ELECTRONIC MATERIALS CHEMISTRY

ELECTRONIC MATERIALS CHEMISTRY

edited by

H. Bernhard Pogge
Microelectronic Division
IBM Corporation
Hopewell Junction, New York

Marcel Dekker, Inc.　　　New York•Basel•Hong Kong

Library of Congress Cataloging-in-Publication Data

Electronic materials chemistry / edited by H. Bernhard Pogge.
   p. cm.
   Includes index.
   ISBN 0-8247-9632-2 (alk. paper)
   1. Electronics–Materials. 2. Chemical engineering. 3. Materials science. I. Pogge, H. Bernhard.
TK7871.E42 1995
621.381–dc20   95-40327
                                                                                      CIP

The publisher offers discounts on this book when ordered in bulk quantities. For more information, write to Special Sales/Professional Marketing at the address below.

This book is printed on acid-free paper.

Copyright © 1996 by MARCEL DEKKER, INC. All Rights Reserved.

Neither this book nor any part may be reproduced or transmitted in any form or by any means, electronic or mechanical, including photocopying, microfilming, and recording, or by any information storage and retrieval system, without permission in writing from the publisher.

Marcel Dekker, Inc.
270 Madison Avenue, New York, New York 10016

Current printing (last digit):
10 9 8 7 6 5 4 3 2 1

PRINTED IN THE UNITED STATES OF AMERICA

Preface

Despite the apparent chemically unassociated terms such as microelectronics, optical and magnetic recording devices, solar cells and fuel cells, lasers and micromechanical devices, semiconductors and superconductors, the success of the electronic industry and its wide variety of products are firmly rooted in the chemical sciences and technologies. Fabrication of these products relies extensively on highly sophisticated chemical processes. Chemistry of electronic materials emerged with the invention of the transistor in 1947 and has served the role of a pioneer for the now rapidly expanding general field of materials chemistry. Although electronic materials have come of age over the past four decades, the field continues to expand in many new directions with youthful energy. One electronic material sector alone—the microelectronic industry (whose growth has averaged about 20% per year since its inception in the mid-1960s)—is poised to become a $2 trillion industry by the year 2000. It will consume more than $2.5 billion worth of gases, $1.3 billion of wet chemicals, $1.1 billion of deposition chemicals, and $0.9 billion of lithographic-masking materials (polymers). Worldwide employment in this area will reach well beyond 10 million.

The increasingly complex technical world creates great pressure for the technical profession of tomorrow to be in tune with respect to the rapid progress occurring in the various technologies. The traditional boundaries between science and engineering have become less defined in most modern technologies, as have the boundaries between traditional disciplines, and even their respective subdisciplines. For example, the area of materials chemistry, and specifically the electronic materials chemistry sector, is both highly interdisciplinary and intradisciplinary. It intersects with various other physical sciences (e.g., physics, material science) and with all the different chemical disciplines. It is therefore of critical importance to provide students of the chemical sciences an educational environment and opportunities to expand their awareness, understanding, and skills in these rapidly growing technology areas.

This text has evolved with these aspects in mind. It is intended for the upper-level undergraduate and graduate student with an appropriate chemistry or chemical engineering background, as well as for the young professional entering the electronic product development and manufacturing industry. The chapters are written by academic and industrial chemists or chemical engineers (several of them having crossed discipline boundaries) to focus on the chemical aspects of the respective technologies. Also, specific effort has been made to balance fundamentals with applications, but to abstain from discussing the latest research activities and results. These are constantly changing and are more appropriately covered by various technical journals (see listing below) and conference reports. Furthermore, the chapter topics are not merely restricted to the traditionally popular silicon microelectronic device technologies, but cover other electronic materials, such as ceramics, superconductors and fuel cells to impart an expanded awareness of the diverse chemical content of other electronic materials and products. However, we could not cover all possible topics (e.g., an inclusive set of chip fabrication processes, or chapters on electromagnetic recording materials, solar cells, aerogels, fullerenes, diamond/SiC, laser materials, conductive polymers, nanomaterials, etc.); the length of the book would have become unmanageable. Yet, the generic process topics that are presented, are as critically important and applicable in the fabrication of the many other materials products.

To assist the reader in a quick overview of the chemical elements relevant to a specific topic, each chapter page includes a periodic table highlighting those elements (nomenclature of heaviest elements as per ACS recommendation). The highlighted elements may be either major or minor components or contributors in a material system. For example, silicon is a basic constituent of silicon epitaxy, whereas it is either an undesirable impurity or a controlled dopant in GaAs epitaxial systems. Also, in material reactions, the presence of oxygen can initiate detrimental corrosion in some material systems, while it creates protective films (passivates) in others. Rather than attempting to differentiate among the many possible roles of an element in various electronic material systems, the highlighted elements are merely intended as an awareness of the many elements from the periodic table associated in some form with the chapter's topic. Each chapter concludes with a short list of references to encourage further reading for the interested reader. A set of exercises has also been included (except for Chapter 1) for further stimulation.

The text's outline, after a product overview (Chapter 1), is organized around two major topics: processes and materials.

In Chapter 2 an introduction is given to the unique processes and process environment so critical to the successful fabrication of any of the many different electronic products. The chapter also briefly outlines the general process flow for microelectronic device fabrication and, for completeness, reviews a number of related processes. The chapter concludes with some device reliability issues.

Chapters 3–7 concentrate on specific, yet widely used process technologies to fabricate electronic device products, using mainly microelectronic devices as a foundation. Each of these technologies has a strong chemical base, and their importance in future device fabrications will significantly increase as the electronic device structures become smaller and smaller and the role of chemical interfaces become more dominant.

Chapters 8 and 9 center on discussions of electronic materials issues in terms of material corrosion and means to generate physical, structural, or chemical analytical data from submicron regions and surface areas. Chapter 9 represents a bridge between the semiconductor processes of the preceding chapters and the subsequent chapters discussing other materials. The latter, however, also rely on the same analytical details of the material and material processing issues in order to develop desirable functional products.

Chapters 10–13 discuss insulating polymers, ceramic packages, superconductors, and fuel cells.

Many individuals, knowingly and unknowingly, have contributed to the germination and completion of this text. My expanded contact with colleges and universities over the past two decades convinced me that students in the chemical sciences are generally not made aware of both the extensive and exciting opportunities offered by the field of materials chemistry. This text can be of help for faculty and student alike. I would like to acknowledge the many discussions, held on various campuses and with various professors, that eventually planted the seed for this undertaking. Also, many thanks are in order to a large number of co-workers and peers in the industry who have helped in keeping my career on a constant high. Their enthusiasm and continuous list of new innovations and capabilities have been constant stimuli for my own contributions. The collective efforts of the authors in this book, who were all willing to take time out of their active schedules, will hopefully instill a similar sense of excitement to the reader. I greatly appreciate their efforts.

My deep appreciation must also go to Mrs. Joyce Knapp, who flawlessly maintained control over the many communications so necessary in not losing sight of the end goal—particularly, in the many instances when I was absent from my office. I also thank her for the various typing activities during the editing phase.

H. Bernhard Pogge

Technical Journals Covering Materials Chemistry Topics

1. *Chemistry of Materials*
2. *CHEMTECH*
3. *Journal of Electronic Materials*
4. *Journal of Materials Chemistry*
5. *Journal of Materials Science: Materials in Electronics*
6. *Journal of Materials Research*
7. *Journal of Micromechanics and Microengineering*
8. *Journal of the Electrochemical Society*
9. *Materials Chemistry and Physics*
10. *Materials Science and Engineering: A Microelectronic Journal*
11. *Nanotechnology*

Contents

Preface iii
Contributors ix
List of Acronyms (Non-Chemical Terms) xi

Overview

1. Electronic Products 1
   Michael A. Fury

Part I. Processes

2. Electronic Processes and Environment 35
   Robert K. Lowry

3. Chemical Vapor Deposition for Epitaxial Growth 83
   Thomas F. Kuech

4. Chemical Vapor Deposition of Crystalline and Amorphous Carbon, Silicon, and Germanium Films 127
   Peter Hess

5. Depositions and Reactions of Metals and Metal Compounds 171
   Roy G. Gordon

6. Photolithography 199
   Alois Gutmann

7. The Chemistry of Plasma Etching 251
   Demetre J. Economou

Part II. Materials

8. Electrochemistry of Corrosion: Principles and Protection   323
   Vlasta Brusic

9. Micro and Nano Analyses of Materials   367
   Homi Fatemi

10. Polyimides as Dielectrics and Optical Interconnects   415
    Tohru Matsuura

11. Ceramic Materials   461
    Charles H. Perry

12. Chemistry of the New Superconductors   501
    Frank J. Adrian

13. Fuel Cells   561
    Nguyen Q. Minh

    Index   609

Contributors

Frank J. Adrian* (*Ph.D. Physical Chemistry–Cornell University*) Applied Physics Laboratory, The Johns Hopkins University, Laurel, Maryland; and Institute for Physical Chemistry, Technical University of Darmstadt, Darmstadt, Germany

Vlasta Brusic (*Ph.D. Physical Chemistry–University of Pennsylvania*) IBM Corporation, Yorktown Heights, New York

Demetre J. Economou (*Ph.D. Chemical Engineering–University of Illinois*) Department of Chemical Engineering, University of Houston, Houston, Texas

Homi Fatemi (*Ph.D. Material Science and Engineering–Stanford University*) Advanced Micro Devices, Sunnyvale, California

Michael A. Fury (*Ph.D. Physical Chemistry–University of Illinois*) Rodel, Inc., Newark, Delaware

Roy G. Gordon (*Ph.D. Chemical Physics–Harvard University*) Department of Chemistry, Harvard University, Cambridge, Massachusetts

---

*Retired: Olney, Maryland

Alois Gutmann* (*Ph.D. Physical Chemistry–University of Innsbruck, Austria*) Siemens Components, Inc., Siemens U.S.A., New York, New York

Peter Hess (*Ph.D. Chemistry–University of Heidelberg, Germany*) Institute of Physical Chemistry, University of Heidelberg, Heidelberg, Germany

Thomas F. Kuech (*Ph.D. Applied Physics–California Institute of Technology*) Department of Chemical Engineering, University of Wisconsin, Madison, Wisconsin

Robert K. Lowry (*M.S. Inorganic Chemistry–Florida Atlantic University*) Harris Semiconductor, Melbourne, Florida

Tohru Matsuura (*Ph.D. Organic and Polymeric Materials–Tokyo Institute of Technology, Japan*) Nippon Telegraph and Telephone Corporation, Musashino-shi, Tokyo, Japan

Nguyen Q. Minh (*Ph.D. Chemical Engineering–University of New South Wales, Australia*) AlliedSignal, Inc., Torrance, California

Charles H. Perry (*Ph.D. Electrical Engineering–Vanderbilt University*) IBM Corporation, Hopewell Junction, New York

---

*Mail address: IBM East Fishkill Facility, Hopewell Junction, New York

List of Acronyms (Non-Chemical Terms)

AF	acceleration factor
AFC	alkali fuel cell
ALE	atomic layer epitaxy
ALET	atomic layer etching
AP-CVD	atmospheric pressure chemical vapor deposition
ARDE	aspect ratio dependent etching
ARL	antireflective layer
ASIC	application specific integrated circuits
ATIR	attenuated total internal reflection
BARL	bottom antireflective layer
BBO	binder burn-out
BCS	Bardeen, Copper, Schrieffer theory
BEOL	back end of line
BICMOS	bipolar complementary metal oxide semiconductor
BSE	backscattering electrons
CAM	chemical amplification
CARL	chemical amplification of resist lines
CCD	charge coupled device
CD	compact disc, also critical dimension
CD-ROM	compact disc read only memory
CMP	chemical mechanical polishing
CRH	critical relative humidity
CVD	chemical vapor deposition
DBS	direct broadcast satellite

DOF	depth of focus
DRAM	dynamic random access memory
DUT	device under test
DUV	deep ultraviolet radiation
E/N	electron field to neutral density ratio
EEDF	electron energy distribution function
EPROM	erasable PROM
ESD	electrostatic discharge
ETF	energy transfer function
EVD	electrochemical vapor deposition
FEOL	front end of line
FET	field effect transistor
FIT	failures in time
Gb	gigabits
GR	groundrules
HDP	high density plasma
HDTV	high density television
HEPA	high efficiency particulate air
HHV	high heating value
HOMO	highest occupied atomic/molecular orbitals
I/O	input/output
IC	integrated circuit
ICP	inductively coupled plasma
IEP	isoelectronic point
Kb	kilobits
L/S	line/spaces
LAP	laser ablation processing
LED	light emitting diode
LHV	low heating value
LMIS	liquid metal ion source
LPCVD	low pressure chemical vapor deposition
LSI	large scale integration
LUMO	lowest unoccupied atomic/molecular orbitals
Mb	megabits
MBE	molecular beam epitaxy
MCFC	molten carbonate fuel cell
MCM	multichip module
MHz	megahertz
MLC	multilayer ceramic
MMICs	monolithic microwave integrated circuits
MO	molecular orbits
MOCVD	metal-organic chemical vapor deposition

Acronyms (Non-Chemical Terms)

MOR	modulus of rupture
MOS	metal oxide semiconductor
MOVPE	metal-organic vapor phase epitaxy
MTF	meantime to failure, also modulation transfer function
MUV	mid ultraviolet radiation
NA	numerical aperture
PAB	post apply bake
PAC	photoactive compound
PAFC	phosphoric acid fuel cell
PAG	photo acid generator
PEB	post exposure bake
PECVD	plasma enhanced chemical vapor deposition
PED	post exposure delay
PROM	programmable ROM
PSB	phase shifting mask
PVD	physical vapor deposition
RC	resistance capacitance
RH	relative humidity
RIE	reactive ion etching
ROM	read only memory
SE	selective epitaxy, also secondary electrons
SIS	superconductor/insulator/superconductor
SN	superconductor/normal metal
SOFC	solid oxide fuel cell
SPFC	solid polymer fuel cell
SQUID	superconducting quantum interference device
SRAM	static random access memory
TAB	tape automated bonding
TARL	top antireflective layer
TCE	thermal coefficient of expansion
TH	temperature-humidity
THB	temperature-humidity-bias
TPD	temperature programmed desorption
TPI	tracks per inch
TSI	top surface imaging
UHV	ultra high vacuum
UHV-CVD	ultra high vacuum chemical vapor deposition
ULSI	ultra large scale integration
VLSI	very large scale integration
WORM	write once, read many

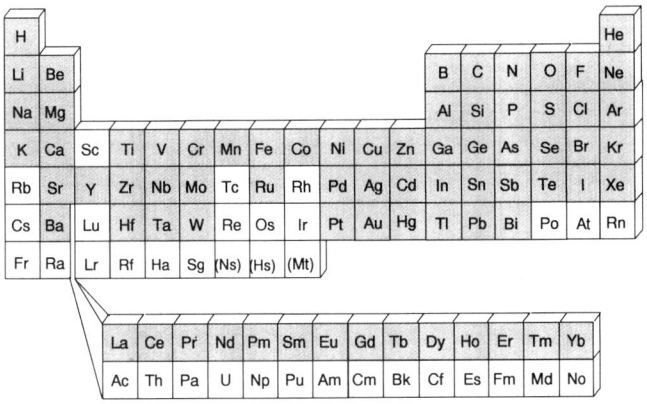

1
Electronic Products

Michael A. Fury   *Rodel, Inc., Newark, Delaware*

I. THE MICROELECTRONICS INDUSTRY

Solid state electronics began in 1947 as an applied research project at what was then Bell Laboratories [1]. The assignment was to develop a solid state device to replace traditional vacuum tubes as switching and amplifying devices. Tubes were becoming too bulky and expensive, consumed too much energy, and had too high a failure rate to be useful in the quantities needed to configure complex communication and computing machines. An alternative technology was clearly required, and solid state electronics held the promise of lower cost and higher reliability. The early work focused on understanding electron transport phenomena that physicists had predicted theoretically. The material selected for this study project, polycrystalline germanium, was chosen based on its position in the periodic table, specifically its electron shell configuration. The resulting device is shown in Fig. 1. The process for forming these *point-contact* bulk transistors involved few steps, and it was carried out in a standard laboratory with no specialty equipment and no stringent specifications on the materials used. While this was adequate for demonstrating the principles at work, the methodology was highly irreproducible, since the electronic properties of these semiconducting devices were very sensitive to chemical and microstructural variations in the polycrystalline material. Other groups of researchers predicted that high-purity single crystals were necessary to ensure stable and reproducible device performance, and so most of the focus shifted to the growth of low-defect single crystals.

In addition to germanium, another element in the same Group IV family, silicon, had also been under investigation as a potential material for semiconducting devices. By the early 1950s, the successes achieved with single-crystal silicon had allowed it to win out over germanium as the material of choice, for several reasons. Its *bandgap*, the electron energy level separation between a conducting state and a nonconducting state, is larger than the bandgap of germanium, making it feasible to operate devices at a higher temperature. Oxidation of silicon produces a water-insoluble oxide that is suitable as both an impermeable diffusion mask and as a high-quality insulator material between con-

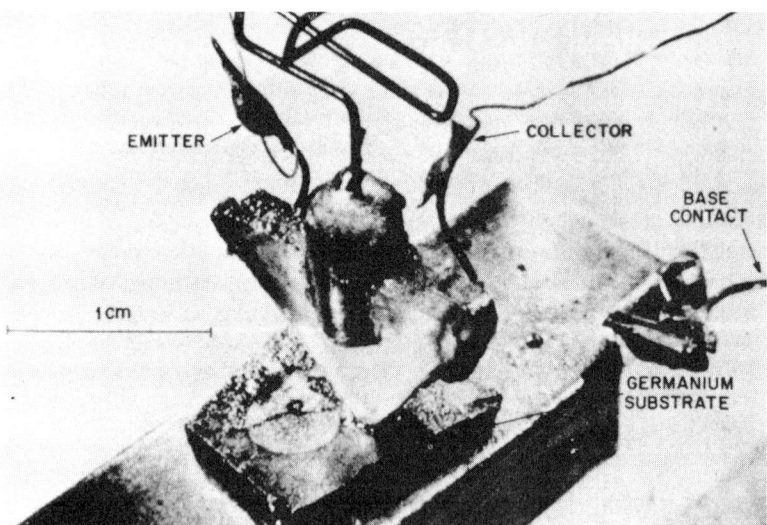

Figure 1 Polycrystalline germanium point-contact transistor, invented in 1947 by J. Bardeen and W. H. Brittain. (From Ref. 1.)

ducting layers; germanium oxide, on the other hand, is water-soluble. Finally, sources of high-purity silicon are readily available: the starting material for semiconductor-grade silicon is beach sand.

Throughout the 1950s, the production of transistors and diodes matured as a manufacturing process. These components were referred to as *discrete devices*, as each unit consisted of a single transistor or diode. The "tin can" transistor with three wire leads became a familiar item in transistor radios and other products of the 1950s and 1960s. The availability of these solid state components enabled the commercialization of a great many new electronic products, and it met well the objectives of lower cost, lower size, lower power consumption, and higher reliability than their vacuum tube predecessors. Product designs evolved to take advantage of the properties of these new devices.

By 1958, the engineering world had come to grips with the commercial limitations of grouping hundreds of discrete transistors into ever more complex circuits. The concern for increasing costs and loss of reliability led to the notion of fabricating more than one transistor on a single piece of silicon, and integrating them into functional circuits [2]. This form of mass production greatly reduced the number of interconnections between unlike material sets, thereby reducing cost and improving reliability. There were simply fewer individual components to fabricate, and fewer places for errors or weak links to occur.

It was at this point that microelectronics, as we know it today, began to assert its unusually demanding requirements on the industrial infrastructure now supporting it. This new infrastructure spawned many companies catering, in whole or in part, to the electronics industry. It included suppliers of specialized equipment for design, fabrication, testing, and packaging of electronic devices. Other companies arose to develop and support computer software used in the design, modeling, and testing of systems, circuits, devices, tools, and processes. Still other companies became suppliers of materials with unique properties demanded by the electronics industry, and of chemicals with specifications critical for, and applicable only to, the successful fabrication of electronic devices. This industry also generated a demand on a wide-ranging set of chemical disciplines to assure continued progress in all phases of development and manufacturing.

By the late 1980s, the electronics industry had become the largest single industry in the world. In contrast to the economic growth of the electronics industry, the microelectronics device dimensions have continuously shrunk. This is driven by the market demand for more function at lower cost and higher speed. Microelectronics is now ubiquitous, with applications that touch every aspect of our lives. These application niches define the different market segments that make up the economic engine of the electronics industry. Industry analysis groups such as Dataquest, Inc., and VLSI Research, Inc., help to standardize nomenclature for these segments and gather information on their growth. The growth of the industry seems to be limited only by the imagination with which new applications are identified and commercialized.

The most visible use of microelectronics is in computing applications. The designs and performance specifications for the chips in desktop and portable systems are derived from the same family as the chips in multimillion dollar mainframes and supercomputers. The differences in system configurations are a matter of satisfying the customer's requirements. There are several types of packaged electronic devices that make up the bulk of those used in computing machines.

The devices that allow a computer to "think" are collectively known as *logic devices*. In large mainframes, these have traditionally been custom designed for a single, specific purpose to service a subset of the processing requirements of the system. Such a design may be optimized for speed and handling large volumes of data, but it is costly to design and requires a large quantity of unique product *part numbers* to make up each machine. The cost pressures induced by the desire to create personal desktop computers fostered the creation of the *microprocessor*, which is made up of arithmetic and logical comparators, hardwired instruction interpreters, buffer storage, input/output ports, and clock timers for synchronization, all on a single chip. The result is a cost-effective computing engine that can be applied to a variety of end uses. Microprocessors were the enabling technology for the personal computer revolution started in the 1980s. Leading the way was Intel Corp., whose president and co-founder, William Noyce, invented the first commercially successful microprocessor, the Intel 4004, in 1971. The end-user demand for greater speed and increasingly complex applications drove the rapid growth and maturing of this technology. Intel led this innovation explosion with a skeleton research and development staff, preferring to base their design and manufacturing commitments on mature unit processes and equipment. For advances in materials and process technology, they relied heavily on published research literature and continuing university research.

In weighing the cost disadvantages and design burden of traditional mainframes against the flexibility and cost advantages of microprocessors, university and industry researchers turned their attention to designing large computers using many microprocessors operating in parallel. These so-called *massively parallel systems* have demonstrated the capability of performing 10^{12} floating point arithmetic operations per second, or *teraflops*, as early as the late 1980s.

There remain some applications, however, for which the general microprocessor is still not the most cost-effective design alternative. This demand sustains the need for what has become known as the application-specific integrated circuit, or *ASIC*, market. The market consists of specialty chips whose design is optimized for a particular function in a particular product. Examples range from high-speed logic processors and cache memory chips for large mainframe computers to control circuits for "smart" home appliances.

In order to realize useful work from the logic elements of the computing system, it is necessary to make instructions and input data available to them electronically, and have a means of manipulating and storing the output data. This element is known as *memory* and it has an evolutionary trail that can be traced from vacuum tubes to magnetic cores to the semiconductor devices of interest in this text. The device that serves as the basic commodity of the microelectronics industry is the *dynamic random access memory*, or *DRAM*. This device makes up the bulk memory of all computers, from laptops to mainframes. Of all semiconductor devices, it is produced in the highest volumes worldwide. As such, it serves some very important functions in the microelectronics industry. As a volume commodity, it must be produced inexpensively with high manufacturing yields, as there is a relatively low profit margin of less than 5% in a highly competitive, volatile marketplace, compared with margins of 30% or more for microprocessors and ASICs. The DRAM manufacturing lines, therefore, are typically the first factories to migrate to larger wafer sizes and to smaller lithography dimensions. The extra expense and risk associated with this technology leadership role must be balanced against the low-cost market position, making the management of this market segment an exceptional challenge.

The DRAM derives the *dynamic* portion of its name from the fact that it stores its memory state (charged or not charged, 1 or 0) as charge in a capacitive storage cell, which leaks and must be refreshed every few milliseconds. While this may seem an inefficient overhead burden to assume, the DRAM has evolved from a multitransistor design per memory bit at its introduction in 1982 as a 256 Kb DRAM (actually 2^{18}, or 262,144 bits) to the single device per bit design, first introduced with the 1 Mb DRAM. The area of a single memory cell has dropped from 100 μm^2 in 1982's 256 Kb DRAM to a projected 0.2 μm^2 in the 1 Gb DRAM, using a vertical trench storage capacitor. This design efficiency allows high-density packing and shrinking of devices on a single chip, and it is an enabling factor in the introduction of gigascale integration ($>10^9$ devices per chip) [3].

The first solid state memory device, however, was the *SRAM*, or *static random access memory*. The SRAM consisted of four or more transistors per memory cell arranged in a flip-flop relay circuit, not unlike that used for vacuum tubes. While the SRAM cannot be fabricated with the same densities per chip as the DRAM, it is capable of achieving higher storage and retrieval speeds than DRAM at the cost of higher power consumption and lower density. This trade-off enables high-speed buffer cache memory designs for servicing microprocessors and other logic engines with frequently needed data.

Both the DRAM and the SRAM are *volatile* memory, losing their stored information when power is removed. A number of nonvolatile memory devices exist, the most basic of which is the *ROM* or *read only memory*. As the name implies, the stored information may be read but new information cannot be written to the device. These are fabricated as hard-wired circuits at the design level, and they are useful for storing basic instructions that do not change over time for computers and other control systems. A number of variations have evolved, starting with the *PROM,* or *programmable ROM,* whose circuits contain a number of fusable elements that can be blown by applying high current pulses in a programming machine. This design allows fabrication of a generic device from a fixed set of lithography pattern masks, while delivering flexibility to update and modify the imbedded code by the manufacturer. Subsequent generations have resulted in the *EPROM, erasable PROM,* which uses fusable links that are reversible between high resistance by application of a high current pulse, and low resistance by exposure to UV light, through a window designed into the device side of the chip. Other variations exist, including electrically alterable ROM, electrically erasable PROM, and extremely-low-power RAM. The latter is a RAM device that can switch automatically to a battery backup to retain its stored information, while consuming so little power as to be stable for a year on a watch battery. Other devices, such as bubble memory and flash memory, are capable of permanent storage in a read–write mode, but they are beyond the scope of this text.

The increases in circuit density, improvements in process control, and the steady rise in manufacturing productivity have resulted in a decrease of 30% per generation in the cost per *bit,* the basic unit of computer memory, since the early 1970s. For this lower cost, the market has also come to expect higher operating speed, lower power consumption, more memory storage capacity, and more computing functionality, and all these expectations have been delivered. These remarkable achievements were possible because of the very accommodating materials set of silicon-based microelectronics. As the industry moves into submicron device dimensions and 100 MHz (10^8 clock cycles per second) operating speeds, the materials and the processes used to fabricate them are being stressed to their physical limits. The ultimate device, which switches states using a single electron, will require structural dimensions of 10 nm (100 Å) or less with control tolerances of less than 1 nm (10 Å). Continuation of the historical decrease in price per bit with each generation is expected to slow, unless there is a significant breakthrough in process, design, or materials technology that creates a discontinuity and places the industry on an entirely new trend curve. Certainly, future progress will incur large development costs. It is possible that a whole new set of materials will be required to achieve subsequent generations of electronic devices.

II. TYPES OF DEVICES

A *device* may be defined as a circuit element that performs a single function by exploiting a single set of physical phenomena. These include transistors, resistors, diodes, capacitors, light emitting diodes, and solar cells. New devices and applications for them are developed as the marketplace demands. At an industry level, there are several extended functions that must be performed in order to achieve a useful product, such as packaging, interconnections, and power distribution.

A. Solid State Electronics

Silicon has four valence electrons, and in its diamond lattice crystal structure, which is essentially two interpenetrating face-centered cubic cells, it shares a covalent bond with each of its four tetrahedrally oriented neighbors. It is possible for some silicon atoms to be substituted by other atoms, intentionally or unintentionally. Intentional substitution by elements with either three or five valence electrons is termed *doping*, typical doping levels ranging from 10^{17} to 10^{22} atoms per cubic centimeter. The Group V elements of phosphorous, arsenic, and antimony are used to bring an extra electron to the crystal lattice, resulting in a *negative space charge* compared with pure silicon, although the net charge is zero. The resulting material is referred to as n-type silicon. Boron is the only three-valence dopant of practical value, since it does not readily diffuse through the silicon lattice at normal use temperatures. Boron is deficient by one electron in the silicon crystal site, resulting in a *positive* space charge and the designation of p-type silicon. The charge carrier in p-type silicon may be thought of as an electron vacancy, or a positively charged *hole*. The charge carrier in n-type silicon is the excess *electron* itself.

1. The Diode

The most fundamental device in solid state electronics is the junction diode. The diode is analogous to a check valve in hydraulic systems, in that it allows current to pass through in only one direction. It consists of adjacent regions of p-type and n-type silicon; the contact area is known as a *pn junction*. When a pn junction has no applied voltage across it, there is a tendency for excess electrons from the n region to migrate toward and fill in the holes of the p region. This migration is counterbalanced by the charge of the dopant atom nuclei, which are locked into their lattice positions. At equilibrium, there is a negative *space charge region* in the p side of the junction, and a positive space charge region in the n side. A stable electrical potential field is formed, and there is no net current flow (Fig. 2a). The size of this region depends on dopant concentrations, and it may range from microns in large discrete devices to only tens of angstroms in integrated circuits of the mid-1990s.

If an external electron source, such as a battery, is applied across the junction so as to move electron current from n regions toward p regions (positive terminal connected to the p material), the junction is said to be *forward biased*. The

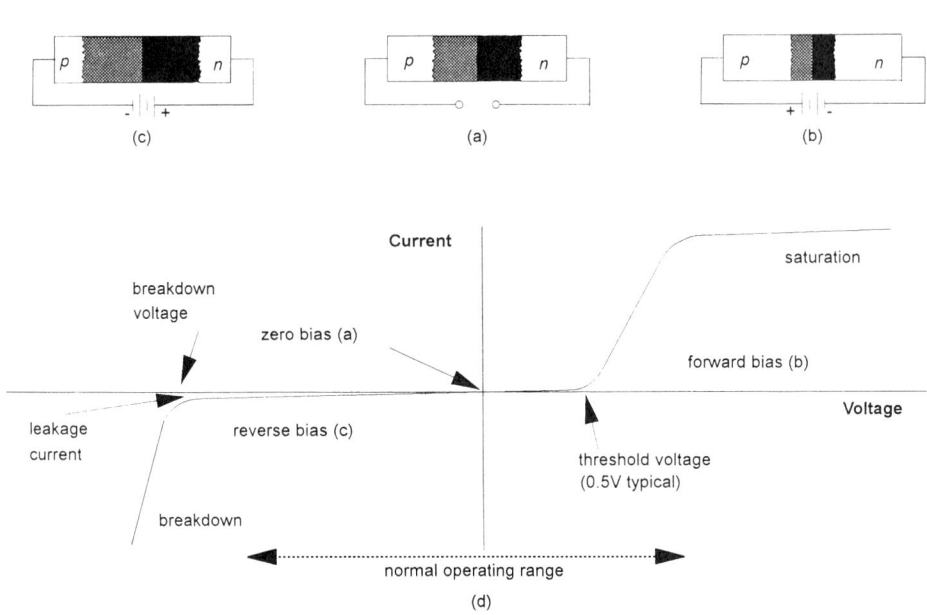

Figure 2 Space charge regions in a pn junction diode for (a) no external bias, (b) forward bias, and (c) reverse bias, and (d) the corresponding current–voltage curve for each condition. Some common terminology is also indicated.

positive terminal attracts electrons across the junction from the n material, just as the negative terminal attracts holes across the junction from the p material. This charge attraction decreases the thickness of the space charge region and reduces the potential barrier to current flow across the junction, as illustrated in Fig. 2b. Current begins to flow proportional to the applied voltage, once the minimum *threshold voltage* is exceeded, and continues to increase until the device is *saturated*. At this point, the junction is unable to carry additional current. As voltage increases further, the diode begins to act as a *resistor*, and it will heat up and eventually destroy itself.

If the battery is applied so as to move electron current from p regions toward n regions, the junction is said to be *reverse biased*. The external electron supply causes the space charge region to increase and raises the potential barrier to current flow (Fig. 2c). Over a range of negative applied voltages, there may be a small *leakage current* on the order of 10^{-12} to 10^{-8} amperes, due to physical imperfections in the lattice, but the current flow is essentially *off*. However, even such low levels of leakage become a concern for advanced high-performance devices, and much effort is spent on assuring high-quality crystals. As the reverse bias is increased further, the space charge region is no longer able to hold back the current flow, and the device is said to *break down*. Some devices, such as the Zener diode, are made to undergo this breakdown in a predictable and reproducible manner, and they will do so repeatedly. Most diodes, however, are irreversibly destroyed by a single breakdown event.

2. The Transistor

If two pn junctions are placed back to back, the resulting three-zone device is a *transistor*. The word itself is a contraction of *transfer resistor*, which is descriptive of the way the electrical signal on the middle zone controls the current flow between the other two zones. It is analogous to a flow regulator in hydraulic systems, although it can be applied more creatively. In analog signal applications, such as a stereo system, a small signal applied to the middle zone can be used to proportionately regulate a large current between the other two zones. As such, the transistor acts as an amplifier of the input signal. In digital applications, such as computers, the only meaningful states are *on* or *off*. In this environment, the transistor acts as a simple switch.

The names of the three zones depend on the type of transistor being described. There are two basic types, the *bipolar* transistor, and the *field effect* transistor (FET). In the bipolar transistor, the center, controlling, terminal is the *base*. The other two terminals are the *collector* and the *emitter*. In the FET, the center, controlling, terminal is the *gate*, and the other two terminals are the *source* and the *drain*. Both transistor types will be described in greater detail in Sec. II.B.1. The operating principles are similar, as illustrated in Fig. 3 for a simple bipolar transistor. When the base is reverse biased relative to the collector, the space charge region essentially fills the base region, and the current between the collector and emitter is blocked by a high resistance. When the base is forward biased relative to the collector, the space charge regions across both pn junctions shrink, and current can flow between the collector and the emitter. The total current that can flow, within normal operating conditions, is a function of the voltage between the emitter and the collector. The proportion of that total current that actually does flow is a function of the voltage between the base and the collector (Fig. 3c).

B. Microelectronics Materials Families

Solid state microelectronic devices can be divided into several broad categories, based on the primary materials from which they are fabricated. The most common of these is single-crystal silicon, making up more than 90% of the worldwide electronics market. Another category critical to high-speed communications and military applications is the compound semiconductors. These families

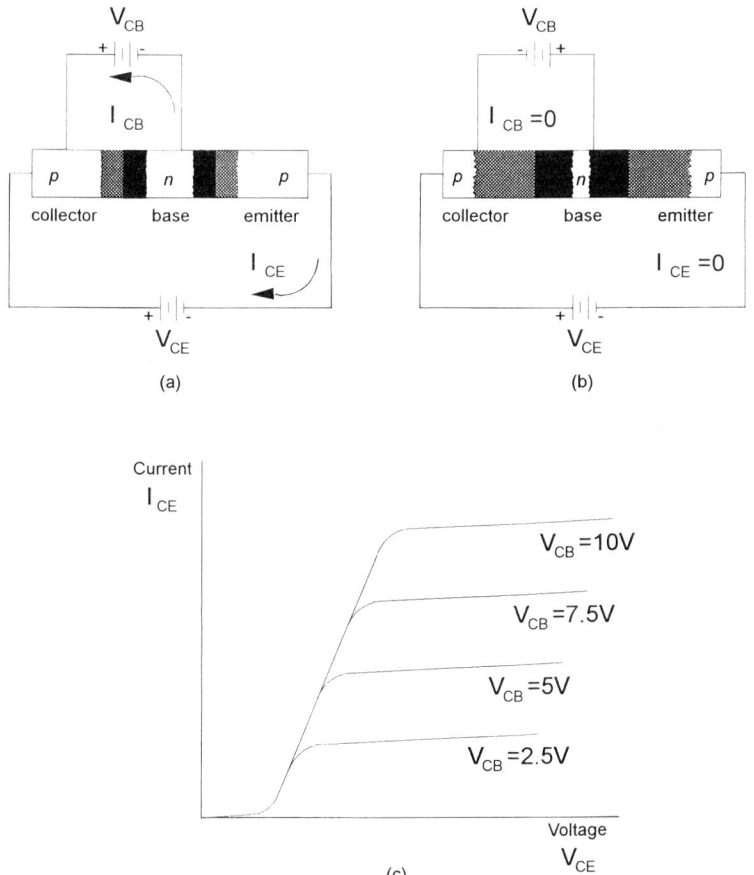

Figure 3 Space charge regions in a pnp junction bipolar transistor for (a) forward bias and (b) reverse bias, and (c) the corresponding family of current–voltage curves for a set of collector–base voltages.

consist of III–V or II–VI pairings of compounds from the periodic table. Amorphous semiconductors, mostly silicon, make up another class of low-cost devices used as solar cells, and transistors for specialty applications.

1. Silicon Materials

Silicon-based microelectronic devices are predominately one of two broad design types: bipolar transistors and field effect transistors (FETs). Although based on the same space charge and mobile carrier principles, the structures and the operating characteristics of these device families are significantly different. The motivations for selecting one over the other for a particular application are governed by the requirements of the end product. In general, bipolar devices operate faster and at higher power levels. These benefits come at the price of higher power consumption and heat generation. Field effect transistors, also known as metal oxide semiconductors (MOSs), experienced significant design improvements in the late 1980s, so that the speed advantage of bipolar devices has been reduced to the point at which the cost and power advantages of MOS devices offer a cost-effective alternative for even the most demanding bipolar applications. These two families of silicon devices share a common materials set, with a set representative of the early 1990s shown in Table 1.

The simple transistor of Fig. 3 is adequate to illustrate the solid state fundamentals of the transistor, but the actual realization of an integrated bipolar transistor design is more likely to resemble Fig. 4. While the single-crystal silicon may be adequately free of defects to allow fabrication of some devices, high performance demands the use of an epitaxial doped silicon layer 0.5 to 5.0 µm thick grown on top of the silicon wafer, in which the active device areas are fabricated. This sequence accommodates the placement of a buried subcollector beneath the device, to enhance the current-carrying capacity and response time of the device. The FET device shown in Fig. 5 is less complex to fabricate, primarily because the source and drain can be fabricated in the same step, whereas the emitter, base, and collector in a bipolar transistor each require separate sequential processing. In order to shrink FET devices and increase their operating speed, it is necessary to reduce the distance between the source and

Table 1 Materials Common to Silicon Devices of the Early 1990s, and Their Applications

Applications	Materials
Substrates	Single-crystal Si, GaAs, Ge, polycrystalline Si, amorphous Si
Contacts, local interconnects	Polycrystalline Si, WSi_2, W
Dopant source materials	PH_3, B_2F_6, P_2O_5, $POCl_3$, AsH_3
Device isolation, insulators	SiO_2, Si_3N_4, SiO_xN_y, polyimides
Intermetal insulators	SiO_2, polyimides
Contact metallurgy	PtSi, CrO_3, $TiSi_2$, WSi_2, $CoSi_2$
Interconnect metallurgy	Al, AlCu, AlSi, AlCuSi, W, Cu
Solder connections	PbSn alloys
Layered isolation between interconnects and solder	Cr–Cu–Au
Ultrasonic wire bonding	Au

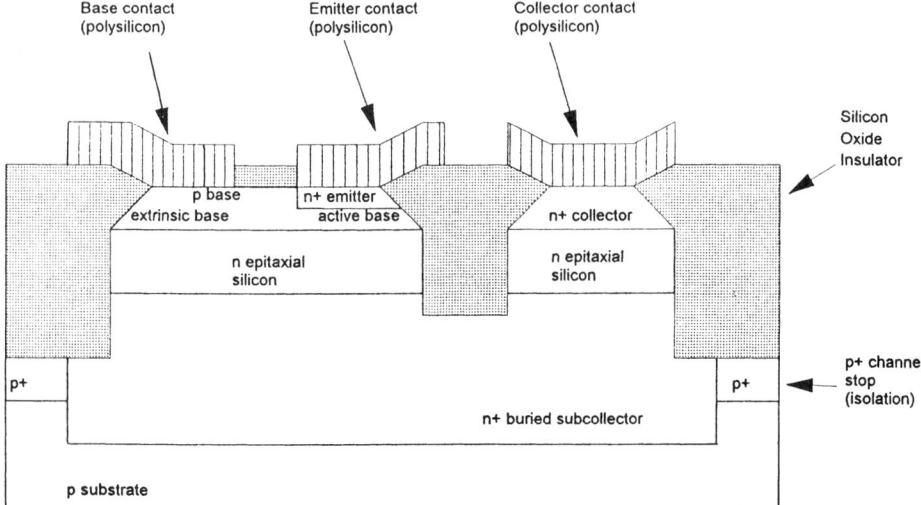

Figure 4 Cross-section diagram of a silicon bipolar transistor.

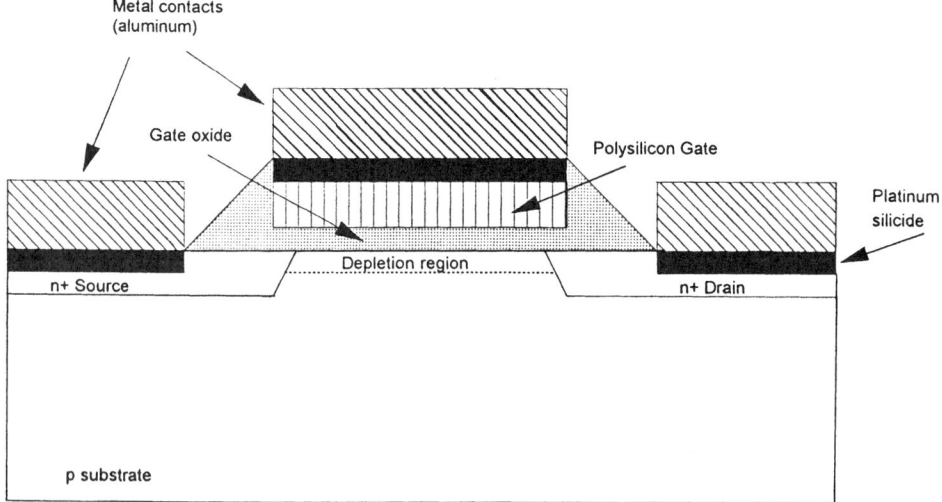

Figure 5 Cross-section diagram of a silicon n-channel FET.

drain doped regions, i.e., reduce the *channel length*, and reduce the thickness of the gate oxide to increase the effective field strength in the channel. Among the technical achievements that have brought these requirements to the level needed for manufacturing are the development of rapid thermal annealing techniques and processes for the deposition of thin, low-defect silicon oxides, as discussed in Chapter 4.

2. Compound Semiconductor Materials

In addition to semiconductors based on silicon, there are several semiconducting compound materials with commercially exploitable properties. GaAs and CdTe are among the more widely recognized compounds, and they typify the III–V and II–VI classes, respectively. These combinations from the periodic table provide the same electron configuration as single-crystal silicon. Differences in lattice spacings and electron energy levels result in unique electrical properties. The nonstoichiometric nature of these compounds also places restrictions on the types of process steps that can be used in device fabrication. For example, it is not possible to form an insulating layer from a compound substrate by thermal oxidation. The chemical and toxological properties of the compound semiconductors create an additional burden of controlling emission of reaction by-products and waste materials to a degree not required for silicon. Such issues are considered in the design of commercial processing systems for this market, adding to their cost and complexity.

Despite these difficulties, GaAs and its derivatives comprise an enabling technology for many applications. Early research in the field was sustained by the need for high-speed digital and analog devices for the military. It has since developed commercial applications in microwave telecommunications and digital switching, which in turn has resulted in many of the new features available for home telephones. The majority of the devices fabricated from compound semiconductors are variations on the FET (see Fig. 5), although bipolar designs (see Fig. 4) exist as well. Compound semiconductors will not be treated in detail in this text. Comprehensive references are available for those interested in additional details [4,5].

3. Polycrystalline and Amorphous Materials

The properties making single-crystal silicon suitable for electronics applications persist to varying degrees in its polycrystalline and amorphous forms. Many semiconducting materials are capable of generating electron–hole pairs upon absorbing light energy, including Si (single-crystal, polycrystalline, and amorphous), GaAs, CuS, and CdTe among those common in commercial applications. Although lattice defects play a significant role in determining electronic behavior, and grain size and grain boundaries become additional materials parameters to be controlled, such grain features have a relatively small effect on the optical properties of silicon, which lends itself to photovoltaic device fabrication from lower-cost polycrystalline forms rather than much more expensive single crystals.

Applications of amorphous semiconductors were first proposed by Chester F. Carlson in 1938, for xerography. This process, using selenium-based materials, was not commercialized until the 1950s. More recent extensions of this technology enable color photocopying based on graded amorphous alloys of selenium, arsenic, and tellurium [6].

Another commercial application of amorphous semiconductors is in solar cells, for the direct conversion of light to electrical energy. Much work has been done in this field in the 1970s and 1980s. As of early 1994, hydrogenated amorphous silicon (a-Si:H) is the most promising material that exhibits the characteristics necessary for this application. These include the ability to absorb a significant fraction of the solar energy, the ability to form a *depletion region*, or space charge zone, and the ability to form both doped region junctions and ohmic contacts with the material.

While FET transistors have been fabricated from a-Si:H, their electronic properties are poor compared with their single-crystal analogs, particularly with respect to their switching time. This is likely due to their distribution of *trapping*

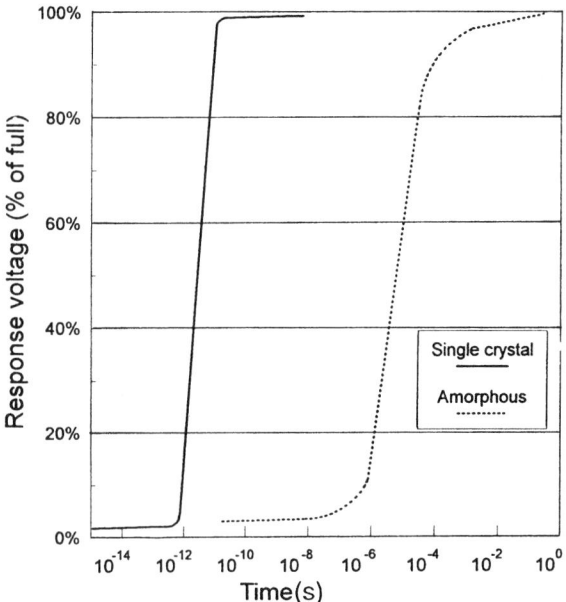

Figure 6 Transition stabilization times for amorphous versus single-crystal transistors.

states, or energy levels and times required for the various charge-sensitive crystal defect sites to stabilize. While their transition from *off* to *on* may be 90% complete in 10 μs or less, there is a tail in the transition that may persist for several milliseconds. In contrast, a conventional single-crystal device can complete its transition to a stable current in 20 ps (10^{-12} seconds) or less (Fig. 6). Amorphous or polycrystalline thin-film transistors can be fabricated with the general configuration shown in Fig. 7. Suitable semiconducting materials include CdS, CdSe, PbS, and CdTe, each material having its electrical and spectral response characteristics optimized for the intended application. These can be configured into transistors, diodes, memory storage cells, and photoconductors for applications such as large-area solid state displays, radiation detectors, television cameras, and electrophotography.

The most common application of non-single-crystal semiconductors is for solar cells. Individual cells are often used as photodetectors in cameras, security systems, and automatic lighting systems. When grouped into arrays, it is possible to generate enough electricity to serve as a primary power source. Such arrays made of single-crystal silicon have been used in spacecraft since the Vanguard 1 space probe was launched in 1958. Earthbound applications include remote radio repeater stations and communications links, where the commercial demand and the lack of alternative power sources balances the high cost of single-crystal solar cells.

A basic solar cell is illustrated in Fig. 8. The thickness of the p+ layer and the depth of the pn junction are important parameters in determining the efficiency with which the cell absorbs and converts solar energy. The optimum thickness varies from 1 μm for GaAs to >10 μm for silicon. The demand for

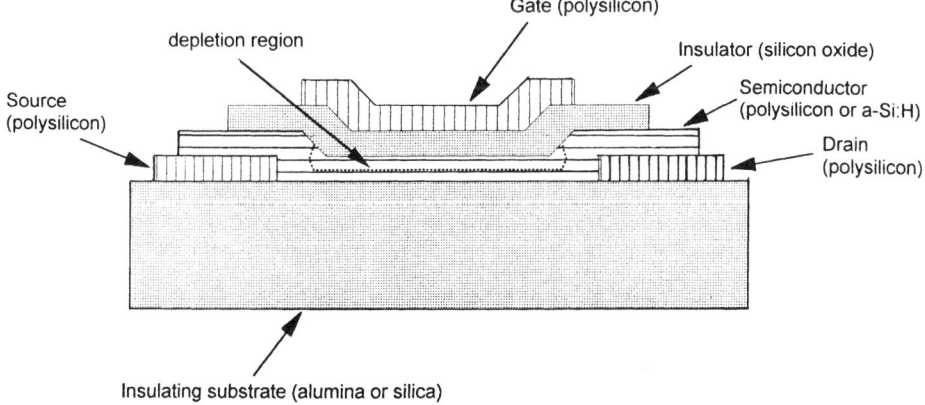

Figure 7 Cross-section diagram of an amorphous or polycrystalline thin-film device.

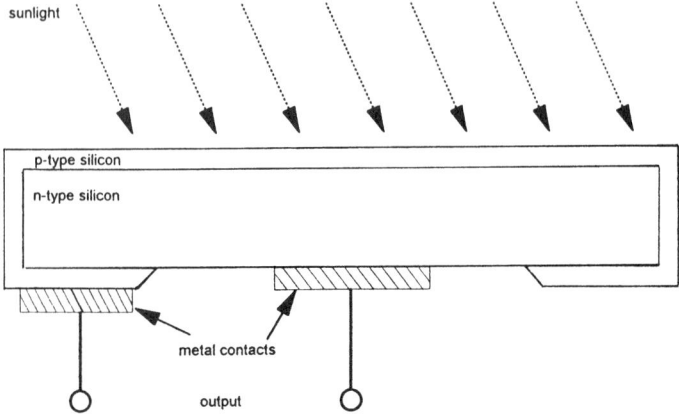

Figure 8 Cross-section diagram of a basic solar cell.

large-area solar cells has prompted the development of manufacturing methods for depositing continuous thin-film sheets of flexible amorphous silicon.

In order to compete with fossil fuel sources, the cost of solar power must be reduced. The two primary means to achieve this have been to reduce the solar cell cost by replacing single crystals with less expensive polycrystalline or amorphous cells, and by increasing the efficiency of light conversion to electricity. Amorphous thin-film solar cells have efficiencies of typically 4–9%, while single-crystal Si normally operates at 10–12%, with experimental cells achieving 18% under ideal conditions. By using light concentrators such as fresnel lenses, laboratory milestones of 20% with Si and 25% with GaAs single crystals have been observed. Another strategy requires that the solar spectrum be split to two separate cells, each optimized for a different band of wavelengths, to achieve efficiencies of 28–35%. Still, these laboratory efforts only serve to demonstrate feasibility and develop some basic understanding of what is necessary to achieve similar efficiencies in less expensive solar cells. The overall motivation for solar energy commercialization lies in the vast amount of solar energy available. The worldwide energy consumption level is about 80×10^{12} kWh per year, while the earth receives 10^{18} kWh of solar energy per year. Theoretically, the planet's entire energy needs could be met with solar cells covering as little as 0.008% of the earth's surface, or 0.08% at 10% efficiency and a low enough cost for the cells, their support structures, and connections to the power distribution network.

4. Interconnects

The manufacture of millions of transistors in a square centimeter of silicon is a remarkable achievement in itself, but it lacks utility until those devices are wired together to form the circuitry necessary to perform the intended design function. Table 1 includes the materials commonly used to form on-chip interconnects. The sequential deposition and patterning of metal interconnects and intermetal insulators creates topography that interferes with the successful processing of subsequent layers, for two major reasons. Metal deposited by sputtering or evaporation processes tends to deposit in a line-of-sight manner from a source above the wafer. Deposition on a flat surface can be very uniform (<5% 3σ variation in thickness across a 200-mm wafer), but the actual thickness over a topography step, measured normal to the step surface at each point, will be thinner than the nominal thickness elsewhere. These thin spots result in higher resistance in the lines formed through them, and they are prone to early-life failures due to resistive heating and *electromigration*, discussed in Chapter 2. In addition, the topographic steps create local variations in the thickness of photoresist applied to the wafer for lithographic imaging (see Chapter 6), resulting in variability in the developed image size versus the intended design image size. These topographical problems limited the number of levels of wiring to two in early integrated circuit (IC) designs, until the introduction of various planarization processes in the early 1980s. In general, these approaches require the deposition of up to 100% excess insulator over the intended final thickness, followed by deposition of sacrificial layers such as photoresist, polyimide, or a different form of silicon oxide, which tend to deposit in a relatively planar fashion. Using this more planar top surface, the wafer is then etched back using a process with a 1:1 selectivity in etch rates between the sacrificial material and the intermetal insulator. The result is a top surface with less topography locally, but still subject to variations in global thickness across the wafer due to the cumulative tolerances of the deposition and etch-back steps. The desire for global planarization was addressed with the introduction of chemical-mechanical polishing for planarization of multilevel interconnects [7], a process that had been used for years in the fabrication of atomically smooth silicon wafers but had not previously been applied to device fabrication. A four-level interconnect structure is illustrated in Fig. 9, showing a materials set typical of 4 Mb and 16 Mb DRAMs and the associated logic generations.

As design dimensions have shrunk, crosstalk between adjacent metal lines and other forms of capacitive coupling delays have become limiting factors in overall circuit performance. The resistance capacitance (RC) time constant for signal transitions becomes large as the distance between lines shrinks, since capacitance is inversely proportional to the separation distance:

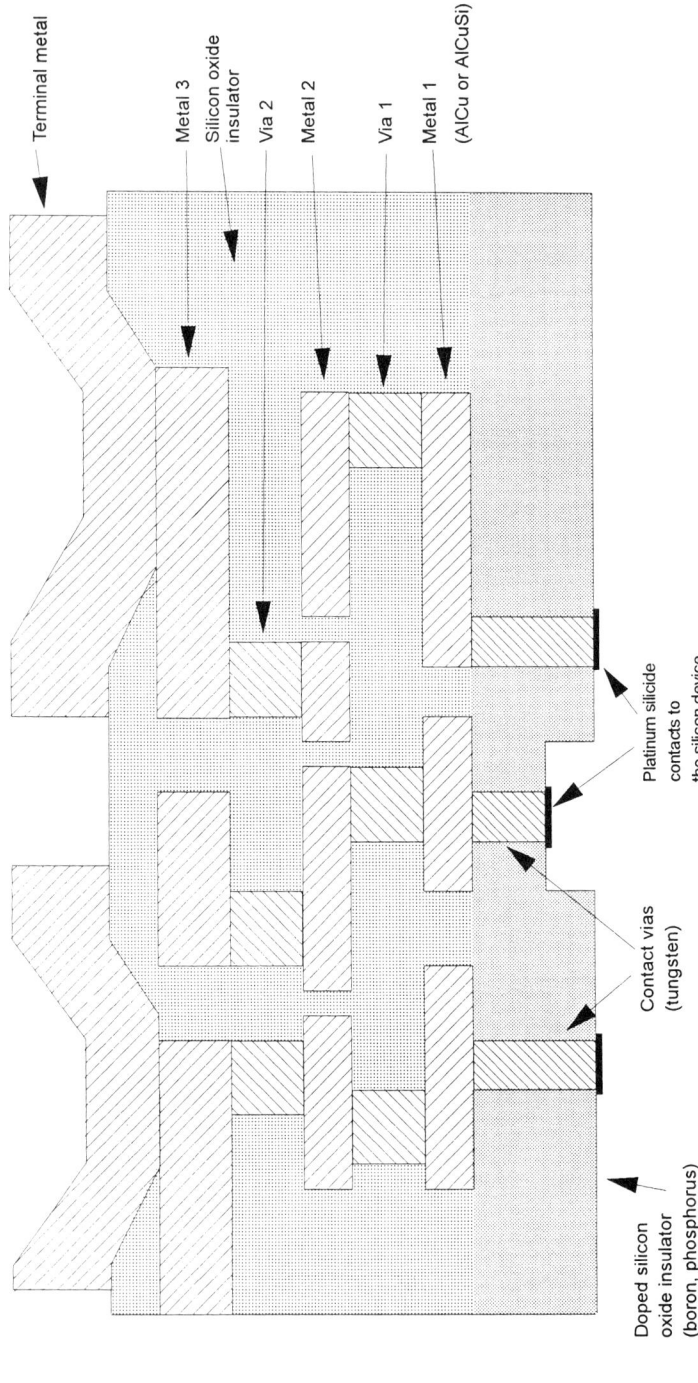

Figure 9 Cross-section diagram of a multilevel metallization stack.

$$C \propto \frac{\kappa A}{t} \tag{1}$$

where

C = capacitance between adjacent lines
κ = dielectric constant of the intermetal insulator
A = cross-sectional area of the lines' overlap
t = thickness of the insulator separating the lines

Also, the cross-sectional area of a line becomes the same order of magnitude as the aluminum grain size, 0.5 to 2.0 µm, increasing its resistance over bulk values. Means to offset the effects of shrinking dimensions can be achieved by decreasing the resistivity of the metal (i.e., using copper instead of aluminum), or by decreasing the dielectric constant of the insulator. Numerous studies have been done of the combined benefits of these two actions [8], pointing toward an interconnect system comprised of copper metal and low dielectric insulator, such as a polyimide. The metallorganic chemical vapor deposition (MOCVD) of copper from both Cu(I) and Cu(II) precursors is established in the literature [9], with resistivities shown as low as 1.8 µΩ-cm, compared with 1.7 µΩ-cm for oxygen-free bulk copper and 2.65 µΩ-cm for aluminum. Although insulators with dielectric constants κ lower than 2.3 are known, compared with 3.9–4.2 for SiO_2, the ability to integrate these successfully into a manufacturing process scheme remains a significant challenge, due to incompatibilities with other processing conditions and chemistries. Some successes have been shown using materials with κ of 2.8–3.2. This barrier will continue to be pushed lower as problems are identified and solutions are engineered into the materials and processes.

Superconductors, discussed in Chapter 12, are materials that have zero resistance to the flow of electric current. One area for very-large-scale integrated (VLSI) and ultra-large-scale integrated (ULSI) applications of superconductors is their deposition as thin films for device contacts and on-chip interconnect metallization. This is done using laser evaporation, sputtering, or chemical vapor deposition (CVD) techniques using a variety of inorganic and organic source materials. The objective is to develop materials and methods that can be both deposited and patterned in a manner compatible with microelectronics processing, without losing their superconducting properties in the process. These materials must achieve and maintain superconductivity at liquid nitrogen temperature while carrying operating currents of several milliamperes over a 10-year lifetime. Sustaining superconductivity under these conditions for prolonged times is proving to be a difficult task, and commercial applications are developing only slowly as a result.

One application that has developed significantly is the *SQUID*, or *s*uperconducting *qu*antum *i*nterference *d*evice, which is based on the *Josephson effect*. In this phenomenon, current through the device passes through a thin barrier junction layer by a quantum mechanical effect called *tunneling*. The junction material determines the device properties, and it includes oxides of V, In, Nb, Pb, or Al, or semiconductors such as CdS, in thicknesses from 100 to 8000 Å. Operating at liquid helium temperatures to minimize thermally generated electron current in the junction, the SQUID is extremely sensitive to fluctuations in magnetic field. It has found widespread use in geophysics, where electromagnetic wave measurements provide information on porosity, salinity, and temperature of earth structures from depths of 100 meters to tens of kilometers for locating oil, water, and geothermal sources. SQUIDs have been configured for surgical insertion into the brain, to measure local neuron activity from the magnetic fields generated by the brain's electrical synapses. They can also be used as ultra-low-noise signal amplifiers in both direct current (dc) and radio-frequency (rf) applications.

Another area of active product development is optical interconnects. When refined to impurity levels of less than 100 ppb, silicon oxide is capable of transmitting light over long distances with little loss. Corning Glass demonstrated such a material in 1970, with a loss of less than 20 db/km. Only 10 years later, the rewiring of the world's telecommunications infrastructure with fiber optics was well under way. By 1990, more than three million kilometers of fiber optic cable with a loss of less than 5 db/km had been deployed in the United States alone. These commercial materials are tailored to applications and operating conditions using a variety of network formers, of which SiO_2 is only one example, to provide the basic glass structure, in combination with network modifiers, as illustrated in Table 2. In addition to the stringent purity levels for the

Table 2 Common Types of Glass Network Formers and Network Modifiers

Network former	Network modifier
SiO_2	K_2O
B_2O_3	MgO
Al_2O_3	CaO
Na_2O_3	PbO
GeO_2	
As_2O_3	

composite fiber, the valence state of the impurities is also important. For example, Fe^{3+} and Cu^+ have absorption bands outside of the spectral range of interest for fiber optic applications, while Fe^{2+} and Cu^{2+} absorb strongly at wavelengths of 1.1 and 0.8 μm, respectively, where they will degrade the optical transmission.

Among the support devices used in this network are optical amplifiers. These consist of segments of erbium-doped fiber fed by a pump laser through a parallel fiber coupler. In this simple device, all wavelengths of signal light are amplified simultaneously as analog data, without the need for conversion to digital electronic form and back to analog. Such technologies are capable of data transmission at 34 billion bits per second.

In addition to the distance transmission needs of the telecommunications industry, the microelectronics industry may realize performance improvements by

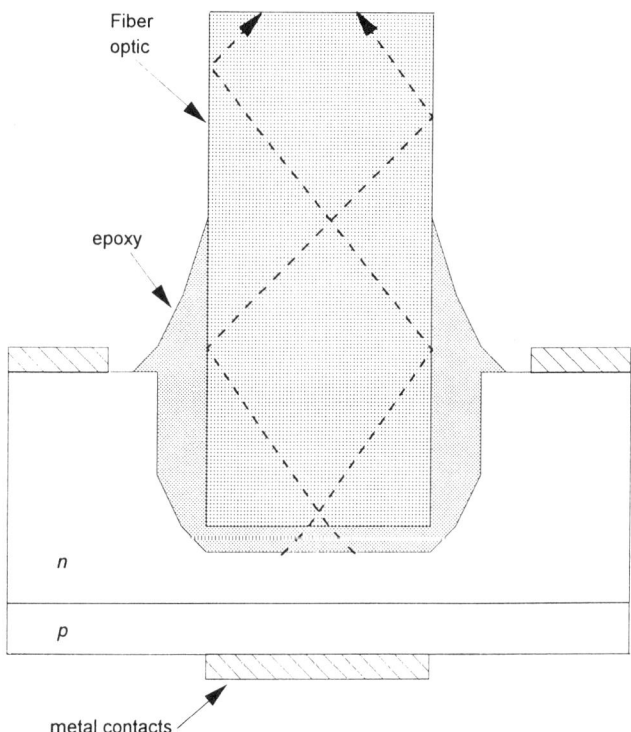

Figure 10 Method for attaching a fiber optic cable to an isolated optoelectronic device.

integrating optical interconnects into package and chip designs. Direct connection of a single device to an optic fiber, as illustrated in Fig. 10, is suitable for isolated or low-density applications, but it cannot replace metal interconnects at submicron dimensions. It also becomes unwieldy in electronic packages containing tens of thousands of interconnects per square centimeter. Optical waveguide technology has developed to address these requirements. This involves the formation of light-conducting pathways in situ on the chip or package surface. Materials range from silicon oxides and nitrides to a variety of polymers such as polymethylmethacrylate (PMMA), Teflon AF, and poly(cyclohexylsilyne). The latter material changes its index of refraction on exposure to UV laser light, which enables direct writing of waveguide patterns using lithographic techniques.

Solving the materials and process issues associated with the changeover from aluminum–silicon oxide interconnects to copper–low κ interconnects, superconductors, or optical interconnects is necessary, but not sufficient to ensure its implementation into manufacturing. Such a materials system is a revolutionary departure from the installed manufacturing base of equipment and processes for Al–SiO_2 interconnects. The performance advantage potential available from this change must be weighed against the costs of new equipment and fabrication facilities, of new reliability studies to ensure its worthiness in the field, and of new yield learning to achieve the productivity level of the prior generations. In addition, when faced with obstacles such as the RC barrier, circuit designers have shown over the years an uncanny ability to find alternative design solutions, extending the lifetime of a particular materials set and manufacturing technology. This scenario for copper–low κ interconnects is typical of the challenges and decisions associated with the introduction of any new materials set or process technology into manufacturing, and it comprises the heart of applied science and engineering in microelectronics.

C. Optoelectronic Devices

The technology of optoelectronics, or photonics, uses pulses of light rather than electron current to transmit information. A photonic system consists of a transmitter, a light-conducting medium, and an optical detector. As in electronic circuits, amplifiers and modulators are also needed to fabricate functional systems. The heart of this technology is the ability of certain compound semiconductors to emit light. The light emitting diode (LED) is a common element of pocket calculators and other consumer electronics. Of more importance for telecommunications is the laser diode, which emits coherent, monochromatic light at higher intensities than LEDs [10].

1. Light Emitting Diodes

An LED is a pn junction whose bandgap between the conduction and the valence bands corresponds energetically to visible or infrared light. As current flows through the junction under forward bias, electrons migrate from the n side to the p side, where they recombine with hole carriers. This recombination reduces the energy level of the electron, with the excess energy given off as photons. This energy level can be varied by adjusting the doping levels in the device, giving rise to the range of commercial devices in common use today. Because the photon emission is governed by the bandgap energy levels, the light is essentially monochromatic. Some LED designs are shown in Fig. 11.

Gallium arsenide is the most common semiconductor substrate for LEDs, and it can be modified with various dopant levels to emit light from the visible green (below 530 nm) to the infrared (>900 nm) regions. Doping GaAs with Si can push this into the 940 nm range, while In doping extends it to 1060 nm. Several materials capable of serving as LED substrates are listed in Table 3. Some LEDs based on GaP can emit red and green light simultaneously, with the relative intensity of each being a function of the diode current. This single device can emit a continuum of shades of red, orange, yellow, and green. The red (650–680 nm) numerical readouts of pocket calculators and digital watches are probably the most familiar example of LED applications. Infrared (IR) LEDs have become commonplace for applications such as remote controls for televisions and stereos, and they have been used for wireless keyboards in personal computers.

2. Laser Devices

In an LED, the recombination of conducting electrons and holes and the emission of photons are random events. Each photon is capable of stimulating the recombination of other electron–hole pairs as it passes through the device, but random events are inefficient for this purpose. The LED device can be constructed so that the end faces of the pn junction are planar, parallel, and highly reflective. In this case, the initial photons are reflected back through the junction, stimulating further emission, which itself reflects back into the junction in an increasing cascade effect. Below a certain threshold current, the junction has more electrons in the valence band than in the conduction band, and the device behaves like an ordinary LED. During a brief pulse of high current, however, the number of conduction electrons is large, as is the number of holes, making the opportunity for electron–hole recombination very great. The capture and internal reflection of the initial photons triggers a coherent photon pulse that exits one device face, which is, by design, less than 100% reflective (Fig. 11c). For the laser LED to be efficient, the pn junction itself needs to be extremely planar and free of light-scattering defects that would divert photons from the intended path.

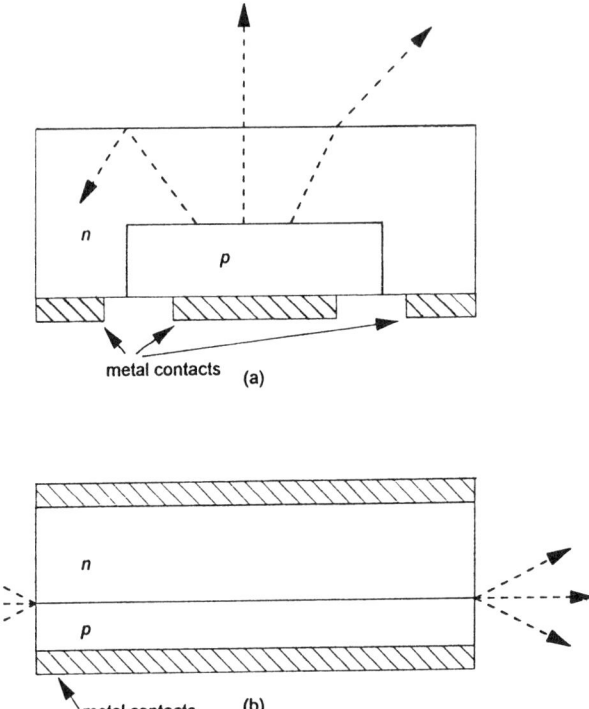

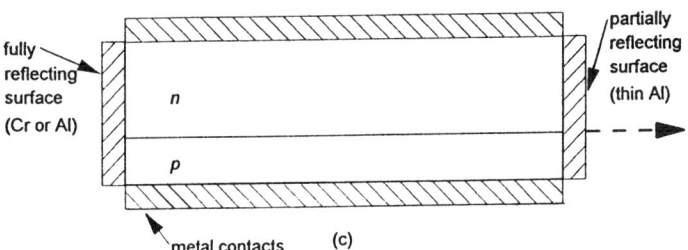

Figure 11 Cross-section diagrams of light emitting diode devices: (a) surface emitter; (b) edge emitter; (c) edge-emitting laser.

Table 3 LED Materials and the Resulting Colors

Material	Color	Wavelength (nm)
AlGaN	Violet	360
AlGaAs diode with LiNbO$_3$ waveguide	Blue	840 with frequency doubling to 420
GaN; SiC	Blue	440
ZnMnSe	Blue	450 at 5 K
ZnSSe	Blue	475
GaP; GaN	Green	520–550
InGaP	Yellow/green/orange	570–590
SiC	Yellow	580
GaAsP	Amber to red	610–700
CdZnTe	Red	620
AlGaInP	Red	640–680
AlGaAs	Red	675–680
GaP:ZnO	Red	690
GaAs	IR	900
GaSiAs	IR	940
GaInAs	IR	1000
PbS	IR	2500
PbSe, PbTe	Far IR	10,600–30,000

Because these devices are integrated into optoelectronic systems, the interconnection technology requires elements in addition to the metal and dielectric materials found in conventional electronic systems. Fiber optic lines may be integrated to the light source for communications off the chip, as discussed in Sec. II.B.4.

3. Compact Disc Technology

The use of vinyl records for distribution of recorded music ended swiftly in the early 1990s. While cassette tapes had co-existed with records for a number of years, compact discs (CDs) rapidly displaced records as the cost of CD players dropped to commodity levels. The superior quality of digitized sound was a boon to the consumer, and the cost effectiveness of reproducing CDs was a boon to the manufacturers. This leveraging of CDs' high data density and low cost paralleled the growth of an industry for computer data storage and distribution known as CD-ROM (read only memory). A single CD-ROM has a typical capacity of 600 MB (megabytes, 10^6 characters), equivalent to 275,000 pages of text. This enables entire encyclopedias to be stored on a single disk, complete with color pictures, audio clips of music, and video demonstrations of selected topics. Such devices are published media; like a book, they are printed once and can be read but not changed.

The CD-ROM itself consists of a clear polycarbonate disk that is stamped with a metal master plate containing the digitized data for the entire disk. This data consists of a series of oval pits of variable length 0.6 µm wide and spaced 1.6 µm apart in a continuous spiral track from the center to the edge of the disk. This is a density of more than 5900 tracks per centimeter (15,000 tracks per inch, TPI), compared with 53 tracks per centimeter (135 TPI) in high-density magnetic floppy disks. Aluminum is then evaporated onto the disk to increase its reflectivity, specifically the contrast between the high and low portions of the pits in the data track. The aluminum is covered with plastic acrylic that is hardened by exposure to UV light. The CD-ROM is a passive electronic storage device, with the active data reading being done by an infrared laser diode and a photodetector that interprets the modulated reflections off of the spinning disk as digital data. A CD-ROM reader will pass this information directly to a computer, whereas a music CD player will first pass the data to a digital-to-analog converter.

Read/write CD technology combines the advantages of high CD data storage density with user selection of the data to be stored. Intended for archiving large quantities of data, this technology writes data onto a blank CD that initially has no stamped data pits except those needed to guide the read/write mechanism. The reading mode is similar to CD-ROM, in that the data is inferred from reflectivity contrasts created during the writing mode. There are several methods for achieving this contrast, which is an active area of research and development in the field of optical data storage. All the methods use the same basic substrate, a polycarbonate disk with stamped tracking guides, a reflective layer to enhance signal-to-noise discrimination, and a photoactive layer. Some active-layer materials can undergo an irreversible phase change on exposure to sufficient photon energy, and thus they can only be written on once. This technology is called WORM, for *write once, read many*. Other materials can transform from an amorphous to a polycrystalline phase reversibly, depending on the intensity and duration of the radiation to which they are exposed. This results in the read/write capability familiar in magnetic storage devices. Some materials are capable of changing the polarization of reflected light, depending on their state of magnetization. These give rise to *magnetooptical* storage technologies. In addition to storing data by spatial resolution of different material phases, it is possible to modify the reflection spectrum in well-defined regions of the material. This technique, called *photochemical hole burning*, does not actually burn the material in the sense of oxidation. Rather, it selectively modifies the material to create gaps of variable wavelength in the spectrum of light reflected from each data storage "pit." This technique can achieve data storage densities of 10^{11} bits/cm^2, compared with 5×10^7 bits/cm^2 in conventional CD-ROM. Many of the materials under investigation for these various applications are listed in Table 4.

Through a combination of low defects in written data, and software error checking techniques, the error rate of CD-ROM is less than 1 in 10^{12} bits read. In order to perform reliably in this failure-sensitive application, several criteria must be satisfied. The data storage material must be optically stable over a range of temperature environments including, for example, a closed car on a hot day, and cannot be sensitive to sunlight or indoor lighting. The layer must be applied with uniform thickness so as to provide the same optical response across the entire disk write area. The application method must ensure that there are no pinholes, bubbles, or inhomogeneities across the entire disk surface. Finally, in order to become commercially successful, the materials must be readily available at a cost appropriate for the customer's application.

III. OTHER APPLICATIONS

In addition to the types of devices already described, there are a great many other electronic solid state devices that carry a significant share of the electronics economy. Read/write heads for magnetic disk storage are fabricated on a silicon substrate using techniques similar to those used in IC on-chip interconnects. Liquid crystal diodes and displays combine solid state electronics with a physical

Table 4 Optical Storage Photoactive Materials

Applications	Materials
WORM: write once, read many	Photoabsorptive dyes, naphthalocyanines, TeSePb
Reversible phase change for multiple read/write cycles	GeTeSn, GeSbTeS, TeGeSnO, SnTeSe, GeTeTi, GeAsTe, GaTeSe, InSePb, SbSeBi, InSe, TbFeCo, AuSnTeGe, PdSnGeTe, InSeTl, GeSbTe, Sb_2Se, InSeTlCo, GeTe, SbTe
Magneto-optical layered films	Fe/Cu, Fe/Ag, Fe/TiN, TbFeCo/GdFeCo/GdTbFeCo
Photochemical hole burning	DAQ/PHEMA, DAQ/PMMA, ABDAQ/PMMA, PBDAQ/PHEMA, where DAQ = 1,4-dihydroxy anthroquinone, ABDAQ = 4-amino-2,6-bis(4-butylphenoxy)-1,5-dihydroxy anthroquinone, PMMA = polymethylmetacrylate, PHEMA = poly-2-hydroxyethyl methacrylate

phase change in response to an electric field, resulting in the ability to display information. Fuel cells, discussed in Chapter 13, are not electronic devices in the traditional sense, but they represent an important growing technology for replacement of fossil fuels in an increasingly sensitive environment. Several economically significant applications markets are discussed in the following.

A. Communications

More than 15% of the electronics components consumed worldwide are used for telecommunications applications. The distinction between communications and electronic data processing is becoming ill-defined, with the widespread use of modems, fax machines, and computer networks that share the same infrastructure as voice communication. A phone system that once ran entirely on copper wire now consists of fiber optic lines, microwave transceivers, and orbiting satellite links to enable rapid routing of calls anywhere in the world, and rerouting of calls in the event of subsystem failures. While the electronic devices used in this network are not fundamentally unique, they do place a particular emphasis on *bandwidth*, or the ability to carry many signals simultaneously on a single carrier. It is this appetite for bandwidth that motivates the migration from conventional circuitry and wires to optical components. Silica glass has the highest transparency of any known material to electromagnetic radiation, in the wavelength range of 1.0 to 1.5 µm. This facilitates the use of carrier frequencies on the order of 300,000 GHz, several orders of magnitude higher than is achievable in practical applications of electrical wiring. On the other hand, the use of analog-to-digital conversion of voice transmissions as well as digital transmission of computer data and live compressed video signals has given rise to an infrastructure for higher-frequency digital phone lines for videoconferencing and for secure, encrypted voice communications. The switching network that connects one phone to any other phone in the world (or cable TV signal, videoconference, etc.) has evolved from hard wires plugged in by an operator to a computer database that associates a virtual address, the phone number, with a physical address, the wire leading to the phone. Services such as call forwarding become a relatively simple matter of temporarily changing the virtual address.

The high-frequency capability of GaAs devices has given rise to a specialty application technology known as *monolithic microwave integrated circuits* (MMICs). These circuits are used by the military in phased array transmitters and receivers for both communications and rapid-scan radar systems. Very low noise amplifiers are particularly useful in satellite telecommunications and control. Commercial applications are likely to flourish in the areas of direct broadcast satellite (DBS) television reception and tuners for high-definition television (HDTV). In order to be able to handle the information included in such broadcasts, these receiving systems will operate in the range of 12–13 GHz and extract the individual signal channels at 360–860 MHz.

B. Automotive

One particularly visible niche market for microelectronics is the automobile industry. The electromechanical ignition system of points, rotor, and distributor, for example, has been largely replaced by electronic ignition systems. While comprising only 5% of the worldwide electronics market, the automotive electronics market is expected to grow at a rate 10 to 15 times higher than the growth in car sales, expected to be under 1.5% per year. During this time, the cost of electronic components per car will more than double from $100 per vehicle to more than $210. This growth is driven by a combination of government regulations requiring lower emissions and higher fuel efficiency, as well as customer preference for air bags over seat belts alone. Many of the functions that have migrated or are migrating to microelectronic control are listed in Table 5.

The circuits used in automotive applications are designed with a combination of elements from VLSI technology and power devices. They must be particularly resistant to vibration, engine heat, environmental temperature extremes, and corrosive lubricants and exhausts. The spark plug wiring system, with its 10,000-volt pulses, serves as an rf broadcast antenna that can induce erroneous signals in the interconnects. Isolation of the automotive ICs must be thorough as well, since leakage of even a small portion of those pulse signals into the chips can destroy them. At the same time, the overall system reliability is critical, given the casual reliance of people on their vehicles, often without the benefit of the recommended maintenance schedules.

C. Sensors

The mechanical properties of the materials used in electronic devices may be exploited to construct electromechanical sensors for a variety of applications. Because of its purity and crystalline perfection, single-crystal silicon's mechanical properties can be very precisely and uniquely defined anisotropically along each crystal axis. Silicon is comparable in mechanical strength with other engineering materials, shows no appreciable plastic flow, is a good conductor of heat, and its brittleness enables precise fracture along crystal axes to separate fabricated devices on a wafer.

Piezoelectric transducers measure the change in oscillation frequency of a crystalline material as a function of mechanical strain. Zinc oxide, ZnO, is one such material that can be sputter deposited with a preferred crystal orientation, and patterned to form device structures. By fabricating this oscillator on a silicon diaphragm, as shown in Fig. 12, changes in the ZnO frequency can be used to

Table 5 Automotive Applications

Emissions diagnostics	Air bag impact sensors
Simplified wire harnesses	Multiplexed wiring
Ignition	Fuel injection
Active suspension	Traction control
Power train management	Antilock braking systems
Collision avoidance systems	Heads-up windshield displays
Power windows	Power door locks
Power seat positioning	Temperature sensors
Stereo radio/tape/CD	Keyless entry
Navigation systems	Noise cancellation
Cellular phones	Pressure sensors
Hall effect sensors	Accelerometers
Power rearview mirror positioning	

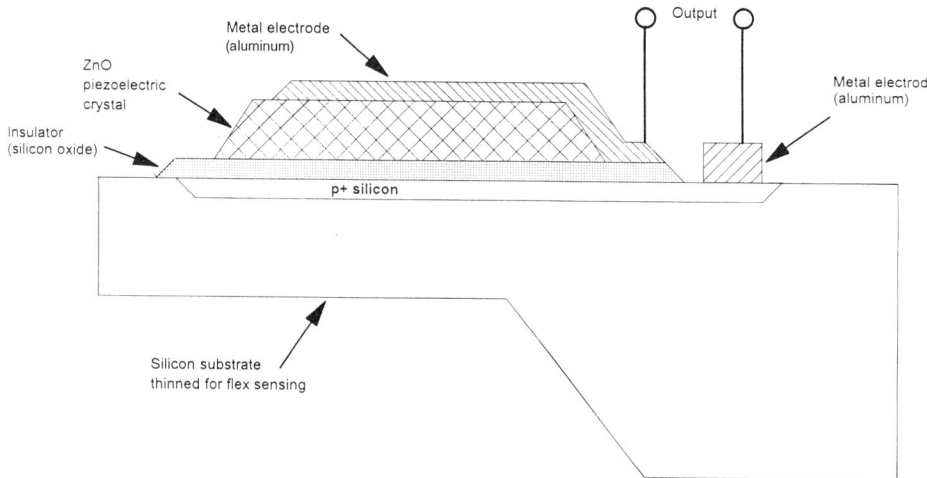

Figure 12 Cross-section diagram of a peizoelectric transducer.

detect pressure changes across the diaphragm or other mechanical causes for stress through the crystal.

A pressure sensor can be fabricated by recognizing that the capacitance between two parallel plates is a function of the distance between them (see Eq. 1). A common configuration is shown in Fig. 13. The top electrode assembly is fabricated on the silicon wafer using a series of additive depositions and subtractive etchings, which result in a sealed, hollow cavity. The back side of the wafer opposite the cavity is then etched to make a diaphragm only 10–30 µm thick for improved sensitivity. Measuring the capacitance between the electrodes can provide an absolute pressure value over the range of calibration.

Accelerometers are used in navigational systems, aircraft and space vehicles, and for active suspension and air bag deployment in automobiles. Solid state devices consist of a seismic mass suspended on a relatively weak hinge, oscillating between two parallel plates (Fig. 14). Deflection of the mass changes the oscillation signal, which is interpreted as a change in acceleration. Operational ranges are determined by appropriate selection of the seismic mass and the strength of the hinge. In an alternative design, the acceleration may be sensed by fabricating a piezoelectric strain gauge across the hinge, rather than using parallel plate capacitance.

By extending the principles used in the devices described here, it is possible to fabricate more complex sensors and even micromechanical devices such as gears, slides, and electric motors. These micromachines are finding early applications in the medical field, often in procedures that obviate the need for traditional surgery.

D. Power Devices

Power devices are those designed specifically for operating conditions that require high current, high voltage, and sometimes high speed. Typical applications of such devices would be signal amplification, switching, and systems control, in which the power device is used to regulate power to other devices. Design considerations that distinguish this application from other devices include high current densities, high field strengths, and high operating temperatures due to resistive heating. Not only must the device be modified to accommodate these conditions, but the packaging must also be more rugged, especially with respect to heat dissipation.

Power devices are fabricated from the same materials sets as more conventional devices, with specialty materials added when required. While VLSI devices are measured in tenths of microns, power devices are more likely to be measured in tens to hundreds of microns. To carry the additional currents required, power devices are often designed with multiple contacts in a single device, all connected in parallel, to distribute the current without overloading

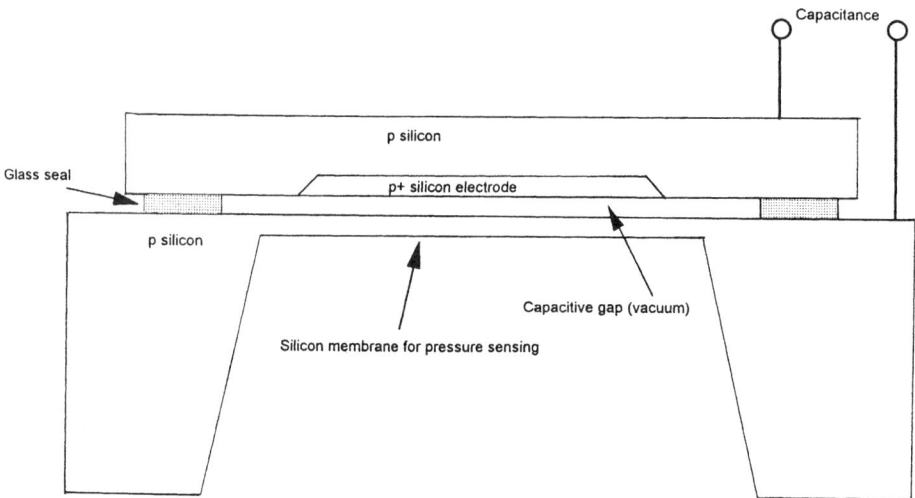

Figure 13 Cross-section diagram of a capacitive pressure sensor.

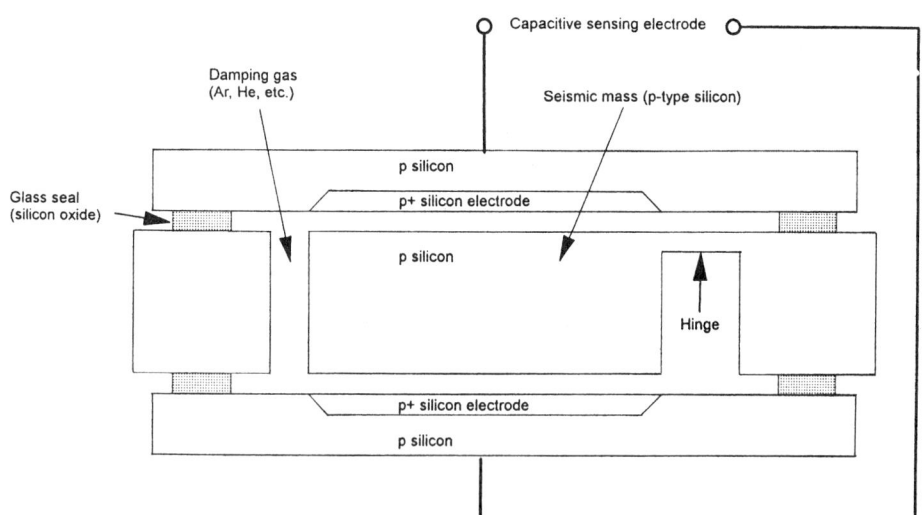

Figure 14 Cross-section diagram of an accelerometer.

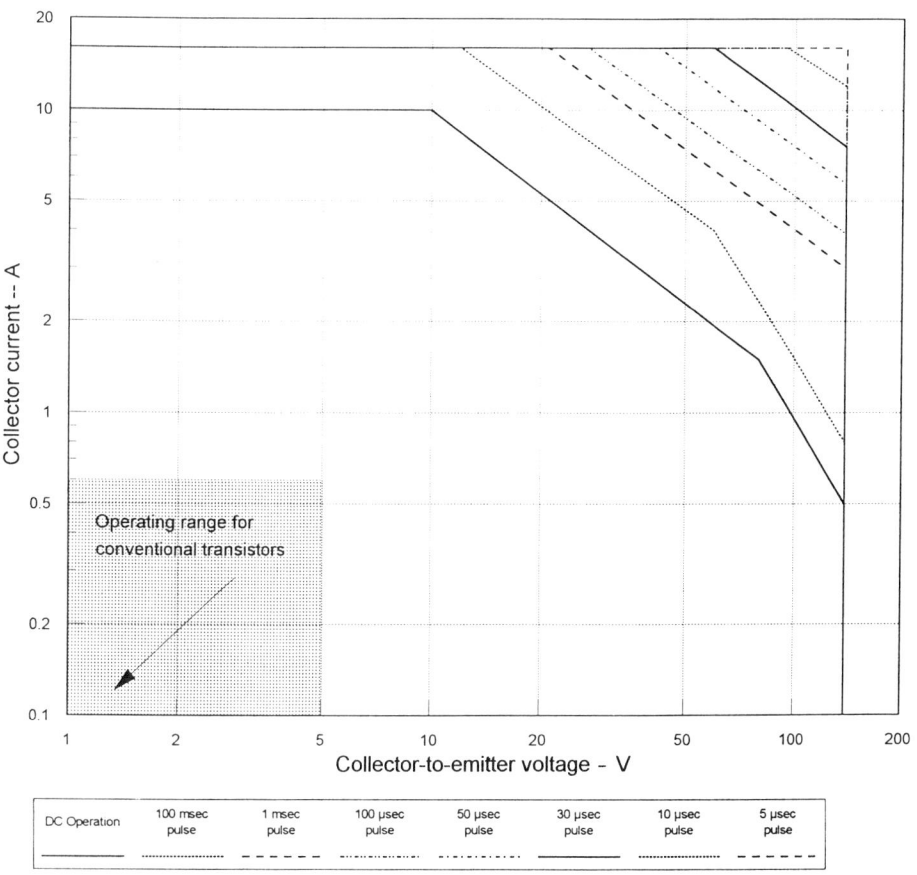

Figure 15 Current–voltage operating regions for a typical power transistor and for conventional transistors.

any one junction. These considerations enable the device to operate at current and voltage levels far above conventional devices, as shown in Fig. 15. Device specifications identify two distinct regions for applications, continuous operation and pulsed operation, where the highest currents and voltages can be applied only for the shortest pulses.

IV. SUMMARY

In addition to the variety of devices discussed in this chapter, there are several other technology families that play a significant role in the electronics industry.

Magnetic storage disk technology has evolved from 60 Kbytes of data on an 8" (200 mm) floppy disk to 2.88 Mbytes on a 3.5" (87 mm) disk for portable data storage, and to over 1 Gbyte of storage capacity in a 3" (75 mm) cube of stacked fixed disks. The magnetic read/write head fabrication process combines thin-film semiconductor process technology with micromachining techniques to continually increase data storage density. The liquid crystal diode technology developed for wristwatches has lead to flat panel liquid crystal displays for portable computers and for overhead projection of computer screens. Photodetectors have evolved from crude light beam sensors for security systems to high-resolution optical elements for video cameras and microscopes.

The demand for electronic components shows no signs of weakening; for each application that becomes obsolete, hundreds more are invented to demand still higher performance and lower cost. At the same time, thousands of convenience applications are brought to market using the lowest-cost, older technology components, giving them a second product life cycle. Each day of production brings with it new lots of source chemicals, new drifts in process equipment components, and new conditions that give rise to corrective actions on the manufacturing floor, each one requiring the skills of trained microelectronics engineers and technicians. The problems solved today become the new base for teaching and for designing processes and equipment tomorrow, a cycle that drives the constantly changing technology front.

REFERENCES

1. J. Bardeen and W. H. Brittain, *Phys. Rev.* 74:230 (1948).
2. J. Kilby, *IEEE Trans. Electron. Devices ED-23*:648 (1976).
3. Y. Kawamoto and S. Tachi, *JEE*, December 1991, p. 32.
4. Cheng T. Wang, *Introduction to Semiconductor Technology. GaAs and Related Compounds*, John Wiley & Sons, New York, 1990.
5. Sandip Tiwari, *Compound Semiconductor Device Physics*, Academic Press, Boston, 1992.
6. N. Goto, Y. Isozaki, K. Shidara, E. Maruyama, T. Hirai, and T. Fujita, *IEEE Trans. Electron. Devices ED-21*:622 (1974).
7. W. L. Guthrie, W. J. Patrick, E. Levine, H. C. Jones, E. A. Mehter, T. F. Houghton, G. T. Chiu, and M. A. Fury, *IBM Journal of Research and Development* 36(5): 845–858 (1992).
8. R. M. Geffken, Proceedings of the 179th Meeting of the Electrochemical Society, Washington, DC (1991).
9. A. E. Kaloyeros and M. A. Fury, *MRS Bulletin* 18(6):22–29 (1993).
10. D. N. Buckley, *The Electrochemical Society "Interface,"* Winter 1992, p. 41.

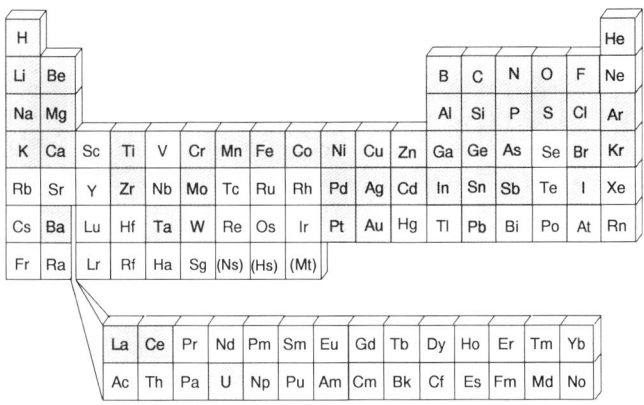

2
Electronic Processes and Environment

Robert K. Lowry *Harris Semiconductor, Melbourne, Florida*

I. INTRODUCTION

Semiconductors have been called the engines of the information age. The advent of small-scale electronic devices was the enabling technology of that information age, which has brought global social, political, and commercial change through advanced telecommunications and information handling. Small-scale microelectronics have enabled new triumphs in medicine and space exploration. Semiconductors continue as the heart and brains of ever-smaller, ever-faster, and ever more powerful computing and data-handling equipment, as well as in myriad other electronic and optical devices. They also comprise a continuously increasing share of the electronics for the highly mobile military forces of the future.

They are a growing component in industrial and consumer electronic equipment, especially for industrial control, communications, and automotive applications.

Microelectronic devices, like semiconductors, tend not to be thought of as chemical products. It is easy to think of everyday household or personal items like drain cleaner, dish soap, and shaving cream as chemical products. Even things less obviously "chemical," like foodstuffs, building materials, and clothing, can still be recognized as chemical products, as they are all intrinsically "chemical." However, to the casual observer, computer chips, or solid state fuel cells, or light emitting diodes, are not viewed as chemical, but as "electrical," because their functions are electrical. Indeed, it is electrical engineering, not a materials-based physical science discipline, which focuses on conceiving, designing, and using these in electrical products.

Semiconductors, though, are as much a chemical product as bug spray or swimming pool supplies. Fabricating them has spawned an entire specialized chemical process industrial sector, much of it in microminiature. Chemistry and the other physical and materials science and engineering disciplines are essential at every step, from making the substrates on which devices are built to enclosing them in protective packages.

Microelectronic products are so pervasively "chemical" that more than two-thirds of the elements in the periodic table are either part of their composition or are otherwise significant to their fabrication. Periodic tables at the beginning of each chapter indicate chemical elements important to that chapter's subject matter, whether as an element or in an alloy or compound essential as part of the device structure, essential to fabricate the device structure, or essential to be avoided as a crippling impurity.

A global perspective of the entire microelectronics manufacturing process is shown in Fig. 1. The process starts with the growth of single-crystal material. Wafers for use in manufacturing are obtained by cutting slices from the crystal rods. Batches of wafers are then sequenced through as many as dozens or even hundreds of process steps, using many gallons of liquid process chemicals, thousands of cubic feet of gases, and hundreds of thousands of gallons of ultrapure water, to create in and on the surface of each wafer a large array of localized individual regions. Each region is a congregation of materials comprising interconnected (integrated) circuits designed for specific electrical functions, and it is termed a "die," since after electrical testing the wafer is sawed up or "diced" to separate them. The individual die (called chips) vary in size from less than 1 mm^2 to more than 4 cm^2. Individual chips are finally "packaged" into enclosures that protect them from the outside environment and enable them to be installed on circuit boards or in other equipment.

"Bulk" materials manufacturing measures its product in meters and centimeters, kilograms, and liters. The "macro" aspects of microelectronics manufacturing operate in these regimes, since process materials are used in large

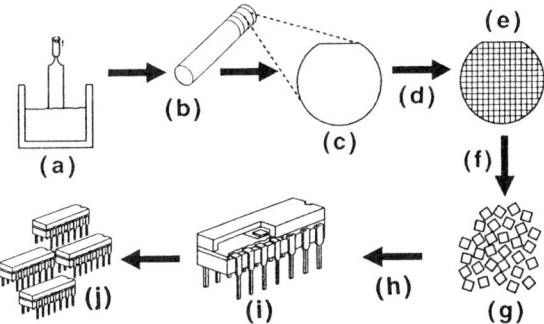

Figure 1 Semiconductor manufacturing process overview: (a) growth of single crystal in rod form; (b) single-crystal rods, sliced into wafers; (c) substrate wafer ready for fabrication process; (d) wafer fabrication; (e) completed wafer with integrated circuit die; (f) electrical test, wafer scribe and break; (g) chips; (h) assembly; (i) packaged chip; (j) product ready for shipment.

quantities. However, physical product features on individual chips are described in dimensions that are smaller by orders of magnitude, i.e., nanograms, microns, nanometers, and Angstroms. It is the diminutive amounts of material and sizes of physical features on a die, and the processes to form them, that are the challenges for microelectronics fabrication.

Figure 2 contrasts a progression of magnifications down to micron dimensions of the setae (hairs) on the mosquito's legs and the chip's metal conductor linewidths. Mosquitoes, and individual chips, are easily seen with the naked eye. But individual small features of each one, like metal conductor lines and setae, are so small that they can only be seen with high-power electron microscopes. The small sizes of these features attest both to nature's wonders and to the industry's ability to fabricate tiny features. However, there is a further hundred- to thousandfold shrinkage in electronic device dimensions in the z (height) dimension. Certain thin-film materials essential to circuit operation, such as thin oxide dielectric films and metal alloys used for resistors, are only 100–300 Å thick. Contaminant layers or subtle variations in surface texture only a few tens of Angstroms high can profoundly affect device performance. Figure 3 compares and contrasts microelectronics feature sizes (bold print) over seven orders of magnitude to some familiar everyday items (light print).

The complexity of microcircuits is evident in the small dimensions and variety of structures and materials present in small spaces, as shown in Fig. 4. This example represents a cross section of only a small portion (i.e., a doped wafer substrate and a number of metal structures whose complete set of components are not all shown) of a bipolar complementary metal oxide semicon-

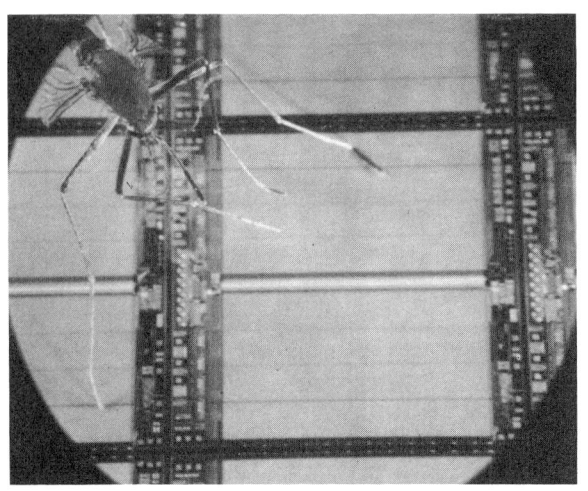

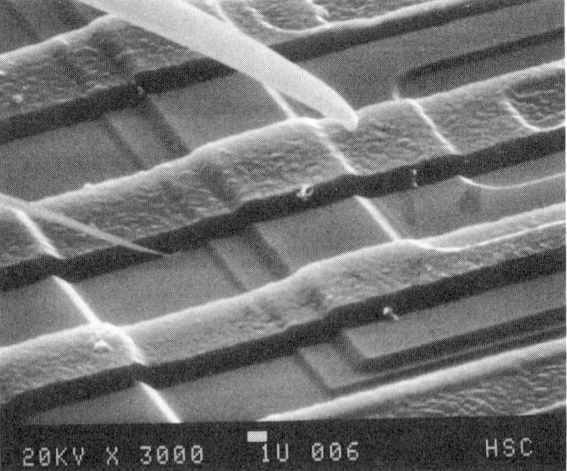

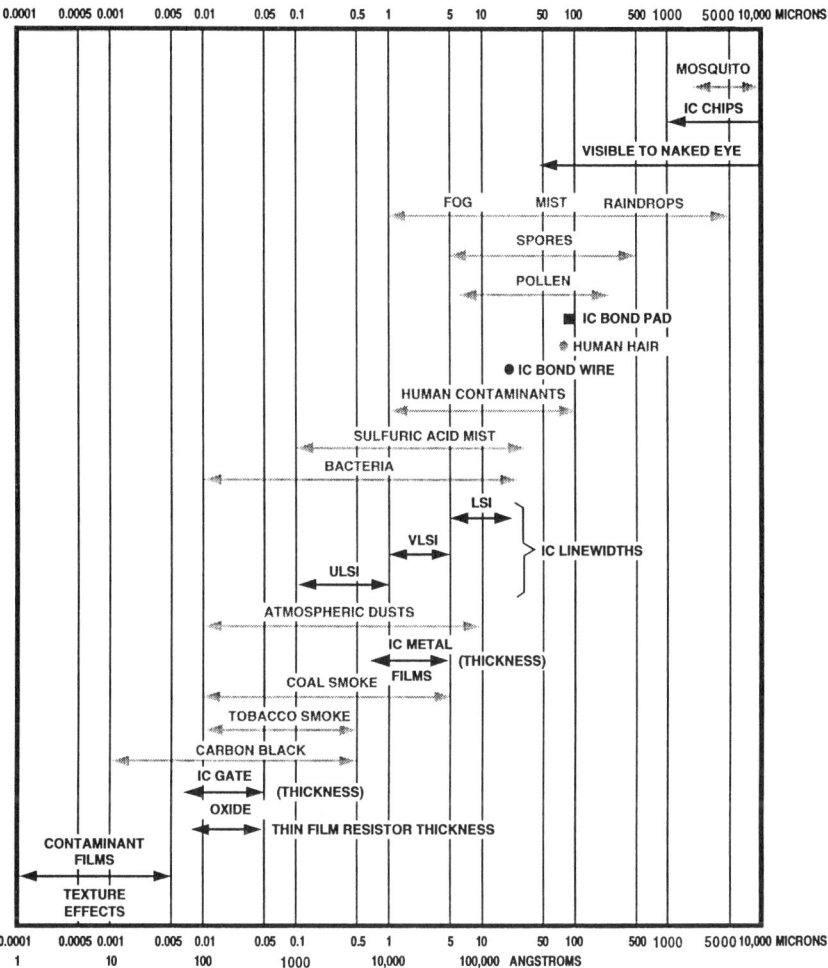

Figure 3 Dimension perspectives.

ductor (BiCMOS) device structure. The biggest feature is the silicon region doped with 10^{17} atoms of phosphorus/cm^3 (Region 10) and represents only 25 µm. The smallest vertical feature is the NiCr resistor in region 13, which is only 125 Å thick and therefore cannot be accurately depicted in this schematic. Making these complex device structures, of which there are more than a mil-

Figure 2 Mosquito on a chip.

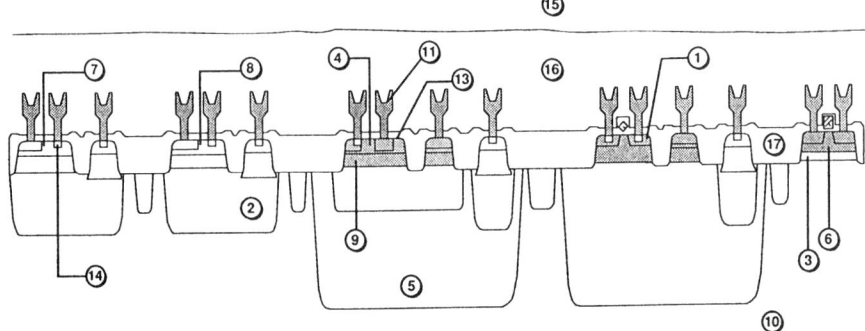

Figure 4 Materials in a bipolar complementary metal oxide semiconductor integrated circuit:

1. Silicon doped with 10^{20} atoms As/cm³ Si
2.   "   "   " 10^{19} "   As   "
3.   "   "   " 10^{15} "   As   "
4.   "   "   " 10^{18} atoms P/cm³ Si
5.   "   "   " 10^{17} "   P   "
6.   "   "   " 10^{16} "   P   "
7.   "   "   " 10^{18} atoms B/cm³ Si
8.   "   "   " 10^{17} "   B   "
9.   "   "   " 10^{16} "   B   "
10.   "   "   " 10^{15} "   B   "
11. Al with 2%Cu/1%Si
12. TiW
13. NiCr (too thin to depict in diagram)
14. PtSi
15. SiO_2
16. SiO_2 with 3%P
17. SiO_2 with 3%B/4%P

17 substances
12 chemical elements
10 kinds of doped silicon, 3 kinds of dopant elements
7 metallic elements
4 kinds of metal alloys
3 kinds of silicon dioxide
2 kinds of elemental silicon (epitaxial, single-crystal)

lion arrayed across a 200 mm silicon wafer, requires well over 200 production steps.

Chemistry is important in all steps in the global perspective in Fig. 1, but the wafer fabrication portion of the process (step e) is especially chemical-intensive (Fig. 5). This chapter describes substrate material formation, explores environments for and some of the major steps in wafer fabrication, and gives material perspectives on packaging and product reliability. Further technical detail on many aspects of these subjects is then developed in subsequent chapters.

II. THE ELECTRONIC DEVICE MANUFACTURING DISCIPLINE

Relentless shrinkage of the physical size of microcircuit device features, coupled with the economic imperative of making them profitably, places increasingly

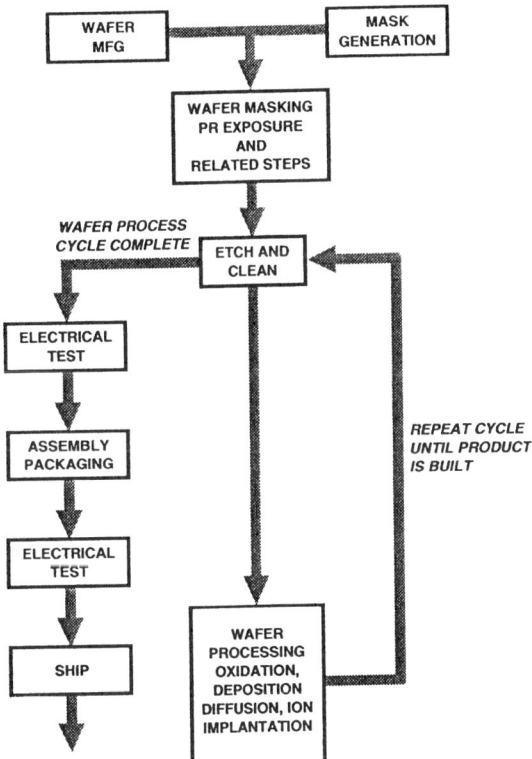

Figure 5 Wafer fabrication and assembly process overview.

Table 1 Microelectronic Surface Contaminants

Type of contaminant	Sources	Effect on product
Particles	Process equipment; Process materials; Process ambient; People	Structural flaws, device killers
Metals	Process equipment; Process chemicals	Defects in Si lattice; Charge traps in SiO_2
Cations	Poor wafer cleans; Process chemicals; People	Anomalous charge effects; Electrical parameter shifts
Inorganic anions, residues	Poor wafer cleans; People	Corrosion of device materials
Hydrocarbon residues	Poor wafer cleans; Poor air control	Loss of film adhesion; Electrical contact resistance

stringent demands on manufacturing materials and technologies. Miniaturization demands increasing accuracy and firm control of the sequential growth or deposition, and subsequent pattern formation, of the many insulating, semiconducting, and conducting materials that comprise an integrated circuit.

Accordingly, the associated environmental requirements for particulate and ionic cleanliness in manufacturing are extraordinarily demanding. To make a complex integrated circuit, such as a dynamic random access memory (DRAM) device, with linewidth definitions ≤0.8 µm and three or more layers of metallic conductor lines "integrating" all the electronic functions together into a circuit, may require, as mentioned, more than 300 individual manufacturing process steps. Clearly, few faults can be tolerated in such a complex procedure. A 99% success rate at each step of the 300-step process would ultimately provide only a 4.9% yield at process completion. Rigorous engineering, superb process equipment (tools), and control of the cleanliness and consistency of process materials are absolute imperatives for successful device fabrication.

"Cleanliness" in a device fabrication environment means the virtual absence of physical particles and ionic, metallic, atomic, and molecular species from the process surroundings. It also means the effective removal from the process surroundings of any such species created as by-products of processing. Table 1 is a general summary of types and sources of contaminants and how they affect microelectronics. Figure 6 shows how these can affect device surfaces, which must be critically clean throughout the manufacturing process.

tures. Accordingly, before use in electronics processes, water is sent through purification sequences including reverse osmosis, cation/anion/mixed bed deionization, chemical or ultraviolet sterilization (to kill bacteria), and ultrafiltration through filters with pore sizes of only hundreds of Angstroms. Ultrapure water must be absolutely sterile, as colonies of bacteria are potential particulate defects in the photopatterning process. Microbiological DNA testing methods can detect as little as a single colony forming unit, i.e., a single bacterial cell, per one liter of ultrapure water.

There are various guidelines for water purity depending on the complexity of devices being fabricated. Table 3 summarizes guidelines for process engineers for pure water to be used in four-megabit dynamic random access memory (DRAM) manufacturing. Like cleanroom facilities, the production and delivery of ultrapure water has also become an important "auxiliary" technology in semiconductor fabrication. Every semiconductor plant relies on its own ultrapure water systems and specialists to operate them.

2. Process Chemicals and Gases

Suppliers of chemical commodities for wafer processing have made great technical strides in producing, containing, and delivering ultrahigh purity materials not only to the manufacturing cleanroom, but also ultimately to the surface of the wafer itself. An example quality specification and measured values for the purest grade of hydrofluoric acid readily available for semiconductor processing are given in Table 4. Remarkable levels of purity are achieved, with 0.1 part per billion (100 parts per trillion) the maximum concentration for each of 26 metal impurities. Even more important than the specification limits are actual concentrations, which for metals are actually less than 100 parts per trillion— less than 1 impurity atom in more than 10^{10} atoms of material, or <100 impurity atoms per trillion atoms of material. Given that mineral acids are derived from ores mined from the earth, the ability to produce, contain, and deliver to the wafer surface a liquid chemical with less than 2 impurity atoms in 10^{10} atoms of material is exceptional.

Gases are also available in ultrapurity, being easier to purify since volatility-based separations are easier to carry out in the gas phase. Unstable reactive gases like silane and arsine, which are produced by liquid-phase chemical reactions, can also be prepared in ultrapurities. As with process liquids, challenges for contamination control are greatest in the containment and delivery of pure gases. Special progress has occurred in producing and maintaining high-pressure cylinders, regulators, fittings, and tubing with ultraclean internal surfaces that minimize particle generation and volatile impurities, such as water vapor. Water is a substance of great contrast in processing, as huge volumes are used in cleaning, yet only microliters of vapor evolved from newly installed or inadequately prepared internal surfaces of gas tubing can cause haze and other defects

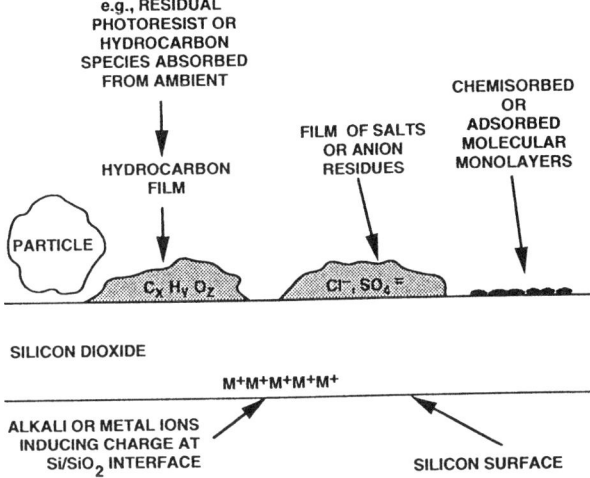

Figure 6 Surface and interfacial contamination on wafers.

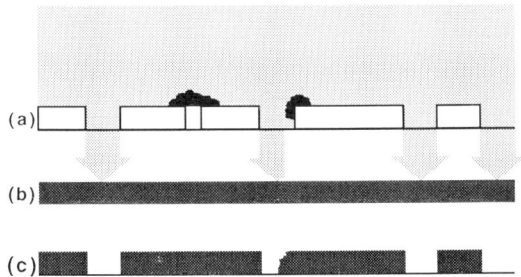

Figure 7 Defects caused by a particle in the patterning process: (a) plate containing master pattern of circuit; (b) film to receive pattern from (a); (c) result of patterning–particles cause ragged line definition or inaccurate feature replication.

A. Particle Control

Consider the effect of one small physical object, a discrete, unyielding, tightly adhered particle (Fig. 7) on a patterning process. Such a particle on a master circuit pattern will print as a repetitive and potentially device-destroying defect in the same location on the product every time the mask is used to generate a device pattern. If the particle is on a substrate surface, it will cause a correlated defect in the patterned material.

One such particle per circuit structure on each die of a wafer has a high probability of destroying functionality of every die. Even a few hundred particles randomly distributed across a wafer's surface (hypothetically leaving one particle to each circuit die) would also result in nonfunctional chips from that particular wafer, i.e., zero yield. Yet wafers can be exposed not to hundreds but literally to millions of particles during the sequential processes of fabrication. Thus, the challenge for process engineers is to assure ultracleanliness of both the production environment itself and the materials entering that environment.

This need for cleanliness becomes increasingly acute with shrinkage in geometries of the features being patterned, since fewer and smaller particles can be tolerated with smaller features. A rough guideline is that no particle larger than 10% of the minimum planar geometric dimension being patterned can be present. If 0.8 µm is the design linewidth for a device, this guideline would prohibit any particle larger than 0.08 µm in the process environment. This excludes virtually every kind of naturally occurring particle; see Fig. 3.

Cleanliness needs dictate that semiconductor production be done in space that is ultraclean, where air flow is filtered to eliminate every possible particle from the ambient air. Clean rooms for semiconductor production are designated as "Class 10," "Class 1," etc. Table 2 contrasts the maximum number of particles allowed by Federal Standard 209E for different cleanroom categories with the numbers of particles generated by human activity. Removal of such huge numbers of particles from cleanroom air to attain the required particle levels is achieved by passing air through "High Efficiency Particulate Air"

Table 2 Numbers of Particles in Various Environments

Type of motion	Particles >0.3 µm generated per minute			
Gymnastics	20,000,000			
Walking fast	10,000,000			
Walking slow	5,000,000			
Sitting with slight body and foot motion	1,000,000			
Sitting with slight head and hand motion	500,000			
Sitting motionless	100,000			
Cleanroom environment	Particles per cubic foot[a]			
	≥0.5 µm	≥0.3 µm	≥0.2 µm	≥0.1 µm
Class 100	100	300	750	—
Class 10	10	30	75	350
Class 1	1	3	7.5	35

[a]Maximum number of particles of each size for each class.

(HEPA) filters. Fiberlike filter media are fitted into filtration units in an accordion-fold array, presenting an extremely high surface area to air flow. The media trap airborne particles without a major reduction in the volume of air passing the filter.

Cleanliness management extends to the supplies and personnel entering cleanrooms. Typical industrial laboratory workers wear laboratory coats and other personal protective equipment to guard them from the materials with which they work. In contrast, in a semiconductor cleanroom, special clothing is worn to protect the materials from the workers, because the human body is a prolific source of chemical and particulate contamination. An initial level of protection for wafer fabrication facilities involved in making complex devices is to employ only nonsmoking employees for positions inside the cleanroom because of the huge particle numbers on smokers' breath, as shown in Fig. 8. However, even the hundreds of particles per cubic foot from a nonsmoker's breath and the fact that the rest of the human body is an abundant particle generator (hair, dandruff, skin flakes, and other effluvia; see again Table 2) require that the product be protected from the people. In cleanrooms of Class 10 or better, people are gowned head to toe in special clothing, called bunnysuits, made from non-particle-shedding fabric. Where these kinds of precautions are not taken, e.g., where cleanroom garments provided inadequate protection or were improperly worn, device contamination will occur. Figure 9 shows people-sourced contamination on circuits, which occurred when lack of face shielding (probably temporarily removed!) allowed spittle to deposit directly on a chip.

Advanced cleanliness management minimizes or even eliminates people by fabricating circuits in cleanrooms "staffed" by robotic equipment. However, cleanrooms that fully surround all process equipment and human or "robotic" personnel are increasingly expensive. The cost of a classical modern cleanroom

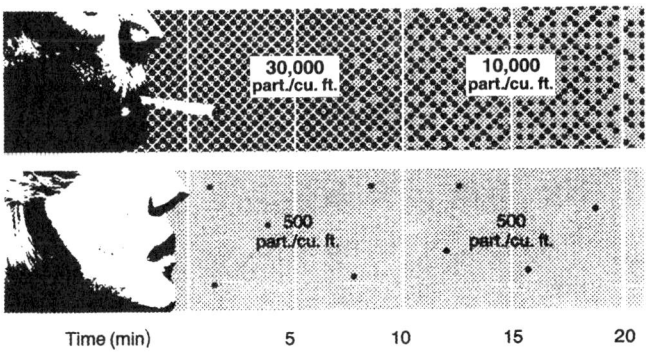

Figure 8 Particle generation by smoker versus nonsmoker.

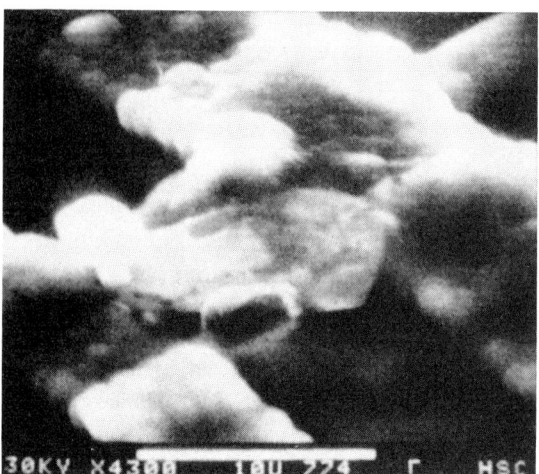

Figure 9 Salt crystal: human-sourced contamination on a chip.

for one megabit (one million components per chip) DRAM production approaches one billion dollars. To reduce this cost, new technologies in device fabrication use modular ultraclean containers in which product wafers are moved into and out of process tools in a manner such that the process equipment and the travels of the transporting containers need not be surrounded by an ultraclean environment. Design and construction of cleanroom facilities and equipment have become an essential "auxiliary" technology in the semiconductor industry.

B. Ionic and Metallic Contamination Control

Ideally, no substance exposed to circuits during device fabrication should add any chemical species that could reduce process yield or degrade operational reliability of devices. As previously summarized, the awareness of the role of chemical species in electronics manufacturing extends beyond its direct use, to concerns for process and product contamination.

For example, the alkali elements in Group I of the periodic table are easily ionized and have relatively low mass. These ions are typically very mobile in device structures. They become trapped at interfaces and can set up anomalous charge fields, shifting the electrical operating parameters of devices outside the intended range. Sodium is of greatest concern because of its ubiquitous nature.

Hydrogen behaves similarly, and its presence is of particular concern in metal-oxide-semiconductor (MOS) structures. Hydrogen, as an impurity, is more likely to be a reaction by-product of chemical processing (Chapter 3, Sec. IVB) than an impurity species in process materials.

Other impurities to be avoided are the transition metals. Iron, copper, etc., often remain as residual impurities even in ultrapurified process chemicals at levels as high as parts per billion, or they may be transported to wafer surfaces by various chemical reaction mechanisms from process equipment. If incorporated in the silicon lattice, metals occupy lattice sites and, with their varying numbers of d-shell electrons, impart extra electrons (conductivity) to a semiconducting structure or become sites with "dangling bonds" that are electrically active device defects.

Residues containing anions such as chloride, phosphate, and sulfate must be avoided. Just a few monolayers of residual moisture can leach these anions, producing a chemically corrosive, galvanically active "salt solution." Merely nanoliters are sufficient to corrode device metallization structures to failure. Chloride ion is well known as a corroding species. Metal corrosion can occur during manufacturing, detracting from production yield, or later on finished devices, causing failure in customers' systems.

Molecular-scale impurities such as hydrocarbon residues left behind by poor cleaning are also a concern. Such deposits can interfere with adhesion of deposited films, or they can cause such high electrical contact resistance at metal contact sites that electrical function is impaired.

Dopant atoms are "purposeful" impurities, placed in certain regions of device structures to impart particular electrical conductivity to specific device regions. When the spatial distribution and concentration (typically on the order of 10^{14}–10^{20} atoms of dopant per cm^3 of silicon) of dopant atoms are properly controlled in both lateral and vertical space, intended device properties are realized. But if dopant atoms are improperly distributed, by being either too shallow or too deeply distributed by moving sideways into unintended regions of a device, or by being too high or too low in concentration, then the "purposeful" dopants become destructive impurities that cause electrical malfunction.

C. Process Materials

1. Ultrapure Water

Ultrapure water is the common denominator in wafer cleaning. It is a diluent for some solvents, the medium for acids, bases, and oxidants, and usually the final substance contacting wafers in any unit process. This is a processing chemical often taken for granted, although it is used in larger amounts than any other liquid chemical. As many as 500 gallons of ultrapure water (18.35 megohm-cm resistivity, meaning theoretical purity, or virtual lack of conducting ions in solution at 25°C) can be used for every wafer processed.

Tap water and even the best distilled water are unsuitable for microelectronics processes because they contain dissolved ions and salts, organic carbon, dissolved and suspended solids, and bacterial growth that endanger device struc-

Table 3 Guideline for Ultrapure Water Quality for Production of Four Megabit Dynamic Random Access Memory Devices

Resistivity @ 25°C, megohm-cm	18.2
Total organic carbon, ppb	<5
Dissolved silica, ppb	<1
Residue, ppb	<0.1
Boron, ppb	<0.5
Ions, ppb, by ion chromatography	
$\quad$ Na$^+$	0.05
$\quad$ K$^+$	<0.1
$\quad$ NH$_4^+$	<0.1
$\quad$ Ca^{2+}	<0.05
$\quad$ Mg^{2+}	<0.05
$\quad$ F$^-$	<0.1
$\quad$ Cl$^-$	0.1
$\quad$ Br$^-$	<0.05
$\quad$ NO$_3^-$	<0.05
$\quad$ HPO$_4^{2-}$	<0.05
$\quad$ SO$_4^{-4}$	0.1
Metals, ppb	
$\quad$ Al	0.05
$\quad$ Ba	0.01
$\quad$ Cr	0.03
$\quad$ Cu	0.05
$\quad$ Fe	0.10
$\quad$ Li	0.03
$\quad$ Mg	0.02
$\quad$ Ni	0.05
$\quad$ Na	0.06
$\quad$ Sr	0.01
$\quad$ Zn	0.06
Particles, per liter, by on-line laser counter	
$\quad$ 0.05–0.1 μm	<100
$\quad$ 0.1–0.2 μm	<50
$\quad$ 0.2–0.3 μm	<20
$\quad$ 0.3–0.5 μm	<5
$\quad$ >0.5 μm	<1
Bacteria, per 100 ml	
$\quad$ by culture	<1
$\quad$ by EPI	<30

Source: Balazs Analytical Laboratory, *93 Pure Water Specifications and Guidelines for Facility and Fabrication Engineers.*

Table 4 Specification for Purest Grade of Hydrofluoric Acid Commonly Available for Electronics Processing

Impurity		Specification maximum	Typical amount present
Anions:	Chloride, Nitrate, Phosphate, Sulfate	10 ppb each	<10 ppb each
Metals:	Al, Ag, As, Au, B, Ba, Ca, Cd, Co, Cr, Cu, Fe, Ga, Ge, K, Li, Mg, Mn, Na, Nb, Ni, Sb, Sn, Sr, Ti, Zn	0.1 ppb each	<0.1 ppb each
Particle	≥0.2 μm	100	<100
	≥0.5 μm	10	<10

Source: Ashland Chemical Company.

on silicon wafers. Most process gases are available in purities of "6-nines" or better, meaning that impurities including water vapor total less than one part per million.

Despite the progress in making inert materials for drums, gas cylinders, and other containers to minimize particle and ion contaminants during transportation and storage of raw materials, the relentless drive for still higher purity has engendered technologies where chemicals are made on demand by in situ reaction and delivered short distances to the wafer surface. This eliminates storage and reduces the potential for brief exposure to contamination from walls of the reaction vessel and the delivery tubing. Inert gases (nitrogen, argon, oxygen) are made cryogenically by distilling liquid in air separation plants located adjacent to wafer fabrication buildings. Ammonium hydroxide is made for wafer cleaning by reacting ultrapure ammonia gas and water from the semiconductor plant's ultrapure water system in equipment directly adjacent to the cleanroom, delivering ultrapure ammonium hydroxide to wafer surfaces on demand. Technology trends will continue to favor on-site production of process chemicals from basic raw materials rather than storage and delivery to the process from bulk containment.

III. BASIC SEMICONDUCTING MATERIAL—THE SUBSTRATE

Making a semiconducting device requires a substrate "platform" with suitable chemical, physical, and electrical properties, on and into which electronic devices are built. Figure 10 shows the electrical conductivity in (ohm-cm)$^{-1}$ of various conductors, semiconductors, and insulators. Semiconducting materials include silicon, germanium, sapphire (aluminum oxide), and compounds such as gallium arsenide. Many types of substrate material can be used, and the generation aspects of these materials are quite similar to each other. The present discussion will focus on silicon as the main substrate for microelectronic products.

A. Producing Semiconductor Grade Silicon

Silicon used for microelectronics manufacturing comes from common sand. To achieve electronically acceptable grades of silicon, sand undergoes a series of purification processes. Mineral sand is first reacted with carbon, obtained from coke or coal, to make metallurgical grade silicon:

$$SiO_2 + 2C \xrightarrow{2000°C} Si + 2CO \tag{1}$$

A small portion of this metallurgical silicon is further purified by reaction with hydrochloric acid, yielding a mixture of chlorosilanes:

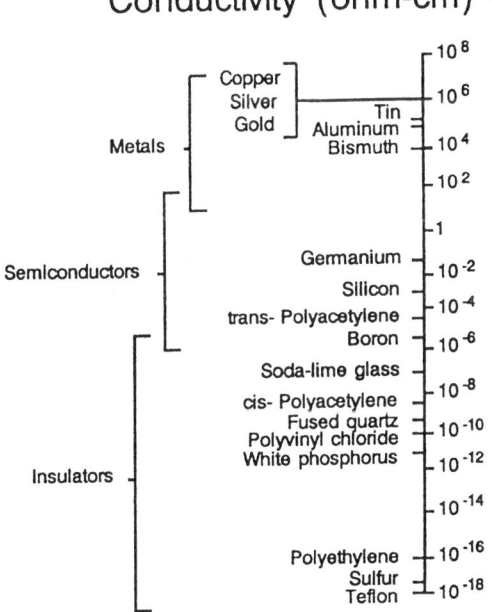

Figure 10 Materials conductivity.

$$Si + xHCl \xrightarrow{\approx 1250°C} Si H_{x-4} Cl_x + H_2 \tag{2}$$

Trichlorosilane is separated from this mixture and is reduced with hydrogen to form ultrapure silicon:

$$SiHCl_3 + H_2 \xrightarrow{1100°C} Si + 3HCl \tag{3}$$

Silicon produced with this sequence of reactions has only trace impurities, typically below the parts per trillion range, and it becomes the source material for growing single-crystal silicon for use in device manufacture. There are two methods of silicon crystal growth: Czochralski (CZ) and float zone (FZ).

Czochralski silicon is manufactured by melting the silicon produced from Eq. (3) in a resistant or radio-frequency (rf) heated quartz crucible. A seed of desired crystallographic orientation (100 or 111) is dipped into the molten silicon and slowly withdrawn with a rotational motion, resulting in a long rod of single-crystal silicon.

The FZ technique requires no crucible. For this case, the silicon rod is formed in an rf heated field between a polycrystalline silicon source and the single-crystal seed. The zone passes around the rotating polycrystalline rod, melting it to yield single-crystal material upon refreezing.

At the conclusion of these processes, the single-crystal silicon rod is cooled, and high-speed mechanical saws slice about 20 mil thick individual wafers of pure silicon from the rod. Further mechanical and chemical processes prepare the product wafers for semiconductor fabrication. These include grinding, lapping, chemical etching for surface impurity removal, polishing, and final cleaning, to achieve a wafer with exact diameter, flatness, co-parallel front and back surfaces, and surface smoothness.

The size of substrate wafers is generally dictated by the economics of the manufacturing process. As customers demand more and faster device functions on a chip at lower cost (resulting in miniaturization), it becomes necessary to make more devices in each process step. This has driven the industry to larger and larger diameter wafers, with wafer diameters routinely in production ranging from 125 to 250 mm.

The crystalline structure of the silicon wafers must attain a high degree of both physical and chemical perfection. Physical lattice imperfections such as dislocations or slip planes, or alkali or transition metal impurities, will alter the semiconducting properties of a silicon wafer. Various methods to control the amount of oxygen incorporated into the crystal are used to "getter" both physical and chemical impurities from the silicon lattice.

B. Adjusting the Electrical Properties of Silicon

At a temperature of absolute zero, the silicon crystal lattice can be depicted as in Fig. 11a. Every electron involved in bonding the atoms together in the lattice is fixed in a perfect uniform crystal structure. Every single lattice site is occupied by one silicon atom, each electronic orbital of which is fully occupied by its resident electrons.

Lattice imperfections naturally occur at temperatures above absolute zero. Some electrons leave their assigned orbitals and spend time closer to adjacent silicon atoms, as in Fig. 11b. Silicon atoms that acquire extra electrons have a negative charge, and silicon atoms that lose electrons have a positive charge. The electron movements impart current-carrying property to the silicon lattice, by electron conduction (moving negative charges), or by "hole" conduction

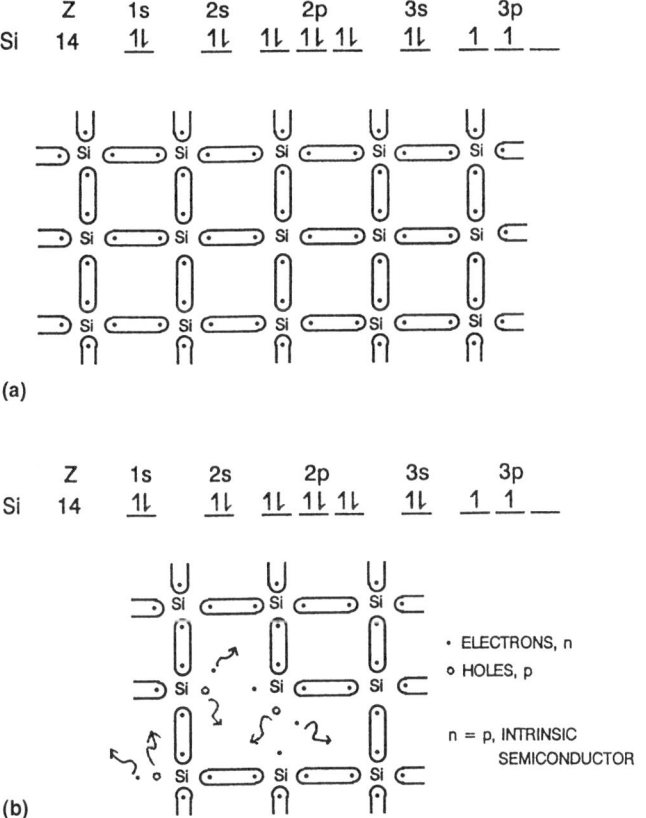

Figure 11 (a) Silicon lattice at 0 K. (b) Silicon lattice at >0 K.

(moving positive charges). When a silicon lattice has an equal number of hole and electron lattice defects, it is defined as an intrinsic semiconductor. The formula for electrical conductivity, σ, is

$$\sigma = eN\mu \tag{4}$$

where

- e = electronic charge
- N = number of ionized impurities per unit volume
- μ = drift mobility of the current carrier

From Fig. 10, the conductivity of intrinsic silicon is about 10^{-3} (ohm-cm)$^{-1}$.

Most silicon crystals are electrically modified by adding selected chemical elements during the crystal growth phase. The elements become incorporated at substitutional lattice sites in the silicon crystal in a process called doping. Adding dopants alters the intrinsic electrical conductivity by increasing N in Eq. (4), the characteristics depending on the electron configuration of dopant ions relative to that of silicon. There are two types, those that impart negative charge (n-type dopant) and those that impart positive charge (p-type dopant).

Phosphorus is a common n-type dopant. It has one more outer shell electron than silicon. When occupying a lattice site, ionized phosphorus supplies one more electron (negative charge) to the lattice than that of a silicon atom. The extra electron becomes a charge carrier, and as an electron supplier, an n-type dopant like phosphorus is called a donor. Arsenic and antimony have similar outer shell configurations compared with silicon and are also n-type dopants.

Boron is a common p-type dopant. It has one less outer shell electron than silicon. When occupying a lattice site, ionized boron supplies one less electron (positive charge) to the lattice than a silicon atom, causing a "hole." P-type dopants forming holes are called acceptors, since they "accept" an electron to neutralize their positive charge.

Electron configurations for phosphorus and boron used for extrinsic silicon doping are shown in Fig. 12, illustrating the n-type/donor, p-type/acceptor doping concept.

Figure 13 plots electrical resistivity as a function of depth for silicon doped with phosphorus and boron. Note that equal concentrations of phosphorus and boron produce unequal silicon resistivities. This is because electrons move through the silicon crystal lattice with less energy (and therefore lower resistivity) than holes. Silicon starting material is available over a wide range of electrical resistivity, and the device manufacturer chooses wafers with resistivities appropriate for the circuits to be built.

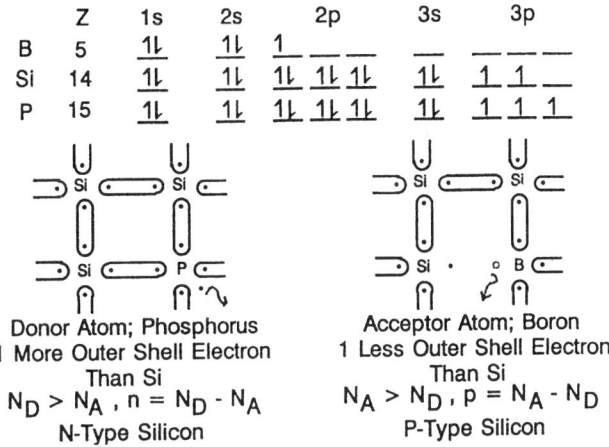

Figure 12 Electron configurations of phosphorus and boron dopants.

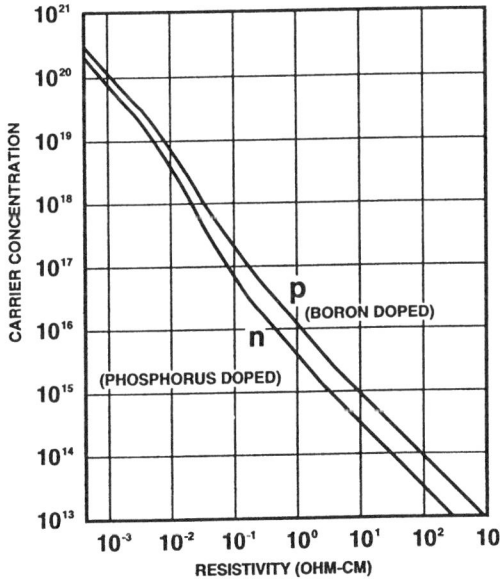

Figure 13 Electrical resistivity of n- and p-type silicon as functions of phosphorus and boron concentration.

IV. A SUMMARY OF SELECTED DEVICE FABRICATION PROCESSES

Much of the success of the microelectronics industry resides in its ability to continually improve chemical techniques and processes to make circuit structures with smaller and smaller geometries. Further dimension shrinkage is propelled by customer demand for more computing power at higher speeds and lower power consumption, all at lower cost, and these factors continue to drive the diverging trends of shrinking circuit geometries and expanding wafer diameters. Shrinkage of feature size, or miniaturization, is key to industry competitiveness.

Figure 14 shows the progress made over several decades in device integration and the resulting increase in microcircuit complexity. The industry is well into the era of ultra-large-scale integration (ULSI). The most advanced dynamic random access memory (DRAM) integrated circuits of the mid-1990s contain 64 million components (64 Megabits) per chip with lateral feature widths on the order of 0.4 µm. Technology projections show geometries continuing to shrink and densities continuing to increase, to 4 billion bits per chip (4 Gigabits) and minimum feature sizes approaching 0.1 µm, during the next decade.

There are three main steps in the wafer fabrication (step e) phase of the manufacturing process in Fig. 1: (1) selective area doping to produce junctions (regions of different electrical conductivity in the wafer), (2) physicohemical formation of various films from 100 Å to ≥1 µm thick (to be part of device structures or to be used in fabrication and then removed), and (3) patterning (precise definition of regions of doped silicon and film materials to form device

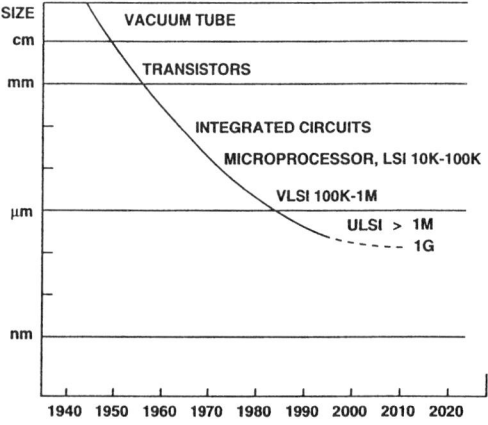

Figure 14 Microelectronics geometry shrinkage.

structures). Hundreds of unit processes incorporating these basic steps are applied to batches (process lots) of wafers. As Fig. 5 indicates, certain wafer fabrication sequences are repeated multiple times as the device layers are built up, until structures like those in Fig. 4 are attained. Many of the processes are closely tied to various film etches and chemical cleaning of wafers, which will be discussed next, followed by a brief review of the major process steps to set the stage for detailed descriptions in later chapters.

A. Wafer Cleaning

Subsequent to silicon wafer production, and dispersed between most unit process steps, wafers are cleaned. As mentioned in Sec. II, pristinely clean wafer surfaces are essential throughout fabrication. Every contaminant (see Fig. 6 and the discussion on process environment) that might be present on a wafer surface must be removed whenever possible.

Cleaning is accomplished by immersing wafers in tanks of liquid chemicals at various temperatures, in which cleaning action can be enhanced by applying ultrasonic energy. The wafers can be passed under cascades of liquids, exposed to high-pressure sprays, or treated in equipment where they are flooded with chemicals while being scrubbed with brushes.

Wafer cleaning chemistries are complex, and they involve an interesting contrast. Hundreds of liters of liquid are used to treat surface increments of just a few Angstroms, albeit over very wide areas. A batch of 24 200-mm wafers has 0.75 m^2 (over 8 ft^2) of critical surface area to be cleaned.

Chemicals effective for removing one contaminant can leave behind another, which must then also be removed. For this reason, cleans are often sequenced, where the chemicals and their order of use are selected for maximum effect to ultimately provide clean surfaces. A common approach to sequenced cleans is to remove organic contaminants first, oxides second, and metals and ions last. These steps are frequently interspersed with and always followed by rinsing in copious volumes of ultrapure water.

Organic (hydrocarbon) residues can be removed by a variety of chemistries. One is a mixture of sulfuric acid and hydrogen peroxide, with the fitting nickname "Piranha" etch. Others include catechol (a noxious odor), and solvents like isopropyl alcohol (flammable). Inorganics like carbon dioxide crystals remove hydrocarbons as the cold solid CO_2 sublimes them from the wafer surface. CO_2 also effectively removes adhered particles, when solid crystals are blown across wafer surfaces. UV-ozone and rf excited plasma discharges using CF_4/O_2 gas mixtures provide a nonliquid dry chemical environment for removing hydrocarbons and oxidizable species from surfaces. Drawbacks of these "dry" methods include the toxicity of ozone and the tendency of some plasmas to introduce defects in thin oxide films by radiation damage (Chapter 7).

Removal of unwanted silicon dioxide is often the second major step in a cleaning process. Hydrofluoric acid (HF), or hydrofluoric acid buffered with ammonium fluoride, is effective for oxide removal:

$$SiO_2 + 2NH_4F + 4HF \rightarrow (NH_4)_2 SiF_6 + 2H_2O \tag{5}$$

The reaction dissolves oxide, but the ammonium fluosilicate by-product must be removed from the vicinity of the wafer surface lest it precipitate as a particulate contaminant. A thorough water rinsing is effective since fluosilicate is water-soluble. Hydrofluoric acid is used to strip oxides to bare silicon just prior to doping operations. However, metal contaminants such as gold (a dopant) easily chemiplate from HF onto silicon surfaces. The HF-treated silicon is prone to adsorb carbon from the ambient, which converts with subsequent heating to silicon carbide particles on the wafer surface. Buffered HF solutions with high hydrofluoric acid:ammonium fluoride ratios impart Angstrom-scale roughness to silicon, the microroughness being a "physical" contaminant adversely affecting growth of thin gate oxide dielectric films a few hundred Angstroms thick. Oxide removal chemistries must therefore be carefully designed to avoid roughening freshly exposed silicon surfaces. Similarly, cleaning solutions must be carefully balanced with regard to concentration of basic chemicals and/or HF.

As a final step, metals and ions are removed with chemistries using aqueous mineral acids, bases, and oxidants. Examples of sequenced chemical cleans are shown in Table 5. Mix ratios and other conditions such as time and temperature are frequently varied to achieve best results.

Table 5 Chemicals for Sequenced Cleans

(a) Removal of Organics, Oxides, and Metals/Ions			
Step	To remove	Volume ratio	Chemicals
1	Organics	4:1	$H_2SO_4:H_2O_2$
2	Oxide	1:10	HF:UPW[a]
3	Metals/ions	1:1:5[b]	$HCl:H_2O_2:UPW$

(b) Removal of Particles, Metals/Ions, and Oxides			
Step	To remove	Volume ratio	Chemicals
1	Particles	1:1:5[b]	$NH_4OH:H_2O_2:UPW$
2	Metals/ions	1:1:5[b]	$HCl:H_2O_2:UPW$
3	Oxide	1:10	HF:UPW

[a]UPW = ultrapure water.
[b]Decreasing the volume ratio of base and peroxide to the 0.1–0.5 range in these formulas lessens the amount of roughening done to the cleaned surfaces.

As device geometries shrink, the expectations for water and chemicals purity shown in Tables 3 and 4 will become more demanding.

B. Selective Area Doping of Silicon

Besides doping bulk substrate material to the desired electrical resistivity, additional localized, selected area doping of wafers is also performed to make dissimilarly doped regions, called junctions. In Fig. 4, for instance, the interface between the region containing 10^{18} atoms phosphorus/cm^3 silicon and the adjacent region containing 10^{16} atoms boron/cm^3 silicon is a junction. Junction regions of different electrical resistivity are interconnected to make functional electronic structures.

To form junctions, dopants are introduced locally into the wafer at the desired dose, i.e., concentration of dopant ions (charges) per unit of silicon (atoms/cm^3), and the desired profile, i.e., concentration of dopants as a function of depth. Table 6 summarizes nomenclature for ranges of dopant concentrations.

1. Diffusion

In this method, dopant atoms are introduced into silicon from a chemical source made available at the wafer surface. Source materials can be solid, liquid, or gas. Dopant atoms are thermally induced to leave the source material and diffuse into the silicon lattice. They diffuse from regions of high concentration to regions of low concentration. Diffusion rates, and the distances that dopant atoms travel, are a function of temperature. Typical diffusions are performed at temperatures above 900–1000°C.

Table 7 lists some chemical compounds used as dopant atom sources. Gaseous sources are ultrapure gases fed into a diffusion tube containing the wafers, where the prevailing temperatures generate elemental atoms that are adsorbed on the wafers' surfaces. Solid sources are available in the form of thin disks placed on top of wafers in the diffusion tubes. Liquid chemical diffusion sources are a blend of volatile organic solvents and a compound containing the dopant

Table 6 Nomenclature for Dopant Levels as Function of Atomic Concentrations

Dopant concentration (atoms/cm^3)	n-type	p-type
$<10^{14}$	Intrinsic	Intrinsic
10^{14}–10^{16}	N$^-$	P$^-$
10^{16}–10^{19}	N	P
$>10^{19}$	N+	P+

Table 7 Dopant Source Materials

Type	Dopant	Gases	Liquids	Solids
N	As	AsH_3, AsF_3	As in $(CH_3CH_2)_4SiO_2$	$AlAsO_4$
N	P	PH_3, PF_3	$POCl_3$	$NH_4H_2PO_4$
N	Sb	SbH_3	Sb_3Cl_5	Sb_2O_3
P	B	B_2H_6, BF_3	BBr_3	BN

element, such as phosphorus oxychloride ($POCl_3$) or boron tribromide (BBr_3). These are applied by directing a stream of the liquid onto a wafer while it is spinning. The spinning action distributes the liquid evenly across the wafer surface. The coated wafer is baked at a low temperature to drive off the volatile organic carrier solvent, leaving a glassy oxide film of the dopant on the wafer surface:

$$4POCl_3 + 3O_2 \xrightarrow{200°C} 2P_2O_5 + 6Cl_2 \tag{6}$$

$$2BBr_3 + \frac{3}{2}O_2 \xrightarrow{200°C} B_2O_3 + 3Br_2 \tag{7}$$

Coated wafers are then heated in diffusion furnaces ≥900°C, volatilizing the glassy oxide film and driving metal atoms into the silicon lattice.

A drawback to diffusion doping is its high temperatures. Because these occur relatively early in the fabrication process, significant stresses can be induced in silicon wafers, thereby limiting the amount of thermal energy that can be used in subsequent processes. Also, the quartz glass tubes used to contain wafers in diffusion furnaces must be scrupulously clean to avoid introducing unwanted dopants or contaminants. Diffusion doping often involves the use of toxic substances, and it can produce waste or scrap materials that are environmentally difficult.

2. Ion Implantation

Implanting dopants is a desirable alternative to diffusion because it can be done without heat and has less potential for introducing unwanted dopants or other contaminants to the device. Implantation also enables a sharper, higher precision profile of dopant concentration with depth in silicon for better device performance. In implantation, ions are accelerated at high energy onto the wafer surface and penetrate into the silicon lattice. Many of the gases used for diffusion (Table 7) are used in implanters to provide source atoms for the ion beam.

Typically, both ion implantation and diffusion are carried out through "windows" in an overlying layer of material that acts as a mask, so that dopant atoms enter silicon only in targeted regions. Overlying mask materials may be oxidized silicon or a polymerized organic film (photoresist). These are patterned in the desired geometry so that the doped region in the wafer replicates the mask pattern and the design of the specific circuit components being built.

Figure 15 shows a depth profile of phosphorus dopant concentration in atoms/cm^3 as a function of depth into silicon in microns. Diffusions and implantations are tailored to achieve desired electrical characteristics, and complex devices may contain several different dopant species in varying concentration profiles.

Implantation is accompanied by some inelastic collisions, shoving silicon atoms out of their normal lattice positions. This causes structural defects in the lattice, and it leaves the implanted species themselves as nonionized atoms and electrically inactive. A postimplantation thermal anneal at temperatures ≥600°C restores the lattice structure by returning displaced atoms to their proper positions, and it allows the dopant atoms in substitutional lattice sites to become

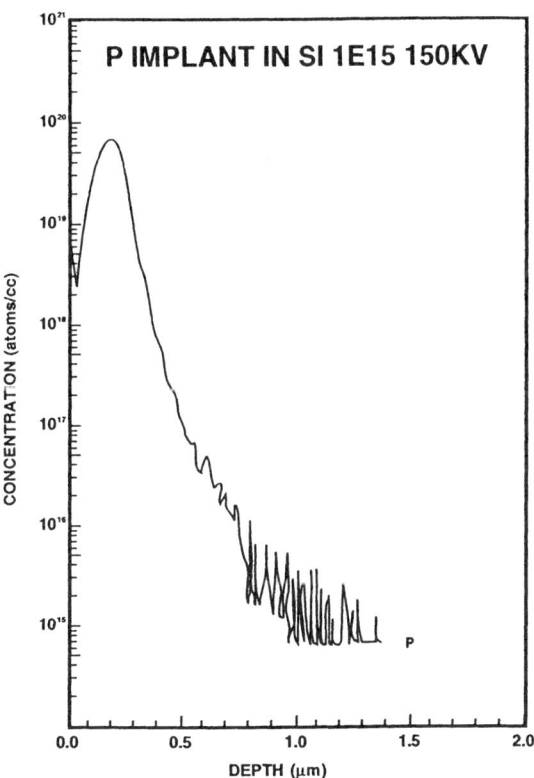

Figure 15 Phosphorus dopant concentration depth profile.

electrically active. The implanter's ion beams are also significant sources of particles, and therefore its operation must be optimized to prevent particles from showering onto product wafers during implants.

C. Physicochemical Thin-Film Deposition

Films of various materials must be created on wafers at multiple points in the fabrication process. Some of these films are sacrificial, i.e., used as masking agents in a process step and then removed from the wafer. Other films become an integral part of finished device structures. Hence, the types of film materials (which can be pure elements, compounds, or mixtures) and their chemical and physical properties are dictated by their role in the circuit structure. Films can be produced by oxidation of silicon substrate or by chemical or physical deposition from gaseous reactants in the vapor phase.

1. Oxidation

Building electronic devices in silicon requires that selected areas or entire wafer surfaces be oxidized. Oxidized silicon can be a structural part of devices in the form of dielectric films that insulate conductive parts of circuits, or they can serve as mask layers during patterning.

Silicon is oxidized by two different chemical processes:

$$Si + O_2 \xrightarrow{700-1200°C} SiO_2 \tag{8}$$

$$Si + 2H_2O \underset{steam}{\xrightarrow{700-1200°C}} SiO_2 + 2H_2 \tag{9}$$

Overall factors affecting wafer oxidation include temperature, pressure, availability of oxidizing species in proper chemical stoichiometry to assure complete oxidation, and surface energy of the wafer where oxidation occurs. Reactant oxygen atoms diffuse through just-grown SiO_2 film to further oxidize underlying silicon atoms, and so the interface between oxidized silicon and underlying unoxidized silicon progresses down into the wafer as oxidation proceeds. Figure 16 shows the thicknesses of oxide obtained as a function of time for various pyrogenic steam oxidations of silicon.

Silicon is often oxidized with chlorine present, which helps stabilize the Si/SiO_2 interface. The electronegativity of chlorine inhibits the mobility of positively charged ionic impurity species by trapping them. Chlorine is sourced from gaseous hydrochloric acid or organochlorine compounds admitted into the furnace during the oxide growth process. Organochlorine compounds like 1,1,1-trichloroethane have been replaced with compounds like oxalyl chloride that have no ozone depletion potential.

2. Chemical and Physical Vapor Deposition

Chemical vapor deposition (CVD) is a process by which pure gases dissociate or react in a thermal and/or radio-frequency (rf) plasma discharge to initiate a

Electronic Processes and Environment 63

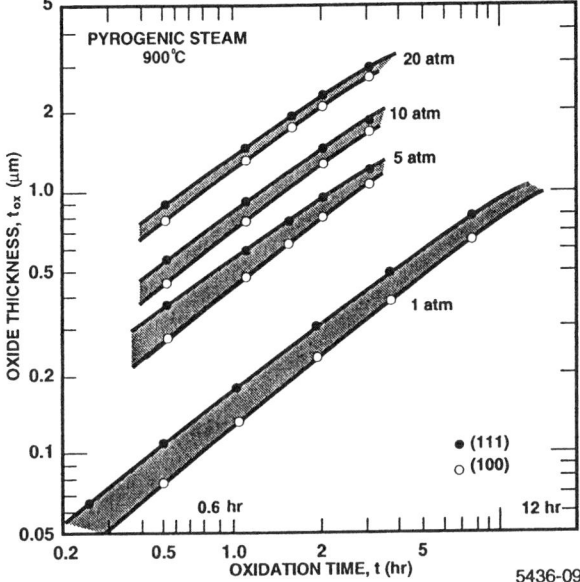

Figure 16 Pyrogenic steam oxidation of silicon at 900°C and 1, 5, 10, and 20 atm.

chemical reaction either just above or directly on the wafer surface. The chemical reaction deposits film material on the wafer. Several types of device-significant films can be made by CVD. The deposition technologies are developed in more detail in Chapters 3–5. Unlike oxidation and CVD, which rely on chemical reactions, the physical vapor deposition process (PVD) is based on a physical phenomenon (i.e., sputtering or evaporation) to produce a chemical film.

D. Photopatterning

Once a "blanket" film is on a wafer, the next step is to make a precise pattern in it, which (in combination with preceding and ensuing patterning steps) gives the intended circuit structure. The process of thin-film patterning is called photolithography. In simplest terms, it involves the following steps:

1. Coating wafers with a film of photoactive polymer in a solvent vehicle, a material called photoresist
2. Placing a mask containing the master pattern to be made in the thin film over the wafer
3. Exposing the masked wafer to photon, X-ray, electron beam, or other energy (depending on the photo resist being used)

4. Using appropriate chemistry to remove either the exposed or the unexposed portion of the photoresist film
5. Using the appropriate chemistry to remove the portion of the underlying film being patterned that is uncovered when overlying photoresist is removed
6. Using the appropriate chemistry to remove remaining photo resist

These steps are then repeated until the desired patterned structures are obtained. Film removal can be done by either wet (liquid chemical baths) or dry (rf excited gas plasma) chemical processes. The chemistry chosen is dictated by the composition of the film to be etched, the geometries to be made, the ease with which contaminating by-products can be removed, susceptibilities of existing device structures to the energetics of the process, and process economics. Details of photopatterning and film etching are covered in Chapter 6 on lithography and Chapter 7 on plasma etching.

E. Chip Passivation

Wafer fabrication nears completion when the full repertoire of unit processes has delivered production lots of whole wafers whose die contain the designed device structures repetitively across the surface. Typically, a 1–2 µm thick film of silicon dioxide, silicon nitride, or a multilayer "sandwich" of these two films is then conformally deposited over each entire wafer. These passivating film layers provide mechanical protection from physical contact and entry by particles or other contaminants to the device structure. The passivation film is chemically patterned via a final lithography step to selectively remove it from areas where it overlies the ≈ 4 mil^2 areas of the conductor metal formed during wafer fabrication and arrayed around the periphery of each die. These exposed metal surfaces are "bonding pads"; the contact points for ≈ 1-mil diameter aluminum wires that will connect the integrated circuit to external electrical systems. At this point, wafer fabrication is complete, and the manufacturing process enters a distinctively different phase, that of test and assembly.

V. POST–WAFER FABRICATION

A. Electrical Test and Wafer Separation

When integrated together, the fabricated devices yield hundreds of circuit die arrayed across each silicon wafer. Each die is potentially a product device, but it may or may not have good electrical functionality if affected by contamination or other defects during processing. Hence, a post–wafer fabrication test first verifies electrical performance of good and bad die on the wafer.

The testing is done by contacting bare metal test points in test circuits on the wafer, and/or the metal bond pads around the periphery of the die, with an array of fine-pointed probes, typically made of chemically sharpened tungsten wire. Through this array of probes, computer programs (chips testing chips!) test each die through a range of electrical biases and other conditions to verify functionality. Nonfunctional die, or those operating outside the specified electrical parameters, receive an ink dot so they can be discarded later.

Tested whole wafers are scribed with a saw blade or laser beam in avenues along the periphery of the individual die. Wafers are then broken ("diced") into the individual circuit die, now called chips. The scribe and break operation uses ultrapure water for washing the die to remove silicon dust particles and other contaminants during the electrical testing, wafer scribing, and breaking operations.

After separation and washing, functionally good die are assembled into protective ceramic, metal, or plastic housings called packages. There are multitudes of package configurations dictated by the end-use application for the product. Packages are classified as single-chip or multichip depending on whether they contain one or multiple chips, respectively.

B. Single-Chip Packaging

Most semiconductor chips are sealed individually in plastic or hermetically sealed glass–ceramic or metal packages, which can then be mounted on circuit boards for installation into electrical systems.

1. Package Parts and Preparation

Single-chip assembly uses a huge variety of plastic, metal and metal alloy, glass, ceramic, and organic materials.

Before assembly begins, items like metal package parts and frames made by mass-production stamping or chemical plating operations are cleaned of residual oils, plating salts, or other residues. Historically, the electronics industry has universally used chlorofluorocarbon (CFC) vapor degreasing for package and circuit board cleaning. Because of their ozone depletion potential, CFCs have been replaced by other materials, some of which are listed in Table 8. Non-CFCs are generally more difficult to use than the one-step CFC degreasing process. However, many cost less and clean equally well.

2. Die Mounting

Each electrically good die is mounted to either a lead frame or a package base, depending on the assembly process. Frames have leads that, outboard of the package body, will plug into board sockets, and that within the package are attached by thin metal wires to the bond pads on the chip. Frames are usually iron/nickel alloys or copper. Their cleanliness is essential since the ultimate seal

Table 8 Non-CFC Cleaning Chemistries

Hydrochlorofluorocarbons (HCFC; still ozone depleting but much less so than CFCs)
Isopropyl alcohol (can be aqueous blends)
Long-chain alcohols (can be aqueous blends)
Other aqueous and semiaqueous blends
Aqueous compounds followed by thermovacuum drying
Terpene solvents
Fluorochemicals
Oxygen, argon, or other dry plasma methods
Carbon dioxide
UV-ozone

integrity of the finished package depends in part on intimate, void-free contact of sealing material to frame metal surface.

Each chip is attached with metal alloy solders or organic adhesives to the center part of the frame or to a package base cavity. Frame surfaces for chip mounting are plated typically with nickel, gold, silver, or aluminum. Chemical and physical criteria for die attachment materials include good adhesive properties, freedom from voids, and good thermal conductivity to conduct heat away from the chip to help prevent thermal stresses and premature failure. Lead–tin or lead–tin–silver solders and organic adhesives filled with 70–80 wt % silver provide good heat sinking.

Other die attachment materials used in package base cavities are glass (often also filled with silver) or a eutectic mixture of gold and silicon. Eutectic die attach is made with a small piece of foil, called a preform, typically containing 98% gold and 2% silicon. The back surfaces of the die to be eutectically mounted have a thin film of pure gold applied by CVD or sputtering during wafer fabrication. The preform is sandwiched between the back of the die and frame surface. The combination is placed on a heater block, mild pressure is applied to the die (scrubbed), and the gold "dissolves" into the silicon, forming a strong gold–silicon eutectic attachment.

3. Wire Bonding

After the die is attached to the frame or package base cavity, wire connections are made. High-purity aluminum wire ≈ 25 μm in diameter, which can contain $\leq \approx 1\%$ of silicon, is used to connect the chip to the lead frame. This is done via a high-speed automated operation using pattern recognition technology. A bond wedge, fed with wire from spools, contacts metal bonding pads on the chip and aluminized or gold-plated posts on each lead frame tip, connecting them with a looped length of wire. Thermocompression, thermosonic, or ultrasonic energy joins the wire to the pad. The first method brings wire and pad together for milliseconds at a controlled temperature and pressure, forming a bond by interdiffusing the metals. Thermosonic bonding combines controlled temperature and ultrasonic energy to form a bond; it is the method most widely used for plastic packages. Ultrasonic attachments are made when the bond tool briefly applies a burst of high-frequency energy (40 KHz), forming a metallurgical weld between wire and bond pad.

4. Seal

Subsequent to wire bonding, the frame/chip assembly is sealed into an enclosure. This envelops the lead frame and its attached chip, sealing the chip away from the external environment, with pins protruding for system installation. This is the completed package, familiar to all users of electronic equipment.

C. Principal Materials for Single-Chip Packaging

There are two main types of single-chip packages, each using different materials and processes.

1. Plastic

More than 95% of integrated circuit (IC) products are molded into plastic, which conformally "shrink-wraps" the chip and its wire connections to the lead frame, encasing them in a solid block of plastic.

Raw plastic mold compound is delivered to IC assembly facilities by the ton in the form of pelletized powders. Table 9 gives the principal ingredients in mold compounds. Major components are cross-linked epoxy resins, typically Novolak resins (20–30%), inorganic fillers (70–80%), and additives such as bromine or brominated organics for flame retardancy (<1–2%).

In plastic packaging, strips of frame assemblies containing wire-bonded die are placed in a mold. Mold compound, softened by preheating, flows into the hot mold, where it surrounds the chip and frame, leaving the lead frame pins

Table 9 Ingredients in Plastic Molding Compounds

Thermosetting polymer compound
Filler, usually fused silica
Fire retardant; brominated resins and Sb_2O_3
Accelerator, to catalyze cross-linking
Hardener
Release agent, to facilitate removal from the mold
Coupling agent, to bond epoxy to the filler
Coloring, usually to render the finished package black

protruding. As the mold compound cools, it shrinks slightly, compressing the entire assembly and forming a tight seal around the package pins to stop penetration by moisture and other contaminants during the operating lifetime of the product. After removal from the molds, units are heated (cured) to ensure complete cross-linking of the mold compound. Figure 17a shows a plastic-encapsulated microcircuit.

Rigorous expectations for materials purity and mechanical properties of plastic packages are shown in Table 10. Plastic formulations to optimize all of these is always a trade-off, where best performance with respect to some of the properties must be sacrificed to optimize others.

2. Hermetic

The term *hermetic* means airtight. Hermetic packages enclose the wire-bonded chip and frame in a sealed cavity, surrounding the chip and its wire attachments in a gaseous minienvironment that cannot be penetrated by external ambients.

Glass-sealed hermetic packages have a layer of vitreous lead oxide/boron oxide glass fired onto a piece of alumina ceramic, called the base. The base is placed on a heater block to soften the glass, and the lead frame is pressed into the molten glass. Upon cooling, the lead frame is solidly embedded on the base by this glass. The die is attached in the base cavity using eutectic metal, glass, or organic adhesive materials, and the wire bond attachments are made. A ceramic lid glazed with the same glass is placed over the base. The assembly is placed on a belt and passed through a furnace, where temperatures ≥350°C cause the glass to soften and flow slightly. The sealing furnace is fed with either dry ambient air or with a "synthetic" air mixture of dry nitrogen and oxygen from ultrapure high-pressure cylinders. Oxygen in the sealing zone of the furnace ensures thermosetting of the vitreous glass. A hard glass seal forms around the pins of the lead frame, encasing the chip in an enclosure with the pins protruding externally from the body of the package. The chip is inside its own environment, sealed away from dust, moisture, and other external contaminants.

Metal-sealed hermetic packages have a metal lid, usually gold, gold-plated, or electroless nickel-plated, which is brazed or welded at relatively low temperature onto the base. Another metal lid type uses the chip mounted with eutectic or epoxy onto a round header with a surface of gold or nickel, capped with a lid of stamped nickel attached to the header, making a hermetically sealed unit known as a "can." Figure 17b shows some hermetic package styles.

After sealing, pins protruding from the package body are coated with a solder-compatible finish, so that packages can be attached to boards or other mounts. Tin and tin–lead solder are the most common lead finishes, applied in high-volume plating or immersion operations. Chemical control of stannous sulfate, pH, tin levels, tin ion ratios, brighteners, and other additives in baths is essential to good plating. Plated tin or solder must have <0.05% carbon, and

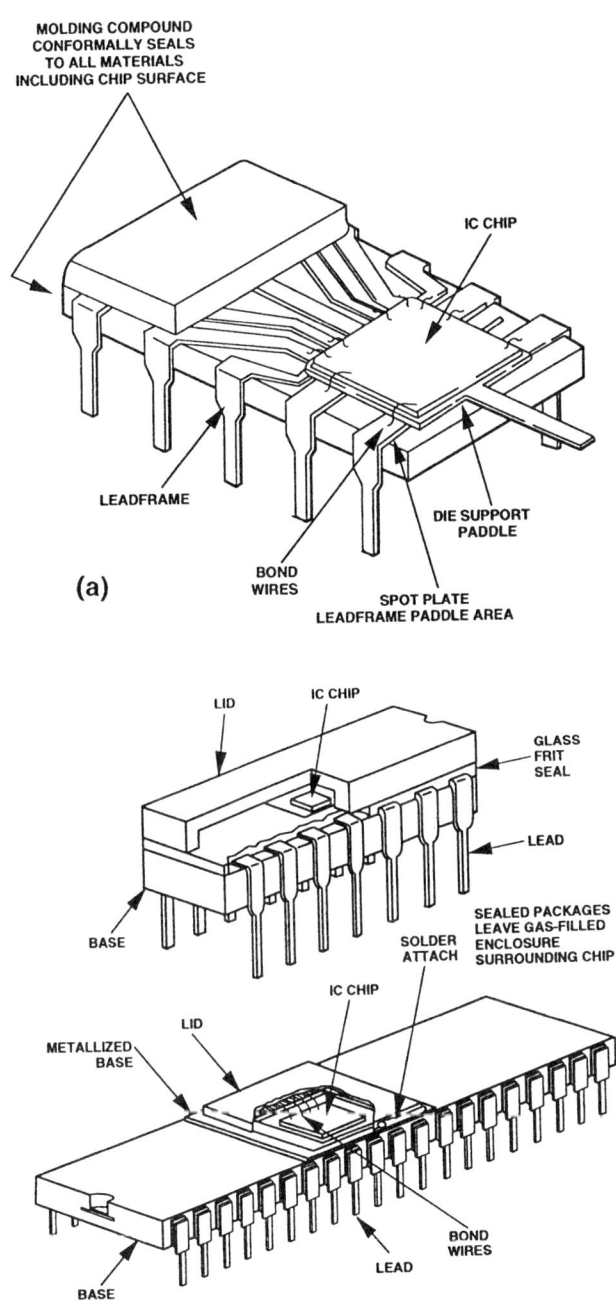

Figure 17 (a) Plastic package construction. (b) Hermetic package construction.

Table 10 Properties of High-Quality Plastic Encapsulant

High cross-linking, long-term stability
Glass transition temperature >150°C
Thermal expansion coefficients <20 ppm
Ionic impurities <10 ppm
Moisture content <0.2%
Moisture uptake <0.2%
Nonflammable in N_2/O_2 mixture up to 34% O_2
Good adhesion to package surfaces
Able to withstand pressure cooker testing for thousands of hours without device failure

no intermetallics or surface layers that would inhibit attachment with lead/tin solder and new lead-free solders that are being developed.

3. Limitations of Single-Chip Packages

All package technologies seek to minimize cost, maximize use of physical space, and reduce distances current must travel from one component to another, so as to maximize speed of the electrical functions being carried out. Single-chip package styles developed to improve these characteristics include leadless chip carrier and flatpacks, which minimize heights and lead lengths by being very thin, with lead terminations metallized onto external edge surfaces of the package. These designs enable higher-density package mounting on boards and reduce the distances current must travel. Despite innovations, single-chip packages are not desirable where physical space is restricted or operational speeds must be fast.

D. Hybrid Packages and Multichip Modules

The hybrid and multichip module (MCM) packages are assemblies of two or more chips on a common substrate, forming a multiple chip unit and performing unique electrical functions. These packages enable the chip manufacturer to offer a broader scope of "solutions" for system designers than the single-chip products. The distinction between hybrids and MCMs is somewhat blurred, but MCMs usually contain only active semiconductor devices, while hybrids contain a mixture of active devices and passive components such as capacitors and resistors. Hybrids and MCMs can be quite large. In contrast to single-chip packages, whose internal cavity volumes are typically less than about 1 cm^3, hybrids and MCMs have internal cavities of 10–100 cm^3.

Hybrids are generally built on multilayer ceramic substrates, with much larger metal interconnection pathways than MCMs. Die can be attached on multichip package substrates using the flip-chip method, wire bonding, or in an automated operation using polyimide films called tape automated bonding (TAB). Materials considerations for each are unique, and details of ceramic materials development for these packaging systems are further explored in Chapter 11.

VI. CHEMISTRY AND PHYSICS OF MICROELECTRONICS RELIABILITY

A. Device Failure

Today's competitive business climate demands dependable microcircuit products that customers can rely on. Reliability means that chips in protective packages (often called ''parts'') will operate under given conditions for a specified length of time.

One might think of a 99% reliability rate, e.g., having 1 bad part in 100, as being good performance. However, "99% good" corresponds to 5000 incorrect surgical operations per week, no electricity for about 7 hours per month, or about two landings missing the runway at every major U.S. airport every day. Hence, a 99% rate of reliability in everyday life, and in electronic parts, is entirely unacceptable. The semiconductor industry has achieved such remarkable levels of performance that reliability is expressed as failures in time (FITs). One FIT is defined as one device failure in 10^9 operating hours (a failure every 114,155 years, or a failure every 114 millennia!). This seems like an astoundingly high level of quality. However, in a production lot of 10,000 units of a given chip type known from qualification testing to have a 10 FIT failure rate, one part will fail every year. Customers do not want that one part on a multimillion dollar spacecraft, in a weapons or missile system essential for national defense, or for that matter, in their personal computers or automobile ignition systems. So the relentless drive for more computing power at lower cost is matched by a similarly relentless drive for lower product failure rates.

There are two kinds of failures. In catastrophic failures, the part stops working completely. In parametric failures, the part still functions, but outside the intended operating parameters. Device failure behavior is typically described by the ''bathtub'' curve shown in Fig. 18, with three distinct regions of failure behavior. A relatively high portion of units fail early in life due to process or material weaknesses, few units fail during the typical operating lifetime of a part, and failure rates increase as parts ''wear out'' near the end of their useful lifetime. Even the slightest defect in device materials can shift or terminate the functionality of a device, causing failure in any of the three regions of the bathtub curve.

With FITs in millennia, determining useful failure rates can only be done by accelerating the mechanisms so they occur in practical time spans. Many failure

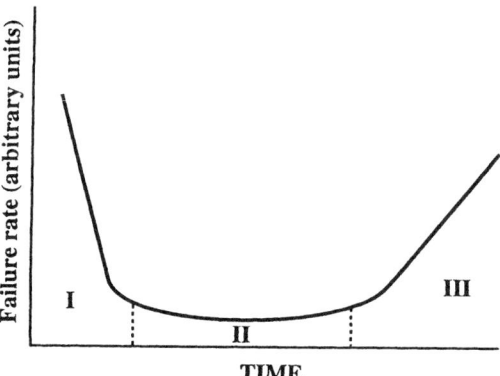

Figure 18 "Bathtub" failure rate curve.

mechanisms are physical or chemical phenomena that are thermally activated. These can be described by the general Arrhenius equation, which relates reaction rates to temperature by

$$\tau = A(T) \exp \frac{-E_a}{kT} \tag{10}$$

where

τ = the rate at which the process occurs
A = a constant
E_a = the activation energy, in electron-volts
k = Boltzmann's constant (8.6×10^{-5} eV/K)
T = temperature, K

Activation energy is nominally a constant, typifying a given failure mechanism (Table 11). Failures are not always as simple as just one mechanism. Two or more mechanisms can operate interdependently, or simultaneously and independently. By operating parts at extreme temperature, current, or voltage, and/or high humidity, latent failure mechanisms can be accelerated so as to compile reliability information for a family of parts in hundreds or thousands of testing hours (weeks to months), rather than years.

B. Overview of Failure Mechanisms

Despite all efforts of the manufacturing disciplines to make functionally perfect parts, failures can and do occur. At the package and wire-bond level, failure may be a relatively macro event, like physical separation of a package pin from

Table 11 Some Experimental Activation Energies for Selected Failure Modes

Failure mechanism	E_a (eV)
Metal migration, contact degradation	1.8
Charge injection	1.3
Gold/aluminum intermetallic growth	1.0
Ionic contamination	1.0
Electromigration	0.5–1.2
Degradation in 80% relative humidity	0.9
Time-dependent dielectric breakdown	0.35
Electrolytic corrosion	0.3–0.6

solder or breakage of a wire bond attachment. At the device level (on/in the silicon chip), failures are more often due to thermally induced stress, wear-out mechanisms of device materials, or incredibly small amounts of contaminants. In many mechanisms, failure is a subtle, hard to characterize, sometimes ephemeral, localized submicron or even Angstrom-sized anomaly deep in the structure of an individual circuit. Failure modes can be obviously chemical, such as corrosion of metal. They can be less obviously chemical (with respect to visual inspection), though no less chemical with respect to the materials. For example, electrical parameter shifts can be caused by positive ions like Na^+ concentrated in a 10–20 Å increment of thickness at an Si/SiO_2 interface, or by oxygen atom deficiencies in an SiO_2 dielectric layer, which trap electrons and shift performance characteristics. These will never be found by chemical analysis, but the cause of failure is still a "chemical" irregularity.

Tables 12–14 summarize chemical or material-oriented causes for some well-known failure modes of plastic and hermetic packages, and for die-level failure not primarily related to package anomalies.

C. Analyzing Failures

Because of the demand for better reliability, analysis of part failures occurs during process and product qualification to identify causes of poor production yield, and to proactively build better reliability into products and processes. It also occurs when products have failed in field usage and are returned for problem diagnosis. Failure analysis begins with electrical testing. Underlying every electrical fault, however, there is a chemical or physical irregularity. While a malfunction can often be categorized from electrical data, physical and/or chemical analysis to verify root cause of failure must often follow electrical tests. Accordingly, in addition to their roles in fabrication, chemistry and materials science are also essential in investigating product failure.

Table 12 Product Quality/Reliability Concerns for Plastic Packages

Failure mode	Material factors; root causes
Bond wire disattachment	Excessive force (sweep) on wire by plastic during molding; Poor wire attach metallurgy
Die lifts, die cracks	Die attach voiding; Thermal stresses
Die-level thermal effects	Die attach voiding; Poor heat sinking
Pin-to-pin electrical leakage	External surface contamination; Plating salt deposits
Poor solderability	Poor lead finish chemistry
Corrosion of bond pads on die, or of wire bond attachments	Water permeation; High halide ion content
Memory errors	Fillers with Th or U emitting α radiation
Illegible brands or package markings	Contamination on lid surface

Table 13 Product Quality/Reliability Concerns for Hermetic Packages

Failure mode	Material factors; root causes
Bond wire disattachment	Poor wire attach metallurgy
Die lifts, die cracks	Die attach voiding; Thermal stresses
Die-level thermal effects	Die attach voiding; Poor heat sinking
Pin-to-pin electrical leakage	External surface contamination; Plating salt deposits
Poor solderability	Poor lead finish chemistry
Corrosion of metal on chip, or of wire bond attachments	Water sealed in cavity; Ionic residues on chip
Soft errors in memory product	Seal glasses with high Th or U levels emitting α radiation
Illegible brands or package markings	Hydrocarbon or oxide contamination on lid surface
Particles loose in cavity	Poor environmental control in assembly; Unclean piece parts

Examining a failed part, known as the device under test (DUT), begins with optical inspection of external package surfaces. Contaminants or other external anomalies can be detected, followed by chemical or microbeam analysis.

The DUT is then "delidded" to reveal the chip and its wire connections for further examination. For hermetic packages this is just a physical removal of the lid from the package to expose the chip. Uncovering the chip in plastic packages is more involved. The part is immersed in a bath of red fuming nitric

Table 14 Product Quality/Reliability Concerns at the Chip Level

Failure mode	Mechanism	Material factors
Electrical opens	Electromigration	Metal atom movement opens physical void in conductor line
Electrical opens	Corrosion	Electrochemical dissolution of conductor metal lines
Electrical shorts	Metal hillock growth	Electromigration; Metal hillocks short adjacent metal conductor lines
Electrical shorts	Moisture	Surface electrical leakage
Electrical shorts	Metal particles	Metals short adjacent conductor lines
Electrical shorts	Oxide defects	Pinhole defect in oxide film
Functional	Inversion	Alkali ions at film surfaces; charge spreads in field oxide
Functional	Surface conduction	Accumulation of electrical charge on thin gate oxides
Electrical leakage	Material defect	Carbon or incorrect amount of oxygen in silicon crystal
Leakage, Functional	Radiation damage	Energetic particle rays cause ionization or leakage sites
Opens, shorts, leakage, etc.	Electrostatic discharge	Unintended impulse of energy to device
Opens, shorts, leakage, etc.	Electrical overstress	Excessive electrical currents melt metal, break down oxides, etc.

or fuming sulfuric acid, which chemically digests the cross-linked polymers of the mold compound. In a more refined technique, a fine stream of the fuming acid is aimed at the DUT to expose certain portions of or all of the chip. This can continue until the chip is entirely uncovered, or it may be stopped short of the chip surface and the remainder of the molding compound ashed in a CF_4/O_2 plasma, to help preserve any chip surface phenomena for further study. Refer to Chapter 9 for a full discussion of chemical and structural analytical methods for electronic materials.

D. Some Specific Failure Modes

1. Electrostatic Discharge

Electrostatic discharge (ESD) is the most common cause of device failure. Acquisition of electrical charge by friction between two materials is called triboelectric charging. The triboelectric series in Table 15 indicates the charge that materials acquire relative to each other; e.g., aluminum acquires a negative charge with respect to air and a positive charge with respect to silicon. Microelectronic materials occur throughout the triboelectric series.

A person walking across a carpet generates about 15,000 volts of static charge, and about 5000 volts when walking across a linoleum cleanroom floor. Circuit packages passing through mechanical handling equipment, or sliding out of shipping tubes, can generate 500–20,000 volts. Yet, many device structures are susceptible to ESD damage at levels as low as 100–500 volts. Static electricity control is therefore imperative in both the semiconductor plant and at the customer's facility to prevent ESD failure. The physical symptoms of an ESD failure range from catastrophically blown out aluminum conductor lines to stress-induced reductions in the operating lifetime of a part. The key to pre-

Table 15 Triboelectric Series

POSITIVE (+)	Air
	Glass
	Mica
	Hair
	Nylon
	Wool
	Fur
	Lead
	Silk
	Aluminum
	Paper
	Cotton
	Steel
	Hard rubber
	Nickel
	Brass
	Noble metals
	Acetate rayon
	Polyester
	Celluloid
	Saran
	Polyurethane
	Polyethylene
	Polypropylene
	PVC
	Silicon
NEGATIVE (−)	Teflon

venting ESD failure is for personnel handling product to be grounded at all times, and to surround product with conductive materials that drain away static charge during shipping and storage.

2. Adhesion Failure

The successful attachment of wire bonds depends on pad and post surfaces being free of contaminants. For aluminum this also means not having an excessive buildup of aluminum oxide, which hardens the metal surface and makes it more difficult to accomplish wire attachment.

Die can separate by delaminating from their substrate when an extremely thin layer of contamination such as a hydrocarbon film (or a fingerprint!) prevents good adhesion of attachment material. More common than complete separation is voided regions in the attachment material, which can be caused by lack of adhesion to die or frame surfaces. Voids cause incomplete heat sinking and chip malfunction.

3. Package Hermeticity Failure

A failure mode in hermetic packages is loss of seal integrity, allowing ambient air ingress to bring moisture into contact with the chip in the cavity. Seal integrity is tested by pressurizing parts in helium or krypton-85 gas and then placing them in an evacuated chamber with a mass spectrometer to see if He or Kr is exiting the package cavity through leaks in the seal. Package leak rates can be detected as low as 1×10^{-8} cm^3 atm/s. A relatively small-volume hermetic package leaking at 1×10^{-6} cm^3 atm/s will leak up to regular air composition in just a few days. Entering air brings humidity into the package cavity, the possibility of condensed liquid water at cold temperatures, and corrosion of aluminum metal.

4. Aluminum Corrosion

In general, moisture access to chips must always be prevented because of the susceptibility of aluminum to chemical and galvanic corrosion:

$$Al + 3H_2O \rightarrow Al(OH)_3 + \frac{3}{2} H_2 \qquad (11)$$

$$Al + 3HCl \xrightarrow{H_2O} AlCl_3 + \frac{3}{2} H_2 \qquad (12)$$

$$Al + 3H_3PO_4 \xrightarrow{H_2O} Al(PO_4)_3 + \frac{9}{2} H_2 \qquad (13)$$

For example, only nanograms of dissolved ions like chloride, a ubiquitous contaminant, need to leach from plastic mold compound or residual contaminants in the region of the chip into a few microliters of water. Microvolumes of hydrochloric acid corrode aluminum chemically or galvanically, and only nanograms of metal need to corrode from a submicron metal line to cause circuit failure. This failure will not happen to metal conductor lines passivated with high-integrity (defect-free) silicon oxides or nitrides, but precautions against corrosion are still essential because bond pads and wire connections are not protected by the passivation films. Refer to Chapter 8 for a full discussion of electrochemistry and metal corrosion.

Aluminum corrosion seldom occurs in hermetic packages due to tight controls on sealing methods and careful engineering of the package part materials to preclude water vapor from the package cavity. Hermetic packages have avoided corrosion failure events by remaining leak-tight for decades.

Aluminum corrosion failure also seldom occurs in plastic parts in field applications. Failures are usually only observed under conditions of extreme humidity, pressure, and temperature, purposely applied to obtain lifetime reliability data. Figure 19 shows an example of the corrosion resistance of plastic parts exposed to these conditions, at lower temperatures, parts do not suffer failure due to corrosion for hundreds of years. Improvements in resin purities have greatly lowered the amount of leachable or hydrolyzable chloride and other conductive ions in plastic mold compounds. Use of mold compounds with high-purity resins, along with defect-free passivation films, has greatly increased the service reliability of plastic parts, even in moist environments, to where their reliability is comparable with that of hermetically sealed parts.

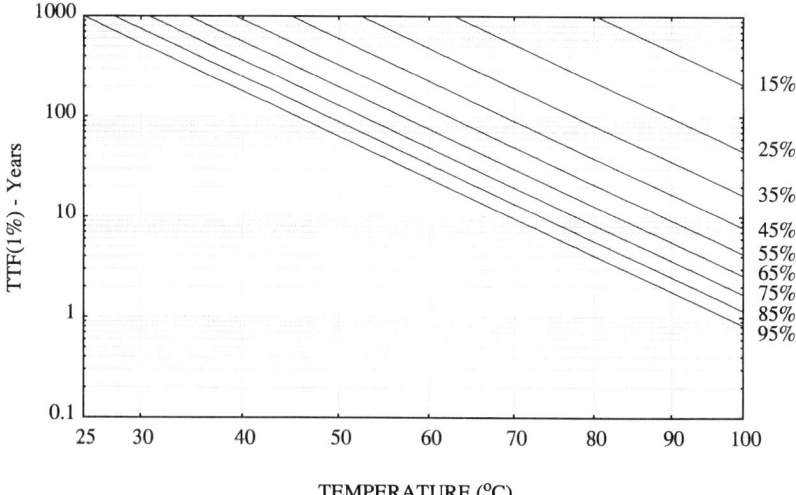

Figure 19 Reliability data for plastic packages.

5. Electromigration

A weakness of aluminum as the main conductor metal on devices is its tendency to form microvoids during high-current operation. Although only nanoamperes of current may actually be flowing, the density of the current in narrow conductor lines can be extremely high. Current densities $>10^5$ amperes/cm^2 cause Joule heating of the metal. Passing electrons acquire enough kinetic energy to physically move aluminum atoms out of lattice positions, resulting in a local microvoid. Microvoid formation often initiates at grain boundaries or grain boundary intersections. Void formation in one portion of a conductor line is generally accompanied by metal hillock growth elsewhere in the line (the metal from a voided area must go somewhere). Continued high current density will cause microvoids to continue to form, and to migrate along the conductor stripe. Significant numbers of microvoids may eventually coalesce together, so that the aluminum comes entirely apart at some point, causing an "open" in the line and device failure. Minor impurities such as silicon up to 1% and/or copper up to 2–4% are usually "doped" into the aluminum during deposition, congregating in the metal grain boundaries and relieving stresses in the metal to help suppress electromigration.

6. Radiation

Certain circuit types are vulnerable to reliability degradation due to the effects of energetic photons such as X-rays, γ-rays, and protons, neutrons, and electrons. Radiation damages devices by ionization and by inducing lattice defects. When passing through a device junction, radiation ionizes the space charge region so as to change the conducting state of a cell from "off" to "on," causing a memory device to store incorrect information. The device will store this incorrect data until reset. Such a condition is called a soft error. Memory circuits are vulnerable to this failure mechanism even in the normal flux of cosmic radiation reaching the earth's surface.

VII. SUMMARY

This chapter has taken an extremely broad-based look at some basic chemistry and materials considerations of silicon semiconductor technologies. Much of what has been described here applies to electronic materials technologies other than silicon, which have similar processes and similar process demands. The purpose here has been to convey an appreciation for the chemical-intensive nature of silicon devices, the processes used to make them, and some of the material-based mechanisms by which they fail. The stage is set for more rigorous descriptions of several of these manufacturing technologies and associated new material developments.

EXERCISES

1. What is the shrinkage factor from 4K LSI devices to 4G ULSI devices? What would the *additional* shrinkage factor be if atomic-scale device structures could be made with 10-Å feature size?
2. Write the three chemical equations for producing semiconductor grade silicon from silicon dioxide. Why might FZ silicon contain fewer crystal lattice impurities than CZ silicon?
3. Draw the electron configurations for phosphorus, boron, antimony, and gold. Identify each as a donor or acceptor.
4. What is the difference in dopant concentration between an n-type and a p-type wafer, each having a resistivity value of exactly 1 ohm-cm? How can the same concentrations of different dopants produce wafers of different resistivity?
5. Identify and discuss three means of maintaining a Class 1 environment in a semiconductor manufacturing cleanroom.
6. Compare and contrast wet chemical versus dry plasma wafer treatments with respect to process effectiveness and the types of contamination each can cause.
7. How much more contaminated than ultrapure water containing the maximum allowable concentration of 11 metallic contaminants (Table 3) is ultrapure hydrofluoric acid containing the maximum allowable concentration of the same 11 metallic contaminants (Table 4)? Speculate as to why etching or cleaning processes using the relatively impure acid do not seriously degrade process yields.
8. What four important thin films can be obtained by deposition using silane as a reactant? What is the difficulty (a) with silane as a raw material, and (b) with the by-product derived from it in each of the chemical reactions used to produce these films?
9. Give two reasons why aluminum is used instead of gold for circuit interconnection.
10. Some substances are simultaneously essential both for microcircuit manufacture *and* to be avoided as yield- or performance-limiting contaminants. Explain with respect to (a) water, (b) phosphorus, (c) chlorine or chloride, (d) iron, and (e) oxygen.
11. Consider a failure in which a gap 0.5 μm wide corrodes open in an aluminum conductor line 8000 Å thick and 1 μm wide. What is the mass of aluminum corroded from the line? How much water is required for this reaction? How much chloride is required if HCl is the corroding species? Use 2.2 g/cm^3 as the density for aluminum.
12. Discuss the advantages and disadvantages of plastic versus hermetic packages as they pertain to microcircuit reliability.

13. Identify three ways in which human-sourced contaminants can cause product failure.
14. Identify and briefly discuss failure mechanisms associated with the following elements or ions: (a) sodium; (b) iron; (c) hydrogen; (d) phosphorus.
15. Write the general Arrhenius equation used for failure rate prediction. Why is temperature such an important factor in determining product reliability?
16. What practical limitation arises in determining FIT values for product via classical life testing as actual failure rates decrease below 1 FIT?

SELECTED READING

1. J. Brodie and J. J. Muray, *The Physics of Microfabrication*, Plenum Press, 1987.
2. *The Chipmakers*, Time-Life Books, 1990.
3. A. S. Grove, *Physics and Technology of Semiconductor Devices*, John Wiley and Sons, 1967.
4. C. R. M. Grovenor, *Microelectronic Materials*, IOP Publishing, Ltd., 1992.
5. D. S. Gupta, ed., *Semiconductor Processing*, ASTM STP 850, 1984.
6. D. S. Gupta and P. H. Langer, *Emerging Semiconductor Technology*, eds., ASTM STP 960, 1987.
7. R. S. Muller and T. I. Kamins, *Device Electronics for Integrated Circuits*, John Wiley and Sons, 1986.
8. E. H. Nicollian and J. R. Brews, *MOS Physics and Technology*, John Wiley and Sons, 1982.
9. W. R. Runyan, *Silicon Semiconductor Technology*, McGraw-Hill, 1965.
10. D. N. Schmidt, ed., *Contamination Control and Defect Reduction in Semiconductor Manufacturing*, Proceedings, Symposium of Electrochemical Society, 1994.
11. *Semiconductor Technology Handbook*, Technology Associates, 1982.
12. S. M. Sze, *Physics of Semiconductor Devices*, John Wiley and Sons, 1981.
13. P. Van Zant, *Microchip Fabrication*, Semiconductor Services, 1986.
14. S. Wolf and R. N. Tauber, *Silicon Processing for the VLSI Era*, McGraw-Hill Series, 1987.

3
Chemical Vapor Deposition for Epitaxial Growth

Thomas F. Kuech   *University of Wisconsin, Madison, Wisconsin*

I. INTRODUCTION

The formation of electronic and optical devices is largely based on the formation and patterning of thin-film layered structures. Many of the semiconductor processing techniques are associated, directly or indirectly, with processes on a single-crystal semiconductor substrate wafer. For example, ion implantation, etching, diffusion, and deposition allow for various modifications of the near-surface region of the semiconductor surface; these allow for the change in chemical composition, as in the case of ion implantation and diffusion, or in physical structures through the deposition of new films or the accurate removal of thin layers of semiconductor to create patterned structures. The latter is key to the definition of the semiconductor device on a wafer surface. Many of these mainstream processes involve high temperatures. Solid state diffusion requires high temperatures in order to increase the atomic mobility of an impurity atom. Ion implantation creates damaged regions within the material that must be thermally annealed in order to reconstruct the damaged regions. Despite the benefits of high temperatures, however, they can nevertheless ultimately degrade the materials through defect formation and atomic redistribution through diffusion. Effective low-temperature process techniques are therefore at the forefront of many materials development efforts. One such process is the fabrication of the electronic device through the controlled deposition of materials.

Deposition processes can be used to create layers of most elements and compounds in a variety of forms. Semiconductor-based devices contain numerous layers of different metals, insulators, and semiconducting materials, each of which must exhibit specific mechanical and electrical properties (Chapter 2), and most of which are related to the specific physical structure of the layers. This physical structure refers to the actual arrangement of atoms within the growing layer and, depending on the presence of any long-range crystallographic order, can be either amorphous or crystalline, as seen in Fig. 1. The relative orientation of the small crystallites within the crystalline films can have a variety

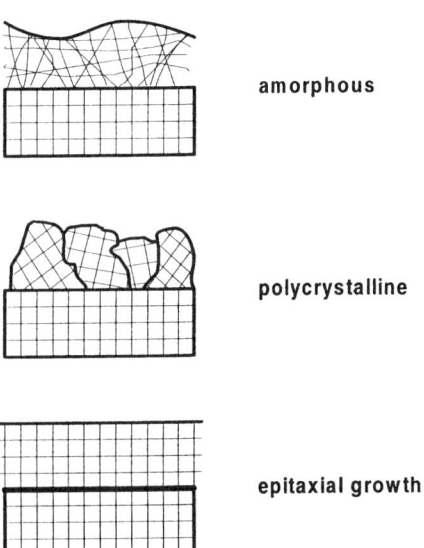

Figure 1 The character of the deposited layer depends on the growth temperature, chemistry, and choice of substrate. In special cases, the atomic arrangement within the deposited material can have a direct relationship to that of the substrate, known as epitaxial growth.

of microstructures. A large number of small crystallites or grains can have important effects on the materials properties of deposited films. While these small crystallites may be free from defects, such as vacancies (missing atoms), dislocations (missing rows of atoms), or stacking faults (extra or missing partial planes of atoms), a large density of defects may exist at the interfacial regions where two small grains meet. The rows of atoms between the grains are in disregistry, and their broken and incomplete bonding arrangements at the grain intersection lead to the presence of electrically active defect states. These defect states can severely degrade the performance and lifetime of the electronic devices and, depending on the function of the material in the device structure, are to be avoided. However, these features may also be quite beneficial. Examples of amorphous films, which are incorporated into modern integrated circuits, are discussed in Chapter 4. Under appropriate process conditions, defect formation in the growing layer can be avoided, and these layers can assume or replicate the physical structure of the underlying substrate. This replication of the crystalline arrangement of the substrate is known as epitaxy. Epitaxial growth offers the opportunity to achieve structurally perfect films.

In the epitaxial growth process, the deposited atoms arrange themselves on the growing surface, bonding to the underlying atoms of the substrate, thereby determining the subsequent arrangement of the next atoms in the growing film. Thus, the resulting film is a continuation of the atomic structure of the single-crystal substrate. Ideally, this film should be as structurally perfect and free from defects as the substrate itself. Since the type of the deposited atoms can be altered through the deposition process, the composition of the growing film can be controlled, in the growth direction, during the deposition. It is possible through several different deposition techniques to produce layered epitaxial structures in which the individual layers are less than a nanometer thick and the interfaces between layers are essentially atomically abrupt.

Epitaxial growth techniques have been developed for all of the major semiconductor device materials: metals, semiconductors, and insulators. The most widely encountered epitaxial growth process is the formation of Si layers on an existing Si wafer or substrate. The deposition of Si on the Si substrate is used to form thin layers of Si in which the electrical properties of the growing layer can differ substantially from those of the underlying layers due to the addition of a small amount of dopant atoms to the growing film, allowing for change and control of the electrical properties of the Si layer. With appropriate process control, atomically abrupt junctions between different doping regions are possible. Transistor structures and diodes can be formed where the change in electrical properties, e.g., a transition from p- to n-type, can occur over a narrow region. The ability to grow various active regions of a device structure, as opposed to the just-mentioned processing of an existing single-crystal substrate, has opened the range of materials structures available for device design.

The epitaxial growth of dissimilar materials, such as Si_xGe_{1-x} on a Si substrate is referred to as heteroepitaxial growth. These heterojunctions provide a difference in both chemical properties and electronic structure and form the basis of many new electronic devices. Besides epitaxial silicon films, growth of heteroepitaxial compound semiconductors is one of the most developed materials system. Quantum well lasers, high-performance heterojunction transistors, and multilayer photodetectors are all products based on heteroepitaxial growth.

There are many issues in the chemical vapor deposition of epitaxial films. As stated, the structure of an epitaxial deposited layer or layers will have a definite relationship to the substrate. Epitaxy describes the overall physical nature of the film; the details of the physical structure or perfection of the epitaxial layers are also important. In order to be further processed, the film must have several other properties bearing on its utility in the formation of a useful device. It must be uniform in thickness, composition, and chemical, electrical, and perhaps optical properties. Variations in film properties lead to changes in the device properties over the wafer and from process or production run to run. In general, the surface must also be very smooth in order to allow subsequent photolithographic patterning. Surface defects, if they occur within the device region itself, will reduce the effective usable area of the deposited film and, therefore, the overall yield of devices and circuits generated from the material. Defects can also occur within the film, in the form of missing atoms (vacancies), rows of atoms (dislocations), extra or missing planes of atoms (stacking faults), as schematically shown in Fig. 2. Their formation must either be prevented or, at the minimum, controlled. Such structural defects can be electrically active and interact with the electronic or optical device formed from the epitaxial layers.

There are several techniques used in the formation of epitaxial layers, the choice of which depends strongly on the required materials and the desired material structure. Physical deposition has been used in the growth of many films. In particular, molecular beam epitaxy (MBE) has been used with great success in the fabrication of very-thin-layer device structures. In this technique, a heated substrate is exposed to a flux of growth nutrients, usually elemental sources, within an ultrahigh vacuum (UHV) environment. Material film growth from the gas phase at higher pressures is far more common and is referred to as chemical vapor deposition, or CVD. Chemical vapor deposition techniques utilize high-vapor-pressure compounds of the elements comprising the film. The volatile source compounds, for example, SiH_4 in the growth of Si, are transported to the growth front, where they react and are incorporated into the growing layer. In all cases, the deposition of the film and the formation of the epitaxial structure proceed through a series of elementary steps or processes (Fig. 3): (i) transport of the growth nutrients to the growth front, (ii) their decomposition at or on the growth surface, (iii) surface migration of the deposited

CVD for Epitaxial Growth

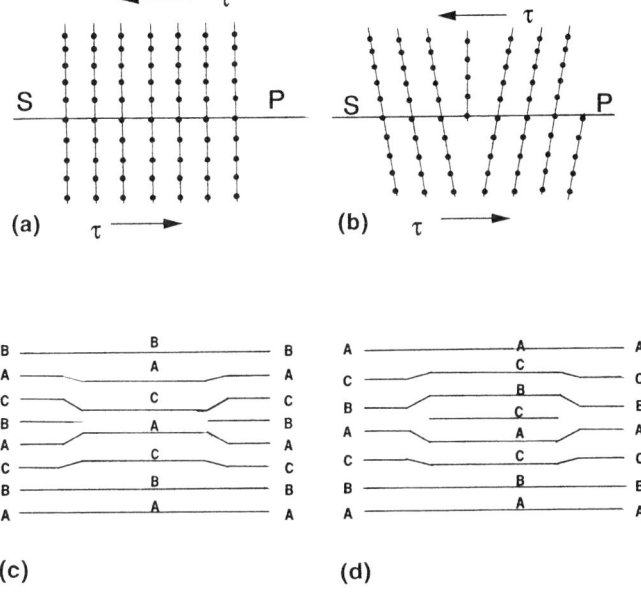

Figure 2 Many types of defects can arise in the growth of the epitaxial layer. Dislocations, such as the edge dislocation in (a), often are due to lattice mismatch between the film and substrate. These defects can move under an applied shear stress, τ. Stacking faults are partially missing (c) or extra (d) sheets of atoms in the crystal, which disrupt the normal ordering of the arrangement of atoms.

species, and (iv) the subsequent bonding into the growth front. The slowest of these elementary processes becomes the rate-limiting step in determining the growth rate of the film, as shown in Fig. 3. These elementary processes take place within the CVD reactor and affect the formation of the epitaxial structure in the deposited layer. The reactor design and the macroscopic growth conditions of temperature, pressure, gas-phase composition, and residence time of the gases within the reactor all, in turn, affect the flux of nutrients to the surface enabling these elementary processes to take place.

The elementary CVD epitaxial film formation steps take place within a chemical reactor. The process itself is conceptually simple; it consists of injecting reactive gases into the reactor wherein the substrates are placed. The gas is transported through the reactor, undergoing gas-phase and surface reactions that result in the deposition of the desired films. The final step is the transport of the reaction by-products and the unused reactants from the system. Energy is provided to the reactor to initiate the chemical reactions producing the films, and this energy input is most commonly in the form of heat, with the reactor

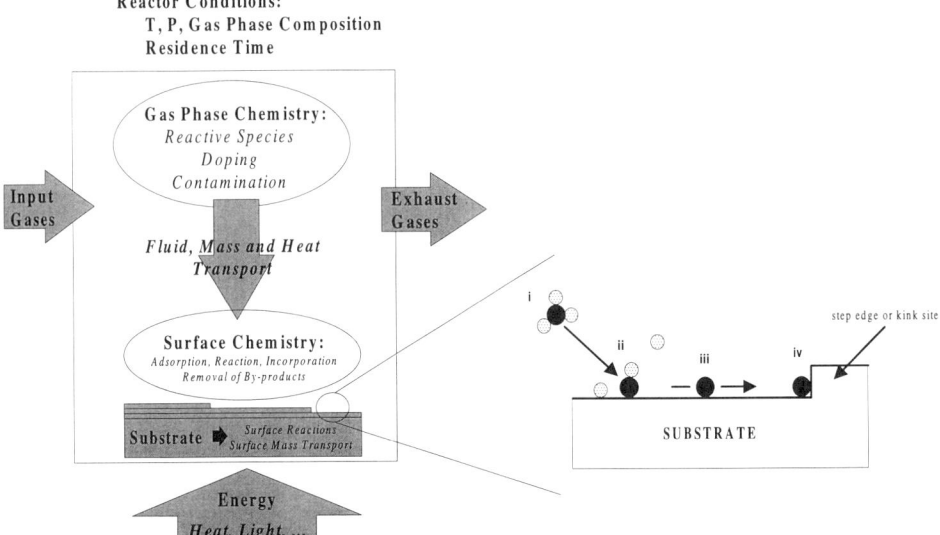

Figure 3 Both gas-phase and surface reactions within the reactor are important in most growth systems. The relative importance of these effects depends on the gas-phase and surface temperatures, reactor pressure, and gas-phase composition. The epitaxial growth of the film on the growth surface proceeds through a series of elementary steps. (i) The source molecule must be transported to the substrate. (ii) The source molecule must adsorb and decompose on the surface with the undesired by-products being transported from the surface. (iii) The surface-adsorbed film constituent must be transported on the surface to an appropriate atomic site. (iv) Lastly, in many cases the atom is transported to a favorable surface site, such as a step edge or kink, and bonds to the surface, resulting in the final incorporation into the surface.

residing inside a furnace or the substrate placed on a heated surface or platen. A variety of reactor geometries are illustrated in Chapter 4. Both gas-phase and surface chemical processes interact with the system parameters to yield a deposit, as shown in Fig. 3. The development of the important materials parameters, which were listed earlier, will depend on the details of the reactor and growth system. The growth chemistry is, however, the main determinant of the materials parameters in the deposited layer. The basic thermodynamics of the CVD process is initially explored to determine the feasibility of a chemical reaction as well as the maximum growth rate expected in the process under thermal energy conditions.

II. BASIC CHEMICAL ELEMENTS OF GROWTH AND EPITAXY

The study of most chemical systems starts with an equilibrium thermodynamic analysis of the overall process. The term *equilibrium* implies, of course, that the system is unchanging in time, which is clearly not the case for any film growth process. However, thermodynamics can provide some useful and important information, such as whether a reaction is energetically possible and, if it is, what is the expected maximum extent of that reaction. All CVD and epitaxial reaction processes require a driving force for the deposition of material, derived from the free energy change of the overall chemical reaction responsible for the net deposition of that material, and generally initiated by thermal stimulation. The deposition and its rate are further affected by a wide variety of process elements, the major ones being mass transport limitations, chemical kinetic limitations, and low supersaturation of the gas phase. The mass-transport and kinetic limitations will be considered in a later section.

The issue of supersaturation, or the driving force for the deposition reaction, can be addressed through several traditional chemical equilibrium concepts. Deposition reactions take place in a hot reactor or on a heated substrate surface, and the rate of reaction depends primarily on the temperature. The deposition reactions can also be initiated by the application of special external energy sources, such as ultraviolet (UV) radiation or an applied plasma. Such reactions are inherently endothermic. These heating sources are typically used to overcome any kinetic limitations in the reaction or decomposition of the chemical sources. The kinetic limitations are generally very small in epitaxial growth systems, thereby driving the reactions rapidly to their thermodynamical end with the external thermal sources. In such cases, the reaction process can often be accurately described by thermodynamic concepts.

A. Thermodynamics of Growth

The starting point of the thermodynamic description of a CVD process is the determination of the free energy of reaction, ΔG_r. For simple reactions, ΔG_r can be determined directly from the free energy of formation for the reactants and products. In the case of GaAs, the film growth can be accomplished by reacting trimethyl gallium [$(CH_3)_3Ga$] and AsH_3. The overall reaction can be written as

$$(CH_3)_3Ga + AsH_3 \Leftrightarrow GaAs + 3CH_4 \tag{1}$$

A gauge of how favorable the reaction can be is determined by its heat of reaction, which is calculated in the conventional manner using the heats of formation at 298 K:

$$\Delta G_r^0 = \sum_{\text{products}} \Delta G_{f,i}^0 - \sum_{\text{reactants}} \Delta G_{f,i}^0 \qquad (2)$$

In the present case, the values listed in Table 1 are used to determine the energy released in this reaction, leading to a $\Delta G_{f,i}^0 = -612$ kcal/mole. This implies that the reaction is very favorable, at least at room temperature, and the reactants will be consumed irreversibly in forming GaAs. GaAs films are of course not grown from these compounds at room temperature, due to kinetic rate limitations; they require elevated temperatures. The calculation of ΔG_f^0 at the growth temperature is more complicated, although it is covered in most standard chemical textbooks. This simple analysis indicates whether the reaction is possible, but not whether it could indeed occur under the conditions employed.

As already stated, a thermodynamic analysis is useful to determine the feasibility of film growth, as well as the maximum growth rate achievable in a particular system. Such analysis is often performed before the development of a new CVD system, and it is based on the thermodynamic calculations of the maximum available gas-phase concentration of the reactants and products present under a given set of reactor conditions. The maximum growth rate will be achieved for a given set of reactor conditions (e.g., temperature, pressure, gas-phase composition) when all available reactants are consumed in the process. The following example illustrates the application of conventional equilibrium thermodynamics to the problem of a flowing gas reactor system, using the case of Si deposition from $SiCl_4$.

Epitaxial silicon can be deposited by the decomposition of any of the chlorosilanes, $SiCl_xH_{4-x}$. Most commonly, dichlorosilane [$SiCl_2H_2$] or silane [SiH_4] are used. The difference between the growth of Si from SiH_4 versus $SiCl_4$ lies in the heat of reaction.

Si growth takes place by the thermal decomposition of $SiCl_4$ in a CVD system according to the reaction

$$SiCl_4 + 2H_2 \Leftrightarrow Si + 4HCl; \quad \Delta G_r \qquad (3)$$

Table 1 Free Energy of Formation

Compound	ΔG_f^0 (kcal/mole)
GaAs	−294.6
AsH_3	+289.3
$(CH_3)_3Ga$	−9.4
CH_4	−12.5

where ΔG_r is the free energy of reaction at the temperature of interest. This reaction is described by the chemical equilibrium constant, K_{eq}, defined in terms of the heat of reaction:

$$K_{eq} = \exp\frac{-\Delta G_r}{RT} \tag{4}$$

and

$$K_{eq} = \frac{P^4_{HCl}}{P^2_{H_2} P_{SiCl_4}} \frac{1}{P_T} \tag{5}$$

where P_i is the partial pressure of one of the reactants, measured in atmospheres, and P_T is the total reactor pressure which is taken here to be at 1 atmosphere pressure. To calculate the final composition of gases, the input concentrations are defined as $P^0_{SiCl_4}$ and $P^0_{H_2}$ and the output partial pressures as P_{SiCl_4}, P_{HCl}, and P_{H_2}. In general, growth will occur when the partial pressure of the reactants, which are input into the reactor, exceeds the amounts determined by the equilibrium relationship of Eq. (5).

The maximum growth rate in this system is calculated by the amount of $SiCl_4$ leaving the system compared with the input amount to the reactor. The difference between these two values, $P^0_{SiCl_4}$ and P_{SiCl_4}, will be the amount of Si deposited in the reactor, ΔSi. Keeping track of the reactants and products is accomplished through the use of mass balances for the various elements involved in the reactor, given below in terms of the partial pressures of the gases in the reactor and the volumetric flow of gas into and out of the reactor, $\dot{V}_{in}$ and $\dot{V}_{out}$. The volumetric flow rates are different due to the change in the number of moles of gas-phase constituents, as the reaction occurs. The reactor is held at a constant pressure, and a steady state is assumed to be achieved within the reactor. Several relationships are known that define the mass action relationships within the reactor. The RT factor in these equations changes the partial pressures into molar quantities:

Mass Balances for Si Deposition from $SiCl_4$

Moles in = Moles out

$$\text{Cl:} \quad 4P^0_{SiCl_4}\frac{\dot{V}_{in}}{RT} = (4P_{SiCl_4} + P_{HCl})\frac{\dot{V}_{out}}{RT} \tag{6}$$

$$\text{H:} \quad 2P^0_{H_2}\frac{\dot{V}_{in}}{RT} = (P_{HCl} + 2P_{H_2})\frac{\dot{V}_{out}}{RT} \tag{7}$$

Amount of Si left in reactor as a deposit = Moles$_{in}$ − Moles$_{out}$

Si: $\quad \Delta Si = \dfrac{P^0_{SiCl_4} \dot{V}_{in} - P_{SiCl_4} \dot{V}_{out}}{RT}$ (8)

Constant Pressure Relationship

$$P^0_{SiCl_4} + P^0_{H_2} = P_{SiCl_4} + P_{H_2} + P_{HCl} = P_T \tag{9}$$

The input partial pressures of $SiCl_4$ and H_2, ϕ_{H_2} and ϕ_{SiCl_4}, are simply given in terms of the flow rates into the reactor, and the partial pressures are given by $P^0_i = [\phi_i / \Sigma_i \phi_i] P_T$. The problem of determining the thermodynamic equilibrium among all the species then requires the simultaneous solution of Eqs. (5), (7), and (8) for the three unknowns, P_i, for $SiCl_4$, H_2, and HCl. The factors in the equilibrium reaction, Eq. (2), can be replaced to obtain a single equation, whose solution yields P_{SiCl_4} as a function of the input parameters and temperature, through the equilibrium coefficient, K_{eq}. The heat of reaction for Eq. (2), needed for this calculation, is given as [1]

$$\Delta G_r(T(K)) = 283.5 - 0.155T \quad kJ/mole \tag{10}$$

These equations, once solved, yield the partial pressure of $SiCl_4$ leaving the reactor at equilibrium and establish the amount of Si deposited in the system through the Si mass balance. The fraction of Si in the gas phase, which is available for deposition, changes with the growth temperature, resulting in the curve presented in Fig. 4. As the growth temperature increases, the amount of available Si for deposition increases for a typical set of input parameters: $P^0_{SiCl_4} = 0.01$ atm, $P_T = 1$ atm, and the temperature $T = 600–1200°C$. This available Si would deposit on the hot interior surface of the reactor, as well as on any Si wafers placed within the reactor.

B. Growth Kinetics: Gas-Phase Considerations

Thermodynamics is a powerful tool in the initial analysis of a growth system, and its application can lead to some unexpected results. The deposition of the Si from SiH_4, in a similar fashion to the foregoing results, is such an example. In this case, the overall growth reaction would be described by

$$SiH_4 \Leftrightarrow Si + 2H_2 \tag{11}$$

with the heat of reaction [1]

$$\Delta G_r(T(K)) = 24.26 - 0.098T \quad kJ/mole \tag{12}$$

This heat of reaction leads to the conclusion that SiH_4 will spontaneously decompose under most conditions to form Si and H_2. SiH_4 is, however, stable at room temperature and above, with no detectable rate of reaction. In order to have the reaction proceed, the reactants must first have sufficient kinetic energy

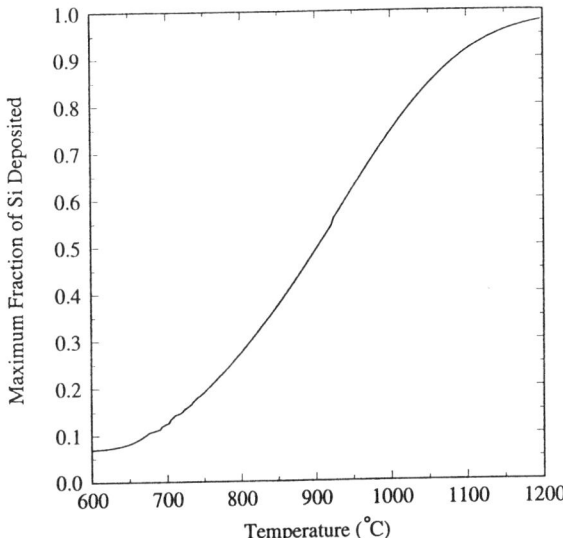

Figure 4 The maximum amount of Si deposited through the use of $SiCl_4$ can be determined utilizing equilibrium thermodynamic calculations. With a higher growth temperature, the amount of available Si for deposition increases.

to overcome any barriers preventing spontaneous decomposition. These kinetic steps can and do occur in the gas phase and at surfaces. The slowest of these steps will determine the characteristic temperature and pressure dependence of the growth process.

The equilibrium coefficient is the balance between the forward and reverse reactions as schematically shown by

$$\left. \begin{array}{c} A + B \xrightarrow{k_f} C \\ C \xrightarrow{k_r} A + B \end{array} \right\} A + B \Leftrightarrow C; K_{eq} \qquad (13)$$

where

$$\begin{aligned} \frac{d[A]}{dt} &= -k_f[A][B] + k_r[C] \\ \frac{d[B]}{dt} &= -k_f[A][B] + k_r[C] \\ \frac{d[C]}{dt} &= k_f[A][B] - k_r[C] \end{aligned} \qquad (14)$$

The rate constants, k_f and k_r, describe the rate of production of C from A and B or A and B from C, respectively. The equilibrium coefficient is the ratio of the forward and reverse rate constants, representing the steady-state relation at equilibrium of the rate of change of A, B, and C (e.g., $d[C]/dt = 0$):

$$K_{eq} = \frac{k_f}{k_r} = \frac{[C]}{[A][B]} \tag{15}$$

The steps involved in the forward and reverse reactions in Eqs. (13) and (14) have a characteristic energy associated with them. These energies, called activation energies, represent the difficulty in the molecular interaction and coordination of motion required for a specific chemical reaction to occur. The temperature dependence of the rate constants are similar to that of the equilibrium coefficient:

$$k_f = k_{f_0} e^{-\varepsilon_f/RT} \quad \text{and} \quad k_r = k_{r_0} e^{-\varepsilon_r/RT} \tag{16}$$

where

$$K_{eq} = \frac{k_f}{k_r} = \frac{k_{f_0}}{k_{r_0}} e^{-(\varepsilon_f - \varepsilon_r)/RT} = e^{-\Delta G_r/RT}, \quad \text{with } \Delta G_r = \varepsilon_f - \varepsilon_r \tag{17}$$

The relationship between these coefficients is schematically shown as a function of energy in Fig. 5. Any vapor-deposited film will therefore be the result of a complex set of chemical reactions, which are not necessarily in equilibrium. Thus, gas-phase chemistry, surface chemistry, and surface transport and incorporation can all be kinetically limiting the growth of the film. Of the various steps, the gas-phase chemistry is, perhaps best understood. In the case of Si

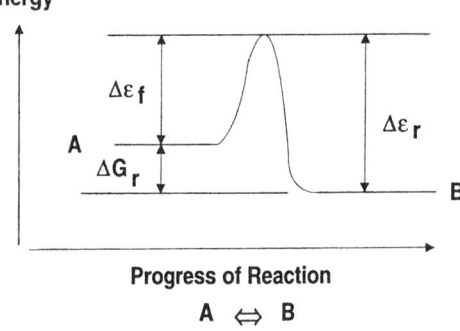

Figure 5 Kinetic factors can limit the approach to equilibrium in gas-phase and surface reactions. The factors describing these kinetic rate constants are the activation barriers for the forward and reverse chemical reactions.

deposition, SiH_4 or some other compound will enter the reactor. The high interior temperatures of the reactor can cause some gas-phase decomposition. SiH_4 and Si_2H_6 decompose in the gas phase to produce a variety of compounds: SiH_2, Si_2H_6, Si_3H_8, and H_2. The temperature and initial composition of the gas phase will determine the overall vapor-phase composition of the system, and the Si-bearing species arriving at the growth front will be a mixture of all these compounds. Not all of the Si bearing species will, however, contribute to growth, as summarized in Table 2. The growth of Si can result from the direct interaction of SiH_4 or Si_2H_6 with the surface. The gas-phase formation of SiH_2 results from the unimolecular gas-phase decomposition of SiH_4 or Si_2H_6. Unimolecular decomposition is the decomposition of a gas-phase species through collisions with the surrounding gas molecules, which results in the transfer of energy. This means, that given sufficient energy, the molecule will, by itself, decompose in the gas phase. Table 2 indicates that even in a simple chemical system there can be a great number of gas-phase species formed. Their gas-phase concentrations are coupled together through the various rate equations describing the consumption and generation of these potentially very reactive species. The importance of these species will depend on their relative concentrations, which are a function of their residence time in the reactor. These concentration calculations can be solved using a variety of numerical techniques, the results of which are

Table 2 Silane Decomposition Kinetics: Simplified Scheme

Reactions to be considered	
Silane decomposition (unimolecular decomposition)	$SiH_4 \Rightarrow SiH_2 + H_2$; k_1
Disilane decomposition (unimolecular decomposition)	$Si_2H_6 \Rightarrow SiH_2 + SiH_4$; k_5
Silane formation (gas-phase reaction)	$SiH_2 + H_2 \Rightarrow SiH_4$; k_4
Disilane formation (gas-phase reaction)	$SiH_4 + SiH_2 \Rightarrow Si_2H_6$; k_2
Silylene decomposition (surface reaction)	$SiH_2 \Rightarrow Si_s + H_2$; k_3
Silane surface decomposition (surface reaction)	$SiH_4 \Rightarrow Si_s + 2H_2$; k_6
Disilane surface decomposition (surface reaction)	$Si_2H_6 \Rightarrow 2Si_s + 3H_2$; k_7

Rate equations
$\dfrac{d[SiH_4]}{dt} = -k_1[SiH_4] - k_2[SiH_4][SiH_2] + k_4[SiH_2][H_2] + k_5[Si_2H_6] - k_6[SiH_4]$
$\dfrac{d[SiH_2]}{dt} = k_1[SiH_4] - k_2[SiH_4][SiH_2] - k_3[SiH_2] - k_4[SiH_2][H_2] + k_5[Si_2H_6]$
$\dfrac{d[Si_2H_6]}{dt} = k_2[SiH_4][SiH_2] - k_5[Si_2H_6] - k_7[Si_2H_6]$
$\dfrac{d[H_2]}{dt} = k_1[SiH_4] + k_3[SiH_2] - k_4[SiH_2][H_2] + 2k_6[SiH_4] + 3k_7[Si_2H_6]$

given in Fig. 6 [2] for the homogeneous gas-phase decomposition of silane. Starting with 0.1 torr of SiH_4 at 625°C, little decomposition of silane occurs (Fig. 6a). In strong contrast, Si_2H_6 under similar conditions (Fig. 6b) decomposes rapidly to form a mixture consisting principally of SiH_4, Si_2H_6, and Si_3H_8. The decomposition of Si_2H_6 reaches a steady-state gas-phase composition after only ~1 ms at this temperature. These differences therefore have a significant impact on the growth of Si from even these simple Si hydride sources.

The differences in gas-phase decomposition indicate that there is a potentially wide range of species that will interact with the growth surface during CVD-based epitaxy.

The residence time of the gas in the reactor can be quite short, on the order of a few milliseconds to several seconds. The kinetic rate coefficients describe the speed at which the reactions can proceed to their thermodynamically determined conclusion. These rate coefficients are generally a function of temperature and reactor pressure. In particular, the unimolecular decomposition of the reactants can be a strong function of reactor pressure, as well as temperature, especially for small molecules. Their decomposition is enabled through the transfer of thermal energy from the other gas species due to gas-phase collisions—the more gas-phase collisions, the greater the chance that the molecule will have sufficient energy to decompose. Experimental data obtained at high reactor pressures do not easily extend to low-pressure growth regimes. Lindemann and co-workers derived a function form for the rate constant that accounts for this collisional activation [3]. The unimolecular rate constant generally takes the form

$$k_{unimolecular} = \frac{k_1 P_T}{1 + k_2 P_T} \tag{18}$$

where both k_1 and k_2 follow the temperature dependence given in Eq. (16), and again P_T is the total pressure within the reactor. At high reactor pressures, this expression simplifies to a constant value, $k_{high\ pressure} = k_\infty = k_1/k_2$, while the low-pressure value of the rate constant is proportional to P_T: $k_{low\ pressure} = k_1 P_T$. The rate constant is therefore determined by the reactor pressure. Exact pressure dependences of important rate constants can be determined experimentally or through the use of calculational programs that incorporate the quantum mechanical details of the decomposition. An example of the pressure-dependent rate constant for the unimolecular decomposition of SiH_4 into SiH_2 and H_2 (the first reaction in Table 2) [4] is given in Fig. 7 as a function of the ratio of the rate constant at a given pressure to the high-pressure limiting value, k_∞. At low pressures, the SiH_4 decomposition rate is proportional to the reactor pressure, as given by the simple Lindemann theory.

Based on the pressure dependence of the gas-phase rate constants, epitaxial growth systems relying on gas-phase reactions will benefit from low growth

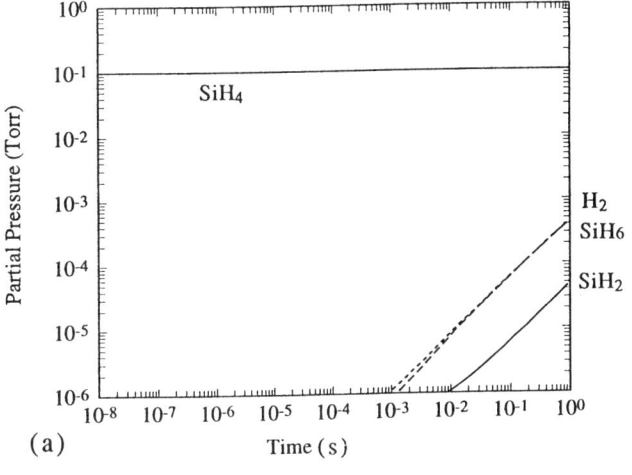

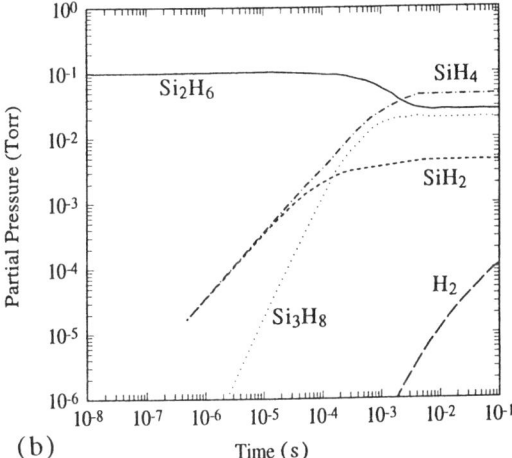

Figure 6 The gas-phase decomposition of (a) SiH_4 and (b) Si_2H_6 are favored at elevated temperatures. The rate of decomposition is limited by the kinetic rate factors governing these decomposition processes. At 650°C, the gas-phase decomposition of SiH_4 occurs slowly, while Si_2H_6 rapidly decomposes. If the time spent over the heated regions of the reactors is short, surface reactions will dominate the growth process from these reactants. Times longer than 1 s are required for substantial SiH_4 decomposition, while only 1 ms is needed for Si_2H_6 to fully decompose in the gas phase. (From Ref. 2.)

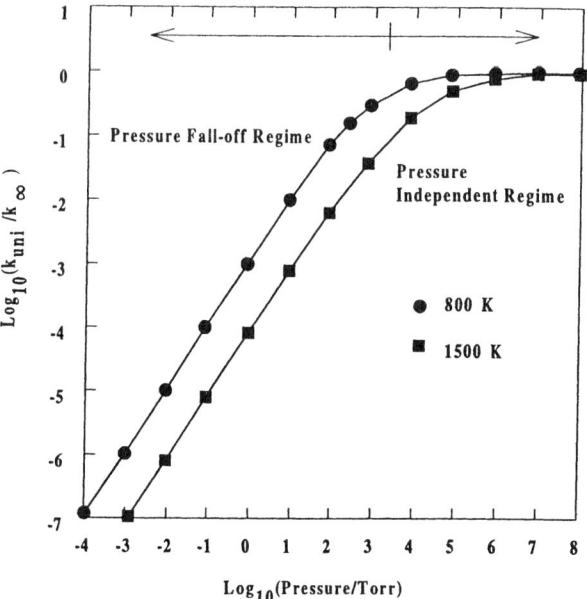

Figure 7 The rate of gas-phase decomposition of small molecules, such as of SiH_4 shown in this figure, is dependent on the gas pressure in the reactor. The gas-phase chemistry can be effectively quenched, at a given temperature, by utilizing low reactor pressures. At high pressures, the reactions become pressure-independent. (Data from Ref. 4.)

pressures. Such process conditions eliminate the undesirable gas-phase reactions. Surface reactions then become more important in the film growth. This simplification of the growth chemistry has motivated many growth CVD systems to be run at low reactor pressures.

C. Growth Kinetics: Surface Considerations

The nature and concentration of species arriving at the growth front are determined by the gas-phase reactions and, in turn, are a function of the gas-phase kinetics and thermodynamic factors. The subsequent reactions and incorporation of material into the growth surface will depend on the detailed chemistry of the surface. The atomistic modeling of the growth surface is very difficult; yet, we need the ability to parameterize the experimental observations so they may be useful in predicting future trends. These models include the basic steps of adsorption, decomposition, and incorporation into simple models.

The adsorption of a reactive species depends on several factors: the energy required for, or released by, the adsorption of reactants, the availability of an adsorption site, and the concentration of gas-phase species that are competing for these sites. The concentration of reacting species on the surface will be the balance between the arriving species, A_{gas}, interacting with vacant surface sites, S, and those once adsorbed available for desorption, A_S, as schematically shown in Fig. 8,

$$\text{adsorption:} \quad A_{gas} + S \xrightarrow{k_{ad}} A_S \tag{19}$$

$$\text{desorption:} \quad A_S \xrightarrow{k_{de}} A_{gas} + S,$$

where k_{ad} and k_{de} are the rate constants for the adsorption and desorption processes. The adsorption of gas-phase species is often modeled via simple Langmuir adsorption isotherms, a model also based on thermodynamic equilibrium. The fraction of available adsorption sites is given in terms of the partial pressure of

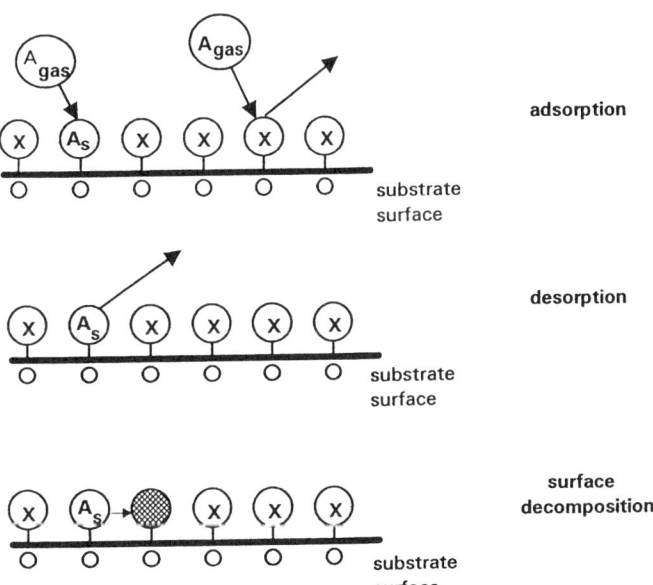

Figure 8 Surface reactions important in vapor growth proceed through a set of elementary steps. The source molecule, A_{gas}, must chemically adsorb, or bind, to the surface on an available surface site, becoming a surface adsorbed species, A_S. The reactant may desorb, leading to no growth, or undergo later surface decomposition and incorporation into the film.

reactant in the gas phase [5]. The fraction of available adsorption sites, θ, covered by given reactive species is

$$\theta = \frac{\alpha P}{1 + \alpha P} \tag{20}$$

where P is the partial pressure of the reactive species over the growth surface, and α is typically written in the Arrhenius form as $\alpha = \alpha_0 \exp(-Q/RT)$, with Q being an activation energy associated with the process, and α_0 a constant. This simple equation allows for a description of the temperature dependence and the partial pressure dependence of the surface concentration of reacting species and can often (despite its simplicity) describe the features of many adsorption processes. If several species are competing for the same adsorption sites, this equation is modified to express the site fraction of each of the species:

$$\theta_i = \frac{\alpha_i P_i}{1 + \Sigma_j \alpha_j P_j} \tag{21}$$

where the sum is over all the types of species (i or j) available for adsorption. The actual surface concentration in this case is complicated by many species competing for a limited number of adsorption sites on the growth front—a common situation during growth. In the case of SiH_4-based Si growth, SiH_2, SiH_4, and H all compete for the same sites on the surface.

Once adsorbed on the surface, these species can further react and decompose. The rate of these subsequent surface reactions will depend on the species concentration, as well as neighbor molecules with which it may react. Many simple reaction schemes can be postulated for any decomposition reaction, and models for rate expressions can be derived to initially characterize the growth behavior. To complete the model of the growth chemistry, the gas-phase chemistry is coupled with the surface or heterogeneous reactions. One such model assumes that the growth of Si from SiH_4 is due to the gas-phase formation of SiH_2 and its subsequent decomposition on the surface. In actuality, at low growth temperatures, the direct reaction of SiH_4 plays an important role. A reduced set of reactions describing the growth of Si from SiH_4 are

$$\begin{aligned}
&\text{Reversible gas-phase reactions:} && SiH_4 \xleftrightarrow{K_{gas}} SiH_2 + H_2 \\
&\text{Reversible process of adsorption:} && SiH_4 \xleftrightarrow{K_a} SiH_4 \cdot S \\
&\text{Surface reaction:} && SiH_2 \cdot S \xrightarrow{k_S} Si(s) + H_2(g)
\end{aligned} \tag{22}$$

The equilibrium coefficients, K_{gas} and K_a, and the rate constant, k_S, are functions of the temperature in the system. These reactions can be combined through the application of the equilibrium thermodynamics to form a model of the Si growth rate. If the partial pressure of SiH$_4$ is assumed to be approximately the same as the input value into the reactor (i.e., $K_{gas} \ll 1$), the partial pressure of SiH$_2$ can be found through the law of mass action:

$$K_{gas} = \frac{P_{SiH_2} P_{H_2}}{P_{SiH_4}} \frac{1}{P_{Total}} \tag{23}$$

Similarly, the surface concentration of adsorbed SiH$_2$, [SiH$_2 \cdot$ S], can be determined through the equilibrium condition:

$$K_a = \frac{[SiH_2 \cdot S]}{S_{Total}} \frac{P_{Total}}{P_{SiH_2}} \tag{24}$$

where S_{Total} is the total concentration of possible surface sites. Si growth is due to the final surface reaction in (22):

$$\text{Growth rate} \propto \frac{d[Si]}{dt} = k_s [SiH_2 \cdot S] \tag{25}$$

Combining these relationships, we can express the growth rate of the Si within the context of this simple model as

$$\text{Growth rate} \propto \frac{d[Si]}{dt} = k_s K_a K_{gas} \frac{P_{SiH_4}}{P_{H_2}} \tag{26}$$

The resulting growth model indicates that the growth rate would be exponentially dependent on temperature due to the temperature dependence of the reaction rate and equilibrium constants. The growth rate dependence on the partial pressure of reactants in the system can be a direct test of the growth model. The development of such simple reaction schemes associated with the epitaxial growth of these materials potentially allows for the prediction of the growth behavior over a wide range of reactor conditions. These models are limited, however, by the accuracy of the assumptions associated with the chosen chemical kinetic scheme and the surface-reaction limitation to growth.

D. Surface Transport and Reactions

The preceding discussion focused on the chemical reactions required for atoms to be deposited on the sample surface. Once the atoms are deposited, the specific atomic motion and the energetics of the growth front will dictate the physical arrangement of these deposited atoms. The epitaxial growth of a semiconductor requires that the deposited atoms assume an orientation that is directly related to the underlying substrate. The nature of the bonding across the interface and any difference in the lattice parameters (interatomic spacing) between the deposited layer and the substrate will determine the ultimate structure of the epitaxial layer. The simplest case of epitaxial growth is homoepitaxial growth, which is the deposition of a material upon a substrate of the same material. Homoepitaxial depositions are used to provide a difference in the optical or electrical properties of the film from that of the underlying substrate. Since there is no difference in the physical and chemical properties, this growth can proceed in a well-controlled manner.

The growth or addition of atoms to the surface can proceed in two different growth modes or behavior: step-flow growth or layer-by-layer growth. The two-dimensional growth of the crystal is often referred to as Frank–van der Merve growth. In the step-flow mode case, atoms deposited on the growth front can diffuse to naturally occurring step edges. Since these step edges and step kinks, shown in Fig. 9a, provide several atoms for the migrating atom to attach to, migrating atoms will naturally bind there and be incorporated into the growing film. At high growth temperatures, the atoms have sufficient mobility to migrate

a) Frank-van der Merve Mode: Layer-by-Layer Growth

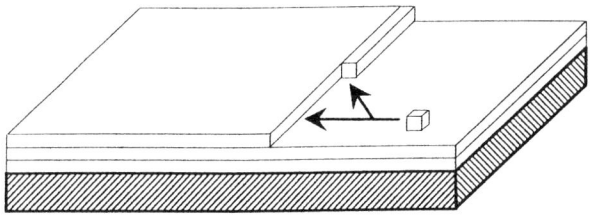

b) Stranski-Krastanov Mode: Finite Layer plus Island Growth

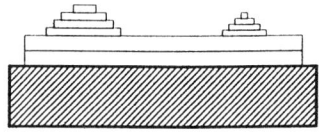

c) Volmer-Weber Mode: Island Growth

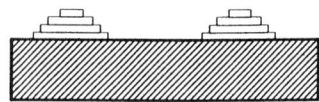

Figure 9 The deposited atoms can undergo a variety of growth behaviors as schematically shown here. The growth mode is dependent on the chemical and physical differences between the substrate and deposited layers as well as the growth temperature.

across the surface before encountering other adatoms. The growth then proceeds with the step flowing across the growth front, leaving a very smooth atomically flat surface with a terracelike structure. A low flux of atoms to the growth front or a low growth temperature leads to a slow surface diffusion of adatoms across the growing crystal. In this case, adatoms will encounter other adatoms on the growth front. Some of these atoms will bind together, resulting in the formation of a new layer of the crystal. In all cases, the added atoms will maintain an epitaxial relationship with the atoms of the underlying surface. This latter form of growth mode can result in rougher surfaces.

A more complicated and yet more interesting case is that of heteroepitaxial growth of a material. In this case, the deposited atoms differ in chemical composition from the substrate. The best-known heteroepitaxial semiconductor systems are Al_xGa_{1-x}. As grown on GaAs, and Si_xGe_{1-x} on Si. Heteroepitaxial growth has several distinct advantages over homoepitaxy, as well as several complicating features. The wide range of compound semiconductor alloy systems can result in an extremely large range of possible bandgap or band structure combinations. Since the chemical composition is changing, the physical and chemical characteristics of the deposited epitaxial material also change. The change in the strength and nature of the chemical bond between the substrate atoms and the deposited epitaxial layer, and the lattice parameter, will contribute to the nature of the deposited film. In the case of $Al_xGa_{1-x}As/GaAs$ materials, their combination forms a natural ''lattice-matched'' system. Other binary compound semiconductors containing Al and Ga, and which possess a common anion, i.e., AlAs and GaAs, AlSb and GaSb, and AlP and GaP, have all nearly identical lattice parameters, as shown in Fig. 10. The constant lattice parameter allows for the growth of an arbitrary composition material on a readily supplied substrate of a binary material.

Lattice-matched heteroepitaxial semiconductor growth can proceed quite similarly to the homoepitaxial case. Layer-by-layer and step-flow growth are observed in many systems in which the chemical bonding and lattice parameters are similar between the two materials. The alloy semiconductor (e.g., $Al_xGa_{1-x}As$) will, in most cases, grow as a random alloy system with the metal atoms randomly distributed over the available cation sites in the crystal. There are more cases, however, where the desired film has neither the same bond strength nor lattice parameter. The heteroepitaxial growth of these generally lattice-mismatched systems can exhibit additional growth modes or behaviors. The appearance of a particular type of growth behavior will depend on the chemical bonding, usually characterized by the interfacial energy associated with the epitaxial layer–substrate interface and the difference in lattice parameters. The interfacial energy determines the adhesion of the epitaxial layer to the substrate, while the lattice parameter difference determines the amount of stress that will be present in the film if it maintains perfect registration with the underlying

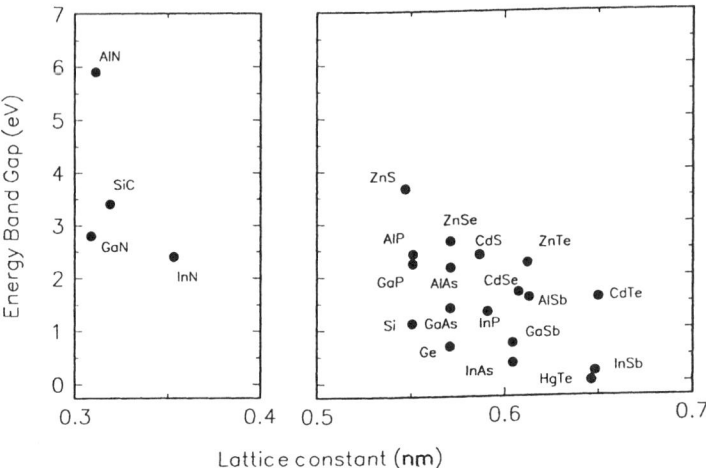

Figure 10 The ranges of available bandgaps and lattice parameters are quite wide for binary compound semiconductors. The small covalent radius of nitrogen leads to a substantial shrinkage of the lattice and widening of the bandgap.

substrate. Many materials combinations will allow for the deposition of one or more monolayers in perfect registry with the substrate. Once the strain in the deposited film becomes too large, the film relaxes and three-dimensional islands form on the surface. This growth mode, schematically shown in Fig. 9b, is referred to as a Stranski–Krastanov growth mode, or a three-dimensional island growth. The resulting film, while maintaining an overall epitaxial relationship to the substrate, will have many defects in the form of dislocations and stacking faults.

The growth of the non-lattice-matched composition generally results in the formation of a network of "misfit" dislocations, required to accommodate the difference in lattice parameters with a subsequent disregistry in the atomic arrangement across the heterointerface. However, the growth of defect-free, generally lattice-mismatched, materials can also be achieved, but only at a specific composition at which the lattice parameter matches that of the substrate, as shown in Fig. 11. There is an intermediate growth regime for lattice-mismatched materials in which defect-free, but highly strained, thin layers of material can be grown on a substrate. This growth mode is often referred to as pseudomorphic growth. Heteroepitaxial growth initiates by the adsorption of atoms, their bonding to the substrate surface, and the transmission of the epitaxial "footprint" to the subsequently deposited atoms. The epitaxial growth will typically proceed with the deposited crystal maintaining the lattice parameter of the substrate par-

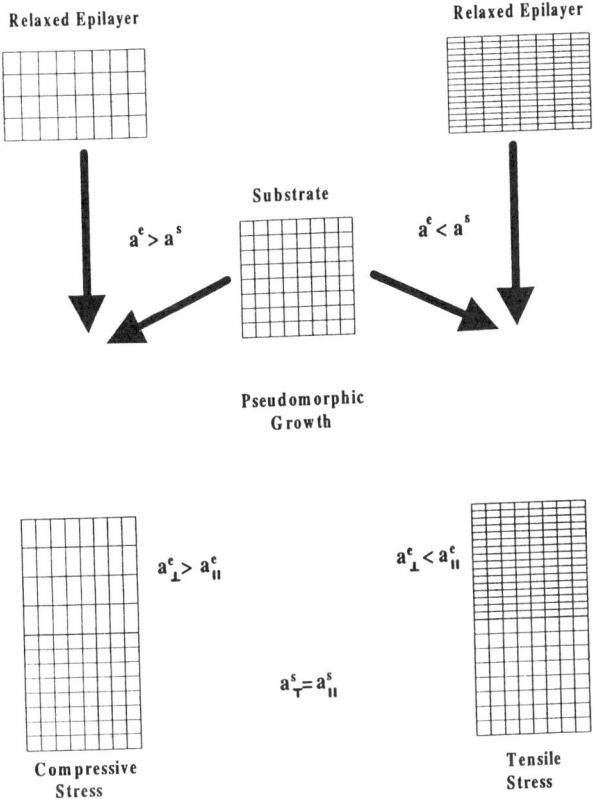

Figure 11 The growth of heteroepitaxial films is complicated by a difference in lattice parameter. The crystal lattice of the overlying film can conform to the in-plane atomic spacing of the substrate, $a_\parallel$. At a larger film thickness, the deposited layer relaxes to its bulk lattice parameter through the generation of lattice defects such as dislocations. The thickness at which the film relaxes is referred to as the pseudomorphic limit.

allel to the surface despite a difference in bulk lattice parameters. The growing layer will tetragonally distort from its bulk structure as a result of the initial ''locking'' to the substrate lattice spacing. The lattice parameter of the epilayer, perpendicular to the surface, will become distorted from the bulk value as the in-plane lattice parameter strives to match the lattice parameter of the substrate. This distortion will build up a strain in the material with a large amount of energy stored in the structure, the amount of which eventually exceeds a critical value required for the creation of strain-relieving defects, such as dislocations. It is therefore possible to grow highly mismatched materials, without extended defects, to a limited, or "critical," thickness. The exact maximum thickness prior to the formation of extended defects, commonly referred to as the critical thickness, is very material-specific. Many device structures are composed of several very thin layers, which often do not exceed this pseudomorphic limit. This allows for the incorporation of a wide range of materials into device structures despite the lack of a bulk lattice-matched condition.

III. CVD CONSIDERATIONS: FLUID FLOW AND MASS TRANSPORT

The CVD growth environment is chemically and physically complex. The gases being transported through the reactor often exhibit complex fluid flow patterns. There are often high thermal gradients in the reactor due to the presence of a cold wall and a heated substrate (Chapter 4, Fig. 6). Other reactors, referred to as "hot-wall" systems, are contained within an isothermal furnace. Chemical vapor deposition reactors work at atmospheric pressure as well as at very low pressures. The choice of reactor and operating conditions depends on the particular growth chemistry employed in the system. In all cases, the growth of an extremely uniform film is desired, where the "uniform" refers to the growth thickness, chemical composition, and electrical and optical properties. If the sample surface is maintained at a constant and uniform temperature, it is critical that a uniform flux of reactants to the growth surface be achieved over the entire deposition area. There are generally two regimes in which the reactor may operate. In the first regime, the surface reactions are very slow and the mass transport of reactants to the surface does not influence the growth rate. In this case, the entire surface has the same concentration of reactants in the gas phase above it, leading to a uniform surface reaction rate. At the other extreme, the surface reactions are very fast and the reaction proceeds to near completion at the growth surface. The transport of nutrients from the gas phase to the surface becomes rate limiting, and the growth becomes mass-transport-limited. These two regimes are typically distinguished by their differing dependence on the growth temperature. Mass transport is influenced by the growth temperature through its weak temperature dependence on the gas-phase diffusion coefficient, $D_{gas} \propto T^{-n}$, where $n \approx 1$–2. Growth taking place within the surface-reaction-limited regime has the strong dependence on temperature characteristic of the surface reaction rate coefficient, $k_S = k_0 e^{-\epsilon/RT}$. The gas-phase transport therefore plays a key role in determining the uniformity of deposition.

Although the nature of the gas-phase transport can be modeled and predicted from the principles of fluid flow, the combined and coupled thermal, momentum, and mass transport mechanisms make the numerical solution of the basic transport equations very difficult. Accurate models of the CVD environment require large numerical simulations often beyond the capabilities of desktop computers. The specific flow regime of the CVD environment can be characterized in terms of the Knudsen number, which is based on the ratio of the mean free path of the gas molecule to the typical physical dimension of the reactor:

$$K_n = \frac{\lambda}{L} \tag{27}$$

where the mean free path of a gas molecule λ of diameter d at a gas density of n is given by

$$\lambda = \frac{k_B T}{\pi \sqrt{2}\, d^2 P} \tag{28}$$

The mean free path of a gas molecule at room temperature is approximately given by

$$\lambda\ (\text{cm}) = \frac{0.66}{P} \tag{29}$$

where P is the pressure in Pascals (Pa). At atmospheric pressure ($\sim 10^5$ Pa), the mean free path is about 7 nm, while at a low pressure of ~ 100 Pa (10^{-3} Torr) the mean free path is ~ 0.7 mm. The flow regimes are then classified by the magnitude of the Knudsen number:

$$\begin{aligned}&\text{viscous flow} \Rightarrow K_n \ll 1 \\ &\text{transition flow} \Rightarrow K_n \approx 1 \\ &\text{molecular flow} \Rightarrow K_n \gg 1\end{aligned} \tag{30}$$

The viscous regime is characterized by low temperatures and high pressures. Most CVD systems operate at moderate or near-atmospheric pressures. In this pressure regime, the fluid transport can be understood through the application of traditional fluid transport theory based on a continuum description of the fluid. While the complete solution of the transport equations is generally difficult, several simplifications have arisen that provide a heuristic model of the growth environment. The design of CVD systems within this pressure regime focuses on the development of a laminar flow profile within the reactor. Laminar flow is characterized by the gas flowing smoothly across the surface, without turbulence. Turbulence or irregular mixing can lead to a high degree of non-uniformity in the film growth, since the gas flow and hence the flux of nutrients to the surface is changing with time. Laminar flow is readily established in most systems over a short entrance length of the reactor. Once laminar flow is established, the resulting gas-phase fluid transport across the growth front is often characterized by the use of boundary layer theory. A boundary layer is a hypothetical gas-phase region near the growth surface over which the gas velocity is zero. This stagnant region of gas allows for the easy solution of the diffusion equation across this boundary layer, the extent of which depends on the gas velocity and viscosity [6]. In practice, the assumptions of the boundary layer theory are not met for most reactors; yet, this simplification can predict some of the general features of the growth process. The flux to the surface is found by postulating a growth reaction at the surface in series with the gas-phase transport to the growth front, as seen in Fig. 12. At steady state, the flux through the gas phase and the reaction rate at the surface are equal, allowing for the assignment of a nutrient partial pressure adjacent to the surface to be used in the model development (see exercise problem 6). Several reviews and discussions of the modeling methods and considerations can be found in Ref. 7.

Reactors operating in the molecular flow regime, $K_n \ll 1$, are often used when the growth rate is limited by the surface reaction rate of a growth species. In these low-pressure reactors, the details of the gas transport of the reactants are not important due to the slow reaction rate of the film formation; yet, these reactors are useful in obtaining a high throughput due to a large-number wafer capacity. Since mass transport is fast compared with the surface reaction rate, the gas-phase transport between the wafers is sufficiently rapid to supply the required growth nutrients.

The fluid flow conditions within the reactor have a direct bearing on the growth rate and uniformity of the films being deposited. It can also influence the detailed properties of the growing film. In particular, morphology of the growing film and the impurity content of the film will be affected by the fluid-thermal environment. The formation of recirculation cells in the fluid stream will allow reacted nutrients and their by-products to interact again with the growing surface. Often this leads to the incorporation of deleterious impurities. An additional effect, particularly in higher pressure reactors, is the phenomenon

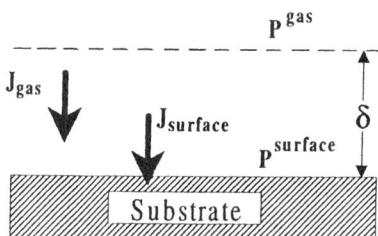

Figure 12 The gas-phase transport of nutrients and the surface reactions are two processes in series in the overall growth of the thin film. The actual film growth will be dominated by the slower of the two serial processes.

of *autodoping*, the transport of previously incorporated dopants into growing regions. Autodoping on uniformly doped substrates refers to the outdiffusion of impurities from substrate into a lower-doping-concentration epitaxial layer. This process is a solid state diffusional process and is well described by conventional diffusion theories [8]. In many applications, epitaxial growth is undertaken on wafers that have selectively doped regions, formed by patterned solid state diffusion or ion implantation. At high growth temperatures and high reactor pressures, these dopants can evaporate from the growth surface, transport laterally in the fluid boundary layer near the growth surface, and be reincorporated into adjacent regions, as schematically shown in Fig. 13. This form of autodoping leads to the lateral expansion of the localized doping region and the incorporation of dopant into the epitaxial layer. Low growth pressure enhances the rapid transport of the evaporated dopants from the growth front, and lower growth temperatures reduce the tendency for evaporation, both parameters thereby improving the development of an epitaxial region with well-defined impurity profiles.

IV. SPECIFIC SYSTEMS

Based on the foregoing discussion, it becomes obvious that there are a wide variety of growth considerations and conditions that can be employed in the epitaxial growth of elemental and compound semiconductors. Two specific materials systems and growth processes will be presented in detail, highlighting some of the particular problems encountered in CVD: the metal–organic vapor-

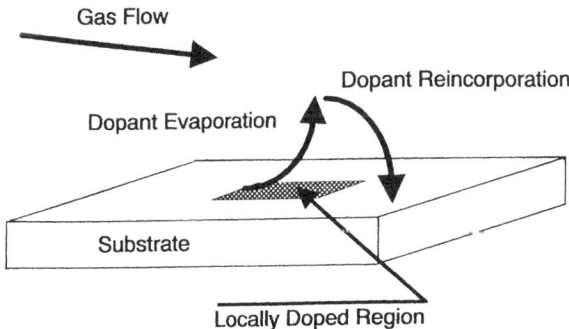

Figure 13 Autodoping refers to the redistribution of incorporated impurities into adjacent growing epitaxial regions. Lateral transport of the impurity occurs when the dopants in the substrate reevaporate and are transported to adjacent regions of the surface, to be reincorporated into the growing layer.

phase epitaxy (or MOVPE) of GaAs, and the very low pressure growth of epitaxial Si.

A. The Metal–Organic Vapor Deposition of GaAs and Related Materials

The MOVPE growth of compound semiconductors was developed in the 1970s and 1980s and is widely used in the growth of semiconductor heterostructures. (There have been several reviews of the MOVPE process. See Refs. 9 and 10.) The growth chemistry is based on the use of volatile metal–organic compounds, such as R_xM, where typically, $R = CH_3, C_2H_5, \ldots$, and $M = Al, Ga, Zn, \ldots$, in combination with Group III hydrides (AsH_3, PH_3, H_2Se, ...) or organic-based compounds (($C_4H_9)AsH_2$, $(C_4H_9)PH_2$, $(C_3H_7)_2Te$, ...). These compounds share several characteristics. They are all quite reactive and decompose at moderate temperatures. The low decomposition temperatures of these compounds, many with decomposition temperatures below 400°C, necessitate the use of a "cold-wall" reactor. The substrates on which the deposition is to occur reside on a heated pedestal. The remainder of the reactor is not directly heated. The reactive gases flowing through the reactor only encounter high temperatures at or near the substrate surface. It is within this local heated gas-phase region and on the hot substrate surface that the precursors react, leading to the deposition of adatoms, which migrate on the surface and are finally incorporated into the growing layer. The temperatures encountered in the growth of GaAs from $(CH_3)_3Ga$ and AsH_3 are typically 600–800°C. Since As has a very high vapor pressure over GaAs, compared with Ga, an excess of the As precursor is always used, with As/Ga ratios of 10–100 in the gas phase commonly employed. These source compounds react very rapidly at these typical growth temperatures. As a result, the slow step in the growth of the film over most of the temperature range is generally the transport of the nutrient from the bulk of the flowing gas to the growth front. The growth rate data for GaAs MOVPE growth, shown in Fig. 14, have been studied over a wide temperature range and indicate three separate temperature regimes [9,10]. At low temperatures, the growth rate is strongly temperature-dependent. Over this range, the temperature-dependent decomposition of the $(CH_3)_3Ga$ is the rate-limiting growth step. As the temperature is increased, its reaction rate increases until the rate of transport of material to the growth front can no longer keep up with the consumption of the growth nutrient [$(CH_3)_3Ga$]. As a result, the growth rate now becomes limited by the gas-phase diffusion of the reactants to the surface. The weak temperature dependence seen in this temperature region (600–800°C) is characteristic of the gas-phase diffusion coefficient of $(CH_3)_3Ga$. At even higher temperatures (>900°C), the growth rate decreases once again, possibly due to thermodynamic influences coming into play. Both Ga and As will have moderate equilibrium

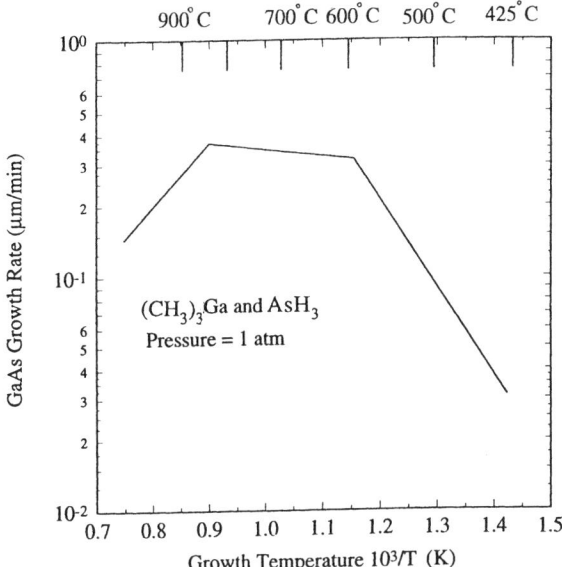

Figure 14 The growth rate in the metal–organic vapor-phase epitaxy of GaAs exhibits three separate growth regimes. At low growth temperatures, the growth rate is dominated by the surface decomposition of the reactants. The nearly temperature-independent growth regime is limited by the gas-phase transport of nutrients to the surface. The decrease of the growth rate at elevated temperatures has been attributed to both the desorption of the nutrients at the surface and parasitic reactions in the gas phase.

vapor pressures over GaAs at these high temperatures. Hence, the partial pressures of Ga and As in the gas phase, required for thermodynamic equilibrium, are now significant, and since most of the growth nutrients are no longer available for incorporation into the growing film, the maximum achievable growth rate decreases as a result.

The growth of MOVPE GaAs typically takes place in the middle temperature regime. Over this temperature range, almost all available Ga in the gas phase can be incorporated into the film. Additionally, the temperature is high enough that surface mobility sufficient for the lateral transport of atoms on the surface to step edges is present. The growth proceeds in the step-flow mode of layer-by-layer growth as seen in Fig. 9a. This mass-transport-limited growth also simplifies the formation of alloy-based semiconductors. All available constituents are consumed at the growth surface, i.e., Al and Ga in the growth of $Al_xGa_{1-x}As$, since As is in excess. The specific alloy composition in the growing layer will be controlled by the relative arrival rates of the two species, since the gas-phase diffusion coefficients for most of these species are very similar. The ratio of the alloy constituents in the solid, e.g., $Al/Ga|_{solid}$, will be equal to the ratio of partial pressures of the precursors in the gas phase, $(CH_3)_3Al/(CH_3)_3Ga|_{gas}$. The mass-transport-limited growth therefore greatly simplifies the formation of multicomponent materials, which has led to the rapid implementation of this technique. While the mass-transport-limited growth does allow for the easy determination of the alloy composition, the uniformity of the growing film thickness will be quite sensitive to the details of the fluid flow within the reactor. Regions of the wafer where the gas-phase transport impedes the mass transport of nutrients will exhibit a lower growth rate.

B. Silicon Growth at Very Low Pressures

The near-atmospheric-pressure growth, AP-CVD, of Si has been used to form a wide variety of Si structures. The high growth temperatures used in this process, in excess of 800°C, can lead to the redistribution of impurities or dopants within the film. The extension of this type of Si growth to the growth of heteroepitaxial systems, such as Si_xGe_{1-x}, has been difficult. The low melting point of Ge, ~935°C, and the large difference in lattice parameters of germanium and silicon (Fig. 10) lead to the formation of a large number of defects in alloy layers. A low-temperature growth technique is therefore required in order to develop these new materials. The lower limit placed on the growth temperature in AP-CVD is due to the incorporation of oxygen. Oxygen is present in all gases, at some level, due to the present inability to sufficiently purify these gases. For example, a gas that is found to have only 10 ppb of impurities, within an atmospheric pressure reactor (760 torr), would have a partial pressure of these impurity gases of 0.001 Pa, or 7.6×10^{-6} torr. A calculation of the rate of impingement of impurities onto the surface at this partial pressure leads to the equivalent of several monolayers of impurities a minute. If only a fraction of these impurities remain on the surface and are incorporated into the film, the film would be very contaminated. One way to reduce this effect is to lower the total pressure within the reactor. If this same gas was flowing into an ultrahigh vacuum (UVH) a reactor at 0.001 Pa total pressure (10^{-8} atm), the partial pressure of impurities within the reactor would correspondingly drop to 10^{-11} Pa, or $\sim 10^{-13}$ torr, resulting in a very low impingement and incorporation of impurities into the growing film. In particular, the equilibrium between SiO_2 and O_2 within the reactor can be shifted to favor an oxygen-free surface. SiO_2 and the high-temperature monoxide maintain an equilibrium with O_2:

$$SiO_2(\text{solid}) \Leftrightarrow SiO(\text{gas}) + \frac{1}{2}O_2(\text{gas}) \tag{31}$$

The low partial pressure of water in the reactor favors the formation of the volatile SiO. The Si surface then remains free of O-based contaminants, allowing epitaxial growth at significantly lower growth temperatures.

Nevertheless, the growth of Si in UHV-CVD [11] is complicated by several factors. The growth takes place under a molecular flow regime given the low operating pressures, and the growth reaction is based on the adsorption and surface decomposition of SiH_4 on the growth surface. At the low temperatures, the surface of the growing film is covered with hydrogen produced from the growth of Si, a by-product of SiH_4 use. The rate-limiting step to the growth of the Si film is the desorption of the hydrogen from the surface allowing for the subsequent adsorption and decomposition of the SiH_4 source. The Si is incorporated into the growing surface through the subsequent removal of the hydrogen atoms from the $SiH_x(ad)$ adsorbed species. Since mass transport limitations are absent in this growth technique, the overall reaction scheme is quite simple, based on the rate of hydrogen removal from the adsorbed SiH_x and the desorption of H_2 from the growth surface, allowing additional SiH_4 to adsorb. The epitaxial relationship is maintained through the lateral diffusion of the deposited Si atoms over the surface after decomposition to an appropriate Si surface site.

The growth of mixed composition systems, such as Si_xGe_{1-x}, is accomplished through the addition of GeH_4 to the reactor. The growth rate of the alloy system is greater than that of Si alone for the same total pressure in the system. The SiH_4 and GeH_4 compete for the adsorption sites on the growing surface. If the ability to adsorb on the growth front were the same for both these sources, a constant growth rate would be assumed. The rate-limiting step, however, is again the removal of hydrogen from the growth front. The Ge–H bond is much weaker than the Si–H bond (20 vs. 38 kcal/mole). The presence of Ge on the growth surface, when growing Si_xGe_{1-x}, leads to the enhanced desorption rate of hydrogen from the surface. Hydrogen attached to the surface Ge is easily desorbed, again opening new adsorption sites. Additional nutrients are then adsorbed, and the growth rate increases. This chemical modification of the surface kinetics leads to the increased growth rate of the Si_xGe_{1-x} alloy system.

The low-pressure environment presented by the UHV-CVD approach leads to several other important consequences. The gas-phase chemistry plays a small role in the growth of the film. The activated gas-phase reactions due to gas-phase collisions are quenched as the pressure is reduced, while surface reactions become the dominant influence on the growth of the film. The growth is typically conducted at low growth temperatures in order to minimize any gas-phase reactions, thereby affecting the growth rate or uniformity. The almost complete coverage of the growing surface with atomic hydrogen leads to a form of passivation or protective covering of the sample surface, which kinetically inhibits the adsorption of other impurities, such as oxygen, by not providing a suitable adsorption site for them. This high hydrogen surface coverage allows epitaxial growth at low temperature (as low as 450–550°C) without the loss of epitaxy due to the inability to prevent oxygen or other impurities from incorporating into the growth surface.

V. SPECIAL CONSIDERATIONS

The underlying chemistry of the deposition process can be modified to change the growth behavior and provide additional utility to the growing layer. Most semiconductor structures require the controlled introduction of specific impurities to provide unique electrical and optical properties to the films. This introduction of impurities, or doping, is generally limited, however, to the modulation of composition and doping in the *growth direction*, and it is discussed briefly in terms of chemical considerations. Another topic, selective epitaxy, relates to the control over the deposition and composition of lateral growth and into localized areas. Such growth methods are used effectively in the formation of new and advanced device structures, possessing unique electronic properties.

A. Doping: Controlled Impurity Introduction

The control of impurities in the growing layer is the focus of much of CVD development. Unintentional or background impurities place a limit on film purity. High-purity semiconductors are essential to the fabrication of high-speed devices, but they often require the controlled introduction of intentional impurities. These are typically electrically active as shallow donor or acceptor states. The incorporation of impurities from the gas phase during CVD growth follows several trends. Their specific incorporation behavior is generally influenced by the operating parameters within the reactor. Each doping precursor exhibits a different rate of incorporation based on its chemical characteristics and relationship with the epitaxial base material. As with many chemically active systems, small amounts of impurities can substantially alter the chemical reactivity of the growing surface. Although several system parameters do affect the growth and doping, the growth temperature has the greatest influence. In contrast to the dopant influencing the growth system, the film composition, growth rate, and other gas-phase constituents can also influence the impurity incorporation.

The limit of the epitaxial film purity (see Chapter 2, Sec. II.C) is ultimately determined by the growth reaction itself. The actual chemical composition of the growth precursor and its kinetic pathways can lead to unintentional incorporation of impurities. In CVD, since the growth nutrients are transported to the growth front in the form of a specific chemical reactant, it is possible that part of the growth source constituents can be incorporated into the growing film. Perhaps the most obvious example of the importance of the growth reaction chemistry on the impurity incorporation is the unintentional incorporation of carbon during MOVPE growth. Carbon incorporation is the ultimate limitation to the purity of such MOVPE-based semiconductor films. The MOVPE technique utilizes organic ligands to create volatile, reactive growth precursors. A necessary condition for the suitability of a possible growth precursor is the efficient removal of these ligands during growth. The growth source, i.e., $(CH_3)_3Ga$ and similar compounds, has carbon-bearing ligands. Since carbon is an electrically active shallow acceptor in most III–V compounds, unintentional incorporation of carbon can lead to degraded electrical and optical properties. The relatively weak metal–carbon bonds in most Group III compounds allows the organic ligand to be removed at low temperatures through a variety of chemical paths. Despite the facile removal of these ligands from the metal atom, there is, however, a small probability that the organic ligands will eventually contribute carbon to the growing film.

Two primary growth sources are available for the MOVPE-based growth of GaAs: $(CH_3)_3Ga$ (TMG) and $(C_2H_5)_3Ga$ (TEG). Although similar, these molecules exhibit significant differences in carbon incorporation, which in general depends on the reactor flow conditions, the growth temperature, substrate orientation, the gas-phase composition, and the specific growth chemistry. Several mechanisms have been proposed for the incorporation of carbon into the growing layer.

The most studied system for carbon incorporation is the growth of GaAs from trimethyl gallium and AsH_3 in an H_2 carrier gas. Growth temperature strongly affects the unintentional carbon incorporation, which increases at the extremes of the growth temperatures, either very high or low. The vapor-phase ratio of $AsH_3/(CH_3)_3Ga$ also alters the carbon incorporation by several factors. A high V/III ratio $[AsH_3/(CH_3)_3Ga]$ within the reactor decreases carbon incorporation.

The conclusions reached from these observations indicate that the removal of carbon from the growth front is limited by a surface reaction, primarily by the hydrogen elimination reaction between the methyl groups on the TMG and the hydrogen on the AsH_3:

$$GaCH_3(ad) + AsH_x(ad) \Rightarrow GaAs(s) + CH_4 \qquad (32)$$

While the carbon removal reaction pathway is understood, the reaction mechanism by which the carbon (from the methyl groups) is actually incorporated has been harder to ascertain. Again on the surface, the successive removal of hydrogen from the adsorbed methyl groups leads to the formation of the reactive carbene, CH_2, through a reaction, such as

$$CH_3(ad) + CH_3(gas) \Rightarrow CH_2(ad) + CH_4(gas) \qquad (33)$$

Once formed, the attachment of this species to the GaAs surface somehow leads to the final incorporation of carbon into the crystal. The level of carbon in GaAs is very low under all usual growth conditions, less than 1 ppm in the crystal.

This level is, however, still high enough to affect the characteristics of some of the devices fabricated from the epitaxial layers.

The chemical nature of the Ga source also strongly affects the carbon levels. The highest-purity GaAs has been produced from the reaction of TEG with AsH_3. The TEG decomposition follows a different set of reaction paths in that the gas-phase reactions appear to be more important in the TEG decomposition, as opposed to the surface reactions just described for TMG. In gas-phase pyrolysis studies, the TEG is found to decompose at much lower temperatures than the TMG. It decomposes at typical growth temperatures through a β-hydride elimination reaction, which results in the fissure of the Ga–C bond with the replacement of the carbon by a hydrogen obtained from the ethyl ligand:

$$(C_2H_5)_2GaC_2H_5 \Rightarrow (C_2H_5)_2GaH + C_2H_4 \tag{34}$$

The co-reaction of the Ga source with the AsH_3 is not required. The carbon-bearing parts of the molecule can be removed in the gas phase before reaching the growth surface. The presence of a unimolecular reaction, combined with the lower pyrolysis temperature of TEG, reduces the amount of carbon reaching the surface, resulting in significantly lower residual carbon incorporation.

The intentional incorporation of impurities is used to impart specific properties to the growing film. In many cases, the addition of impurity-containing molecules, such as SiH_4 or Si_2H_6, to a GaAs growth system leads to the controlled introduction of Si, which is a donor in GaAs. Gas-phase and surface chemistry play a role in these processes; however, the incorporation of impurities appears frequently to be independent of the film growth reaction. The impurity concentration realized in the growing film would simply be the product of the atomic concentration of the host materials and the ratio of the relative deposition rates of the impurity and host materials, R_i:

$$[\text{Impurity}] = N_{\text{host}} \times \frac{R_{\text{dopant}}}{R_{\text{host}}} \tag{35}$$

The addition of impurities to a surface can influence other reactions within the growth system. For example, the addition of PH_3 or AsH_3 during the growth of Si from SiH_4 can lead to a dramatic drop in the growth rate of the film. As and P are both donors in Si, and their hydrides can adsorb strongly on the Si surface. PH_3 will decompose slowly on the surface relative to SiH_4 and contributes, along with the strong adsorption behavior, to the accumulation of PH_3 on the growth front. This prevents SiH_4 from being adsorbed, and hence its growth rate drops significantly. Such behavior is similar to some types of surface "poisoning" often seen on catalyst surfaces. The Group III hydride B_2H_6 can result in a very different effect. Boron is the most common acceptor in Si and Ge. The addition of B_2H_6 to an SiH_4-based growth system leads to an increased growth rate of the Si layer. While not well understood, the added dopant in the p-type Si is thought to lead to a high concentration of holes in the near-surface regions. These holes, representing empty valence band states, could migrate to the surface, leading to their insertion into the Si–H bond at the surface, which is weak; this facilitates the desorption of H from the surface, thereby leading to the opening of a new adsorption site for the SiH_4. The rate-limiting step to the growth of Si from SiH_4 is the desorption of H from the growth front. These two cases, PH_3/AsH_3 and B_2H_6, indicate the strong influence that impurities can have in surface-controlled reactions.

B. Selective Epitaxy

Epitaxial growth techniques are generally used for the controlled uniform deposition of materials over the entire semiconductor surface. These processes are limited, however, to the modulation of composition and doping in the growth direction. If the underlying substrate was covered by a surface layer, e.g., SiO_2 on Si, the deposition films would not be epitaxial, since it has no access to the structural arrangement of the substrate. For many devices and applications, a selectively deposited material only at specific areas on the wafer surface would be preferred, and such lateral definition of the growing layer can be accomplished by choosing the appropriate growth chemistry. Opportunities for significant advances in device structures are anticipated with the effective development of epitaxial techniques that simultaneously also provide direct control over the lateral dimension of the structure, such control being otherwise exceedingly difficult to achieve during the traditional growth of an epitaxial layer. Localized, or selective, area epitaxy offers the opportunity to form such structures.

Effective localized growth on a substrate surface can eliminate process steps, provides added functionality to the structure, and allows the incorporation or embedding of the masking material into the device structure. Selective epitaxy (SE) has been accomplished in several different growth systems. It has been investigated as a path to achieving three-dimensionally integrated structures in the Si-based materials. The investigations have focused primarily on Si chemical vapor deposition using dichloro- and tetrachloro-silane ($SiCl_2H_2$ and $SiCl_4$) as growth sources. While still at a very early stage of development, the results indicate the feasibility of such three-dimensional integrated circuits. These applications require control over both the growth characteristics of the SE material and the electrical properties of the interface between the SE and substrate regions, referred to as the regrown interface.

Perhaps the most intriguing use of an SE technique is the possible integration of several different materials, and hence devices, on a single integrated chip. The ability to controllably produce high-quality regions of both well-defined composition and structure allows for the development of multifunction device structures that are not attainable by traditional planar processing technologies.

Several approaches have been used to accomplish this goal. Localized heating, or photolysis through an externally imposed light source (most commonly a laser), has been used to produce localized regions of growth [12,13]. The lateral resolution of these regions is a function of the wavelength of light, the substrate thermal behavior, and the duration of exposure. Regions as small as several microns have been produced and utilized in the demonstration of this growth technique. However, these features are still somewhat large for the growth of the desired smaller device structures. Alternatively, either a masking layer is used to prevent epitaxial growth, resulting in the deposition of polycrystalline materials, or growth conditions are chosen to eliminate deposition on the masked regions entirely, as shown in Fig. 15. The latter technique obviously avoids any need for subsequent removal of polycrystalline deposition from the masked field regions. The specific choice of the appropriate growth chemistry and process conditions are central to any effective selective epitaxial growth.

The SE of Si has been studied for many years. Typically, the areas for the selective growth are defined by a masking layer opening to an underlying Si substrate. SiO_2 and Si_3N_4 are preferred as masking layers. The selective area growth of Si relies on the suppression of the growth on the masking region relative to the openings, and several approaches are employed. Any nucleation and growth of Si on SiO_2 would require that the concentration of the surface Si atoms on the mask reach a high enough concentration (critical coverage) to initiate the nucleation of Si crystallites from the adsorbed molecules. The object of the SE process is therefore to prevent this critical concentration from being realized on the mask region. The Si exposed in the mask openings to the growth

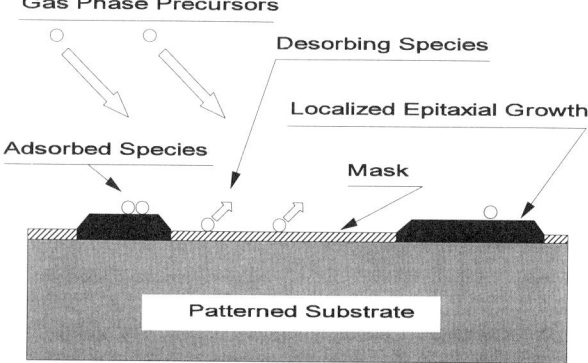

Figure 15 Highly selective growth occurs when the adsorption of the growth nutrients readily occurs on the unmasked regions but is inhibited on the mask itself. These selectively grown regions are typically characterized by faceted growth on unmasked regions next to the mask boundaries.

ambient will not experience any substantial nucleation barrier, since no new phase is being formed.

Thus, the prevention of nucleation on the mask, ensuring selective area growth, is accomplished primarily by the choice of growth chemistry (Si precursor). SiH_4, $SiCl_4$, and $SiCl_2H_2$ have been used as Si sources. The respective growth rate behavior of these precursors is shown in Fig. 16 [14]. The Si chemical source must either be able to interact with the surface-masking material chemistry to prevent the adsorption of the precursor on the growth surface, or chemically provide for species that will assist in that process. SiH_4 works well at low growth pressure, due to low adsorption of SiH_4 on typical mask surfaces, while the much stronger adsorption of SiH_4 on the exposed Si surface leads to deposition. Eventually, however, nucleation of Si will occur on the mask region after some time; the period prior to the time required for such initial nucleation of the Si on the mask is referred to as the incubation period. The incubation period will depend primarily on reactor temperature and pressure and the characteristics of the masking material. It is typically sufficiently long for the desired Si layer to be deposited in the mask opening.

An alternative approach relies on the presence of additional chemical reactions in the growth environment to assist in removing Si nuclei from the masking regions. The presence of Cl in the growth environment provides such a reaction

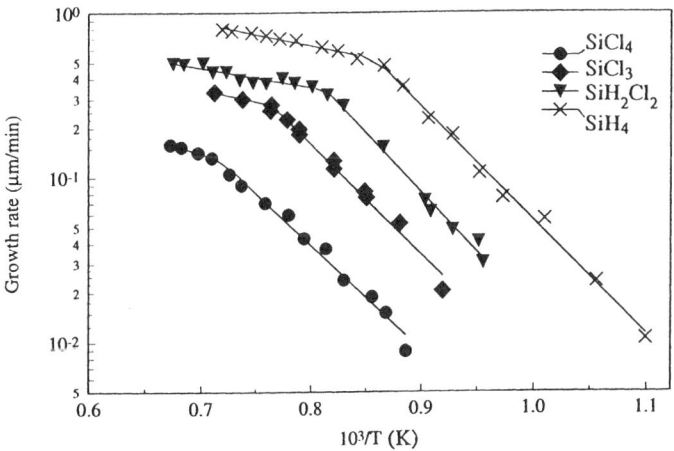

Figure 16 The vapor-phase growth of Si from the chlorosilanes, SiH_xCl_{4-x}, exhibits a strong dependence on the chemical nature of the growth reactants. Lower decomposition temperatures can be used as the ratio of H/Cl in the precursor is increased. (Data from Ref. 14.)

pathway whereby deposited Si is removed from the mask region through the reformation of chlorosilanes:

$$SiCl_4 + 2H_2 \Leftrightarrow Si + 4HCl \tag{36}$$

Hence, both etching and growing can occur, the rate of these reactions being sensitive to the local concentrations of HCl and $SiCl_4$ immediately above the growth surface. Fluid flow, temperature, and pressure are used to adjust this process so that the balance between growth and deposition favors growth on the exposed Si surface and etching on the mask regions. At high reactor pressures, the growth reaction by-products over an exposed region are transported across adjacent regions, and this local transport will affect the local growth rate across the wafer, and hence the deposition uniformity.

The deposition uniformity and edge morphology of the SE film regions play a central role in the utility of these structures for a particular application. The uniformity of the deposited material is measured on very disparate length scales. The larger length scale (in centimeters) involves the overall growth uniformity of the selectively grown regions over the entire wafer. The finer length scale, typically on the micron level, is dominated by very local surface and gas-phase diffusion processes. At this level, the growth morphology of the small structures is characterized by the crystallographic planes forming the SE structure boundary. The facet formation is dominated by the surface and gas-phase transport, but further complicated by the adjacent masked and exposed regions.

The edge morphology of selectively grown regions exhibits a complex dependency on the materials and growth system parameters. They are dominated by the appearance of slow-growth crystal planes, as shown in Fig. 17 for the case of GaAs SE. These features develop as a result of both the gas-phase and surface diffusion of deposited atoms across the developing flat to an exposed surface. Since there are many crystal planes that possess comparable surface energies, variations in the growth habit will appear with growth temperature changes. The prediction of the edge morphology is generally very difficult due to this complex growth behavior. A predictive model of the surface diffusion processes and knowledge of the structural nature of the growth surface are required to fully develop an understanding of this transport and growth behavior. Because of the presence of the masked regions, the bounding edges of the selectively grown regions have a larger net deposition volume of material due to the effects of the gas phase and from the adjoining mask areas. This additional flux of material results in growth nonuniformities on the distance scale comparable with a gas-phase or surface diffusion distance. Several theoretical modeling efforts have investigated the effects of either or both of these diffusion effects. The establishment of an effective surface-reaction-rate-limited SE growth chemistry would mitigate these transport-related edge morphologies.

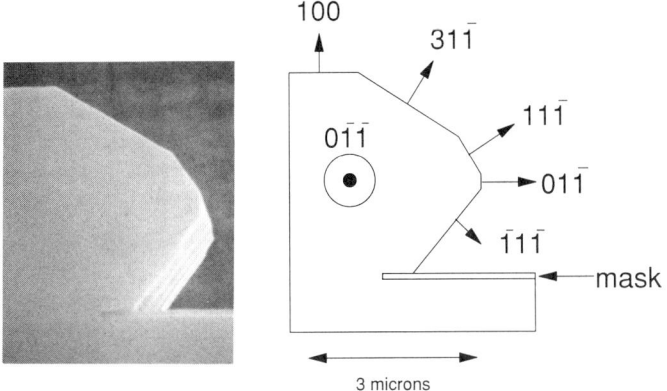

Figure 17 Crystal facets tend to form at the edges of the selective growth regions. This is shown here for the SE of GaAs via the MOVPE process. The facets are slow-growth planes that will bound the growing crystal.

VI. SUMMARY

The process of epitaxial growth by chemical vapor deposition will continue to be one of the principal tools used in the formation of semiconductor devices. As the device dimensions continue to shrink, the importance of the new low-temperature growth processes will continue to grow in importance. Growth based on CVD offers the potential for a high throughput, and hence a highly manufacturable process. Despite the current, and even the anticipated increasing importance of this technique, basic understanding of many of its underlying chemical and physical issues remains lacking. The atomic-level picture of the growth process will be needed as device dimensions approach the nanoscale. The motion of the atoms on the surface, the specific chemical reactions occurring at selected atomic sites on the growth front, and the modification of the growth reactions to control feature shape and dimension are areas of continuing research. While an atomic-level picture of the growth process is emerging, the development of new CVD technologies is also moving forward. Whereas the device dimensions shrink, the wafer diameter is projected to increase to 300–350 mm. The traditional use of multiwafer batch CVD reactors will shift to single-wafer CVD systems. The large wafers will place great demands on the ability to uniformly control the thermal and fluid environments for the wafers in order to achieve the exacting growth of the thin layers and small structures needed in future device generations. The rational design of these complicated system awaits the further development of new computational approaches to modeling the thermal, radiation, and fluid environments. The answers derived from these challenges will certainly open other doors in the processing of advanced materials and lead to more and more control over the atomic-level structure of materials.

EXERCISES

1. Si growth can take place by the thermal decomposition of $SiCl_4$ in a CVD system according to the reaction:

    $SiCl_4 + 2H_2 \Leftrightarrow Si + 4HCl; \quad \Delta G$

    The growth rate of the Si can be modified by the addition of HCl to the reactor feed. In this case, the Si deposition will be reduced. The feed concentrations of $SiCl_4$ and HCl are $P^0_{SiCl_4}$ and P^0_{HCl} as dilute mixture in an H_2 carrier gas, with concentration $P^0_{H_2}$. What is the dependence of the Si growth rate on the input partial pressure of HCl at a constant temperature? What input partial pressure of HCl is required to just prevent Si deposition? Use the mass action relationships to obtain an equation relating the input feed gases to the Si growth rate at a constant temperature.

2. In ultrahigh vacuum, almost no decomposition of SiH_4 occurs. The silicon growth may go through successive hydrogen extraction:

    $SiH_4 + S \Leftrightarrow SiH_4 \cdot S$
    $SiH_4 \cdot S \Leftrightarrow SiH_3 \cdot S + H$
    $SiH_3 \cdot S \Leftrightarrow SiH_2 \cdot S + H$

    where S represents a surface site. For this problem, however, assume that SiH_4 is the adspecies mainly responsible for the growth, so that the following growth steps are obeyed:

    $SiH_4 + S \overset{K_A}{\Leftrightarrow} SiH_4 \cdot S$

    $SiH_4 \cdot S \overset{K_B}{\Leftrightarrow} Si(c) + 2H_2$

    Derive a simple model of the growth of Si that incorporates these basic reaction steps.

3. A CVD reactor has a simple tube geometry with cross-sectional area A. The gas flowing into the tube has an initial reactant concentration c_0, flowing at a velocity v_0. The film growth rate follows a simple rate power law: growth rate, $GR = k_s c^n$. The rate constant, k_s, follows the usual Arrhenius temperature dependence, $k_s = k_0 \exp(-\varepsilon_a/k_B T)$. The temperature is varied along the reactor tube in order to achieve a uniform growth rate. Find the functional form of the temperature profile, $T(x)$, along the length of the reactor, in terms the reactor temperature at the inlet, T_0, and the parameters given,

that would achieve uniform growth rate along the reactor length. Assume the growth is only occurring along the walls of the tube. What is, if any, the maximum reactor length?

4. The growth of high-purity GaAs for detectors and other devices can be based on the halide vapor-phase epitaxy process. This process can be modeled using equilibrium thermodynamics. This process entails the flow of $AsCl_3$ into the reactor in a carrier gas of H_2. The $AsCl_3$ breaks down irreversibly into As_4 and HCl upon entering the reactor. The HCl reacts with a GaAs-crusted Ga source (which can be understood to be just a chuck of GaAs) to produce GaCl. The GaCl, As_4, HCl, and H_2 then flow down the tube to a cooler region, whereupon GaAs is re-formed or deposited on the GaAs wafer due to the change temperature.

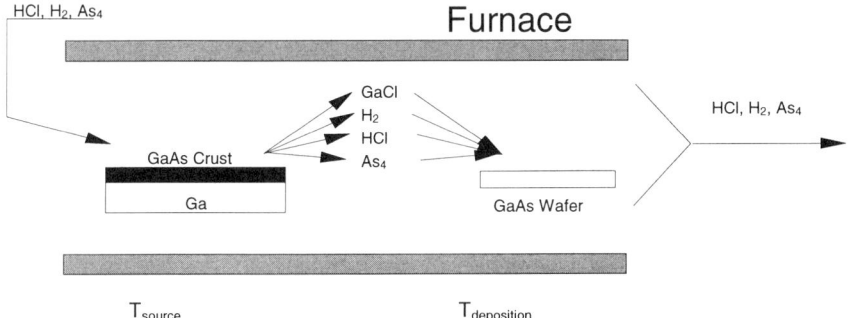

The source temperature is 700°C, and the reactor is kept at a pressure of 1 atm. Find the maximum growth rate of GaAs at the deposition zone under two conditions:

a. The flow rate of $AsCl_3$ is constant at 20 sccm (standard cm³/min volume) and $T_{deposition}$ is altered from 0 to 50°C below T_{source}.
b. With a fixed $T_{deposition}$ of 650°C, as the $AsCl_3$ flow rate is changed from 0 to 50 sccm. The reactor has a constant input flow of H_2 of 1000 sccm.
   Assume all the deposition is occurring on a GaAs wafer 25 mm in diameter and that the growth rate is uniform over the surface. Provide graphs of the maximum growth rate in microns of deposited GaAs per minute on the wafer in each case as a function of the changing variable. Do not consider in your model the equilibrium between As_4 and As_2. Use the following data:

	$\Delta H^0_{f,298\,K}$ (kcal/mole)	$S^0_{f,298\,K}$ (e.u.)	Heat capacities $C_p = a + bT + cT^2$ (e.u.)		
			a	$b \times 10^3$	$c \times 10^5$
GaAs(s)	−19.8	15.34	10.46	2.80	0
HCl	−22.06	44.64	6.18	1.34	0.35
H_2	0	31.21	6.58	0.65	0.12
As_4	34.5	75	19.81	0.023	−1.18
GaCl	−19.1	57.4	8.93	0	−0.38

5. The growth of Si from SiH_4 can be due to the direct reaction of SiH_4, SiH_2, and Si_2H_6 on the growth front. The Si_2H_6 and SiH_2 species are a result of the gas-phase decomposition of the SiH_4 as shown in Fig. 8 for the temperature of 625°C. The ''reactive sticking'' coefficient describes the probability of the molecule, upon reacting with the surface, will adsorb, decompose, and incorporate an atom on the surface. You are using SiH_4 in a reactor in which you may change the residence time of the gas phase, as in Fig. 8. Plot and compare the growth rate of Si in the surface reaction rate limit. Assume the reactive sticking coefficients of SiH_2, SiH_4, and Si_2H_6 are given as 1, 10^{-5}, and 10^{-3} respectively. What impact does the residence time have on the growth uniformity in a CVD using this growth chemistry?

6. The growth rate in many CVD systems is limited, in part, by both mass transport and surface reactions. These basic steps are in series, as shown in Fig. 8. The mass transport to the surface will be driven by the concentration gradient across the boundary layer. The surface reaction is determined by the rate constant, k_S, the inlet partial pressure, P^{gas}, the gas-phase diffusion coefficient, D, and the boundary layer thickness, δ. Write an expression for J_{gas} and $J_{surface}$ in terms of these quantities and the unknown surface partial pressure, $P^{surface}$. At steady state, these two fluxes to the surface will be equal to the overall observed growth rate:

$$J_{growth} = J_{gas} = J_{surface}$$

Show that the growth flux can be expressed as

$$J_{growth} = \frac{k_S P^{gas}}{1 + \dfrac{RT\delta k_S}{D}}$$

REFERENCES

1. J. A. Dean, ed., *Lange's Handbook of Chemistry*, McGraw-Hill Co., New York, 1972.

2. B. S. Meyerson, B. A. Scott, and R. Tsui, *Chemtronics 1*:150 (1986).
3. K. F. Roegnigk, K. F. Jensen, and R. W. Carr, *J. Chem. Phys. 92*:4254 (1988).
4. B. S. Meyerson and J. M. Jasinski, *J. Appl. Phys. 61*:786 (1987).
5. A. W. Adamson, *Physical Chemistry of Surfaces*, Wiley Interscience, New York, 1990.
6. A simple model based on a boundary layer theory is given by F. C. Eversteyn, P. J. W. Severin, C. H. J. v. d. Brekel, and H. L. Peek, *J. Electrochem. Soc. 117*:925 (1970).
7. See, for example, as a introduction to the description of many semiconductor processes, D. W. Hess and K. F. Jensen, eds., *Microelectronics Processing*, Advances in Chemistry Series, American Chemical Society, Washington, DC, 1989.
8. B. Tuck, *Introduction to Diffusion in Semiconductors*, Peter Peregonius Ltd., 1974.
9. T. F. Kuech, "Metal–Organic Vapor Phase Epitaxy of Compound Semiconductors," Materials Science Reports, vol. 2, 1987.
10. G. B. Stringfellow, *Organometallic Vapor Phase Epitaxy, Theory and Practice*, Academic Press, San Diego, 1989.
11. B. S. Meyerson, *IEEE Proc. 80*:1592 (1992).
12. N. H. Karam, H. Liu, I. Yoshida, B.-L. Jiang, and S. M. Bedair, *J. Crystal Growth, 93*:254 (1988).
13. Q. Chen, J. A. Osinski, C. A. Beyler, and P. D. Dapkus, *Appl. Phys. Lett., 57*:1437 (1990).
14. L. J. M. Bollen, *Acta Electronica, 21*:185 (1978).

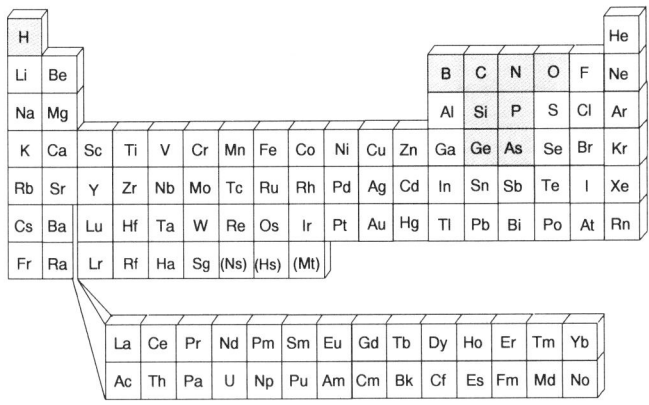

4
Chemical Vapor Deposition of Crystalline and Amorphous Carbon, Silicon, and Germanium Films

Peter Hess *University of Heidelberg, Heidelberg, Germany*

I. INTRODUCTION

A. General Considerations

Chemical vapor deposition (CVD) is a technique for growing solid films on a substrate by gas-phase and surface reactions.

The CVD process proceeds under nonequilibrium conditions—in some cases near to equilibrium (e.g., epitaxy), in others further away (e.g., metastable materials)—by generating chemically active species that induce the growth of the solid network. At the center of this deposition process is the source, or parent, gas containing the element(s) required as film constituent(s). The source gas is chosen for its combination of thermodynamic, thermochemical, chemical, and spectroscopic properties, and it is decomposed by introducing energy into the system, for example, by heat (e.g., hot filament), by electrons (e.g., glow discharge), or by photons (e.g., lamp or laser). The film precursors needed to initiate the growth process are generated either directly or by secondary chemical reactions, as shown in Fig. 1. The chemical reactions occurring in the gas phase and at the surface determine the composition and structure of the solid film.

Chemical vapor deposition has long been known to humankind. When early humans cooked their meals on an open fire, the cooking utensils became covered with a layer of soot. When wood burns, hydrocarbons are generated in the gas phase, which form a carbon film on colder surfaces. Of course, the adherence and mechanical strength of these carbon layers are low due to the uncontrolled fast growth. As we all know from experience, the film properties are more graphitelike than diamondlike in this case.

The goal of modern CVD methods is to provide films with the following characteristics: (1) good thickness uniformity, (2) controlled composition and stoichiometry, (3) high purity and density, (4) good adhesion, (5) a high degree of structural perfection, and (6) good mechanical, optical, and electrical properties. As discussed in this chapter, CVD methods have been developed in recent years that satisfy these demanding criteria.

Chemical vapor deposition is an active field of research developing in many different directions. In this chapter the deposition of crystalline silicon and carbon, the growth of dielectric silicon compounds such as SiO_2 and Si_3N_4, and the formation of amorphous hydrogenated carbon (a-C:H), silicon (a-Si:H), and

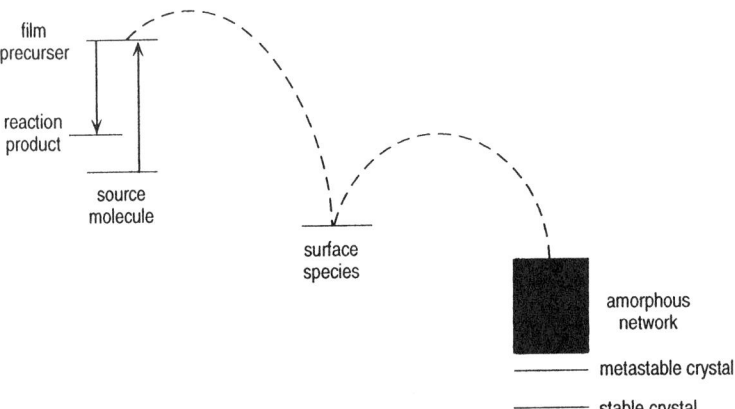

Figure 1 Simplified scheme of the main chemical pathways of a CVD process including gas-phase activation and reaction, chemisorption of film precursors, and activated solidification steps.

germanium (a-G:H) films are discussed in more detail to demonstrate the wide variety of possibilities provided by CVD processing for the synthesis of materials with tailored properties.

Depending on the deposition conditions, the film structure is crystalline, polycrystalline, or amorphous. A great variety of chemical compositions can be deposited, including nonstoichiometric materials. Indeed, most elements of the periodic table and an even larger variety of compounds can be grown today as high-quality films by CVD techniques.

Optimization for specific applications is achieved by adjusting the mechanical, optical, and electronic properties. Strong external activation of the system (e.g., by electrons or photons) makes it possible to drive the chemical environment far from equilibrium and thus synthesize materials that cannot be realized by conventional CVD or molecular beam epitaxy (MBE). This wide versatility is the reason for the growing technological interest in these process techniques.

Several books are available on different aspects of CVD process technologies, describing the deposition of elemental and compound films of insulators, semiconductors, and metals. In particular, Refs. 1–3 concentrate on silicon, Ref. 4 on diamond, Ref. 5 on amorphous materials, and Refs. 6–8 on laser-assisted processes.

B. Applications

Chemical vapor deposition is closely connected with thin-film technology, which pervades our everyday life: hard coatings for tools, decorative coatings on plastics, photovoltaic glass coatings for the generation of solar energy, antireflection coatings for camera lenses, transparent coatings for computer displays, etc. Advances in microelectronic technology and further miniaturization of integrated circuits (ICs) depend in part on the ability to reduce the size of devices containing different high-quality thin-film materials in ever-smaller dimensions.

One of the world's best-studied materials is silicon, the dominant semiconductor material in microelectronics. Its success as a semiconductor is connected not only with its electronic properties but also with its chemistry. It forms an oxide that can be used as a high-quality dielectric layer. The only component still required to make the silicon world complete is an efficient light emitting diode (LED) made of crystalline silicon. Such a device would enable optoelectronic circuitry to be entirely based on silicon, and it would therefore revolutionize very-large-scale integration (VLSI) technology.

Chemical vapor deposition methods play a crucial role in integrated circuit processing to provide high-performance films. High-quality epitaxial and polycrystalline silicon layers can be deposited at relatively low temperatures. As an example, consider a device used for switching purposes. In the so-called MOSFET (metal oxide semiconductor field effect transistor, see Fig. 5 of Chapter 1) the gate consists of polycrystalline silicon instead of a metal. The gate functions as a control electrode to switch the current flow from the terminal source to the terminal drain. Photovoltaic devices, in which light is transformed directly into electrical energy, are another application of polycrystalline silicon.

Attention has also been focused on dielectric layers in the silicon technology. In silicon, the number of bandgap electronic states present at a clean surface can be reduced by 4–5 orders of magnitude in the presence of a high-quality thin oxide layer, which can act as a barrier to further oxidation by blocking the diffusion of the relevant species to the reacting interface.

In fact, the superior features of silicon dioxide and its interfacial properties with silicon are major reasons contributing to the importance of silicon technology compared, for example, with germanium (water-soluble oxide) and GaAs. SiO_2 is chemically very stable, and it is an excellent insulator with high resistivity and a dielectric breakdown strength greater than 10 MV/cm. Such layers can be grown thermally by oxidizing the silicon surface above 800°C in an oxygen-rich environment (i.e., oxygen, wet oxygen, or steam) or at lower temperatures with vapor deposition film growth techniques.

Other dielectric materials used as insulators in the semiconductor industry include silicon nitrides, silicon oxynitrides, and aluminum oxide. The deposition of silicon nitride (Si_3N_4) will be discussed in some detail, because Si_3N_4 serves as an excellent barrier to the migration of alkali ions, is nearly impervious to water, and can be used to make planar structures.

The technique of chemical vapor deposition, and in particular low-temperature CVD, is a nonequilibrium method that allows the synthesis of metastable materials in a wide range of compositions, structures, and morphologies. Depending on the deposition conditions, these materials possess a Gibbs energy that can be much higher than the thermodynamically most stable form of the material. Examples are metastable diamond crystals and metastable amorphous solids. An important class of amorphous solids comprises the hydrogen alloys of carbon (a-C:H), silicon (a-Si:H), and germanium (a-Ge:H). The term a-C:H refers to diamondlike hydrocarbons with up to 60% hydrogen, while a-C contains less than 1% hydrogen. These hydrogenated amorphous alloys of Group IV elements of the periodic table are arousing increasing scientific and technological interest and therefore will also be discussed in more detail.

The synthesis of metastable crystalline materials such as diamond is achieved with optimized low-temperature and low-pressure CVD conditions. The main attractions of such an energizing CVD process are not only that deposition occurs at much lower temperatures and pressures, but also that thin films can be grown on a variety of substrate materials. The technological interest in CVD diamond films as a coating material is due to its outstanding hardness, chemical inertness, low friction, high thermal conductivity, and low optical absorption in a large spectral range, among other properties. Its use as a coating is the most mature application. Especially in optics, there are many examples of applications as protective, wear-resistant, or antireflection coatings due to the low optical absorption. Based on its wide bandgap (5.5 eV), its high breakdown voltage (20–40 times that of silicon), and its high electron and hole mobilities (≈ 2000 cm^2 V^{-1} s^{-1}), higher operating frequencies and power levels can be expected for diamond transistor devices than for those with silicon and GaAs. Such devices have not yet been realized due to several technical problems, one of which is to overcome the nucleation and growth of heteroepitaxial diamond on a single-crystal substrate. Another is the impurity control of homoepitaxial diamond growth for the fabrication of a p-type semiconductor. Since these barriers appear to be surmountable, diamond offers promise for future high-temperature semiconducting devices and for packaging of high-speed, high-power switching devices and logic circuits.

Films of amorphous (hydrogenated) carbon (a-C, a-C:H) are being used for specific applications in which not all the extreme properties of crystalline diamond are needed. In fact, it is possible to optimize the mechanical and chemical properties of an a-C:H film so that it can be used as a protective, wear-resistant coating ("diamondlike film"). Especially in the case of amorphous materials and alloys, CVD processing has great potential for the synthesis of solids with tailored mechanical, chemical, optical, and electronic properties, and for its ability to deposit on a large number of different surfaces without any major nucleation problems.

The main applications of a-Si:H and a-Ge:H are in the area of semiconductors. The germanium materials can be used as low-bandgap semiconductors in tandem solar cells. They may also be used for bandgap tailoring in more complicated alloys such as a-SiGe:H. The most important electronic applications of a-Si:H are in thin-film solar cells, thin-film transistors for liquid crystal flat displays, photoreceptors for electrophotography and laser printing, and image sensors.

II. CHEMICAL VAPOR DEPOSITION

A. Basic Principles
1. Thermodynamics

The source gas molecules employed in CVD processing are metastable with respect to transformation into their elements. For example, silane is unstable relative to solid silicon and molecular hydrogen. The film formed during deposition may be either metastable amorphous, metastable crystalline, or a thermodynamically stable crystalline structure, as shown in Fig. 1. One of the most important problems to be solved is the control of deposition parameters so as to obtain defined metastable phases in CVD processing.

To understand the fundamental requirements of thermodynamics, consider a steady-state deposition process. In this case, the Gibbs formula can be applied to a nonequilibrium open system consisting of k chemical components:

$$dG \leq \delta U + V\,dp - S\,dt + \sum_{i=1}^{k} \mu_i\,dn_i \qquad (1)$$

where δU refers to the work being done on or by the system (δU is not a function of state and depends on how this work, other than work of expansion, is done), G is the Gibbs free energy, S is the entropy, T is the temperature, p is the pressure, V is the volume, μ_i is the chemical potential of the ith component, and n_i is the number of moles of this component.

At steady state, the composition of the system must be constant in time. This means that the transport of species through the open system and the chemical reactions occurring in the system must balance each other. Also, at steady state, the temperature and pressure are constant due to externally imposed conditions. Therefore, dT, dp, and dn_i in Eq. (1) vanish to obtain

$$dG \leq \delta U \qquad (2)$$

In the case where no work is done on the system except work of expansion, δU also vanishes. Thus, one obtains the fundamental law of thermodynamics that only those processes occur in the system which lead to a reduction of the free enthalpy, i.e., $dG \leq 0$. In plasma CVD, where the growing surface may be immersed in the plasma, δU in fact may be positive due to electromagnetic work done on the solid. In this case, a positive dG value is possible. The implications of this law with respect to CVD processing as applied to the synthesis of diamond (a metastable crystal) and other amorphous metastable materials will be discussed later.

Thermodynamics not only determines the processes that can be realized, it can also be used to estimate the equilibrium concentration of species present under certain deposition conditions of pressure and temperature. Such a list of gaseous species is obtained from the thermochemical data by minimizing the Gibbs free energy of the overall system. The equilibrium constant K_p of a reaction can be calculated from the relation

$$\Delta G^\circ (T) = -RT \ln K_p \qquad (3)$$

where ΔG° is the Gibbs free energy of the reaction and R is the universal gas constant.

However, as already mentioned, equilibrium is not established in the CVD reactor. In a practical deposition system, kinetic factors limit the amount of material deposited and thermodynamic considerations set only an upper bound. The kinetic factors depend on operating conditions such as energy supply and gas flow and on the system design, e.g., the reactor geometry. Despite the need for the molecular mechanism for a detailed understanding, useful insights can be gained from quasi-thermodynamic approaches.

2. Chemistry

There are several ways to initiate the deposition chemistry by supplying energy to the system:

1. The source gas is decomposed via the lowest reaction channel ("soft processing"). In this case, the initiation step is relatively simple and corresponds to vibrational excitation in the electronic ground state up to the dissociation limit, as shown in Fig. 2 ("pyrolysis").
2. In energetic processing, more energy is introduced into the system and many channels are open to different reactive species for the decomposition of the source gas. In this way, different radicals or even electronically excited species can be generated by excitation of higher electronic states, as indicated in Fig. 2 ("photolysis").

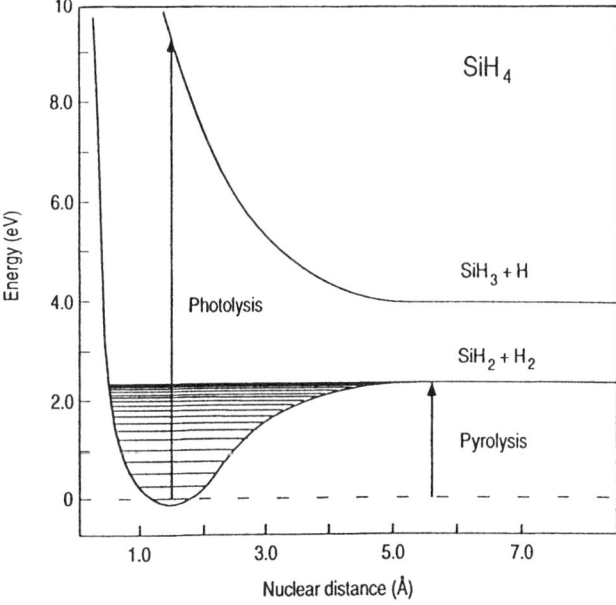

Figure 2 Potential curves illustrating pyrolysis due to strong vibrational excitation of the lowest electronic state and photolysis based on electronic excitation in the case of silane.

3. The ideal deposition scenario would start with a selected source gas yielding the precursor needed for the growth of the film with tailored properties by, e.g., specific photochemical initiation. This precursor reaches the surface without additional transformations (e.g., by secondary reactions), thereby avoiding detrimental effects caused by other species reaching the surface.

The film precursor induces the solidification reactions at the surface. Its reactivity should be not too high, so as to allow enough mobility along the surface. By diffusion of the active species over the surface, ordered layer-by-layer growth can be achieved, generating a high-quality film.

The general principles of growth at the surfaces of tetrahedrally bonded carbon, silicon, and germanium are well understood for deposition from hydrides as source gases. In these systems, the surface dangling bonds are saturated by hydrogen. The surface X–H bonds (X = C, Si, Ge) are weak enough to allow further chemical change by reactions. Growth of the solid network can be achieved, for example, by insertion reactions into the surface X–H bonds or by first creating dangling bonds by hydrogen abstraction with reactive radicals. The important point is that film growth can be initiated by energized chemical processing at temperatures much lower than those needed for thermal dehydrogenation of the surface under equilibrium conditions.

Since no detailed in situ information is available on the different stages of dehydrogenation of the growing surface, the following discussion is based on results obtained for well-defined single-crystal (100) surfaces of tetrahedrally bonded carbon, silicon, and germanium, as shown in Fig. 3. For these surfaces, hydrogen desorption has been studied for the (1 × 1) dihydride and the (2 × 1) monohydride, and it has been found that the recombinative H_2 desorption process is surprisingly insensitive to the specific surface structure. Thus, the results obtained for single crystals also seem to be applicable to the situation during film growth, at least qualitatively.

The binding energies of hydrogen on the carbon (100) surface have been determined by temperature-programmed desorption (TPD) measurements, as shown in Fig. 4. Since there is no agreement on the carbon data in the literature, their values are shown as dashed lines with large error bars. Also in Fig. 4, the arrow indicates typical surface temperatures used for the deposition of high-quality diamondlike films as well as for diamond, demonstrating that diamond can be grown at temperatures where the dihydride group is stable under equilibrium conditions. Obviously, the chemical potential of the species governing the surface chemistry during metastable diamond growth is sufficiently high that hydrogen can be removed and carbon–carbon bonds can be formed. It has been proposed that the chemical potential of atomic hydrogen could be responsible for this processing step; however, other radicals may also contribute.

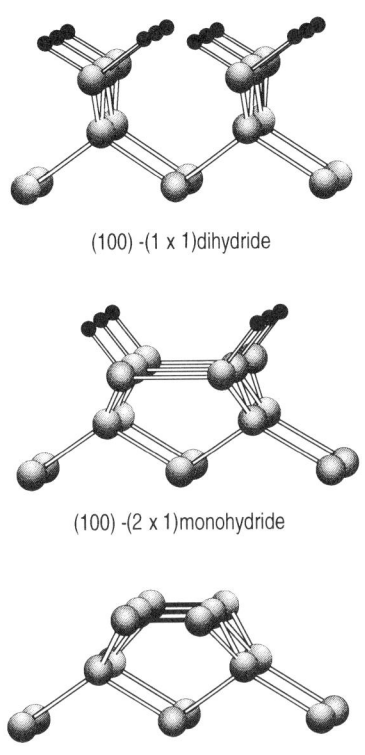

(100) -(1 x 1)dihydride

(100) -(2 x 1)monohydride

(100) -(2 x 1)π bonded

Figure 3 The different stages of hydrogenation of the (100) single-crystal surface of tetrahedrally bonded carbon, silicon, and germanium.

This explains why diamond crystals can be grown by CVD at much lower pressures (<1 bar) and temperatures (<1000°C) than those derived from the carbon phase diagram for high-pressure and high-temperature diamond synthesis (~50,000 bar and ~2000°C). The paradox of low-pressure and -temperature diamond crystal growth can be explained by the argument that not the bulk Gibbs free energy but its surface value governs the CVD growth, and that the surface with the lowest free enthalpy is crucial. Thus, the main role of atomic hydrogen (or other radicals) may not be that it promotes faster etching of graphite but that its chemical potential is sufficiently high to maintain the carbon surface atoms in an sp^3 hybridized state, suitable for growth of crystalline diamond. When the partial pressure of hydrogen atoms is too low, however, the surface carbon will reconstruct into an sp^2 state and form graphitic carbon.

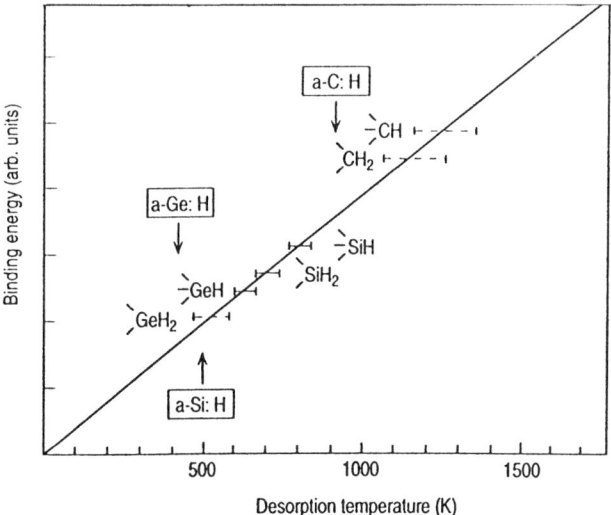

Figure 4 Hydrogen desorption temperatures measured for the XH and XH$_2$ groups at the (100) surface of diamond, silicon, and germanium crystals by temperature-programmed desorption (TPD). Typical CVD processing temperatures are indicated by arrows.

The binding energies of hydrogen on silicon and germanium have also been determined (see Fig. 4), and the corresponding arrows indicate the typical surface temperatures used for the deposition of high-quality a-Si:H and a-Ge:H. Since the incorporated hydrogen should only saturate dangling bonds and thus should be bonded as monohydride in these films, one can conclude that the chemical potential of the species governing the surface reactions must be high enough to destroy polyhydride structures and to drastically reduce the incorporation of hydrogen into the network. It seems that the chemical potential of atomic hydrogen is one of the determining factors for the composition and structure of the deposited film under strongly energized deposition conditions, as in the case of diamond growth.

As stated, low-temperature CVD can be used to deposit a large variety of metastable amorphous networks and alloys. In the case of carbon, an exceptionally large number of amorphous structures can be realized due to different types of bonding, namely sp^3, sp^2, and sp hybridization. The primary interest lies in the amorphous structures with dominant sp^3 bonding ("diamondlike films"). These possess good mechanical, chemical, optical, and thermal properties. The bonding configuration consists in this case of σ states from sp^3 and sp^2 bonding and π states from sp^2 hybridization. The π states lie symmetrically around the Fermi energy at midgap and define the optical bandedges, which depend mainly on the size of the sp^2 clusters, generating some medium-range order in the network. The bandgap, given by the π–π^* separation, varies inversely with the cluster size. A broad range of cluster sizes can therefore explain the small slope of the bandedge, or so-called broad Urbach tail.

Chemical vapor deposition processes for silicon and germanium compounds offer a large variety of amorphous networks, in which the long-range order of the crystal lattice is destroyed. However, the remaining short-range order of the tetrahedral bonding is sufficient to retain the band structure. The transition to the amorphous state is connected with relatively large distortions of the bond angle and relatively small changes of the bond length. Since the mean coordination number falls below four, undercoordinated atoms are created ("dangling bonds"), which increase the density of defect states in the bandgap drastically ("gap states"), as shown in Fig. 5. The incorporation of hydrogen improves the electronic properties of amorphous films by saturating the dangling bonds and thus reducing the number of gap states to a level where the material can be doped. As indicated schematically in Fig. 5, the density not only of gap states but also of defect states at the bandedge ("tail states") is reduced in hydrogenated amorphous materials with small "Urbach energy." The much steeper Ur-

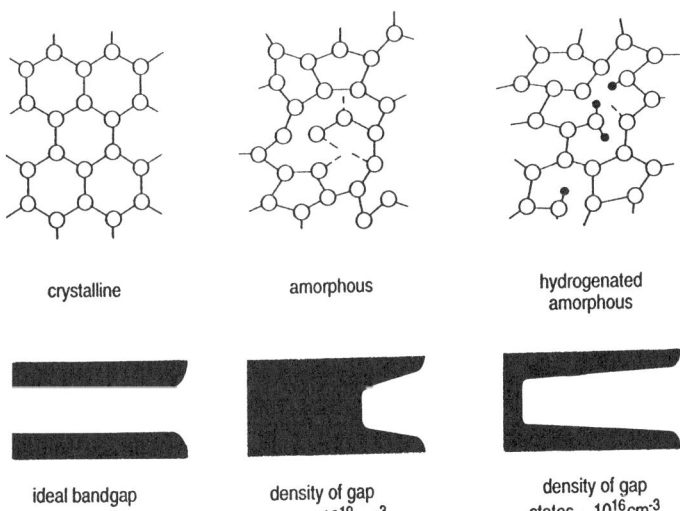

Figure 5 Schematic of the structure and bandgap of crystalline, amorphous, and hydrogenated amorphous materials. The open circles represent silicon atoms, the full circles hydrogen atoms. The full lines indicate chemical bonds, the dashed lines dangling bonds.

bach tail in a-Si:H and a-Ge:H is due to fluctuations in the short-range order of the network and is not caused by π states as in a-C:H.

3. Kinetics

Although the source molecule is thermodynamically unstable with respect to its elements, significant kinetic barriers normally exist to its decomposition and the solidification reactions of the film precursors, as indicated in Fig. 1. Thus energy is required to overcome these barriers and to achieve film growth at an acceptable rate.

If the kinetics is dominated by a chemical reaction step either in the gas phase or at the surface, the activation energy of the rate-determining step can be obtained by the Arrhenius equation,

$$k = k_0 \exp \frac{-E_A}{RT} \tag{4}$$

where k is the rate constant, k_0 is the preexponential factor, and E_A is the activation energy. The value determined by such an analysis may help to identify the rate-limiting step. However, a serious problem in such a mechanistic analysis is that in the usually complicated CVD systems several reaction processes may have comparable rates and thus contribute to the measured kinetics. In addition, the actual temperature range for such an analysis may be rather small since the dominant reaction mechanisms tend to change with temperature.

The kinetics of the different steps such as gas-phase and surface reactions or transport processes is crucial for the overall deposition process. As a general rule, a higher growth rate has the tendency to deteriorate the growing structure by forming a more distorted network. Thus, epitaxial growth of the thermodynamically stable crystal phase generally occurs at lower supersaturation and thus at a lower deposition rate than the formation of a metastable amorphous film.

Device-quality films can usually be obtained only at limited growth rates. A typical rate in many systems is on the order of 1 to 10 monolayers per second.

For deposition processes that operate under reaction-rate-limited conditions, the temperature is a critical parameter due to the exponential dependence of the reaction rate on temperature (see Eq. 4). Thus, uniform deposition requires a constant temperature throughout the reactor.

According to the Arrhenius equation, the reaction rate increases exponentially with temperature. Therefore, at a certain temperature, the reaction rate exceeds the rate at which reactant species arrive at the surface. The reaction rate then becomes limited, this situation being referred to as mass-transport-limited deposition. At this point, temperature control is not nearly as critical; however, it is very important to maintain the supply of the same concentration of reactants over the whole surface. At intermediate temperatures, both surface reactions (low-temperature limit) and mass transport (high-temperature limit) influence the overall deposition process.

4. Transport

The ultimate deposition rate may be limited by the transport of reactant and/or product species. There are two limiting cases in the transport regime: free-molecular and diffusive transport to the surface. If the mean free path λ of the active species is much larger than the dimension L of the reactor, virtually all precursors produced in the gas phase reach the surface without undergoing additional gas-phase collisions. This is called the free-molecular or ballistic transport mechanism and is characterized by $K_n = \lambda/L \gg 1$, where K_n is the Knudsen number. On the other hand, if the precursor will undergo numerous collisions in the gas phase before reaching the surface when the mean free path is much smaller than the reactor size ($K_n = \lambda/L \ll 1$), the transport mechanism corresponds to diffusive deposition. As a general rule, free-molecular deposition depends more strongly on the sticking coefficient of the precursor than diffusive deposition. In the latter case, a species that does not stick immediately at the surface may experience additional gas-phase collisions directing it back to the surface and providing a new chance for sticking.

For free-molecular deposition, the deposition flux can be calculated via geometrical considerations. With increasing total pressure in the reactor, the mean free path decreases according to

$$\lambda = \frac{kT}{\sqrt{2}\, p\pi d^2} \tag{5}$$

where d is the molecular diameter of the active species. At higher pressures, a smaller gas volume (for $K_n \gg 1$ only a very small part of the reactor volume) contributes to the deposition by free-molecular transport, and most of the deposition results from diffusive flux. However, ballistic deposition occurs to some extent at any Knudsen number and thus always makes a contribution to the total deposition rate. This contribution represents the lower limit of the actual deposition rate and depends on the particular Knudsen number and the sticking coefficient.

Diffusive transport dominates for $K_n \ll 1$, and the diffusion equation has to be solved in order to calculate the diffusive flux. Analytical solutions can be obtained under simplified conditions such as constant reactant concentration in the gas phase and zero concentration at the surface and infinitely distant outer boundary conditions. At atmospheric pressure, the mean free paths are on the order of microns, and mass transport can be treated diffusively. In the case of laser microfabrication with small-area reactions, the observed reaction rates are often orders of magnitude larger than those of the corresponding large-area reactions. This can be explained by enhanced mass transport to and from the small reaction zone. Thus, for very small laser spot sizes the spatial scale of the surface reaction zone may reach the mean free path length at relatively high pressures, and ballistic transport has to be considered.

In the transition regime ($K_n \approx 1$), where both the diffusive and ballistic deposition mechanisms contribute significantly to the growth process, a rigorous treatment of the transport problem requires the solution of the Boltzmann equation.

The flow conditions and species transport in a CVD reactor can be characterized by the Peclet number, Pe. Transport of the gas by forced convection can be described by the product of the gas velocity u and the distance l traveled. Transport by diffusion can be characterized by the diffusion coefficient D, which is a function of pressure, temperature, and molecular weight. The Peclet number is defined by the ratio of these two parameters:

$$\mathrm{Pe} = \frac{ul}{D} \tag{6}$$

For values much larger than 1, transport is dominated by convective species transport or laminar flow with minimal mixing. High-growth-rate methods such as plasma jets typically work in this regime. Even in this case, a thin fluid layer where diffusive transport dominates exists near all solid surfaces, since the velocity must go to zero at the surface. In the case Pe $\approx$ 1, slip-type flow is observed with moderate mixing, and for Pe $\ll$ 1, molecular diffusion occurs with efficient mixing. Low-pressure CVD reactors such as conventional hot-filament and microwave systems operate in this regime.

Continuous growth of a homogeneous film requires a continuous supply of the deposition gas mixture controlled carefully by mass flow controllers. It should, however, be pointed out that even under steady-state conditions for the concentration gradients in the gas phase, homogeneous growth may not be guaranteed; especially during the nucleation period and the first phase of film growth, the film properties may change.

In many deposition reactors, diffusive transport occurs across a thin boundary layer, the thickness of which increases with the viscosity of the gas and decreases with the density and gas velocity (square root dependence). The concept of a boundary layer separating the forced convection region from the stationary surfaces is useful for both cold-wall, atmospheric-pressure reactors and hot-wall, low-pressure reactors. The gas velocity within the boundary layer gradually increases from zero at the surface to its bulk value in the forced convection region. This gradient also affects the temperature and the gas-phase chemical concentration, as indicated in Fig. 6 [9]. The concentration gradient provides the driving force for the transport of the reactants through the boundary layer. At atmospheric pressure conditions, the thickness of this boundary layer is in the range

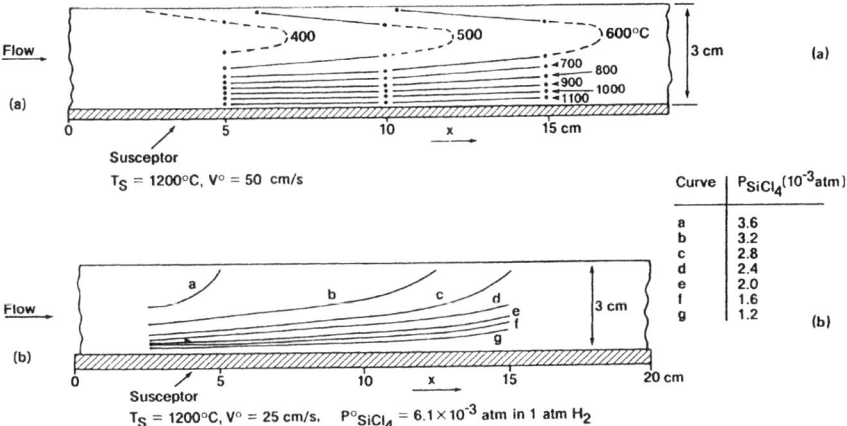

Figure 6 Measured variation (a) of the temperature and (b) of the SiCl$_4$ partial pressure in a horizontal epitaxial reactor. (From Ref. 9.)

of several millimeters. As the pressure decreases by a factor of 1000, in a low-pressure reactor, the boundary-layer thickness increases by a factor of 3–10, causing the diffusion coefficient to increase by a similar factor. The change also tends to shift the reaction into a reaction-limited regime as opposed to a transport-limited process, thereby enhancing deposition uniformity across the full deposition area.

B. Deposition Techniques and Reactors

1. Conventional Chemical Vapor Deposition

The simplest CVD process is achieved by thermal decomposition of the source gas induced by heating. Figure 7a shows a typical hot-wall reactor. Essentially, this is a quartz tube with a heater producing temperatures between 300 and 1100°C. The source gas passes from one end of the chamber to the other. The substrates may be placed parallel to the gas flow or perpendicular, to increase throughput. The gas may be supplied continuously at atmospheric pressure or at reduced pressure, with the consequence of a lower deposition rate.

The hot-wall, low-pressure reactor shown in Fig. 7a is designed for high wafer capacity and thus is commercially important. At low pressures, films with uniform thickness across the wafer are obtained because the reactor is operated in the reaction-limited regime. To reach reasonable deposition rates, little, if any, carrier gas is used. Polycrystalline silicon, for example, is deposited at about 625°C with 100% SiH$_4$ as the source gas in the pressure range of 10–100 torr.

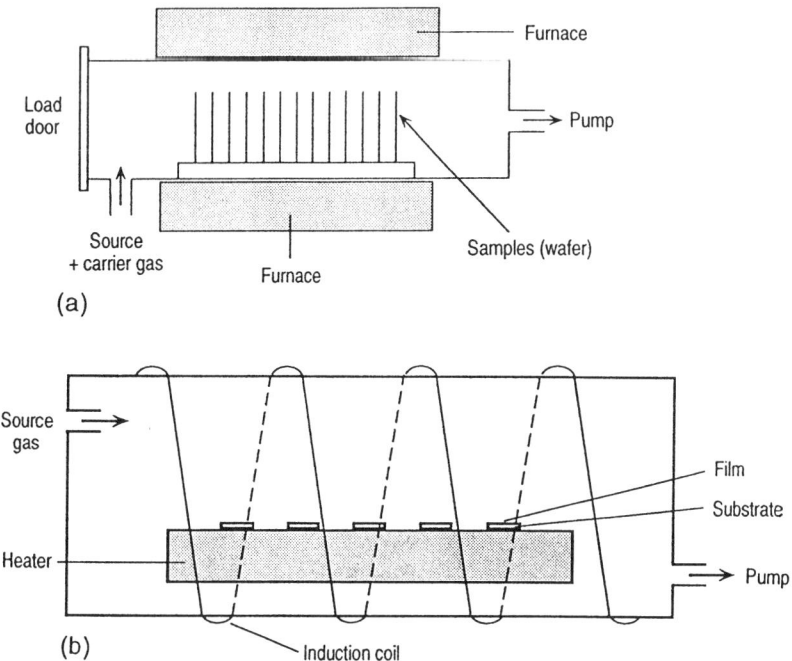

Figure 7 (a) Schematic diagram of a hot-wall reactor for thermal CVD at reduced pressure. (b) Basic configuration of an inductively coupled plasma reactor.

Another type of reactor suitable for polycrystalline silicon manufacture is the cold-wall atmospheric pressure apparatus. The wafers are placed on a flat supporting susceptor that can be externally heated by radio-frequency (rf) induction, as shown in Fig. 7b. Since the rf energy couples to the susceptor, the quartz walls remain relatively cool. The deposition gas mixture, containing 1000 times more carrier gas (N_2 or H_2) than silane, has a pressure slightly higher than atmospheric to push the gas through the reactor. The major portion of the temperature and reactant concentration gradient occurs across the boundary layer and drives the mass-transport-limited deposition.

2. Plasma Chemical Vapor Deposition

The activation of the gas phase in plasma CVD is performed by a direct current (dc), radio-frequency (rf), or microwave (mw) plasma with frequencies up to about 8 GHz. Industrial production of large-area deposition of films is mainly based on plasma techniques. This includes hard and wear-resistant coatings, optical coatings, all kinds of materials for the semiconductor industry, and amorphous hydrogenated silicon.

Due to the practical importance of plasma processing, great efforts have been made to develop in situ diagnostic methods, which are required to increase the fundamental understanding of the complicated processes occurring in the plasma and at the growing surface. Since species exist in the plasma with energies up to several hundred electron volts, essentially all reaction channels are open, including ion reactions. The chemistry is generally extremely complicated and will be discussed, for example, for depositions generated in a rf glow discharge system using capacitive or inductive power coupling.

A capacitively coupled rf plasma reactor is shown in Fig. 8. The plasma of a rf glow discharge consists of a partially ionized gas containing ions, electrons, and neutral species. All of these species have their own characteristic density and energy distribution. The concentrations of electrons and ions are the same since the plasma is electrically neutral. The concentration of neutrals is several orders of magnitude higher than that of the charged species (see Chapter 7). Typical electron energies of 1–10 eV correspond to temperatures in the 10^4–10^5 K range, while ions are thermalized with a temperature near 300 K. These temperatures create strong nonequilibrium conditions in the plasma. For CVD processes, the hot electrons are crucial, since they are able to induce chemical reactions (e.g., by strong vibrational excitation). And since most chemical reactions can occur at energies of 10 eV, the chemistry is rather complex.

As a result of the higher mobility of electrons, all surfaces within the plasma reactor assume a negative potential with respect to the plasma, generating a sheath with an electric field directed perpendicular to the surface. Therefore, positive ion bombardment of the interior surfaces, including the growing film, occurs. Since the energies of the positive ions can be very high (≥ 100 eV),

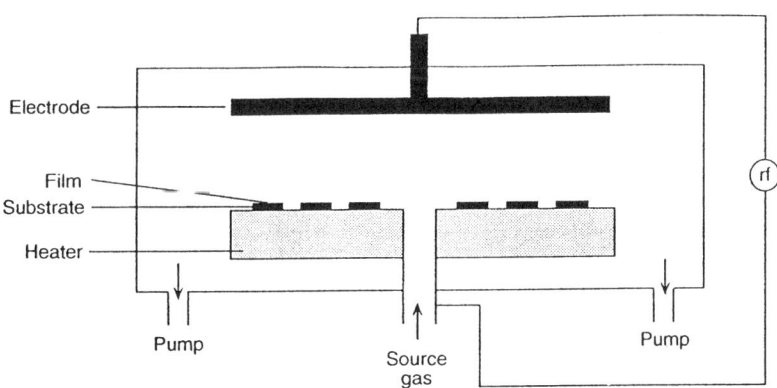

Figure 8 Schematic of the experimental setup used for rf glow discharge deposition.

chemical processes, surface sputtering, and film damage can occur, all of which can affect the film properties.

On the other hand, the electrons reach the surface only during small time intervals within each rf period. They may be largely responsible for the deposition chemistry. The more energetic electrons are capable of dissociating molecules and radicals, and of ionizing molecules, radicals, and atoms. For the development of a realistic model, the energy distribution ("electron temperature") and the electron density have to be known (Chapter 7).

3. Hot-Filament Chemical Vapor Deposition

One of the simplest CVD methods, used frequently to deposit metastable crystalline and amorphous materials, where strong nonequilibrium conditions are needed, is the hot-filament technique illustrated in Fig. 9a. The chemistry is

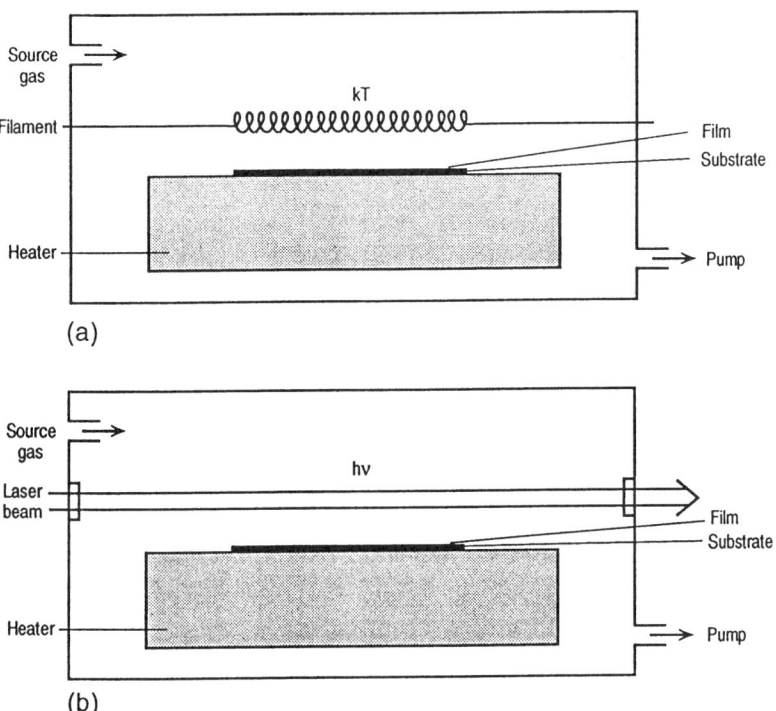

Figure 9 (a) Schematic of the experimental setup used for hot-filament deposition of diamond and amorphous semiconductors. (b) Schematic of the experimental setup used for laser deposition employing parallel illumination. An infrared (IR) laser may be used for pyrolysis and an ultraviolet (UV) laser for photolysis of the source gas.

induced by a refractory metal filament (e.g., tungsten or tantalum) heated to about 2000–2600 K at a distance of 0.1–2 cm from the substrate surface, creating a strong temperature gradient to the much lower surface temperature—e.g., around 1000 K for diamond deposition and around 500 K for the growth of amorphous silicon. A drawback of this method is the strong corrosion of hot refractory metals in an oxygen-containing atmosphere. To increase the area of deposition and the homogeneity of the deposit, an array of hot filaments can be used.

Several model calculations have been performed for such a setup to determine the steady-state concentrations and reaction paths by estimating the chemical gas composition at the wire temperature, using chemical thermodynamics and kinetics. Transport can be taken into account by assuming a one-dimensional gas flow within a given temperature profile, including concentration gradients and thermal diffusion. In a more realistic model, heterogeneous chemistry at the filament and the surface must be included—a rather difficult challenge.

The purely thermal hot-filament process not only gives state-of-the-art films but is also ideally suited for the development of chemical models and mechanistic studies of the complicated nonequilibrium CVD processes.

4. Photochemical Vapor Deposition

In photochemical vapor deposition (photo-CVD), photons from a lamp or a laser are used to initiate the deposition process. Photons provide the ideal energetic control via selection of the primary reaction step by the wavelength of the optical source. The best spatial control of the excitation process is achieved with laser radiation, where detrimental effects of the walls can be minimized. Disadvantages include limited throughput, the required line-of-sight, and the cost of lasers and lamps.

Usually two limiting cases are considered in photoassisted growth. The radiation in a photothermal process provides spatially localized heating of the gas and (or) surface, depending on the laser beam–surface configuration, which may be parallel or perpendicular. In the parallel configuration, the source molecules decompose (pyrolyze) in the heated gas-phase region, and the decomposition temperature can be decoupled from the surface temperature (see Fig. 9b). In the perpendicular configuration, it may not be necessary for the gas to absorb the radiation. Here, the source molecules can be decomposed at the heated surface region, as is the case, for example, in laser direct writing. The ability to ''write'' CVD patterns with a laser, restricting the deposition of material to predetermined micron lines or areas, is a particularly neat trick applied in microelectronic repair and unique device fabrication.

Chemical bonds in the source molecules can be broken directly by the absorption of a single quantum of visible and UV photons of several electron volts energy. Such a process is classified as a photochemical decomposition due to photolysis of the source gas (Fig. 2). Electronic excitation may also lead to direct bond breaking via photolysis.

If the laser beam is not in contact with the surface (in a parallel configuration), the absorption of photon energy in the irradiated gas volume is governed by the Lambert–Beer law at low laser intensities:

$$I = I_0 \exp(-nl\sigma) \tag{7}$$

where I_0 and I are the incident and transmitted optical intensities (in W/cm^2), n is the number density of absorbing species (in cm^{-3}), l the path length (in cm), and σ the absorption cross section (in cm^2). In a parallel photothermal process, heating of the gas column above the substrate initiates deposition, and the CVD process resembles hot-filament CVD. The decomposition temperature obtained at the center of the light beam is determined by the optical absorption coefficient of the gas and the concentration of the absorbing species. A sensitizer with high absorption at the wavelength employed can be used to increase the energy available for deposition.

The reactive species produced by a specific photochemical decomposition of the source gas (photolysis) will reach the surface without secondary collisions and with reactions only at low pressures, i.e., in the ballistic regime. Such a process may be called photochemical vapor deposition and would be the ideal scenario for controlling the deposition chemistry. Since low pressures are needed for such a processing scheme, the deposition rate will be low.

The so-called perpendicular, or normal, incidence configuration is used for deposition with UV and VUV lamps that are available commercially. The advantages of both continuous and pulsed lamps are the high duty cycle and the lower costs. A disadvantage is the low intensity, which limits large-area deposition but also restricts surface heating. A completely different situation is realized when a wavelength is selected that is not absorbed by the gas phase. Then, only decomposition of adsorbed molecules by substrate heating or photochemical dissociation can drive the process. In laser direct writing, spatial localization of the process is achieved by focusing a laser beam onto the surface through a transparent gas.

An important development in the direction of in situ processing is an interlocking multichambered CVD deposition apparatus, where each chamber sees only one specific source gas to avoid contamination. Film fabrication and modification can be achieved by plasma deposition and directed ion-, electron-, or laser-beam processing.

C. Selected Chemical Systems

1. Polycrystalline Silicon

One of the simplest CVD processes is thermal conversion of silane into solid silicon and gaseous hydrogen:

$$SiH_4 (g) = Si(s) + 2H_2 (g) \tag{8}$$

This reaction is exothermic by 34 kJ/mol. Other silicon hydrides are also thermodynamically unstable with respect to their elements. In thermal CVD, the energy required to surmount the activation barrier of reaction (8) is supplied by heating the substrate to temperatures above 500°C.

The deposition of polycrystalline silicon by low-pressure chemical vapor deposition (LPCVD) ($\leq$ 0.4 torr) in the temperature range 550–650°C is an important commercial process. Thin polysilicon films are vital to metal oxide semiconductor (MOS) integrated circuits and interconnects. Most commercial systems use pure SiH_4 and a 10–30°C temperature ramp over the deposition zone with the wafer standing vertically in the heated tube (see Fig. 7a).

Generally, the films are amorphous or polycrystalline below 580°C with a grain size $\leq$100 Å. Around 600°C, a polycrystalline, fine-grained equiax structure is obtained. Between 620 and 640°C the growth becomes polycrystalline with a tendency to a columnar structure and strong (110) texture.

Gas-phase reactions become dominant above 650°C and at pressures greater than 0.4 torr, reducing the film quality. Hence the unique deposition conditions (nucleation, growth, kinetics, etc.) are confined to a narrow range of temperature and pressure, but they are reasonably well understood for a highly productive manufacturing process.

The different structures and morphologies can be explained as follows. In the case of low supersaturation (which means small reactant concentrations) and sufficient temperatures to permit efficient surface diffusion of the film-forming species, a layer-by-layer growth of epitaxial deposits may be achieved (Chapter 4). At higher supersaturation and limited diffusion, randomly oriented fine grains may grow at the surface into a large columnar structure with preferred orientation. Under conditions where surface diffusion is restrained but nucleation occurs continuously due to a sufficient supply of reactants, very fine equiaxed grains are produced. Such a microstructure is sometimes desirable due to its relatively high mechanical strength and fracture toughness.

Thus, changes in CVD deposition conditions allow the realization of very different materials properties ranging from crystalline to amorphous with short-, intermediate-, and long-range order or disorder. Structural variations and morphological aspects (voids, grain boundaries) increase the scope of film materials with potentially significantly different properties than those of the bulk.

2. Metastable Diamond Growth

The synthesis of diamond films at low pressures and temperatures is typically achieved via hot-filament CVD or plasma CVD techniques. The source gas is usually very rich in hydrogen, e.g., $\leq$1% CH_4, and the pressure varies between 1 mbar and 1 bar. Most diamond deposition experiments have been performed at substrate temperatures between 700 and 1100°C; however, depositions at temperatures as low as 300°C have been achieved using thermal CVD and a halogen-containing gas. Typical deposition conditions are listed in Table 1 for hot-filament and microwave-plasma CVD. The growth rates vary between 1–50 Å per second.

To elucidate the chemical mechanism of the gas-phase and surface reactions, in situ diagnostic methods have been applied during deposition. The film precursors for diamond growth remain ambiguous, and in reality several species may contribute to the growth process under the various deposition conditions employed. The main information available on the gas-phase chemistry has been obtained by mass spectrometric analysis, infrared tunable diode laser spectroscopy (TDLS), UV resonance enhanced multiphoton ionization spectroscopy (REMPI), laser-induced fluorescence (LIF), and coherent anti-Stokes Raman spectroscopy (CARS). The species considered as film precursors are either reactive molecules or not-too-reactive radicals. For metastable diamond deposition, the primary species discussed in this respect are the methyl radical (CH_3) and acetylene (C_2H_2). Both species have been detected in situ by several optical techniques and by mass spectrometry. Model calculations including more than 50 reactions and more than 25 species indicate that besides the source gas CH_4 the most prominent carbon-bearing species are CH_3 and C_2H_2.

Most researchers favor the CH_3 radical as being the most important film precursor coupled with the requirement of the presence of a crucial amount of H atoms:

$$CH_4 \; (g) = CH_3 \; (g) + H \; (g) \tag{9}$$

$$S \cdot + CH_3 = S\text{--}CH_3 \tag{10}$$

$$S\text{--}CH_3 + H = S\text{--}C + 2H_2 \tag{11}$$

The first step in the solidification process is the adsorption or chemisorption of the film-forming precursor at a vacant surface site S·. Assume that the CH_3

Table 1 Typical Deposition Conditions for Hot-Filament and Microwave-Plasma CVD of Diamond Films

Feed gas composition	1 vol % CH_4 in H_2
Gas pressure	10–100 mbar
Gas flow rate	~100 sccm
Substrate temperature	700–1100°C
Filament temperature	~2000°C
Microwave power	100–1000 W
Growth rate	1–50 Å/s

radical is such a species. This radical has a planar D_{3h} symmetry in the gas phase and is sp^2 hybridized with an unpaired electron in the unhybridized p orbital lying along the threefold axis. Upon chemisorption at the surface, it must transform into the sp^3 hybridization favored for diamond growth. The CH_3 radical requires a dangling bond for chemisorption, forming a highly hydrogenated surface that consists of trihydride groups. Thus, the solid diamond network is constructed in a second dehydrogenation step (Eq. 11), which determines the cross-linking in the solid and possibly controls the growth kinetics.

The following radical reactions can be considered as crucial for diamond growth to take place at practical rates. One result of the energy supplied to the system is the dissociation of molecular hydrogen in a rather inefficient endothermic process. Atomic hydrogen can remove hydrogen from the surface,

$$S-H + H \cdot = S \cdot + H_2 \tag{12}$$

by a slightly exothermic abstraction reaction because the H–H bond energy is greater than the C–H bond energy (435 vs. 330 kJ/mol). Hydrocarbon radicals R· fill the vacant surface sites to form a new diamond bond,

$$S \cdot + R = S-R \tag{13}$$

suggesting that a partially hydrogenated surface with a sufficient concentration of free radical surface sites is needed for diamond growth.

It is interesting to note that oxygen increases the growth rates significantly and improves the crystalline quality. The removal of graphitic co-deposits by OH radicals may be at least partially responsible for these effects.

According to the C–H–O phase diagram determined for the carbon system, the atomic gas-phase compositions completely define the diamond deposition region and the conditions where amorphous (hydrogenated) carbon is obtained, as shown in Fig. 10. In the pure C–H system a large excess of hydrogen is needed, whereas in the pure C–O system about equal amounts of the two different atoms yield diamond. At the lower boundary of the narrow diamond growth region, the deposition rate approaches zero and thus gives the best diamond material. The upper boundary line describes the transition to amorphous carbon. Therefore, the data suggest that the atomic ratios, and not the molecular structures, are actually important for metastable chemical deposition of diamond.

3. Dielectric SiO_2 and Si_3N_4 Films

As stated earlier, modern high-density device fabrication processes require the availability of much lower process temperature methods. In the case of SiO_2, several film deposition alternatives to the high-temperature (≥800°C) thermal oxidation of silicon exist, most of which exhibit desirable film properties such as low defect densities and conformal step coverage. One such process is the

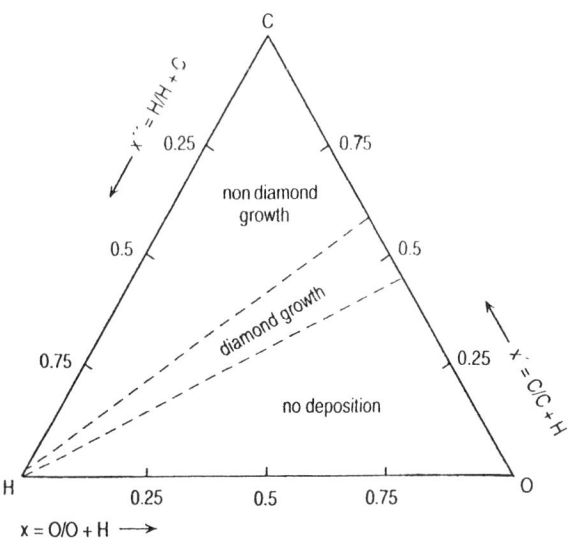

Figure 10 C–H–O phase diagram showing the gas-phase compositions suitable for diamond synthesis and the regions where no deposition occurs or amorphous carbon films are grown.

widespread use of the low-temperature thermal CVD method of reacting silane with oxygen at 300–500°C:

$$SiH_4 + O_2 = SiO_2 + 2H_2 \tag{14}$$

Either nitrogen or argon is used as the carrier gas. This deposited material is not as dense as the thermally grown SiO_2: however, it can be densified by flash heating to 1000°C. The deposition pressure can be reduced by the reaction

$$SiH_2Cl_2 + 2N_2O = SiO_2 + 2N_2 + 2HCl \tag{15}$$

Such low-pressure processes have the advantage of better uniformity of the deposited layer over surface steps (Chapter 5, Fig. 4a).

The deposition pressure and temperature can be further reduced by means of a glow discharge. In such plasma CVD processes, the reactants are introduced at low pressure without diluents. Depositions at 200–350°C have been achieved for

$$SiH_4 + 2N_2O = SiO_2 + 2N_2 + 2H_2 \tag{16}$$

However, incorporation of nitrogen and hydrogen into the film cannot be avoided. The lowest deposition temperatures of silicon oxides and nitrides are achieved with photo-CVD, and these reaction systems yield the most detailed mechanistic information.

Commercial reactors with mercury photosensitization or ArF laser photodissociation are available that allow large-area growth of these dielectric films. These techniques feature depositions with low defect densities and good step coverage. These insulator films are generated by pyrolytic reactions similar to thermal CVD, but induced by photon heating via UV lamps or UV lasers.

High-quality photodeposited films of SiO_2 have been achieved by irradiating a mixture of SiH_4 and O_2 or SiH_4 and N_2O at substrate temperatures below 200°C with a low-pressure Hg lamp (253.7 and 194.9 nm) or an ArF laser (193 nm), as per Eq. (16).

The individual reaction steps of this process are not clearly known. However, it is assumed that in the 193 nm photolysis the reaction starts with the production of oxygen atoms from the photolysis of nitrous oxide, which has a much larger absorption cross section than SiH_4:

$$N_2O + h\nu = N_2 + O(^1D) \tag{17}$$

The $O(^1D)$ atoms are rapidly transformed into $O(^3P)$, which then react with SiH_4 to form SiH_3 radicals and OH, the latter reacting also with SiH_4 to form more SiH_3:

$$O(^3P) + SiH_4 = SiH_3 + OH \tag{18}$$

$$OH + SiH_4 = SiH_3 + H_2O \tag{19}$$

The SiH_3 radicals then react to give SiH_2 by disproportionation:

$$SiH_3 + SiH_3 = SiH_2 + SiH_4 \tag{20}$$

which induces a complex set of reactions forming products such as SiO_mH_n and finally SiO_2.

Silicon nitride can be used as a secondary layer to improve the passivation of microelectronic devices by gettering undesirable impurities, such as alkali ions. It can also act as a trap or barrier to the diffusion of impurities to the silicon substrate.

Thermal CVD of Si_3N_4 in the 700–900°C range is realized by the reaction of SiH_4 with an excess of NH_3 (and H_2):

$$3SiH_4 + 4NH_3 = Si_3N_4 + 12H_2 \tag{21}$$

By introducing an oxidant, oxynitride films are formed. Low-pressure depositions are achieved with the reaction system

$$3SiH_2Cl_2 + 10NH_3 = Si_3N_4 + 6NH_4Cl + 6H_2 \tag{22}$$

Similar to SiO_2, depositions at temperatures as low as 300°C can be realized by processing a mixture of SiH_4 and NH_3 in a rf glow discharge. Low-temperature glow discharge processing in IC manufacture is in most cases executed with a capacitively coupled plasma system. Low operating pressure of a few torr enhances conformal coatings and allows better wafer throughput.

A dielectric film of Si_3N_4 can also be deposited by irradiating a mixture of SiH_4 and NH_3 or Si_2H_6 and NH_3 with UV light. A good-quality film is obtained at temperatures as low as 200°C. The first step of the SiH_4/NH_3 reaction by 193 nm photolysis is probably

$$NH_3 + h\nu = NH_2 + H \tag{23}$$

Silane, which absorbs very weakly at this wavelength, reacts with the reaction products and forms SiH_3:

$$SiH_4 + H = SiH_3 + H_2 \tag{24}$$

$$SiH_4 + NH_2 = SiH_3 + NH_3 \tag{25}$$

As discussed before, the highly reactive SiH_2 radical can be formed by SiH_3 disproportionation, leading to a not-well-understood sequence of intermediates of different aminosilanes, and finally to Si_3N_4.

4. Chemical Vapor Deposition of a-C:H, a-Si:H, and a-Ge:H Films

There are many similarities between the deposition of metastable diamond films and commercially important a-C:H, a-Si:H, and a-Ge:H films. Extensive energization of the gas phase is necessary to create the reactive chemical environment for the growth of these thermodynamically unstable materials.

As mentioned, most of the compositions in the C–H–O diagram in Fig. 10 either lead to amorphous films (a-C, a-C:H) or to no deposition at all. There is a tendency for most of the deposition techniques to deliver graphitelike films with predominantly sp^2 hybridization. Only a few methods produce films with the necessary sp^3 bonding, such as sputtering of carbon, or condensation of carbon plumes produced by pulsed laser vaporization (see Table 2), or from filtered or mass-selected carbon ion beams from a solid carbon source. Hydrogenated amorphous carbon (a-C:H) containing an sp^3 fraction of 0.75 has been deposited from acetylene using a plasma beam source (13.6 MHZ rf powered electrode and a smaller tungsten grid electrode at ground potential). The ion energy for the highest sp^3 fraction is about 100 eV per C atom, similar to that for a-C deposition. The properties of even the best films are, however, still far from the outstanding diamond values, e.g., the density is 10–20% lower and the Young's modulus is lower by a factor of 2–3. This seems to be due to the different kinds of hybridization possible in the carbon system, since high-quality a-Si:H films come much closer to their respective crystal data. Nevertheless, applications are increasing for high-quality diamondlike films because their nucleation problems are less restrictive than those for diamond growth.

Among the different deposition techniques available for the preparation of device-quality a-Si:H, the plasma CVD method is technologically mature and suitable for mass production, although the complex chemistry and kinetics underlying film growth from nonequilibrium glow-discharge reactive plasmas are not well understood. The important externally controllable parameters for plasma CVD are the source gas, gas flow rate, gas pressure, power-source frequency, rf power density, and substrate temperature. Typical parameters for the deposition of a-Si:H by plasma CVD with 100% SiH_4 as a source gas are presented in Table 3.

Important mechanistic reaction information can be extracted from laser CVD experiments. When the source gas is exposed to resonant vibrational excitation with a CO_2 laser, deposition by thermal activation of the lowest reaction channel is possible. For the silane system, this lowest reaction channel is the decomposition of SiH_4 into SiH_2 and H_2. The SiH_2 radical is highly reactive and will

Table 2 Typical Deposition Conditions for Diamondlike Amorphous Carbon (a-C) Films by Pulsed-Laser Ablation

Source material	Pyrolytic graphite
Pressure	High vacuum conditions
Ablation laser	ArF excimer laser (193 nm)
	KrF excimer laser (248 nm)
	Nd:YAG laser
Kinetic energy (ablated species)	Up to 50 eV
Substrate	Cleaned silicon wafer
Bonding structure	>80% sp^3 hybridization

Table 3 Typical Deposition Parameters for the Growth of a-Si:H Films by Plasma CVD

Source gas	100% SiH_4
Gas flow rate	5–10 sccm
Gas pressure	50 mtorr
Rf frequency	4–13.56 MHz
Rf power density	0.01–0.03 W/cm^2
Substrate temperature	220–300°C
Growth rate	0.5–10 Å/s

not reach the surface at higher pressures; it reacts with silane, forming disilane, then with disilane, forming trisilane, and so on ("cluster formation"):

$$SiH_4 = SiH_2 + H_2 \tag{26}$$

$$SiH_2 + SiH_4 = Si_2H_6 \tag{27}$$

$$SiH_2 + Si_2H_6 = Si_3H_8 \tag{28}$$

$$SiH_2 + Si_nH_{2n+2} = Si_{n+1}H_{2n+4} \tag{29}$$

Higher silanes are obviously detrimental to the growth of high-quality films if they are not completely decomposed at the surface. Therefore, CO_2 laser CVD, requiring relatively high pressures (several torr) to reach reasonable growth rates, does not result in desirable films, but it remains useful as a vehicle for basic reaction studies.

High-quality a-Si:H films can be realized with a UV laser beam. The energetic beam not only generates diradicals (such as SiH_2 and Si_2H_4), but also monoradicals (such as SiH_3 and Si_2H_5). Extensive secondary reactions can be avoided in UV laser CVD by decreasing the partial pressure of the source gas down to several µbar. Typical deposition parameters of high-quality a-Si:H, using Si_2H_6 and a F_2 laser (157 nm), are given in Table 4.

Diradicals are much more reactive than monoradicals, and the mono- to diradical ratio has been observed to increase as the laser beam is moved away from the surface. The deposited film properties also improve with the larger beam to surface distance, indicating that the monoradicals are responsible for the better films. Further support for the importance of SiH_3 in film growth has been obtained with threshold ionization mass spectrometry studies of the silane discharge, which identified the SiH_3 radical also as the dominant species in the vapor phase.

Table 4 Typical Deposition Conditions for the Growth of a-Si:H Films by Pulsed VUV Laser CVD

Source gas	Si_2H_6 + He
Disilane flow	0.5 sccm
Helium flow	100 sccm
Disilane partial pressure	5 µbar
Total pressure	1.0 mbar
Laser wavelength	F_2 laser (157 nm)
Laser power	3 W
Substrate temperature	220–300°C
Growth rate	0.2–2 Å/s

The main argument as to why SiH_3 is considered to be the dominant precursor for high-quality films is its relatively low reactivity and high surface mobility. SiH_3 can move across the surface to fill low spots and produce a smooth film on an atomic scale. Contrary to the SiH_3 radical (which needs a dangling bond for chemisorption), SiH_2 or SiH can insert into any SiH bond with nearly gas kinetic collision efficiency. This includes insertion at the surface with high sticking coefficient. From the chemical point of view, the insertion of the SiH_2 radical into a surface-bonded SiH leads to a SiH_3 group similar to the chemisorption of a SiH_3 radical at a dangling bond site. Thus, precursors with different incorporation probability may form the same surface species. However, the resulting surface topography and the film morphology may be different.

Such a radical mechanism (SiH_3 or SiH_2) is connected with the formation of a highly hydrogenated surface, and the solid network is constructed in a similar manner as the metastable diamond synthesis discussed before (Eqs. 10,11).

The deposition of device-quality amorphous hydrogenated germanium (a-Ge:H) has been studied mainly by rf glow discharge from GeH_4 with a hydrogen excess. Considerably more demanding deposition conditions have to be chosen compared with those for a-Si:H film formation. High-energy electrons and an ion bombardment energy above 100 eV are needed. These results are also relevant for germanium–silicon alloy films, which are important for "bandgap engineering." The reasons for the required deposition parameter changes are not well understood.

5. Doping

The controlled incorporation of dopant atoms into the semiconductor lattice is one of the important processing steps in the production of microelectronic devices. Doping requires the control of the final dopant profile and concentration in the semiconductor (Chapter 2, Sec. IV.B).

Traditionally, diffusion techniques are provided either by solid-phase sources such as evaporated, painted-on, or spun-on coatings or by gas-phase compounds that are decomposed pyrolytically at the hot surface. The ion implantation technique, where ions are generated from a gaseous source, allows precise control of the depth of penetration as well as the dose of implanted atoms. However, the high ion energy of up to 500 keV leads to crystal damage and the formation of amorphous regions. Both of these traditional doping methods need an additional high-temperature processing step. In the case of ion implantation, high-temperature treatment is essential to reestablish a crystalline structure with the dopant atoms placed at electrically active interstitial sites. In the diffusion process, the high-temperature drive-in cycle drives the dopant to the required depth and realizes the necessary concentration level.

The foregoing techniques are employed to generate very specific doped regions with very specific depths and impurity profiles. Other requirements call for uniformly doped films of various thicknesses. For example, the electrical resistivity of undoped polycrystalline silicon is several ohm-cm, but it may be shifted by several orders of magnitude with appropriate doping. This is traditionally achieved with in situ doping during the CVD process.

Polycrystalline silicon films can be considered as an array of crystallites whose surfaces or boundaries can reversibly trap or release charge carriers or dopant atoms. Low dopant concentrations do not increase conductivity as expected, due to trapping of the charge carriers in the crystallite boundaries. As soon as the traps are saturated, the resistivity decreases sharply. At high dopant concentration (e.g., 10^{20}–10^{21} atoms/cm^3), the resistivity approaches that of the single-crystal silicon value, due to a drastic increase in grain size.

Specific dopants behave differently in the crystal matrix. Both arsenic and phosphorus exhibit appreciable grain boundary segregation, while boron does not. Also, the addition of PH_3 or AsH_3 to SiH_4 during the CVD process tends to lower the silicon growth rate, while boron tends to increase it, and the degree of that influence changes with growth temperature. Hence, competitive surface growth kinetics plays an important part in these CVD systems [9].

As an example of the different doping behaviors, Fig. 11 shows the free-carrier concentration as a function of phosphorus concentration for monocrystalline, polycrystalline, and amorphous silicon. Below a carrier concentration of about 10^{18} cm^{-3}, the curve for polycrystalline silicon with a grain size of about 0.1 µm shows a characteristic decrease due to trapping of charge carriers in grain boundary states. The doping behavior of a-Si:H is even more complicated. Below 10^{18} cm^{-3}, the carrier concentration decreases rapidly due to trapping in bandgap states. Above 10^{19} cm^{-3}, the free-carrier concentration remains several orders of magnitude below the doping concentration. Obviously, only a small fraction of the incorporated impurity atoms is activated.

Extremely high heating and cooling rates (up to 10^{15} K/s) via laser doping offer great advantages in large-area (sheet) doping or very localized doping. The dopant source molecules are either pyrolyzed or photolyzed by the laser radiation in the gas phase or at the surface. Detailed information on laser-induced silicon doping is available. P-type doping has been studied, for example, using BCl_3, $B(CH_3)_3$, $B(C_2H_5)_3$, and B_2H_6 in conjunction with excimer laser radiation (193 nm, 351 nm). As source gases for n-type doping, PCl_3, PH_3, and AsH_3 have been used. Normal incidence laser doping may melt the silicon surface, and the transport of the dopant to the surface may be the rate-limiting step, due to the rapid diffusion of atoms into the liquid phase. The diffusion depth can be controlled by the surface temperature (laser fluence) and heating time. Very flat doping profiles and high doping densities can be obtained. Typical values reported for boron doping of silicon are carrier concentrations of 4×10^{19}–5×10^{21} cm^{-3} and junction depths between 0.05 and 1.0 µm.

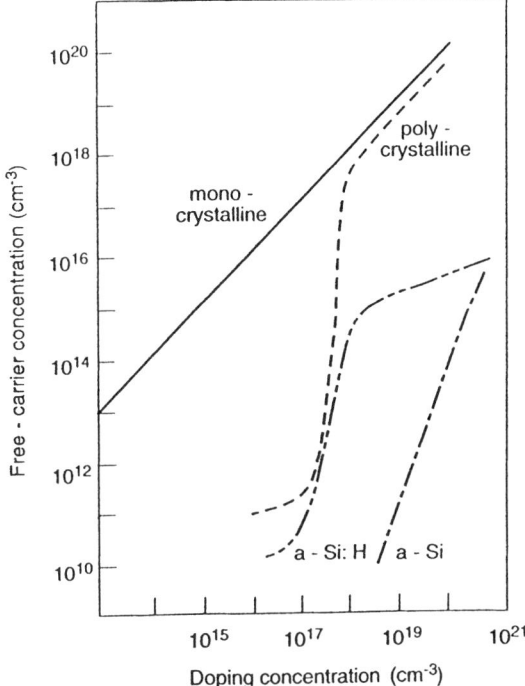

Figure 11 Free-carrier concentration versus phosphorus doping concentration for monocrystalline, polycrystalline, amorphous hydrogenated (a-Si:H), and amorphous (a-Si) silicon. (Adapted from Ref. 10.)

Figure 12 shows the depth profile of boron for pulsed laser doping, obtained by secondary ion mass spectrometry (SIMS) for an n-type (100) Si wafer. The fluence of the ArF laser radiation was 0.6 J/cm^2, and the BCl$_3$ pressure 67 mbar. Near a dopant concentration of about 10^{17} cm^{-3}, the signal-to-noise ratio becomes too small. Extrapolation of the linear part of the boron depth profile to the substrate doping level of 2×10^{15} cm^{-3} yields a junction depth of about 0.35 µm, as indicated by the dashed lines. In experiments using triethylboron instead of BCl$_3$, carbon was found in the doped material due to cracking of the hydrocarbon ligands.

Pulsed laser doping has also been studied for polycrystalline silicon. The required laser fluence must be sufficient to melt, but insufficient to cause optical breakdown, i.e., in the regime 0.2–1 J/cm^2. The results indicate that for boron doping, the grain boundaries present in the polycrystalline material do not alter the dopant distribution. If amorphous silicon films are irradiated under these

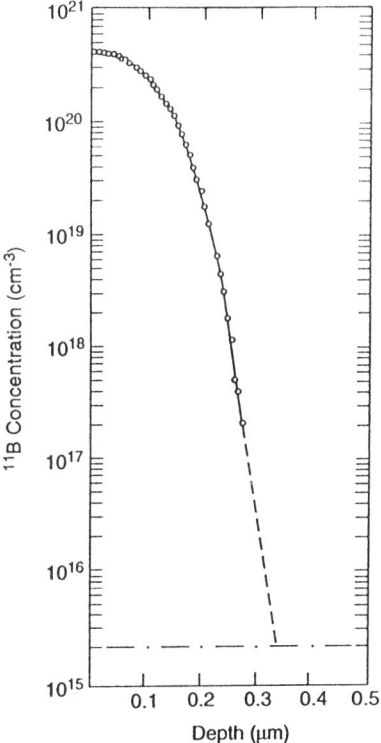

Figure 12 Boron doping profile versus depth in single-crystal silicon obtained by pulsed excimer laser doping at 193 nm and 0.6 J/cm^2 with 67 mbar BCl_3. (Adapted from Ref. 11.)

conditions, simultaneous recrystallization of the amorphous material into fine-grain polycrystalline silicon is observed with the doping process.

Doping has also been studied for a-Si:H and a-SiC:H technologies. For the application of these materials in low-cost solar cells, p-i-n devices are needed, which have been fabricated by plasma CVD and photo-CVD. Higher efficiencies of these devices are achieved by performing the different deposition processes in separate reaction chambers.

III. MATERIALS PROPERTIES

A. Crystalline Materials

Of the different allotropes of carbon (diamond, graphite, and fullerene), the cubic diamond structure has the most outstanding properties and thus is attractive for many applications in materials science. Diamond has the highest atom density of any solid, has the highest bulk modulus (and consequently the lowest compressibility), is the hardest material, and has the highest Young's modulus. At room temperature it has a higher thermal conductivity than any other material, with its value being four times higher than that of copper. Its thermal expansion coefficient is less than the value of Invar. The material is transparent from the ultraviolet through the visible and into the infrared with the exception of two phonon bands between 3 and 6 μm. Due to its high Debye temperature, diamond has one of the lowest heat capacities at low temperatures. Diamond has the same crystal structure as silicon and germanium and is a wide bandgap semiconductor with an indirect gap of 5.45 eV. Most of the diamond properties listed in Table 5 refer to natural diamond crystals.

How well these limiting crystal values can be realized in a deposited polycrystalline film depends on the quality of the CVD process. Imperfections, impurities, lattice disorder, strain, etc., disturb the local symmetry and thus increase optical absorption and decrease the elastic constants and mechanical strength. Nevertheless, CVD technology provides diamond films that approach the single-crystal properties surprisingly well.

For illustration, Fig. 13 gives a comparison of the densities of the crystalline forms of carbon and the range of values observed for polycrystalline and amorphous carbon films. Depending on the nature of bonding (sp^3, sp^2, or sp hybridization) and the structure of the film (monocrystalline, polycrystalline, or amorphous), any value between the different allotrope densities can be realized. A similar behavior is observed for other film properties. Figure 14 shows a comparison of the Young's moduli of single-crystal carbon materials and films. The Young's modulus describes the stiffness of the material and can be easily related to the hardness for certain structures. As can be seen from these two

Table 5 Mechanical, Elastic, Optical, and Thermal Properties of Diamond, Silicon, and Germanium

Property	Diamond	Silicon	Germanium
Density (g/cm3)	3.52	2.33	5.35
Young's modulus (111)(GPa)	1164	169	138
Poisson's ratio (111)(GPa)	0.079	0.26	0.25
Hardness (111)(GPa)	90	9	—
Refractive index (VIS)	2.42	3.99	4.1
(IR)	2.38	3.42	4.0
Bandgap (eV)	5.45	1.11	0.67
Thermal expansion (K^{-1})	1×10^{-6}	3×10^{-6}	5×10^{-6}
Thermal conductivity (W cm^{-1} K^{-1})	6.6	0.84	0.62

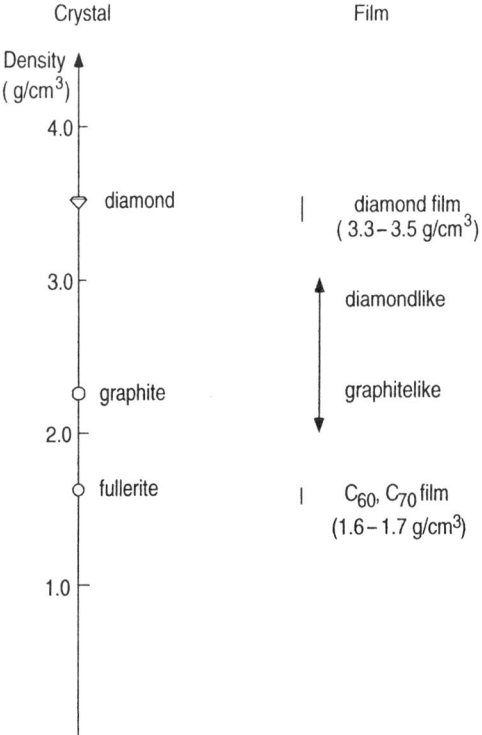

Figure 13 Comparison of the densities of carbon allotropes (diamond, graphite, fullerites) with the values measured for polycrystalline and amorphous films.

figures, polycrystalline diamond films, in fact, approach the diamond values, whereas the amorphous diamondlike films do not reach 100% sp^3 hybridization and thus show larger deviations.

A similar situation is encountered for epitaxial or polycrystalline silicon or germanium films. Table 5 lists the crystal data, and again the properties of a real CVD film may deviate to a greater or lesser extent from these "ideal" values depending on the quality of the film.

The bandgaps of silicon and germanium are much smaller than the diamond value, and the stiffness of the network is also much lower, as can be seen from the corresponding values of the bulk modulus and Young's modulus. For this reason, the carbon materials are preferred for hard coatings.

The polycrystalline silicon used in many microelectronic devices consists of single-crystal regions with an extension of typically 1000 Å ("crystallites" or "grains"). The mechanical, optical, and electronic properties are similar to those

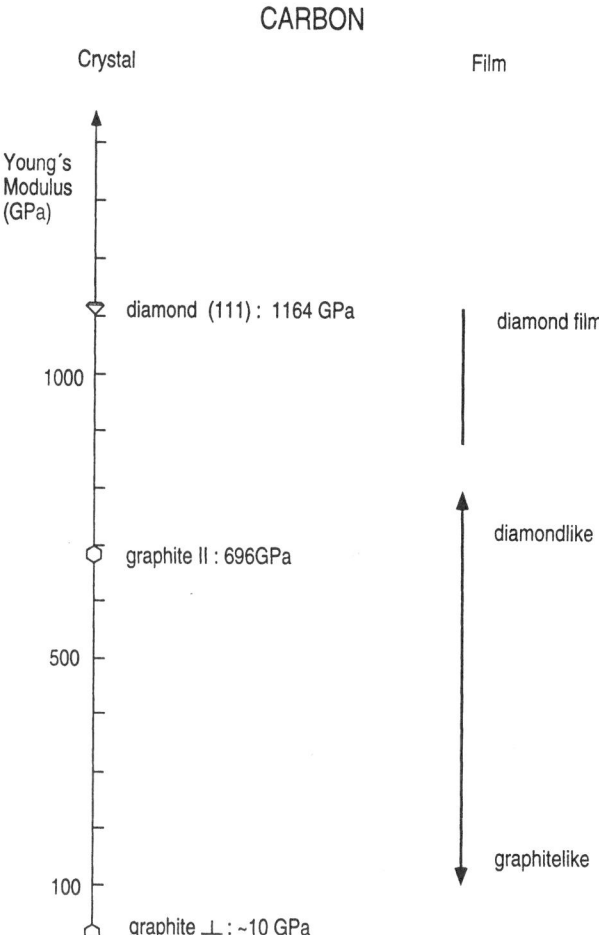

Figure 14 Comparison of the Young's moduli (E-moduli) of carbon allotropes with values measured for polycrystalline and amorphous films.

of bulk single-crystal silicon; however, they do not reach them (e.g., density 2.3 g/cm^3, coefficient of thermal expansion 2×10^{-6} K^{-1}), since only the silicon in the interior of the grains behaves like bulk silicon.

The grain boundaries are disordered atoms containing a large number of defects due to distorted or incomplete bonding. These defects and the lack of periodicity substantially alter the diffusion characteristics and the dopant distribution. As expected, dopant atoms not only segregate to grain boundaries, where they do not effectively produce free carriers, but they also exhibit a significantly higher diffusion coefficient compared with single-crystal regions. With increasing grain size the bulk properties are approached, since the volume increases faster than the surface area.

An example of the use of polycrystalline silicon in a photovoltaic application is shown in Fig. 15. It illustrates the generation of a voltage by absorption of light in a pn junction solar cell. Upon absorption of a photon, an electron–hole pair is created, and these have to be spatially separated and collected in order to capture the energy of each pair. A thickness of several hundred microns is needed to absorb the sunlight for indirect bandgap crystalline silicon, with its relatively low optical absorption coefficient.

B. Dielectrics

The development of advanced IC and VLSI technology requires not only reduced film thicknesses, but in many cases also lower processing temperatures. As shown in Tables 6 and 7, CVD techniques provide acceptable solutions for this problem with insignificant degradation of the corresponding film properties.

The successful integration into the microelectronic and optoelectronic industries of films with acceptable material properties and formed with in situ film-processing techniques depends on a number of factors. In the case of dielectric films, the major parameters for device-quality insulators are (1) density, (2) refractive index, (3) dielectric constant, (4) dielectric breakdown strength, (5) defect density, (6) film resistivity, (7) intrinsic stress, (8) step coverage, (9) stoichiometry, (10) thickness uniformity, and (11) etch rate in various acid solutions of the densified or as-grown film. Besides these film properties, the following deposition parameters are of great interest: (1) deposition temperature, (2) deposition pressure, and (3) growth rate.

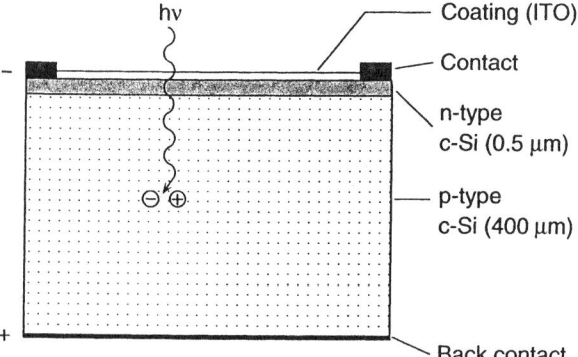

Figure 15 Simplified cross section of a crystalline pn junction solar cell.

Table 6 Chemical Vapor Deposition of Silicon Dioxide (SiO_2) Films

Property	Atm.-press. CVD	Plasma CVD	Photo-CVD
Density (g/cm^3)	2.1	2.2–2.3	2.1
Refractive index	1.44	1.4–1.5	1.45–1.49
Dielectric constant	3.8	4.6	3.9
Dielectric strength (V/cm)	8–10 $\times$ 10^6	4–8 $\times$ 10^6	4–6 $\times$ 10^6
Defect density (cm^{-2})	1–10	≤1	≤10
Film resistivity (Ωcm)	>10^{17}	>10^{17}	—
Intrinsic stress (dynes/cm)	6–8 $\times$ 10^8 (tensile)	2 $\times$ 10^9 (compressive)	—
Step coverage	Poor	Good	Good
Deposition temperature (°C)	300–500	200–350	50–200
Deposition pressure (torr)	760	0.1–5	0.3–1
Growth rate (nm/min)	1	100	15

Table 7 Chemical Vapor Deposition of Silicon Nitride (Si_3N_4) Films

Property	High-temp. CVD	Plasma CVD	Photo-CVD
Density (g/cm^3)	2.9–3.1	2.5–2.8	1.8–2.4
Refractive index	2.0–2.1	2.0–2.1	1.8–2.4
Dielectric constant	6–7	6–9	5.5
Dielectric strength (V/cm)	1 $\times$ 10^7	6 $\times$ 10^6	4 $\times$ 10^6
Defect density (cm^{-2})	—	≤0.5	≤10
Film resistivity (Ω-cm)	—	2 $\times$ 10^6	—
Intrinsic stress (dynes/cm)	1 $\times$ 10^{10} (tensile)	1–8 $\times$ 10^9 (compressive)	—
Step coverage	Poor	Conformal	Conformal
Deposition temperature (°C)	700–900	200–400	100–200
Deposition pressure (torr)	1	0.2–3	0.3–1
Growth rate (nm/min)	10–50	10	5

The previously mentioned polycrystalline silicon gate of the MOSFET device is separated from the silicon substrate by a high-quality SiO_2 insulating layer, whose interface state densities are typically below 10^{10} cm^{-2}. These interfaces have only about one unsaturated bond in every 10^5 interface atoms. Such SiO_2 films can be achieved with CVD techniques or thermal oxidation.

Typical values for SiO_2 films deposited by atmospheric-pressure CVD, plasma CVD, and photo-CVD, and process parameters characteristic for typical reactors are given in Table 6. The great advantage of photo-CVD over the atmospheric-pressure CVD technique is the opportunity of increasing the growth rate from 1 nm/min by a factor of 10–20 at deposition temperatures below 200°C, while still yielding acceptable film properties.

The film properties and deposition parameters for high-temperature CVD, plasma CVD, and photo-CVD of Si_3N_4 are given in Table 7 for comparison. As can be seen, there is no major difference in the Si_3N_4 film properties compared with SiO_2. Also, the deposition conditions are very similar. In terms of photo-CVD equipment, laser-based techniques may have a stronger impact on patterning of thin films and devices, whereas large-area deposition of thin films of dielectrics, insulators, and oxides is dominated by plasma or incoherent lamp systems.

C. Amorphous Materials

The elastic properties of an amorphous network are essentially determined by the average number of atom coordination. The constraint-counting model requires that a mean coordination number of at least 2.4 exists in order to form a rigid network. Strain may influence the network to increase the coordination, although it may be partially relieved via hydrogen incorporation. The average coordination number in amorphous networks of carbon, silicon, and germanium without long-range order is below 4. As seen in Fig. 16, the Young's modulus increases from zero to its bulk value as the coordination shifts from 2.4 to 4.0.

Different hybridization (sp^3, sp^2, sp) is one reason for the extremely large variation of the properties in the a-C and a-C:H systems with deposition conditions, which are more complex than in the a-Si:H and a-Ge:H systems. The mechanical and elastic properties (e.g., density 2.9 g/cm^3, Young's modulus 400

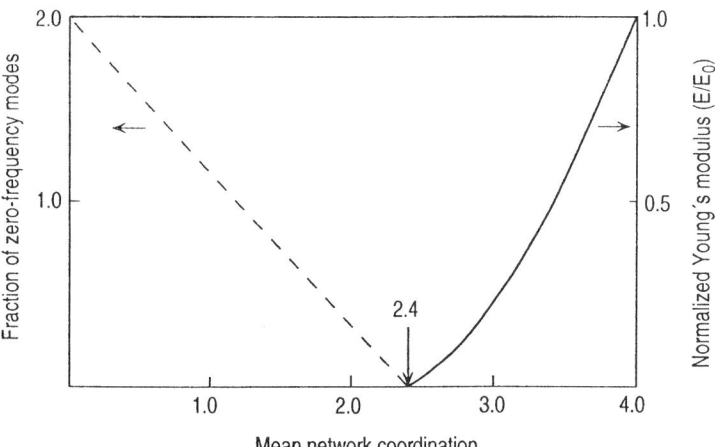

Figure 16 The number of degrees of freedom for an average coordination number below 2.4 and the variation of the Young's modulus between 2.4 and 4.0 for a tetrahedrally bonded random network.

GPa) of the hardest amorphous carbon films still deviate appreciably from the diamond properties (3.52 g/cm^3, 1164 GPa), as can be seen from Tables 5 and 8. Nevertheless, these amorphous carbon films are of interest as coating materials since they are denser and harden than, for example, even the best amorphous silicon films, which approach the bulk crystal data (2.33 g/cm^3, 169 GPa) with their density of 2.3 g/cm^3 and Young's modulus of 140 GPa.

The bandgap structure and electronic properties are also very different in the amorphous material. Figure 17 indicates that the bandgap for carbon may vary in the wide range between 0 and 2.6 eV. Note that in the amorphous carbon systems the gap is much smaller than for diamond (5.5 eV). This seems to be connected with the failure to achieve 100% sp^3 bonding in amorphous carbon networks. As discussed before, sp^2 bonding is also responsible for the high density of defect states at the bandedge, leading to the high Urbach energy of 250 meV compared with 50 meV for device-quality a-Si:H. Thus, the bandgap is not very well defined, and since the density of gap states is also orders of magnitude higher in a-C:H than in a-Si:H, the former material has no electronic applications.

As can be seen in Fig. 17, the bandgaps of a-Si:H and a-Ge:H are widened compared with the respective crystalline values (c-Si, c-Ge). However, only in a very limited range of bandgaps can materials of device quality be obtained for silicon (around 1.7 eV) and germanium (around 1.1 eV) alloys. Nevertheless,

Table 8 Mechanical, Optical, Electronic, and Deposition Properties of a-C:H, a-Si:H, and a-Ge:H Deposited by CVD Methods

Property	a-C:H	a-Si:H	a-Ge:H
Density (g/cm3)	2.4–2.9	2.3	4.8–5.2
Refractive index	1.8–2.8	3.4–3.5	3.4–4.4
Young's modulus (GPa)	300–400	100–140	—
Poisson's ratio	0.2–0.4	0.3–0.4	—
Hardness (GPa)	30–60	6–10	—
ημτ product (cm^2/V)	—	5×15^{-6}	5×15^{-7}
Photo/dark conductivity	—	10^5–10^6	0.1–1
Bandgap (eV)	0–2.6	1.6–1.8	1.1–1.3
Urbach energy (meV)	250	50	50–60
Density of gap states (cm^{-3})	10^{18}–10^{21}	10^{15}–10^{16}	10^{16}–10^{17}
H concentration (atom%)	<30	5–15	2–10
Deposition temperature (°C)	100–200	230	150–200
Deposition pressure (Pa)	0.5–15	10–50	100
Growth rate (nm/min)	6–60	1–10	20

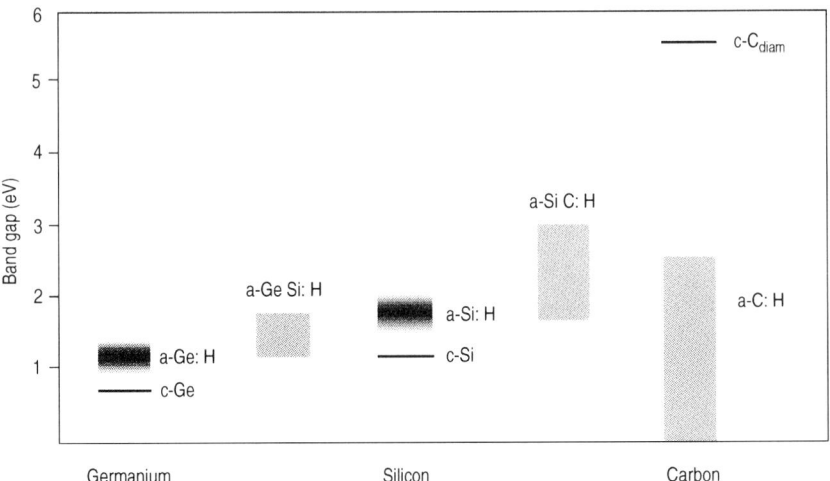

Figure 17 Bandgaps of tetrahedral crystalline solids (c-C, c-Si, c-Ge), the binary hydrogenated amorphous alloys (a-C:H, a-Si:H, a-Ge:H), and the ternary amorphous alloys (a-GeSi:H, a-SiC:H).

the synthesis of amorphous alloys leads to a drastic extension of bandgap values over those available from ideal tetrahedral crystals ("bandgap engineering").

In the silicon and germanium system, the amorphous hydrogenated materials are of interest due to their strong optical absorption and the good electronic properties. Table 8 presents typical values for state-of-the-art films that can be produced by plasma, hot-filament, and photo-CVD. Much better electronic properties (such as Urbach energy and density of states in the gap) are obtained for a-Si:H and a-Ge:H than for a-C:H.

The a-Si:H and a-Ge:H films have a direct bandgap structure with optical absorption coefficients up to a factor of 100–1000 higher than the crystalline material. Since this property is connected with the disorder in the amorphous network, there is a fundamental trade-off between optical absorption and the electronic properties needed in photovoltaic applications, e.g., electron–hole pair extraction efficiency. Figure 18 shows an amorphous p-i-n silicon thin-film solar cell. A thickness comparison with the polycrystalline silicon cell (see Fig. 15) demonstrates the drastic reduction in the amount of semiconductor material needed for such a cell. The basis for this reduction is the high optical absorption coefficient of the amorphous material, which has a direct bandgap. Thus the reduction in the carrier diffusion length observed in an amorphous network is compensated for by the drastic size reduction of this device.

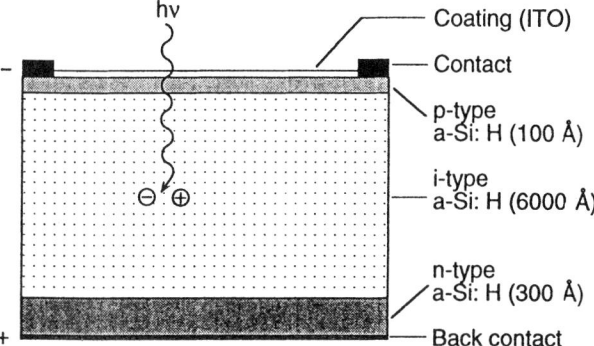

Figure 18 Simplified cross section of a thin-film p-i-n amorphous silicon solar cell.

The stability of the metastable amorphous materials is another important issue, since this material deteriorates during illumination with light ("Staebler–Wronski effect").

IV. SUMMARY

The chemical vapor deposition of crystalline, metastable crystalline, and amorphous materials offers great potential for the synthesis of solids with tailored electronic, optical, chemical, and mechanical properties. Full capitalization of that potential will be based on increasing the knowledge of the molecular processes taking place in the gas phase and at the surface. This includes the series of chemical reaction steps from initiation in the gas phase to solidification of the network, as well as the transport processes taking place under nonequilibrium conditions. It is these strong deviations from equilibrium, which create serious problems for modeling the CVD process, that offer the main potential of the CVD method.

A more detailed analysis of the complicated deposition process by in situ and ex situ techniques will allow a better understanding of the fundamental "chemistry–structure–performance" relationships. These methods will provide the microscopic information for the selection of optimized deposition techniques and conditions. With such molecular guidelines, in fact, a much larger parameter space can be explored and the ultimate goal can be reached—the realization of specific material properties and functions by controlling chemical dynamics, composition, and structure.

EXERCISES

1. Why is it possible to deposit a diamond film by a CVD process at much lower temperatures and pressures than used for conventional diamond synthesis?
2. What are the characteristic features of chemical vapor deposition (CVD) compared with physical vapor deposition (PVD)?
3. Why is a-Si:H more important than a-Si? What are the problems connected with hydrogen incorporation?
4. How can a CVD process be initiated? What are the characteristic features of the different CVD methods?
5. Formulate the kinetic equation of the formation of SiH_2 by dissociation of SiH_4 in a single-photon process. Calculate the probability of forming SiH_2 for an optically thin sample ($\sigma Il \ll 1$ where σ is the absorption coefficient, I the photon flux, and l the optical path length). For a pulsed light source the product of the photon flux and time "It" should be replaced by "nF," where n is the number of pulses and F is the photon fluence per pulse.
6. What fraction of the optical radiation energy will be absorbed over a path length of 1 cm if the absorbing source gas has an optical absorption cross section of $\sigma = 10^{-18}$ cm² (10^{-17} cm²) at 300 K and a pressure of 1 torr?
7. Estimate the growth rate of diamond from the collision frequency of CH_3 with the growing surface. The partial pressure of the CH_3 radicals is approximately 1×10^{-4} bar at 1200 K. The ratio of reactive to hydrogenated surface sites is about 6×10^{-4}. Reported rates for hot-filament deposition of diamond are in the range 0.3–3 μm/h.
8. Use absolute reaction rate theory for a bimolecular addition reaction to estimate the rate constant for methyl radical addition. For a radical addition to a sterically unhindered radical surface site, an activation energy of zero can be assumed. Thus the rate constant varies with the entropy of formation for the transition state. Estimate the value for a "loose" transition state.

REFERENCES

1. S. J. Moss and A. Ledwidth, eds., *The Chemistry of the Semiconductor Industry*, Blackie, Glasgow, 1987.
2. O. C. Gupta, ed., *Silicon Processing*, American Society for Testing and Materials, 1983.
3. S. Wolf and R. N. Tauber, *Silicon Processing for the VLSI Era, Vol. 1: Process Technology*, Lattice Press, Sunset Beach, California, 1986.
4. R. E. Clausing, L. L. Horton, J. V. Angus, and P. Koidl, eds., *Diamond and Diamondlike Films*, Plenum, New York, 1991.
5. J. Pankove, ed., *Semiconductors and Semimetals*, Academic Press, New York, 1984.
6. D. Bäuerle, *Chemical Processing with Lasers*, Springer-Verlag, Berlin, 1986.

7. I. W. Boyd, *Laser Processing of Thin Films and Microstructures*, Springer-Verlag, Berlin, 1987.
8. D. G. Ehrlich and J. Y. Tsao, eds., *Laser Microfabrication: Thin Film Processes and Lithography*, Academic Press, Boston, 1989.
9. S. P. Keller, ed., *Handbook on Semiconductors*, Vol. 3, North-Holland Publishing Company, Amsterdam, 1980.
10. H. J. Möller, *Semiconductors for Solar Cells*, Artech House, Boston, 1993.
11. K. G. Ibbs and R. M. Osgood, eds., *Laser Chemical Processing for Microelectronics*, Cambridge University Press, Cambridge, 1989.

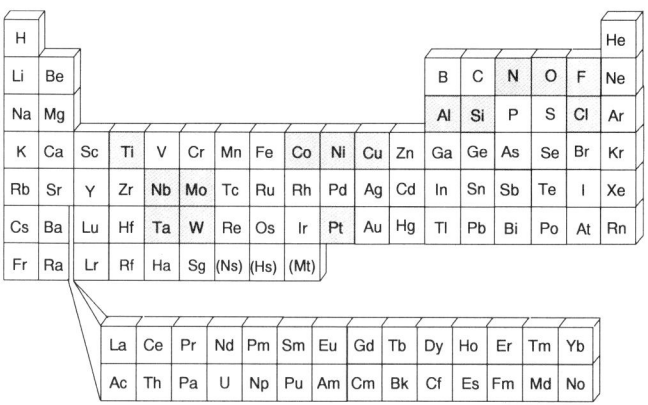

5
Depositions and Reactions of Metals and Metal Compounds

Roy G. Gordon *Harvard University, Cambridge, Massachusetts*

I. FUNCTIONS OF METALS IN INTEGRATED CIRCUITS

Tiny strips of metal form the electrical connections between the millions of transistors on a computer chip. These metals pass the signals from one transistor to another, and to and from the world outside the chip. The name *integrated circuits* is appropriate because these electrical connections are made through connections integrated into the structure, rather than through the external wires used in older electrical devices.

These structures are formed by depositing metal layers and etching them into the intricate patterns required to connect the circuits. The performance and stability of these structures are related to their thermodynamic properties and the interactions at their interfaces.

An interconnect generally consists of an electrically conductive metal film deposited on top of an insulator (usually silicon dioxide, SiO_2), which covers the semiconductor (usually silicon). As sketched in Fig. 1, the metal contacts the silicon through holes in the insulating layer, these holes being termed contact windows. In modern devices, additional layers of patterned insulators and metals are added on top of the layers shown in Fig. 1, to form multilayer interconnections, as shown in Fig. 9 of Chapter 2.

There are several requirements for industrially useful interconnects:

1. "Proper" electrical contact to adjacent layers
2. Low electrical resistance along their length
3. Stability during formation and under operating conditions

For connections to the sources and drains of transistors (see Fig. 5 in Chapter 1), "proper" electrical contacts have low contact resistance, in order to carry the required electrical currents. Gate electrodes, on the other hand, require high contact resistance to the silicon. Gate electrodes are simpler, and so their structures and means of formation will be discussed first, followed by the low-resistance contacts to sources and drains.

II. PROCESSES FOR MAKING METAL SILICIDES

Gate electrodes apply a voltage to a transistor without allowing a significant current to flow. Thus, they require a high-resistance contact to the silicon, which

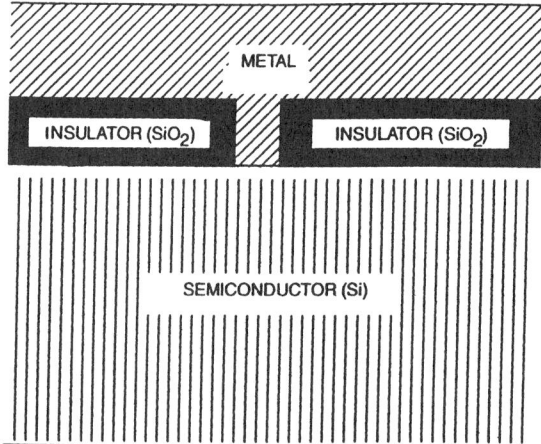

Figure 1 Cross section of a metal contact to a semiconductor device.

is achieved by forming a thin (about 10 to 20 nm) layer of silicon dioxide on the silicon by oxidation at high temperatures, before the gate electrode is deposited. The electrical resistance along the gate electrode must be sufficiently low to maintain a constant voltage over the electrode.

A. Polycrystalline Silicon

Traditionally, gate electrodes in microcircuits have been made from highly phosphorus-doped polycrystalline silicon, deposited by chemical vapor deposition (CVD) onto the crystalline silicon substrate covered with a thin layer of insulating silicon oxide (see Fig. 5, Chapter 1). Doped polycrystalline silicon, however, has a rather high resistivity. Therefore, the resistances along gate electrodes are often reduced by forming more conductive layers on top of the polycrystalline silicon. Metal silicides, such as titanium disilicide ($TiSi_2$), tungsten disilicide (WSi_2), or cobalt silicide ($CoSi_2$), are typically used for this purpose.

The methods used to deposit the metals or their silicides include both sputtering and chemical vapor deposition. Sputtering has been the most commonly used deposition method in the past, but CVD processes have become more popular and are necessary to achieve the smaller dimensions in the advanced microelectronic circuits of the 1990s.

B. Titanium Disilicide

Titanium disilicide is usually formed by depositing titanium metal on polycrystalline silicon and then heating it to a temperature above about 600°C to form the silicide:

$$2Ti + Si \rightarrow TiSi_2 \tag{1}$$

Heating (annealing) to temperatures above 700°C is usually required to transform the resulting titanium silicide from the initially formed, metastable crystal structure (called C49) with a high resistivity (over 60 $\mu\Omega$-cm), into the stable crystal structure (C54) with a lower electrical resistivity (13 $\mu\Omega$-cm).

Titanium silicide can also be formed by a CVD process based on the reaction

$$TiCl_4 + 2SiH_4 \rightarrow TiSi_2 + 4HCl + 2H_2 \tag{2}$$

This reaction deposits $TiSi_2$ rapidly at temperatures of 500–700°C.

C. Tungsten Disilicide

Tungsten disilicide may be made by any one of several CVD reactions. For example, reaction of tungsten hexafluoride with silane,

$$2WF_6 + 7SiH_4 \rightarrow 2WSi_2 + 3SiF_4 + 14H_2 \tag{3}$$

operates at temperatures from about 150 to 500°C. Still lower deposition temperatures (down to about 100°C) can be achieved by using disilane (Si_2H_6) in place of silane (SiH_4):

$$4WF_6 + 7Si_2H_6 \rightarrow 4WSi_2 + 6SiF_4 + 21H_2 \tag{4}$$

The initially deposited tungsten silicide has a relatively high resistivity (often over 600 μΩ-cm). Annealing at temperatures over 900°C causes the resistivity to decrease to values (around 40 μΩ-cm) almost as low as the bulk resistivity (33 μΩ-cm).

During these CVD reactions, some of the fluorine atoms are not removed, and they remain as impurities in the tungsten silicide, typically at a concentration of about 10^{20} cm^{-3}. This fluorine can cause changes in the insulating gate oxide, decreasing the breakdown field (deleterious) and decreasing the interface state density (advantageous).

The dichlorosilane reaction

$$WF_6 + 10SiH_2Cl_2 \rightarrow 2WSi_2 + 3SiF_4 + 3SiCl_4 + 8HCl + 6H_2 \tag{5}$$

produces a purer form of tungsten silicide, with much less fluorine as an impurity (below 10^{17} cm^{-3}), than does the silane reaction (3), perhaps because it operates at higher substrate temperatures (usually 400–600°C). These low levels of fluorine appear, however, to be too small to alter device properties.

D. Cobalt Disilicide

Cobalt disilicide, $CoSi_2$, is usually made by sputtering cobalt onto silicon, followed by heating to about 700°C. The structure of cobalt disilicide is a close match (1.2% difference in size) to the lattice of silicon, and so it can be grown epitaxially on silicon. This opens up the theoretical possibility of growing epitaxial single-crystal silicon on top of metallic cobalt disilicide, to form multilayer semiconductor devices integrated in three dimensions.

Cobalt monosilicide may be deposited from a single source, silyl cobalt tetracarbonyl,

$$H_3SiCo(CO)_4 \rightarrow CoSi + \frac{3}{2}H_2 + 4CO \tag{6}$$

This reaction produces single-phase cobalt monosilicide at about 500°C.

E. Platinum Silicide

Platinum silicide, PtSi, is usually made by sputtering platinum onto silicon, and heating. One CVD source for platinum is the volatile liquid tetrakis(trifluorophosphine)platinum, $Pt(PF_3)_4$, whose vapors decompose at about 400°C to form platinum metal.

III. LOW-RESISTANCE ELECTRICAL CONTACTS TO SILICON

A. Factors that Contribute to Contact Resistance

In many parts of microcircuits, the purpose of the metal contact to the silicon is to provide low electrical resistance between the metal and the silicon. In order to achieve and maintain this low resistance, three conditions must be met:

1. High free carrier concentration near the surface of the silicon.
2. No insulating silicon oxide layer between the silicon and the metal.
3. The interfaces remain chemically and mechanically stable.

A high free carrier concentration (of either electrons or holes) means that many carriers are available to move into or out of the silicon through its interface with the metal. Although there often is an electrical barrier to this movement of the carriers due to carriers trapped at the interface, when the carrier concentration is very high (around 10^{20} cm^{-3}), the barrier is effectively very narrow (a few nanometers), since the other carriers screen the charges giving rise to the barrier. According to the laws of quantum mechanics, the carriers can tunnel efficiently through a narrow barrier.

When exposed to air, silicon quickly becomes covered with a thin layer of insulating silicon oxide. In order for a low-resistance electrical contact to be formed, this silicon oxide layer must be removed or otherwise avoided.

No single metal has been found to meet simultaneously the three requirements just listed. In practice, metal interconnects need to be formed by combining at least three different metal sublayers:

1. A reactive metal, such as titanium, forms the electrical contact layer.
2. A low-resistance metal, such as aluminum, tungsten, or copper, forms the bulk of the interconnect.
3. A nonreactive barrier metal, such as titanium nitride, stabilizes the interface between the other two metals.

Figure 2 indicates this type of triple metal layer, the details of which we will discuss in terms of their respective functions as successful interconnects in modern integrated circuits.

B. Processes for Making Low-Resistance Electrical Contacts

Several processing steps are needed to achieve low-resistance electrical contact between silicon and a metal.

The high carrier concentration in the silicon is achieved by various alternative methods for placing dopants in the silicon layer (diffusion, ion implantation,

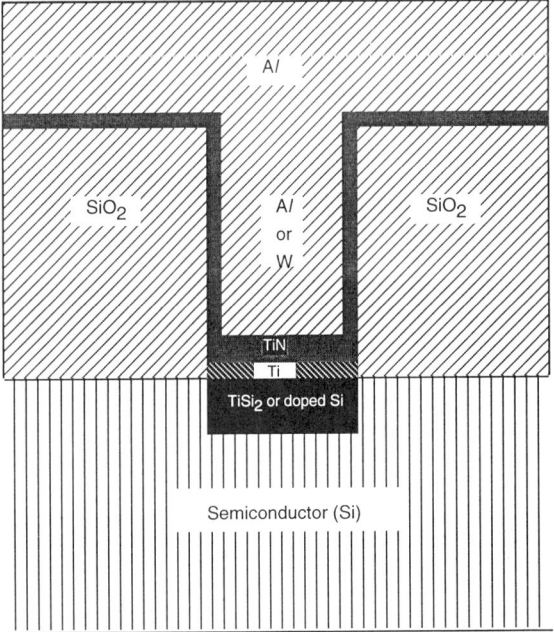

Figure 2 Cross section of a multilayer metal contact.

epitaxial growth), and activating them, if necessary, by an annealing process to remove defects. See Chapters 2–4 for more details about forming a highly conductive silicon layer.

The most effective and reliable way to eliminate the insulating effect of a thin silicon oxide layer on silicon is to convert it with a reactive metal such as titanium into a conducting metal oxide. For example, if a thin titanium layer is sputtered onto a silicon surface that is covered with a thin layer of silicon oxide, the titanium metal removes the oxygen from the silicon, by reactions such as the following:

$$9Ti + 3SiO_2 \rightarrow Ti_5Si_3 + 2Ti_2O_3 \qquad (7)$$

A remarkable feature of this transformation is that the product materials, titanium silicide and titanium(III) oxide, are both metallic electrical conductors. Thus, the insulating silicon dioxide layer has been converted into an electrically conducting layer by this reaction.

Titanium is the *only* metal that can achieve this spontaneous reduction of silicon oxide to highly conductive products. Some metals (such as zirconium) can also reduce silicon oxide by spontaneous chemical reactions, but the product metal oxides (such as ZrO_2) are electrical insulators. Other metals (such as vanadium and niobium) can form conducting oxides, but their reduction reactions with silicon are not as rapid, and the thin films of their silicides have higher resistances than titanium silicide.

Normally, there is only a thin silicon oxide layer (about 2–3 nm thick) remaining on a contact window area of silicon after the thick (about 1000 nm) layer of silicon dioxide is opened by etching. Sputtering from a titanium target is normally used to deposit the titanium metal, to a thickness of 10 or 20 nm. Then the structure may be annealed (heated) in a vacuum or an inert (e.g., argon) atmosphere, in order to induce the reduction reaction (7), with more than enough titanium available to react with the SiO_2. During the anneal, at temperatures of 700–800°C, the remaining titanium reacts with more of the silicon to form more titanium disilicide, by reaction (1).

The top surface of the silicon, which is thus converted into titanium silicide, often contains defects and impurities, which would increase the contact resistance between the silicon and a metal. In the final structure of the contact, the more perfect underlying silicon becomes a fresh surface in contact with the metallic titanium disilicide, forming a very-low-resistance contact. Values of contact resistance as low as 10^{-7} Ω-cm^2 can be achieved in this way. This means that a contact window 0.5 μm in diameter contributes about 50 Ω to the resistance of its circuit.

C. Metals Used for the Electrical Connections

A primary requirement for an electrical interconnection is that it must have a low, stable electrical resistance. It also must remain chemically stable during all manufacturing steps and through long-term use. Aluminum and tungsten are the primary metals used for electrical connections within integrated circuits. Copper is considered a possible future replacement for these metals, for reasons that will be explained. Some properties of these metals are given in Table 1.

1. Silver

Although silver has the lowest resistance for a given size interconnect, it has not been used in integrated circuits because of its electrical instability when too large an electrical current is drawn through it. The current tends to move silver atoms by a process called electromigration, which can cause breaks in the electrical connection. Silver corrodes easily in the presence of many oxygen and sulfur containing materials encountered during the manufacturing processes, as well as in typical computer operating environments. Also, silver very readily diffuses into silicon, in which it catalyzes the recombination of electrons and holes.

Table 1 Properties of Bulk Metals

Metal	Electrical resistivity (µΩ-cm)	Melting point (°C)	Chemical stability	Electrical stability
Ag	1.6	962	Poor	Poor
Cu	1.7	1083	Poor	Good
Al	2.7	660	Fair	Poor
W	5.3	3410	Good	Good
$TiSi_2$	13	1540	Good	Good
$CoSi_2$	15	1277	Good	Good
TiN	22	2930	Good	Good
PtSi	28	1100	Good	Good
WSi_2	38	2050	Good	Good
CoSi	168	1395	Good	Good

2. Aluminum

Aluminum has been and remains the predominant workhorse of interconnect technology. It has a fairly low electrical resistivity (about 1.7 times that of silver). Although aluminum is very reactive, its surface becomes covered with a thin (about 5 nm thick) oxide layer, which generally protects it against further oxidation. Aluminum is easily deposited by several methods, including sputtering, evaporation, and chemical vapor deposition. Patterns can be etched into aluminum by either wet (solution) or dry (gaseous) processes.

Aluminum is also subject to electromigration instability. This effect can be reduced by the addition of small amounts (1–2%) of copper and silicon to the aluminum. Such additives, however, increase the electrical resistance of the aluminum by 10–20%. Even with these stabilizing additives, current densities must be kept below about 2×10^5 amperes/cm^2 in order to avoid electromigration.

3. Tungsten

Tungsten has the highest melting temperature (3410°C) of any metal. This fact illustrates the very strong bonds between tungsten atoms, which also makes tungsten resist electromigration much better than aluminum does. It is possible to deposit tungsten by CVD selectively on the silicon in contact windows, while avoiding deposition on neighboring areas of silicon oxide. This selective deposition process makes it easier to maintain a level topography for deposition of subsequent layers, by forming tungsten "plugs" inside the contact windows. Because of these advantages, tungsten is receiving increasing use in integrated circuits.

A disadvantage of tungsten is that its electrical resistivity is higher than that of aluminum. The circuits of the same size operate more slowly if made with tungsten metallurgy rather than with aluminum.

4. Copper

Although no full-scale use has been made of copper within integrated circuits, it has the potential advantages of lower resistance than aluminum and tungsten, and much better stability toward electromigration than aluminum. Copper can reliably carry current densities up to 5×10^6 amperes/cm^2. The necessary currents could flow through narrower copper interconnects; this advantage will be particularly important for metal linewidths below 0.25 μm. Thus, copper would have a very significant advantage over aluminum or tungsten in permitting size and cost reduction and higher operating speeds (an estimated 50% increase for devices with channel lengths of 0.25 μm).

Copper's disadvantages are similar to those of silver. Copper diffuses easily into silicon, in which it acts as a recombination center. Thus, improved diffusion barriers must be developed for use between copper and silicon. Copper is also easily corroded by water and oxygen, and so greater protection may be needed in handling partially completed devices during the manufacture of a chip. Furthermore, encapsulation of the final devices may need to be improved, in order to protect the copper adequately during long-term use. In particular, plastic encapsulation may not be sufficiently protective for copper circuits, and more expensive inorganic glass encapsulation may be needed.

5. Refractory Metal Silicides

The silicides of refractory metals, such as titanium disilicide (TiSi$_2$), cobalt disilicide (CoSi$_2$), and tungsten disilicide (WSi$_2$), are metallic compounds that have higher resistivities than the elemental metals already discussed. They are extremely stable against chemical attack, diffusion, and electromigration, and they are finding increasing use for short interconnections for which their higher resistances still allow acceptable device performance.

IV. DEPOSITION PROCESSES FOR ELEMENTAL METALS

A. Deposition of Metals by Sputtering

In current technology, sputtering is usually used to deposit aluminum. Sputtering is a physical deposition process carried out in a vacuum chamber. Electrical energy is applied to a low-pressure nonreactive gas, such as argon, within the chamber, forming some positive ions, such as argon ions, and free electrons. On one side of the chamber there is a plate of the metal to be sputtered, called the target. The positive argon ions are accelerated electrically toward the surface of the metal target, from which they dislodge metal atoms. These freed metal atoms travel across the vacuum chamber to the other side, where they coat the substrate with metal.

An advantage of sputtering is its flexibility; almost any metal can be deposited simply by making the appropriate target. Alloy films, such as aluminum stabilized with 1–3% copper and silicon, can be made from targets of essentially the same composition. Metals generally sputter with different efficiencies, and so the first few films made from a new alloy target may be enriched in the element that sputters most easily. However, after a few depositions, the face of the target becomes enriched in the less easily sputtered element, and the deposited films approach the bulk composition of the target.

Sputtering has some disadvantages. Bombardment by ions and electrons from the plasma discharge can generate damage (defects and impurities) in the substrate and on its surface. Sputtering also has some difficulty filling narrow holes and trenches with metal, because the metal atoms tend to stick to the upper part of the walls. As this deposit gets thicker, it can eventually block off the top of the hole, leaving a void in the metal deposit near the bottom of the hole. Stages in this process are illustrated in Fig. 3. This type of deposit represents an example of poor "step coverage." For comparison, Fig. 4 shows a CVD deposit with excellent step coverage, in which the coating thickness is nearly the same on the upper surface, the sidewalls, and the bottom of the contact hole.

The voids left in a deposit with poor step coverage increase the electrical resistance of the interconnect, and they may lead to failure by electromigration of the metal, induced by the high current densities through the adjacent thin metal.

This void formation by sputtering can be avoided by inserting a plate with narrow holes (called a collimator) between the target and the substrate. The collimator only passes metal atoms that are traveling nearly perpendicular to the substrate. In this way, most of the metal atoms are delivered directly to the bottom of the contact window, so that a dense, void-free plug of metal can be formed. A disadvantage of collimated sputtering is a reduced deposition rate, which tends to increase needed maintenance to replace the collimators before deposits on them start to flake off and contaminate the substrates.

Collimated sputtering appears to be adequate for fabricating typical integrated circuits into the mid-1990s. The insulating layers are not likely to be much thinner in the future designs, because adequate separation must be maintained between the different circuits. Although the height (around 1 µm) of the via holes for the metal will remain about the same, their widths will be smaller, and thus the ratio of height to width (called the "aspect ratio") will be larger. In 1994, aspect ratios of 1–2 are typically used, while future devices call for aspect ratios of 3–4 or more. Deposition of metal by collimated sputtering cannot be used to fill holes with these high aspect ratios.

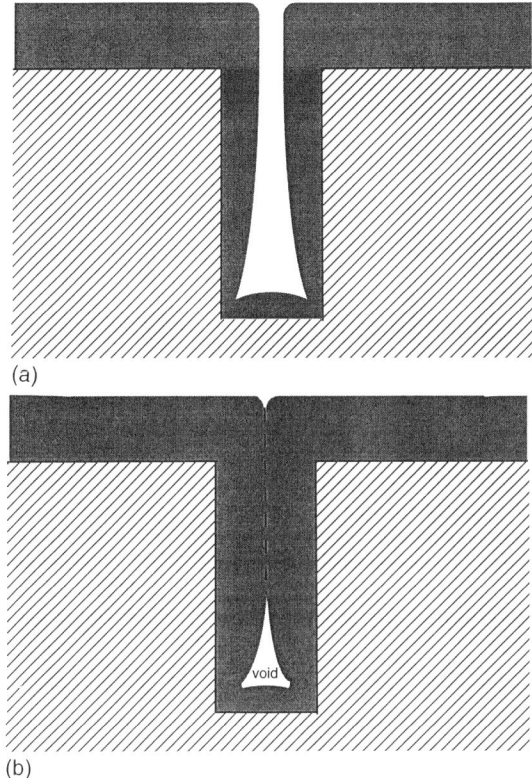

Figure 3 Cross section of a deposit with poor step coverage, leading to a void: (a) early stage of film growth; (b) at end of film growth.

B. Chemical Vapor Deposition of Tungsten

The CVD process has been used since the early 1990s to deposit tungsten into holes with high aspect ratios. General aspects of CVD have been discussed in Chapters 3 and 4. Tungsten CVD is usually carried out at moderately low gas pressures, on the order of 10 torr, although higher pressures, up to 1 atm (760 torr), have also been used successfully. An ideal CVD reactor would provide a uniform flux of reactant gases to all parts of the reactor surface, to give a uniformly thick layer of tungsten. Detailed modeling of the flow, diffusion, and reaction of gases in a CVD reactor are needed in general to determine the flux distribution of the gases (see Chapter 3). Low gas pressure promotes uniform distribution of the gases, because diffusion constants increase with lower pressure. Proper design of the inlet gas nozzle is also very important; often, "show-

Figure 4 Cross section of a deposit with excellent step coverage at (a) an early stage of film growth, and at (b) the end of film growth, showing complete plug filling.

erhead'' designs are used, in which the gas enters the CVD chamber through many small holes. Rotation of the substrate may be used to make the film thickness more uniform. At higher pressures (0.1 to 1 atm), it is usually impossible to distribute the gases uniformly over a large wafer. In such cases, uniform thickness may still be achieved by distributing the gases uniformly from a straight slot, and moving the substrate (for example, on a belt or platform) past the slot.

For tungsten, the most commonly used CVD source is tungsten hexafluoride gas, WF_6. Chemical reactions convert the source gas into the metal to be deposited. For example, hydrogen may be mixed with the tungsten hexafluoride, so that the following reduction reaction can take place:

$$WF_6 + 3H_2 \rightarrow W + 6HF \tag{8}$$

The by-product hydrogen fluoride, HF, is a gas that is pumped out of the deposition chamber and neutralized for disposal. The hydrogen fluoride may also cause problems by reacting with silicon dioxide insulating layers. This substrate etching must be prevented by first depositing a barrier layer on the substrate, as discussed in Sec. V.A.

According to thermodynamics, the hydrogen reduction of tungsten hexafluoride should be spontaneous, since it has a negative free energy change (-138 kJ/mol at 523°C). At temperatures below about 300°C, however, the rate of the reaction is negligibly slow, even though it should happen spontaneously if one waited long enough. In the range 300–450°C the deposition rate increases rapidly with temperature, with an activation energy of about 67 kJ/mol, according to the expression (in nanometers per minute):

$$\frac{\text{thickness}}{\text{time}} \approx 7.8 \times 10^5 \, e^{-8040/T} P_{H_2}^{1/2} \tag{9}$$

in which T is the temperature in degrees Kelvin. The growth rate is proportional to the square root of the hydrogen partial pressure, P_{H_2}, in Pascals, at the surface of the substrate. It does not depend on the tungsten hexafluoride concentration, as long as it is larger than some minimum value (around 10 mtorr).

Above about 500°C, the deposition rate increases only slowly with temperature. In this higher temperature range, the reaction rate is so fast that the slowest step in the process (the rate-limiting step) is the diffusion of the reactants to the surface. Diffusion rates increase only slowly with temperature, and thus the overall deposition rate increases only slowly at high temperatures.

There is, however, evidence that the chemical reaction steps in the reduction of tungsten hexafluoride by hydrogen (reaction (8)) occur on the surface of the deposited tungsten metal, which catalyzes the reaction. (Reactions also occur in the gas phase during most CVD processes.) In general, clean, oxide-free metal surfaces catalyze this reaction, but most insulator surfaces, such as silicon dioxide or silicon nitride, do not promote it.

A CVD reaction has good step coverage when its slowest (rate-limiting) step occurs on the surface and is slow compared with the arrival rate of the reactants. Thus, tungsten deposited by the hydrogen reduction of tungsten hexafluoride in the lower temperature range, 300–450°C, shows good step coverage, because of the slow rate of reaction at the surface. For example, at 430°C, over 80% step coverage has been achieved by this reaction in vias with an aspect ratio of 3:1, at a modest deposition rate of 25 nm/min. As the deposition temperature increases, so does the surface reaction rate, and the step coverage becomes poorer. The good step coverage found for the hydrogen reduction may be attributed to either the slow adsorption of hydrogen or to the slow desorption of HF byproduct from the surface.

Silicon also catalyzes the deposition of tungsten. Initially, the silicon itself is the reductant, according to the reaction

$$2WF_6 + 3Si \rightarrow 2W + 3SiF_4 \tag{10}$$

This silicon reduction of tungsten hexafluoride stops after a thin layer of tungsten is deposited, because the tungsten layer prevents contact between the gas-phase reactant (WF_6) and the silicon surface, which has become covered with tungsten. However, the deposition of tungsten can be continued by shifting to the hydrogen reduction (Eq. 8) catalyzed by the initially deposited layer of tungsten over the silicon. Therefore, selective deposition of tungsten on silicon contacts is possible. The tungsten nucleates on the silicon at the bottom of the contact hole and grows to form a "plug" through the silicon dioxide insulator layer, but it does not grow on top of the silicon dioxide. The resulting structure has a nearly planar surface, onto which a layer of lower-resistance interconnect metal (such as aluminum) can be deposited and then etched into the required pattern of connections.

Although this selective tungsten process has been studied extensively, it has not yet (in 1994) come into widespread commercial use for the production of tungsten plugs in contact holes. One problem is that the removal of silicon by reaction (10) may interfere with the operation of shallow semiconductor devices. Also, practical difficulties have arisen in completely preventing growth on various insulator surfaces. Unwanted tungsten growth may occur on insulator surfaces because of catalytic action of a variety of contaminants, including some less volatile tungsten fluorides that occur as intermediates in the reduction reactions. Use of a cold-wall reactor, in which only the substrate is heated, tends to trap the lower tungsten fluoride by-products and maintain selectivity better than hot-wall reactors do.

Because of the difficulty of reproducing selective tungsten growth, tungsten plugs are usually deposited by the so-called "blanket tungsten process." In this process, tungsten is deposited over the entire wafer surface, and then etched away in all areas except for the thicker regions deposited within the contact holes. Nonselective deposition on all surfaces, including insulators as well as metals and silicon, is achieved by beginning the reaction with silane as a reductant:

$2WF_6 + 3SiH_4 \rightarrow 2W + 3SiF_4 + 6H_2$ (11)

The silane reduction proceeds without the need for any catalytic action by the surface, and it deposits tungsten rapidly and nonselectively on all surfaces. In fact, it can even nucleate spontaneously in the gas phase, yielding undesirable tungsten particles. To avoid vapor-phase nucleation reactions, the silane reduction reaction is normally run with low partial pressures of the reactants (less than 1 torr). At low pressures, the rate of deposition by this reaction is proportional to the flux of silane gas to the surface of the film, and it is essentially independent of the tungsten hexafluoride concentration (provided it exceeds some very small value).

Because of its rapid surface reaction, the silane reduction reaction yields rather poor step coverage. For example, with equal gas concentrations of silane and tungsten hexafluoride, only 25% step coverage is found at 430°C in holes with an aspect ratio of 3:1. The growth rate under these conditions is more than four times higher than achieved by the hydrogen reduction reaction. In practice, because of this poor step coverage, only a thin initial layer is produced by silane reduction. Then, the silane flow is terminated, while the hydrogen flow is initiated. The hydrogen reduction reaction (reaction (8)) is then used to deposit tungsten over the entire surface to benefit from the catalytic effect of the thin tungsten layer already produced by the silane reduction reaction.

Another approach that has been taken to initiate tungsten deposition onto titanium nitride barrier/adhesion layers (see Sec. V.D) is to flow hydrogen gas over the titanium nitride surface prior to beginning the flow of tungsten hexafluoride gas. Presumably, some of the hydrogen chemisorbs onto the titanium nitride surface, leading to almost immediate initiation of tungsten film growth. In contrast, if the tungsten hexafluoride gas is started first or at the same time as the hydrogen, then there is a long delay (many minutes) before the tungsten growth begins, and the interface between the titanium nitride and the tungsten appears to be contaminated by significant amounts of fluorine.

Because the step coverage of the hydrogen reduction reaction is good, "blanket" deposited tungsten fills the contact holes and covers the rest of the insulator surface to a nearly planar level. Then the tungsten is etched by a fluorine-containing plasma, so that only the tungsten within the contact holes remains. Alternatively, mechanical polishing is also used to remove the blanket tungsten layer, leaving contact holes filled with tungsten up to the flat surface of the silicon dioxide insulator.

Stress is another important property of thin films. A film is said to be in tensile stress if it is stretched over the substrate, and in compressive stress if it is squeezed by the substrate. In the case that the tensile stress in a film is too large, cracks may form in it, and/or it may result in delaminations off the substrate. Large compressive stresses may result in delaminations off the substrate in the form of bubbles or blisters. During a CVD process a certain amount of stress is built into the film, with tensile stress being more common than compressive stress. Tensile stress may arise because some of the ligands are lost from the surface layers of the growing film, leaving metal atoms separated by distances larger than usual for the bulk material. Compressive stress is more common in sputtered films, perhaps because of the bombardment of the film surface by fast-moving ions. Stress can also be changed (either increased or decreased) as the film and substrate are cooled from the deposition temperature to room temperature.

Normally, CVD tungsten films have high tensile stress (more than 10^{10} dynes/cm^2). By reducing the concentration of tungsten hexafluoride gas, the stress in tungsten films may be reduced or even made slightly compressive. The lower growth rates under these conditions presumably allow more time for the tungsten atoms to rearrange at the growth surface before being covered by a new layer of atoms. Poorer step coverage may, however, be obtained in tungsten deposited under these conditions.

C. Chemical Vapor Deposition of Aluminum

Several different compounds of aluminum have been used as sources for CVD. Aluminum alkyls, such as triisobutylaluminum (TIBA), allows deposition at substrate temperatures in the range 200–400°C:

$$AlR_3 \rightarrow Al + H_2 + \text{olefins} \tag{12}$$

(R is a hydrocarbon radical, such as isobutyl, C_4H_9.)

These temperatures (200–400°C) are not high enough to break the strong aluminum–carbon bonds (about 280 kJ/mol) directly. Instead, there is evidence that a bond between aluminum and hydrogen forms at the same time that the aluminum–carbon bond is breaking, in what is called a beta-hydride elimination, as illustrated by

$$\begin{array}{c} \text{H–CH}_2 \\ | \\ \text{Et}_2\text{Al–C} \\ \text{H}_2 \end{array} \rightarrow \begin{array}{c} \text{H- -CH}_2 \\ \vdots \quad | \\ \text{Et}_2\text{Al – C} \\ \text{H}_2 \end{array} \rightarrow \text{Et}_2\text{AlH} + \text{CH}_2 = \text{CH}_2 \tag{13}$$

Two more beta-hydride elimination reactions result in AlH$_3$, which quickly decomposes to aluminum and hydrogen at temperatures above 100°C:

$$AlH_3 \rightarrow Al + \frac{3}{2}H_2 \tag{14}$$

The effect of temperature on the deposition rate of aluminum from triisobutylaluminum is shown in Fig. 5. Below 400°C, the rate increases rapidly

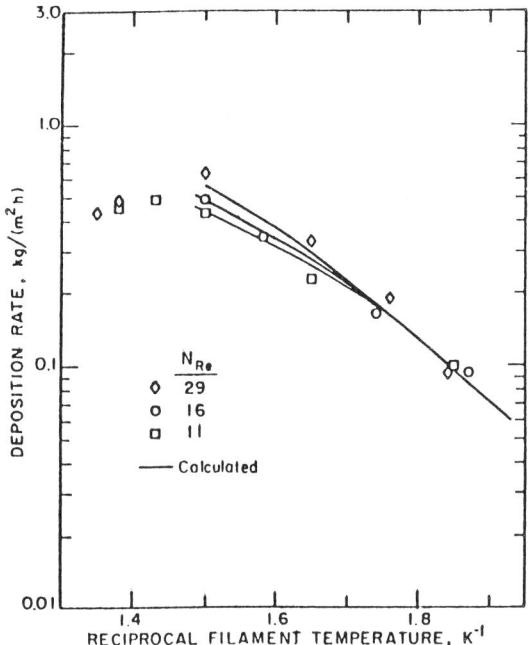

Figure 5 Temperature dependence of the deposition rate of aluminum from triisobutylaluminum. (From A. Malazgirt and J. W. Evans, *Metall. Trans. 11B*:225 (1980).)

with temperature, because of the activation barrier to the beta-hydride elimination reactions. Above 400°C, the diffusion of reactants to the surface becomes rate-limiting, and the deposition rate becomes nearly constant.

Not all of the aluminum–carbon bonds are eliminated during the decomposition of aluminum alkyls, which thus leave some carbon contamination in the aluminum film, particularly at higher substrate temperatures. Another problem with this CVD process for aluminum is that the deposited aluminum film tends to be somewhat rough, because of large crystallites, which can cause problems forming narrow interconnections. Heated liquid aluminum alkyls undergo some beta-hydride elimination of ligands before they are even vaporized, resulting in complex mixtures containing variable amounts of mixed aluminum alkyl-hydride species of lower volatility. Premature source decomposition may be mitigated by using an alkene carrier gas, such as isobutylene. Also, the aluminum alkyls associate in the liquid state by strong intermolecular interactions, which slow down the kinetics of vaporization and make it difficult to obtain reproducible vapor concentrations. Close control of the temperature of the liquid aluminum alkyl is insufficient to ensure a reproducible rate of vapor delivery, because the slow vaporization kinetics results in a less than saturated vapor pressure. Thus, the concentration of vapor delivered depends on the carrier gas flow rate, the volume of liquid remaining, and so on. These difficulties contribute to the slow introduction of the commercial use of aluminum alkyl CVD for use in microelectronic circuit fabrication.

Surfaces of many clean, oxide-free metals, including aluminum itself, catalyze the deposition of aluminum. On the other hand, film growth does not occur readily on some other surfaces, such as oxides. In principle, this distinction could form the basis for a selective aluminum process. However, a reproducible selective aluminum process has been difficult to demonstrate.

Another type of aluminum precursor is aluminum hydride, AlH_3, which was previously mentioned (Eq. 14) as a likely intermediate compound formed during the decomposition of aluminum alkyls. AlH_3 might seem to be the ideal source for deposition of aluminum, because it contains no carbon that could enter the film as an impurity, and it decomposes at a low enough temperature (around 100°C) so that the aluminum crystallites would remain small. The normal, stable form of AlH_3 is a solid material that cannot be vaporized without decomposition. However, when it is complexed with a tertiary amine, in the form of an amine-alane, $R_3N \cdot AlH_3$, it becomes volatile enough to serve as a CVD precursor at low reaction temperatures (100–200°C):

$$R_3N \cdot AlH_3 \rightarrow Al + \frac{3}{2}H_2 + R_3N \qquad (15)$$

R is a hydrocarbon radical, usually methyl (CH_3) or ethyl (C_2H_5).

These amine-alane reagents begin to decompose by breaking the relatively weak bond between nitrogen and aluminum:

$$R_3N \cdot AlH_3 \rightarrow AlH_3 + R_3N \qquad (16)$$

The carbon is removed very completely by this dissociation reaction. The AlH_3 then decomposes according to equation (14), leaving the aluminum film free of carbon contamination. The films are also smoother because of the smaller crystallites formed at these lower deposition temperatures. Trimethylamine-alane (R = CH_3) and triethylamine-alane (R = CH_2CH_3) are solids, which are less reproducible vapor sources than liquids. The mixed alkyl-amine, dimethylethylamine-alane, is a liquid source that tends to give more reproducible amounts of vapor. On storage, however, some of the amine is lost from this dimethylethylamine-alane, and so it may be necessary to add dimethylethylamine vapor to the carrier gas in order to prevent decomposition. Dimethylethylamine-alane is being actively studied for possible application to microelectronics.

D. Chemical Vapor Deposition of Copper

Copper CVD precursors are based on compounds containing the hexafluoroacetylacetonate ligand (abbreviated as hfa):

$$F_3C-\underset{O^-}{C}=\underset{H}{C}-\underset{\underset{O}{\|}}{C}-CF_3$$

This fluorinated ligand yields more volatile precursors than the corresponding ligand with hydrogen in place of the fluorine. At temperatures of about 250–400°C, hydrogen reduces vapors from the solid compound copper(II)-hexafluoroacetylacetonate to deposit copper according to the reaction

$$Cu(hfa)_2 + H_2 \rightarrow Cu + 2H(hfa) \tag{17}$$

A lower-temperature CVD source for copper is the reaction

$$2Cu(hfa)L \rightarrow Cu + Cu(hfa)_2 + 2L \tag{18}$$

in which L is a neutral ligand such as an olefin or trimethylphosphine. Most of these starting materials of the type Cu(hfa)L are solids, but for L = vinyltrimethylsilane, it is a liquid, and thus it is more easily vaporized than are solids.

Resistivities as low as 2 $\mu\Omega$-cm have been obtained, close to that of bulk copper (1.7 $\mu\Omega$-cm). These reactions can produce a high deposition rate of copper, approaching one micron per minute. The temperature dependence of the deposition rate is shown in Fig. 6. At temperatures below 200°C the deposition rate increases steeply with increasing temperature, according to an Arrhenius rate law, probably due to the rate-limiting step being a surface reaction with an activation energy near 96 kJ/mol (23 kcal/mol). Above 200°C, the deposition rate becomes nearly independent of temperature, because the surface reaction is very fast and the rate-limiting step is then the rate of diffusion of reactants to the surface.

Selective growth of copper from reaction (18) has been demonstrated on various metals, including copper itself and metallic titanium nitride, at rather low temperatures of 150–200°C. No growth of copper occurs on dry oxide surfaces, but moisture or other contaminants lead to unwanted spurious growth of copper and loss of selectivity. Better understanding of these and other growth parameter effects is required to achieve a practical selective copper process.

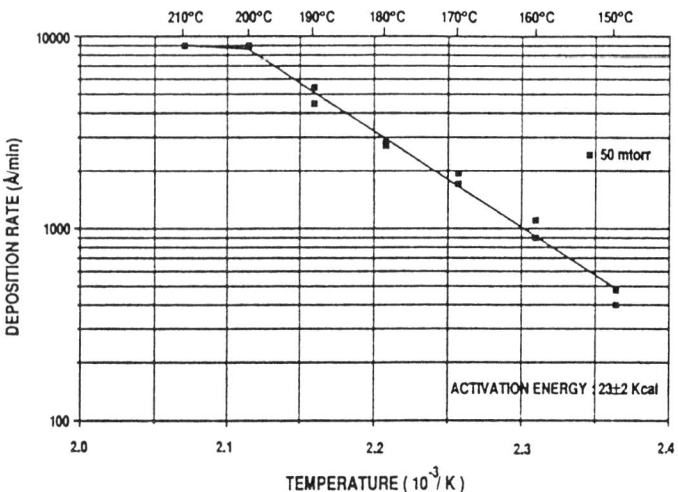

Figure 6 Chemical vapor deposition rate of copper from the precursor (hexafluoroacetylacetonate)copper(2-butyne) as a function of deposition temperature at a precursor partial pressure of 50 mtorr. (From A. Jain, K.-M. Chi, T. T. Kodas, M. J. Hampden-Smith, J. D. Farr, and M. F. Paffett, *Chem. Mat.* 3:995 (1991).)

The likely mechanism of this selective CVD reaction involves dissociation of the precursor on the surface of a metal to form a copper(I) surface species and the free gaseous ligand:

$$Cu(hfa)L(s) \rightarrow Cu(hfa)(s) + L(g) \tag{19}$$

Two adjacent copper(I) species then disproportionate to give the surface-bound product copper and a volatile copper(II) by-product, which sublimes into the gas phase:

$$2Cu(hfa)(s) \rightarrow Cu + Cu(hfa)_2(g) \tag{20}$$

V. BARRIER AND ADHESION LAYERS

A. The Need for a Diffusion Barrier for Aluminum Interconnects

A chemical instability occurs when aluminum comes into direct contact with silicon. Aluminum and silicon react to form a compound (called a eutectic) with a relatively low melting point of 577°C. The interface between aluminum and silicon becomes unstable when heated, even to temperatures somewhat below the eutectic temperature, causing etch pits in the silicon to such a depth that they short out pn junctions near the surface of the silicon. These etch pits can even form through a thin titanium silicide contact layer.

This undesirable interaction is prevented in practice by placing a barrier material on the contact silicide prior to depositing the aluminum. The barrier material must not react with silicon, titanium silicide, or aluminum, and it must not allow the aluminum, titanium, or silicon to penetrate or diffuse through it.

The most commonly used barrier material is titanium nitride, TiN. A TiN layer only 20 nm thick prevents interdiffusion of the contact structure shown in Fig. 2 even when it is heated to a temperature of 500°C for an hour.

B. The Need for a Barrier/Adhesion Layer Under Tungsten

When tungsten is used as an interconnect metal, there are two problems that may occur with a substrate made of silicon and silicon dioxide. The initial reaction of tungsten hexafluoride with the silicon substrate occurs via reaction (10), which consumes some of the silicon surface. Etching of the silicon dioxide can be caused by the hydrogen fluoride by-product of the hydrogen reduction reaction (8), when hydrogen is used as a reactant gas,

$$4HF + SiO_2 \rightarrow SiF_4 + 2H_2O \qquad (21)$$

These reactions remove silicon and silicon dioxide from the substrate, and they can leave etch pits, often described as "wormholes," due to uneven etching across the surface.

Another problem that arises is that the tungsten does not adhere well to the silicon dioxide.

In order to protect the substrate during tungsten CVD and to cause the tungsten to adhere, the substrate must first be protected by a layer that is both impervious to attack by the chemicals present in the CVD chamber and adherent to the tungsten. Titanium nitride has been found to be effective in this role, since it does not react either with tungsten hexafluoride or with hydrogen fluoride, and the deposited tungsten adheres well to the titanium nitride layer.

C. The Need for a Diffusion Barrier Layer Under Copper

Copper diffuses rapidly through many solid materials, including silicon. Inside silicon, copper impurities catalyze the recombination of electrons and holes. Early in the development of solid state transistors, many experimental devices failed because of an unknown contaminant, nicknamed "deathnium," later identified as copper. Thus, if copper is ever to be used as a viable interconnect metal for microcircuits, a very effective diffusion barrier is needed under the copper.

Titanium nitride is effective against the diffusion of copper. Furthermore, very thin layers of amorphous alloys containing tantalum, nitrogen, and silicon have also been demonstrated to be effective as diffusion barriers.

D. CVD of Titanium Nitride Barrier and Adhesion Layers

As indicated, titanium nitride layers serve as barrier films for each of the three common interconnect metals. Titanium nitride layers have generally been made by reactive sputtering of a titanium target in a nitrogen atmosphere. This sputtering process is adequate to fill holes and trenches with aspect ratios under about 2 or 3. However, if the aspect ratio is too large, the sputtered titanium nitride layer is very nonconformal, resulting in only a few tens of nanometers on the bottoms and sidewalls of the holes. Therefore, alternative CVD processes have been investigated, in order to more uniformly fill the narrow holes, or to coat contoured surfaces.

Titanium tetrachloride vapor reacts with ammonia at temperatures above 500°C to deposit titanium nitride with excellent step coverage as per

$$6TiCl_4 + 8NH_3 \rightarrow 6TiN + 24HCl + N_2 \qquad (22)$$

However, it is important that these reactants are not mixed at ambient temperatures, in which case a rapid acid–base reaction precipitates an unwanted salt or adduct, by reactions such as

$$TiCl_4 + 2NH_3 \rightarrow TiCl_4 \cdot 2NH_3 \qquad (23)$$

To avoid reaction (23), the reactants are separately preheated to temperatures around 250–300°C. The mixed reactants then maintain a homogeneous vapor phase sufficiently long to deposit a relatively pure TiN film at substrate temperatures above about 600°C. This temperature is low enough to allow its use in commercial depositions of diffusion barriers on silicon microelectronic devices, at the first level of metallization. The same method is also employed to produce titanium nitride solar control coatings on large areas of window glass.

The deposition temperature for this reaction is, however, too high to be used between upper layers of metals, if some aluminum has already been deposited on the wafer. Another disadvantage of this reaction is that some chlorine remains in the deposited film, where it may later cause corrosion of the metal layers. Also, solid by-products of the CVD reaction, such as ammonium chloride, generate particles, which can contaminate wafers and cause problems in handling the reactor exhaust gases.

An alternative CVD process operating at lower temperatures (typically, 200–400°C) and involving no chlorine is that of tetrakis(dialkylamido)titanium reacting with ammonia as per

$$\text{Ti(NR}_2)_4 + 2\text{NH}_3 \rightarrow \text{TiN} + 4\text{HNR}_2 + \frac{1}{2}\text{N}_2 + \text{H}_2 \qquad (24)$$

in which R represents either methyl or ethyl groups. This CVD process yields better step coverage than sputtering, but it is inferior to that produced by the chloride CVD process (22). However, these films are more effective diffusion barriers between metals and silicon compared with the commonly used sputtered TiN films.

The chemical mechanism of the CVD reaction between tetrakis-(dialkylamido)titanium compounds and ammonia is partly understood. The reaction begins very rapidly as soon as the reactants are mixed in the gas phase. By analogy to similar solution-phase reactions, it seems likely that these initial gas-phase reactions are transaminations:

$$\text{Ti(NR}_2)_4 + \text{NH}_3 \rightarrow \text{Ti(NR}_2)_3\,\text{NH}_2 + \text{R}_2\text{NH} \qquad (25)$$

$$\text{Ti(NR}_2)_3\text{NH}_2 + \text{NH}_3 \rightarrow \text{Ti(NR}_2)_2\,(\text{NH}_2)_2 + \text{R}_2\text{NH} \qquad (26)$$

$$\text{Ti(NR}_2)_2(\text{NH}_2)_2 + \text{NH}_3 \rightarrow \text{Ti(NR}_2)(\text{NH}_2)_3 + \text{R}_2\text{NH} \qquad (27)$$

$$\text{Ti(NR}_2)(\text{NH}_2)_3 + \text{NH}_3 \rightarrow \text{Ti(NH}_2)_4 + \text{R}_2\text{NH} \qquad (28)$$

Flow tube experiments have confirmed that these fast reactions produce dimethylamine (in the case R = methyl), and they have measured the rate constants for these fast transamination reactions. The transamination reactions become slower with bulkier ethyl groups. This mechanism also predicted that the nitrogen in the film comes from the ammonia, rather than from the nitrogen initially bound to the titanium. Isotopic tracer experiments have subsequently verified this source of the nitrogen, giving further support to the transamination mechanism.

The resulting titanium amide, $\text{Ti(NH}_2)_4$, is not a known compound. It is likely to undergo rapid unimolecular decomposition reactions, eliminating ammonia:

$$\text{Ti(NH}_2)_4 \rightarrow \text{HN=Ti(NH}_2)_2 + \text{NH}_3 \qquad (29)$$

$$\text{HN=Ti(NH}_2)_2 \rightarrow \text{HN=Ti=NH} + \text{NH}_3 \qquad (30)$$

The elimination reactions should be somewhat endothermic, since they convert two titanium–nitrogen single bonds into one double bond of lower bond strength. The resulting titanium diimide, HN=Ti=NH, should have a high sticking coefficient on the surface of the growing film, because the titanium atom has only two shielding ligands. Thus, the titanium diimide is a likely growth species, leading to a material with a transition composition of $(\text{TiN}_2\text{H}_2)_n$. Subsequent loss of some nitrogen and hydrogen from the solid would then lead to the observed composition of nitrogen-rich titanium nitride with some residual hydrogen.

Dimerization reactions are also likely for the titanium diimide:

$$\begin{array}{c} \text{HN=Ti=NH} \\ + \\ \text{HN=Ti=NH} \end{array} \rightarrow \begin{array}{c} \text{HN-Ti=NH} \\ |\quad| \\ \text{HN-Ti-NH} \end{array} \quad (31)$$

These are likely to be followed by polymerization reactions such as the following:

$$\begin{array}{c} \text{HN-Ti=NH} \\ |\quad| \\ \text{HN=Ti-NH} \\ + \\ \text{HN=Ti=NH} \end{array} \rightarrow \begin{array}{c} \text{HN-Ti=NH} \\ |\quad| \\ \text{HN-Ti-NH} \\ |\quad| \\ \text{HN-Ti-NH} \end{array} \quad (32)$$

$$\begin{array}{c} \text{HN-Ti=NH} \\ |\quad| \\ \text{HN-Ti-NH} \\ |\quad| \\ \text{HN=Ti-NH} \\ + \\ \text{HN=Ti=NH} \end{array} \rightarrow \begin{array}{c} \text{HN-Ti=NH} \\ |\quad| \\ \text{HN-Ti-NH} \\ |\quad| \\ \text{HN-Ti-NH} \\ |\quad| \\ \text{HN=Ti-NH} \end{array} \quad (33)$$

The polymerization reactions should be exothermic, due to the exchange of a Ti=N double bond for two single Ti–N bonds, the latter of which are each more than half as strong as a double bond. The rates of polymerization are likely to be fast and nearly independent of temperature. Extensive polymerization could lead to the formation of particles. By making the gas residence time in the CVD reactor as short as possible, the growth of particles is limited by reduced polymerization. Gas-phase polymers may also add to the surface of the growing film, resulting in more material of the same composition, $(TiN_2H_2)_n$. With the sticking coefficients of the polymeric species likely to be higher than those of the monomer, the physical properties of the film deposited from polymers may differ from those of films deposited from monomers. Gas-phase polymerization occurring before film deposition will therefore reduce the step coverage, density, and conductivity of films. The best film properties are thus obtained when the reactant gases are mixed as close to the growth surface as possible. Special showerheads have been developed with two interspersed sets of holes to bring the titanium precursor and the ammonia gas in separately, so that they mix close to the wafer surface. Rapid mixing by fans between the showerhead and the substrate has also been used to improve film properties. Better properties (step coverage and conductivity) are obtained from the ethyl precursor (tetrakis(diethylamido)titanium, TDEAT) than from the analogous methyl precursor (TDMAT), presumably because the transamination reactions are slower and subsequent deleterious polymerization reactions are more limited in the ethyl case.

VI. ETCHING PROCESSES FOR METALS

In order to form the intricate patterns of metal lines connecting the various electrical components on a microcircuit, etching processes remove parts of the metal layers not covered by photolithographically produced mask layers (see Chapter 6). The removal of metal can be achieved by appropriate solutions (wet etching), or by reactive gases or by plasmas (dry etching; see Chapter 7). There are several chemical approaches used to etch metals.

A. Wet Etching

Aluminum can be dissolved by aqueous hydrochloric acid, unless its covering aluminum oxide layer is too thick.

One wet tungsten etch is aqueous hydrogen peroxide, H_2O_2. Another is a solution obtained by combining potassium ferrocyanide, potassium hydroxide, and potassium acid phosphate ($K_3Fe(CN)_6$ + KOH + KH_2PO_4).

Copper may be dissolved by an aqueous solution of ferric chloride ($FeCl_3$).

Titanium and titanium nitride may be etched by an aqueous solution containing hydrogen peroxide and ammonia. The solution remains active for a few hours, after which the hydrogen peroxide decomposes. Since this solution does not dissolve titanium silicide, it permits the selective removal of titanium and is the basis of the widely used self-aligned silicide ("salicide") patterning process.

Tungsten silicide may be dissolved in an HF/HNO_3 acid solution. The practical usefulness of this etch is limited because it also dissolves silicon and silicon dioxide.

Wet etching processes are simple to carry out in the laboratory, but they have some practical disadvantages in chip production. In some cases, bubbles of gaseous by-products hinder complete removal of the metal. Rinsing and drying steps are needed to clean the wafers after wet etching. Spent etching solutions must be neutralized, and in some cases toxic metals must be removed before disposal. For these reasons, dry etching processes have often become the preferred means for manufacture of microcircuits.

B. Dry (Gaseous) Etching

Aluminum can be removed by first sputtering in an argon atmosphere to remove the thin surface layer of aluminum oxide, and then applying carbon tetrachloride plasma. It is likely that the CCl_4 plasma converts the aluminum to the more volatile aluminum chloride, $AlCl_3$. The detailed mechanism of this reaction is not known, but it probably involves chlorine atoms and various chlorine-containing free radicals, such as CCl_3.

Tungsten is etched by a carbon tetrafluoride plasma, in which it may be converted to gaseous tungsten hexafluoride.

Chlorine-containing plasmas etch copper if it is held at temperatures above 250°C, so that the by-product Cu_3Cl_3 has a sufficiently high vapor pressure. Copper can also be removed by using the reverse of the CVD reaction (18), using a vapor mixture of copper(II)hexafluoroacetylacetonate (hfa) and vinyltrimethylsilane (L):

$$Cu + Cu(hfa)_2 + 2L \rightarrow 2Cu(hfa)L \tag{34}$$

Titanium and titanium nitride are etched by a chlorotrifluoromethane plasma, which may convert them into titanium tetrachloride vapor.

VII. SUMMARY

Metal interconnections in microelectronic circuits consist on multiple layers. One layer, typically aluminum or tungsten, forms the low-resistance pathway for the electric current to travel between different parts of the silicon semiconductor wafer. Other layers, typically titanium nitride, provide stability to structure by preventing reaction between the metal and the silicon, and by making a stronger bond between tungsten and silicon dioxide insulators. Additional layers, such as titanium metal, are often used to lower the electrical resistance between the metals and the silicon. Very stable metal silicides, such as titanium silicide and tungsten silicide, are being used as gate electrodes that are more conductive than polysilicon, the material traditionally used for gate electrodes.

Sputtering has traditionally been the most common method of applying these metallic layers to silicon substrates. During the 1990s, increasing use has been made of chemical vapor deposition for this purpose, because CVD can fill more completely the very tiny holes and channels in modern microcircuits. But CVD involves complex chemical reactions, which in many cases are only partly understood. Better knowledge of CVD mechanisms is aiding in the optimization and control of known reactions, and in the design of new CVD processes.

EXERCISES

1. Titanium and zirconium are metals in the same group of the periodic table, and they have similar, but not identical, properties. Why do you think that titanium is preferred over zirconium as a contact metal in silicon semiconductor devices?

2. What deleterious effect would be noticed in a microcircuit's operation if the titanium layer in Fig. 2 were omitted from the contact structure? What deleterious effect would be noticed in a microcircuit's operation if the titanium nitride layer in Fig. 2 were omitted from the contact structure?
3. Consider the tungsten CVD reactions (8), (10), and (11). Suggest appropriate chemical reactions that could be used to convert the gaseous by-products from these reactions into safe, nontoxic chemicals suitable for recycling and/or disposal in an environmentally sound manner.
4. Suppose you have carried out a CVD process, and found a void structure in the deposited material, similar to that shown in Fig. 3. What changes in the processing might you suggest to form a dense, void-free deposit?
5. Which of the CVD reactions discussed in this chapter are reversible? Give the conditions that would drive the reaction to the right, and those that would drive it to the left.

GENERAL REFERENCES

M. J. Hampden-Smith and T. T. Kodas, eds., *The Chemistry of Metal CVD*, VCH, Weinheim, 1994.

H. O. Pierson, *Handbook of Chemical Vapor Deposition: Principles, Technology, and Applications*, Noyes Publications, Park Ridge, New Jersey, 1992.

G. K. Rao, *Multilevel Interconnect Technology*, McGraw-Hill, New York, 1993.

J. E. J. Schmitz, *CVD of Tungsten and Tungsten Silicides*, Noyes Publications, Park Ridge, New Jersey, 1992.

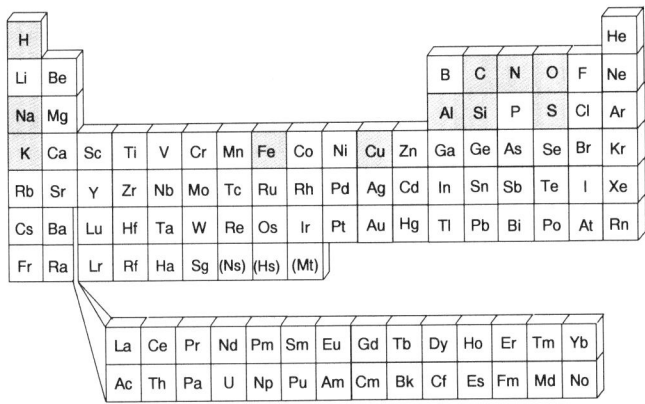

6
Photolithography

Alois Gutmann*   *Siemens Components, Inc., Siemens U.S.A., New York, New York*

I. INTRODUCTION
A. The Principle of Lithography

Lithography is the technique of delineating a pattern into a thin polymer film (called the resist) that is sensitive to electromagnetic or particle radiation. A latent image formed within the resist by irradiation with photons, electrons, ions, or X-rays can be transformed into an actual pattern by selectively removing either the exposed or unexposed regions of the film during a subsequent treatment ("development") in a solvent of suitable chemistry. If the chosen developer removes the exposed areas of the resist, a positive tone image is created. Development of unexposed parts results in a negative tone image. The principal

---

* *Mail address*: IBM East Fishkill Facility, Hopewell Junction, New York

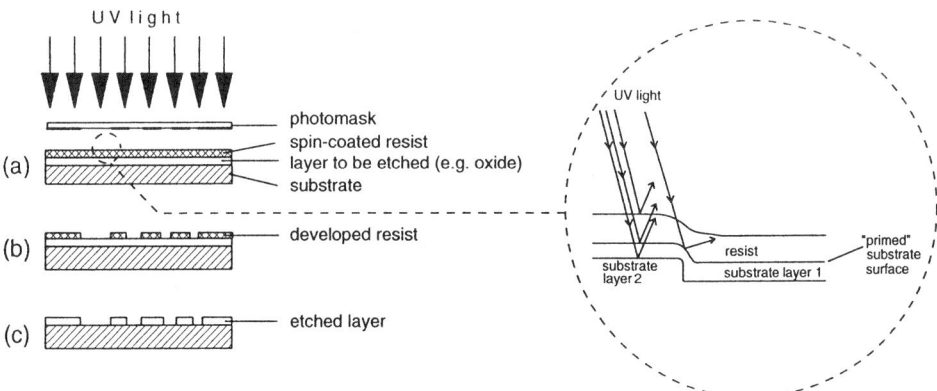

Figure 1 Principle of positive tone photoresist application: (a) exposure of resist coated wafer; (b) after development of resist; (c) after etching and resist stripping.

operation mode (including an etch step subsequent to lithography) of a positive tone photoresist (resist sensitive to light) is shown in Fig. 1.

B. Lithography Options and Application Areas

Lithography is a key technology used in integrated circuit (IC) device fabrication and related electronic technologies. The architecture of an IC device exhibits a series of precisely patterned and positioned layers of dielectric and conducting materials stacked upon each other. Lithographic techniques provide the delineation of the areas for doping and/or internal interconnection, the developed resist patterns protecting part of the underlying substrate during subsequent implantation or dry/wet etch processes. Subsequently, the resist is stripped; it thus has only a temporary function and is not an integral part of the device structure. Lithography has to be carried out repeatedly during IC fabrication. Close to 20 or even more lithographic levels have to be delineated, for example, with most advanced dynamic random access memory (DRAM) generations. The quality of the lithographic process employed, being determined by the overall performance of the exposure tool and resist material under optimized process conditions, therefore has a decisive influence on device performance and production yield.

Photolithography has been the workhorse for IC fabrication since the mid-1970s, and it will maintain this dominant position at least during the 1990s. For the majority of IC devices produced, lithographic levels are patterned by employing exposure tools that utilize the mid-UV (MUV) portion of the spectrum of high-pressure Hg lamps. One-to-one projection systems using broadband exposure in combination with refractive mirror optics have been developed as well as refractive optics depending on monochromatic light provided by high-intensity lines of the Hg spectrum, e.g., the so-called "g-line" at 436 nm and the "i-line" at 365 nm. Photolithography options that use exposure tools and resist materials optimized for application at these wavelengths are correspondingly termed g-line or i-line lithography. Projection systems employing MUV light sources dominate production of devices requiring delineation of feature sizes in the range of 1.0–0.5 µm. Deep-UV (DUV) lithography operating at wavelengths around 250 nm entered the scene in the 1980s and is used to create feature sizes down to 0.25 µm.

The alternatives to optical lithography have remained the same for the past few decades: electron beam (e-beam) and X-ray lithography. E-beam lithography operates without masks ("direct writing"), rastering a sharply focused electron beam across the substrate. The high resolution of e-beam lithography is limited by scattering effects: small-angle forward scattering of electrons in resist and large-angle back-scattering from the substrate (requiring "proximity" corrections of radiation intensity). In spite of excellent resolution, low throughput has remained the main drawback of this technique, limiting its application either to low-volume device production or mask fabrication. X-ray lithography offers resolution down to 0.1 µm. By the mid-1990s, it will enter the pilot line production stage. Its full acceptance as the dominant lithography technique for the most advanced device generation continues to be challenged by enhanced optical lithography. Ion lithography, another method, is not considered a candidate for volume IC production. It has been employed, however, successfully in specialized application areas, e.g., mask repair.

Optical and nonoptical lithography options suitable for volume device production are listed in Table 1, together with corresponding information on the exposure systems employed, practical resolution limits, and application areas. A more detailed discussion concerning nonoptical lithography techniques can be found in the literature [1]. This chapter focuses on optical lithography only, with emphasis on material and processing issues. Over the years, resist designers have proposed a wide palette of materials adapted and optimized to meet the specific needs of a particular exposure technique and process mode employed. For some areas of application—as with respect to MUV lithography—the development has been evolutionary. In other fields more challenging to the resist designer, as for example in the area of DUV lithography, a striking variety of materials and corresponding synthesis options have been proposed. It is beyond the scope of this chapter to detail all the major design strategies employed, many of them losing importance to innovations. Instead, our main focus will be on resist materials/techniques that either have resulted in commercial products that are widely used by the industry or that offer promising potential for the future— although not yet widely used. Single-layer processes employing wet-developable resist materials have been and will remain the preferred approach for production

Table 1 Lithography Options for Volume Device Production

Lithography technique	Exposure method	Exposure wavelength	Production resolution	Application area (example)	Comments[a]
Photolithography mid-UV (MUV)	(Hard) contact printing, proximity printing	Mid-UV broadband	≥3 µm	K DRAM	Flood exposure through mask; hard contact printing prone to defects
"	1:1 projection printing	Mid-UV broadband (350–450 nm)	≥1.0 µm	256K	Wafer and mask are scanned across arc-shaped image field; reflective mirror optics
"	Step and repeat projection printing, 10:1 or 5:1 reduction	436 nm (g-line)	>0.6 µm	1M, 4M	$k_1 \geq 0.75$, NA ≤ 0.55
	"	365 nm (i-line)	≥0.35 µm[b]	16M	$k_1 \geq 0.65$, NA ≤ 0.60
deep-UV (DUV)	"	248 nm (KrF)	≥0.25 µm	64M, 256M	$k_1 \geq 0.60$, NA ≤ 0.50
"	Step and scan projection printing	—			—
E-beam lithography	Gaussian beam, shaped beam	Around 1 nm	<100 nm	Standard process for mask fabrication	Direct writing without masks; low throughput
X-ray lithography	Proximity printing, projection printing	0.4–20 nm soft x-rays, e.g., synchroton radiation	→ 100 nm	In pilot line evaluation phase	Main concerns are mask-related: mechanical stability, thermal expansion, distortion (impact on overlay)

[a] k_1 and NA describe the refractive optics of step and repeat; k_1 describes practical resolution in production (theoretical resolution limit at $k_1 = 0.5$; NA is the numerical aperture of exposure system. See text for details.
[b] Using enhancement techniques.

applications from the standpoint of process complexity and costs. However, more sophisticated techniques—such as the application of antireflective layers, and multilevel and top surface imaging systems—have also been developed to alleviate particular problems inherent to optical lithography caused by high substrate topography or high substrate reflectivity. These techniques can be generally applied, irrespective of the exposure wavelength employed, assuming corresponding optimization of the materials involved.

II. OPTICAL LITHOGRAPHY: GENERAL ISSUES

A. Requirements on a Production Process

A critical prerequisite for the successful application of any lithographic technique is adequate resolution capability. Taking a high performance level of the resist system for granted and assuming application of projection printing (the dominant type of exposure technique, discussed in more detail in Sec. II.C), the ultimate resolution in optical lithography is determined by the numerical aperture (NA) of the exposure tool and the wavelength (λ) of the actinic light. The resolvable minimum dimension under production conditions (d_{prod}) and the depth of focus (DOF) can be estimated using the modified Rayleigh equations

$$d_{prod} = \frac{k_1 \lambda}{NA} \tag{1}$$

and

$$DOF = \frac{k_2 \lambda}{(NA)^2} \tag{2}$$

The k factors are conventionally defined factors, depending on resist system and feature type. A k_1 factor of 0.5 corresponds to the theoretical limit of resolution that can be achieved with high-performance resists patterned over flat silicon substrates. Under the conditions of a production process, however, the practical resolution is reduced, corresponding to k_1 factors of 0.6–0.8 (see also Table 1). Applying a shorter wavelength and increasing the NA are feasible ways to enhance the resolution capability, and both strategies have been employed in the industry: g-line lithography (436 nm) was followed by i-line lithography (365 nm), and finally by DUV lithography (focusing on the 248 nm exposure wavelength). Concomitant development in lens design led to a gradual increase of NA (irrespective of the wavelength), eventually furnishing exposure tools with NA values of 0.50–0.60 (compare Table 1). Providing adequate process windows with respect to variations of processing parameters (e.g., wide exposure dose and defocus latitudes) is another essential requirement for any production process. Increasing NA furnishes better resolution; however, it has two counteracting effects on the available process window. It decreases DOF, according to Eq. (2), but it improves dose latitude (a fact rarely mentioned in the literature but highly appreciated in the production environment). The DOF limitation becomes even more stringent in case of severe substrate topography. Substrate topography and high substrate reflectivity are in general the two major obstacles to obtain adequate control of critical dimensions (CD) and process window. Local resist thickness variations over topography induce variations in absorption (bulk effect) and interference of light. The impact of bulk and interference effects on CD control are illustrated in Fig. 2 (a so-called "swing curve") and Fig. 3 (for a resist line patterned across an oxide step). The swing curve amplitude, an indicator for the expected CD variation over topography, increases with rising substrate reflectivity. In case of high substrate reflectivity (rule of thumb: ≥50%), reflections from sloped substrate regions may further result in severe notching or even complete interruption of resist lines. The influence of the interference effect as well as reflective notching can be reduced by application of dyed resists (see Sec. III.A.5) or antireflective layers (see Sec. II.D.5). Multilayer processes (Sec. III.C) and innovative techniques like top surface imaging (Sec. III.D) try to alleviate topography and substrate reflectivity problems simultaneously. Improved substrate planarization (e.g., via chemical mechanical polishing, CMP) has gained importance with the most advanced chip generations. This has brought some relief to lithography and allowed further reduction of practical resolution limits (further decrease of k_1 factor).

The capability of keeping CDs and overlay within defined tolerances dictates (a.o.) whether a particular device design is feasible or not. Adequate control of critical dimensions to about 15–20% (3 sigma) of the design ground rules (GR), as generally required, has to derive from the overall performance of exposure

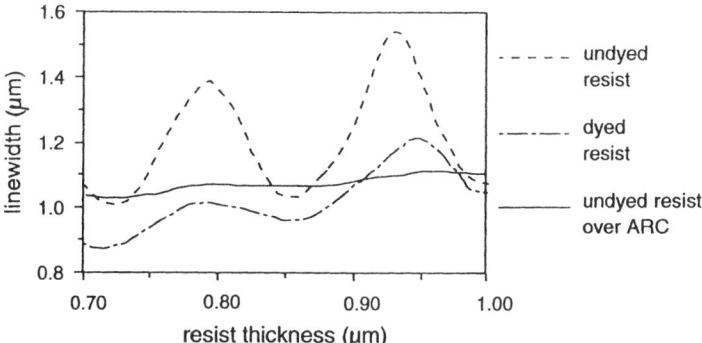

Figure 2 Variation of linewidth as a function of resist thickness (influence of bulk and interference effect).

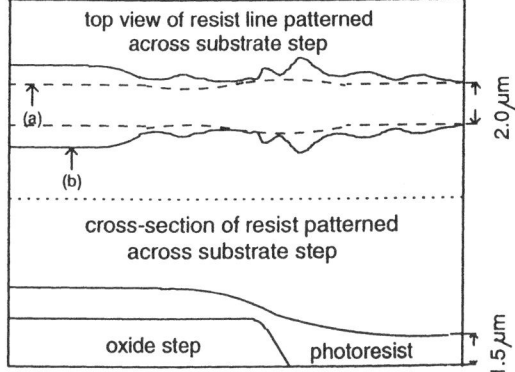

Figure 3 CD variation of resist line patterned over topography: line profile in case of antireflective coating (ARC) application (a) and without ARC (b). (Adapted from Ref. 3.)

equipment and the applied lithographic materials (i.e., primarily resist and developer) under optimized processing conditions. The accurate positioning of individual levels to each other, demanding generally an overlay accuracy ($|x|$ + 3σ) of about 40% of GR, on the other hand, is mainly determined by the quality of the optics and the stage of the exposure tool. Further requirements for resist materials used in production are good adhesion, sufficient thermal stability and etch resistance, strippability, low particle density and low metal ion content in order to prevent degradation of gate oxide quality. Resists and developers, as well as any other material applied, should have reasonable shelf life and good batch-to-batch consistency. Economical issues play—last, but not least—an important role in the decision process regarding a particular lithographic option. Costs for equipment (investment, maintainance), materials, and processing as well as the throughput obtainable with a particular lithographic technique are crucial criteria when considering an application for high-volume production. The trend toward larger wafer sizes is, for example, driven by economic considerations and had a considerable impact on the technological path of exposure tool development, as discussed below.

B. The Interaction Between Light and Resist

The interaction between light and resist, which provides the latent image in the resist prior to development, as well as the interaction between resist and developer converting the latent image into the actual pattern, is of fundamental importance for any lithographic process. The latent image formed after exposure is determined by the type of the applied exposure system (for more details, see

Sec. II.C) and the optical properties of the resist, e.g., refractive index or optical density. The absorption characteristics of a resist material influence the local intensity of light within the exposed resist region, and consequently the concentrations of photoactive compound(s) and photoproduct(s). The chemical nature of the main components of photoresists, as well as the photochemistry of MUV and DUV resist materials, will be described more extensively in Sec. III.

Applying the principles of macroscopic absorption, the dependence of light intensity (I) as a function of the traveling distance of light into the resist (x) can be described using Lambert's law or the Lambert–Beer law. Dill et al. [3] have developed a corresponding mathematical model for conventional nondyed MUV resists, which can also be applied to other resist systems if modified appropriately (in order to simplify the mathematical treatment of the problem, Dill assumed in his model an optically matched substrate. This means that reflections from the substrate do not have to be taken into account):

$$\frac{\partial I}{\partial x} = -I\alpha = -I \sum -a_m \, m_m(x,t) + a_p m_p + a_r m_r + a_s m_s \tag{3}$$

where

α = absorption coefficient
a_m, a_p, a_r, a_s = molar absorption coefficients of nonreacted photoactive compound (M), photoproduct (P), resin (R), and solvent (S), respectively
m_m, m_p, m_r, m_s = corresponding molar concentrations

The variation of the local concentration of nonreacted photoactive compound (PAC)—in the case of conventional MUV resist chemistry also termed ''inhibitor'' because it inhibits dissolution in developer—as a function of time (t) can be expressed by

$$\frac{\partial m_m}{\partial t} = -m_m(x,t) \, I(x,t)C, \qquad C = \text{constant} \tag{4}$$

Using a normalized inhibitor concentration $M = m_m/m_0$, with m_0 being the initial PAC concentration, one can rewrite Eq. (3) and (4) to

$$\frac{\partial I}{\partial x} = -I\alpha = -I(x,t)[AM(x,t) + B] \tag{3'}$$

where $A = (a_m - a_p)m_0$, $B = a_p m_p + a_r m_r + a_s m_s$, and

$$\frac{\partial M(x,t)}{\partial t} = -I(x,t) \, M(x,t)C \tag{4'}$$

Equations (3') and (4') can be solved by numerical integration techniques after the definition of certain boundary conditions. As the portion of nonreacted photoactive compound (*M*) is reduced with increasing exposure dose (*It*) (experimental results confirm the validity of the principle of reciprocity with resist photochemistry, i.e., high light intensity combined with short exposure times is equivalent to low intensity and long exposure times), the absorption coefficient α becomes smaller, as indicated by Eq. (3'). This phenomenon, termed "bleaching," can be seen in Fig. 4, which shows the absorption spectra of a typical MUV resist prior to and after fully completed exposure. The PACs used in conventional MUV resists exhibit suitably high absorption in the wavelength region 330–440 nm, including g- or i-line, whereas the resulting photoproducts are practically nonabsorbing. The so-called Dill parameters *A*, *B*, and *C* appearing in the foregoing equations can be determined from experimental UV spectroscopy data furnishing transmission or absorbance values as a function of exposure dose. *A* and *B* are frequently used in the literature to characterize the absorption behavior of resist materials. With undyed resists, the *B* parameter is mainly determined by the absorption of the resin component. High *B* values reduce the amount of nonabsorbed light arriving at the resist–substrate interface, leading to difficulties in pattern resolution and a deterioration of the wall angle. *B* values ≤ 0.15 are generally required in order to produce almost vertical resist profiles, as demanded for etch processes subsequent to lithography (compare Table 5). This condition can be met relatively easily in MUV lithography. Extending the application of standard MUV resists to the lower DUV wavelength of 248 nm produced, however, inadequate results due to too-high absorption (with optical densities ≥ 0.5).

On the microscopic level, an absorption process involves the absorption of a photon by an atom or molecule, inducing the promotion of an outer electron to

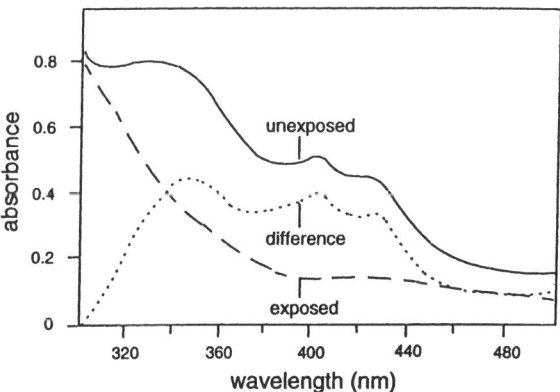

Figure 4 Absorption spectra of nonbleached and bleached MUV photoresist.

a higher level. For diazonaphthoqhinone (DNQ) compounds, the materials of choice for conventional MUV resists, the quantum efficiency remains relatively low—between 0.2 and 0.3. The photosensitivities of MUV resist systems, determined by the properties of resist and developer, have nevertheless been sufficient to furnish acceptable throughput values because the employed mercury illumination sources exhibit relatively high power in the near UV. The power output of these illuminators is, however, distinctly lower in the DUV range. In order to achieve a throughput comparable with that of MUV systems, DUV resists must provide an improvement in photosensitivity of about two orders of magnitude. This technological challenge has led to the introduction of "chemical amplification" (see Sec. III.B.2) as an option for resist design. The overall efficiency of chemically amplified resist is determined by the quantum efficiency of a primary reaction times the chain length of a secondary catalytic reaction.

C. Exposure Techniques/Exposure Tool Options

The evolution of exposure tool development is mainly driven by tightening requirements regarding four key parameters: resolution, overlay, image size, and defect level.

The earliest exposure methods, dominant until the mid-1970s, were contact and proximity printing. These techniques utilize flood exposure of a mask, which is brought in direct contact with or within a close distance of the resist-covered wafer. For these methods, the theoretical resolution limit for equal line and space structures of width d_{min} is given by

$$d_{min} = \frac{3}{3} \lambda \left(s + \frac{1}{2} z \right)^{1/2} \tag{5}$$

where

λ = exposure wavelength
s = gap size between mask and wafer surface
z = resist thickness

Hard contact printing ($s = 0$) provides in principle a resolution capability unmatched by any other optical lithography technique. However, the direct contact between mask and wafer facilitates particle generation, limiting this technique to applications where higher particle counts are still acceptable. Particles cause a gradual degradation in mask quality as well as deviations from flatness of mask and/or wafer, which reduce resolution as well as overlay accuracy. The defect problem can be alleviated by introducing a small gap between mask and wafer surface ($s > 0$), at the expense of resolution (as shown in Eq. (5)). This exposure strategy, termed "proximity" printing, puts high demands on flatness of wafer and masks to obtain a sufficiently constant gap width across the image field. State-of-the art technology allows gaps as small as 10 µm, resulting in resolution down to 3 µm. Proximity printing provides improved overlay accuracy compared with contact printing, because distortions of mask and wafer due to hard contact are avoided.

The simple mode of operation as well as the comparatively low costs have made contact and proximity printing the methods of choice for the fabrication of devices with design rules in the range 3–5 µm. However, the limits of these techniques precluded their general application with further miniaturization, and therefore the needs for improved registration and lower defect densities to achieve the targeted yields were stressed. Thus, by the mid-1970s another exposure technique, namely projection printing, had become dominant in the fabrication of leading-edge devices. The innovative idea behind all variants of this technique— different strategies of projection printing have been developed over the years— was to keep mask and wafer separated while projecting the image of the mask onto the wafer by means of optical systems containing lenses and/or mirrors.

An early, but rather successful approach was scanning projection printing, introduced by Perkin Elmer in 1973 (Micralign systems). Reflecting optics (main elements: two concentric spherical mirrors) are employed with this method. A clever design of the optics (introducing three "folds" in the optical path) allow image projection by simultaneously moving mask and wafer, which are aligned in parallel orientation to each other, through a narrow (about 1 mm wide) annular zone of good correction. In this way, distortions and aberrations of the optical system are minimized. Improvements in optics and scanning stages enabled 1:1 scanning projection printing to delineate feature sizes down to 1.3 µm when employing UV light (350–400 nm), while fabrication of devices with 1.0 µm design rules became feasible when making use of a deep-UV light source.

The fabrication of defect-free masks to be applied at 1:1 magnification becomes, however, very difficult if patterning of submicron structures is required. This problem of 1:1 scanning projection printing was further aggravated with the industry's trend toward larger wafer sizes (from 100 to 150 mm diameter in the mid-1980s, to 200 mm wafer in the late 1980s), which necessitated a corresponding increase of mask sizes. The technological challenges arising from the defect and field size issues drove the development of an alternative exposure strategy, i.e., "step and repeat" projection printing. With this technique, refractive optics with high-resolution lenses are used to project the image of the mask onto the wafer. State-of-the art lens systems providing exposure fields with diameters up to 31 mm (as of 1994) allow the simultaneous patterning of only a few chips (in most cases, 2 or 3) in a single exposure. Hence, in order to expose the whole wafer, it is moved under the projection optics in a stepwise manner by means of a high-precision mechanical stage. Exposure tools operating in this way, termed "steppers," have become the equipment of choice for patterning in the submicron regime. The majority of steppers in use exposure wavelengths at 436 nm (g-line lithography) or 365 nm (i-line lithography). Steppers using deep-UV light at 248 nm entered the scene in the 1980s, favored by the development of KrF excimer laser technology, which offered higher illumination power than Hg lamp sources.

Most steppers are equipped with reduction optics projecting the image of the mask (termed "reticle" in order to distinguish it from a full-field mask) with reduction ratios of 10:1 or 5:1 onto the wafer. In 1994, 2:1 or 2.5:1 reduction i-line steppers with relatively low NAs but very large imaging fields (>40 × 40 mm^2) appeared on the market. These tools were designed with the intention of providing high throughputs for the patterning of less critical levels. Reduction optics diminishes the defect issue drastically. The task of producing a defect-free reticle at economically tolerable yield becomes much easier as the size of the smallest printable defect shrinks. Mask specifications (e.g., CD control or registration on the mask) are also distinctly relaxed when applying reduction optics.

The industry's demand for further enlarged field sizes, as required for leading-edge devices of the 1990s, e.g., 64M or 256M DRAM chips, has become a problem even for step and repeat systems, because under typical production conditions the imaging field of the employed exposure tools should contain not only one chip but at least two chips in order to allow die-to-die comparison regarding defects. This has led to the development of so-called "step and scan" systems, which combine step and repeat systems with scanning technology. Such tools, equipped with reflective optics and deep-UV light sources ($\approx$250 nm), were introduced to manufacturing in the early 1990s. Employing NAs of 0.35—-0.50, 0.40–0.25 µm design rules have been resolved. The optical fields of step and scan systems are defined by the length of the scanning slit and the maximum scan width. A Micrascan II system of SVGL (1993)—operating at 4:1 reduction--provides, for example, an image field of 22 × 32.5 mm. A further expansion to 22 × 50 mm is in principle feasible once larger reticle sizes are used. One of the major technical challenges for the design of this exposure tool was to assure a very precise synchronization of reticle and wafer stage movements at different scan speeds (due to the 4:1 reduction employed).

D. Processing Issues

1. Process Flow for Conventional Single-Layer Lithography

The general description of the process flow for conventional single-layer lithography is reviewed below and summarized in Table 2, followed by a more detailed discussion with respect to some of the individual process steps.

The first three process steps mentioned in Table 2—surface cleaning, dehydration bake, and wafer "priming"—are to provide a clean substrate surface to

Table 2 Process Flow for Conventional Single-Layer Lithography

Process step	Comments
Dehydration bake	Removal of adsorbed water
Wafer priming (e.g., using HMDS)	Prevents lift-off or undercut during development or wet etch process
Resist coating	
Post-apply bake (prebake)	Reduction of solvent content to low and uniform level
Reticle/wafer alignment	In exposure tool
Exposure	See Sec. II.C
Post-exposure bake (PEB)	Prevents standing waves in resist profile
Development	
Postbake (hard-bake)	Further reduction of solvent content, improvement of thermal stability
Resist hardening (optional)	Enhancement of thermal stability (if required)

make it chemically compatible for resist application. Spin-coating has been established as the primary method to produce highly uniform, almost defect-free resist films on the wafer. A post-apply bake (PAB), also termed "prebake" or "soft-bake" in the literature, establishes a uniform level of residual solvent prior to exposure and reduces stress introduced into the polymer film due to shear forces operating during the spin-coating process. For the majority of resists, the prebake is carried out at temperatures between 90 and 110°C. Alignment of the mask to the wafer and subsequent irradiation of the resist-coated wafers take place in the exposure tool. The purpose of the subsequent "post-exposure bake" (PEB) depends on which resist system is applied. With conventional MUV resists it reduces the standing wave effect (interference between incoming light and light reflected from the substrate), which otherwise would lead to rippled resist profiles. The PEB—generally done at temperatures which are 15–20°C higher than the prebake temperature—induces diffusion of the photoproduct by about 50 nm, which flattens the local minima/maxima in the photoproduct concentration along the direction perpendicular to the substrate. The PEB is optional, but it has become practically a standard procedure if resolution of submicron patterns is desired. When chemically amplified resist systems are applied, the PEB is primarily to enhance the catalytical secondary reaction and drive it to a well-defined endpoint. The next step after the PEB is resist development. Immersion, puddle, or spray processes may be applied. A further anneal step, called "postbake" or "hard-bake" removes the major part of the solvent prior to subsequent processing such as etching (see Chapter 7) or implantation. In addition, it improves the adhesion to the substrate. Postbake temperatures are kept a few degrees below the temperature that induces profile degradation and resist flow. Prior to particularly demanding etch or implantation processes, an additional procedure, termed "resist hardening," may be necessary to improve the thermal stability inherent to the developed and hard-baked resist. The preferred means for hardening is irradiation with deep-UV light, inducing crosslinking in the resist, frequently combined with a simultaneous temperature ramp (generally to 140–180°C).

Although batch processing and the use of stand-alone tools for the individual process steps of single-layer lithography is feasible in principle, sequential processing of single wafers is most commonly used. In this case, so-called "wafer tracks" are employed; these consist of modules dedicated to wafer priming, bake steps (on hot-plates), resist coating, and development. The highest level of automation and best reproducibility of processing can be achieved by interfacing wafer track and exposure tool with each other to form a "photocluster."

2. Substrate Preparation

Contamination of wafers by deposition of thin films or of organic or inorganic particulates can occur in many ways during the complex interaction with the environment, processing agents, machinery, and operating personnel. A few examples are airborne contamination, residues of resist, remnants of polymers created during etch processes, oxide films/particles as a result of the reaction with ambient air or residual oxygen in etch chambers, oily films from pump fluids or lubricants, sweat, dandruff, or surface-absorbed water (see also Chapter 2). For those cases where cleaning procedures prior to lithography cannot be avoided, several options are available. Cleaning can be carried out, for example, in aqueous solutions of acids (like HF, HNO_3, H_2SO_4/dichromate) or by using organic solvents like N-methylpyrrolidone. In other cases, high-temperature annealing, plasma etching with O_2 or fluorine-containing gases, or aerosol dry cleaning may be successful. Sometimes a combination of different cleaning steps becomes necessary.

Besides cleanliness, chemical compatibility of the wafer surface to resist coating is also required. Thus, a process called "wafer priming" is generally carried out prior to the coating step, to modify the substrate surface properties so as to prevent lift-off of resist or undercut along the resist/substrate interface, which might otherwise occur during development or wet etching. Wafer priming does not, however, increase the mechanical adhesion between resist and substrate [4] as is frequently assumed. The most commonly used primer is hexamethyldisilazane, HMDS, but other compounds with similar chemical structures like trimethylsilyldiethylamine, TMSDEA, are also viable options. The reaction of HMDS at a Si wafer surface converts polar hydroxyl groups into nonpolar trimethylsilyl groups. A certain minimum coverage of $(CH_3)_3Si$ is required in order to successfully block the penetration of polar liquids along the interface. Trimethylsilyl groups can be attacked by alkaline developers, but the replacement reaction ceases for coverages >80%. Problems may also occur in case of very high $(CH_3)_3Si$ coverage, which may weaken the attractive forces between the primed substrate and resist to such an extent that dewetting of the resist, i.e., a pull-back of the resist after coating, occurs. Overpriming may furthermore facilitate blistering ("popping") in conventional mid-UV resists if the application of relatively high exposure doses leads to a sufficiently high concentration of nitrogen along the resist/wafer interface, which is formed during the photochemical reaction occurring in MUV resists (described in more detail in Sec. III.A.1). All in all, the range of suitable $(CH_3)_3Si$ coverage is rather narrow. The $(CH_3)_3Si$ coverage can be monitored indirectly by determining the hydrophobicity of the substrate via contact-angle measurements of water droplets placed on the wafer. Contact angles in the range 65–75° are generally acceptable. For some resists, however, the process window is even smaller.

3. Resist Coating

Spin-coating is widely accepted as the best method to provide optimum film thickness uniformity. The resist has to be applied in a liquid state, i.e., dissolved in an organic solvent or solvent mixture. Suitable solvents are listed in Table 3. The solvent content of resists varies generally between 65 and 75%. Commercial vendors offer most of their products with different viscosities targeted to cover certain thickness ranges suitable to the customers. For most applications, thicknesses of 0.5–3.0 µm are required, the thickness target depending a.o. on wafer topography, or the conditions of the subsequent etch or implantation process. Spin-coating consumes a few milliliters of resist solution, which is poured onto the static or slowly rotating substrate, per single wafer. Distribution of the material across the whole wafer is achieved during a short acceleration step. A subsequent step at constant spin speed ensures uniform thinning of the polymer film to the specified thickness and the removal of the majority of the solvent. For common wafer sizes of 100–200 mm diameter, good film uniformity is generally obtained for spin speeds of 3000–5000 rpm. Thickness homogeneity across a wafer as well as from wafer to wafer must be typically ±5 nm or better in order to maintain the required CD control. With state of-the-art equipment, a 3-sigma thickness uniformity of 3 nm (even for 200 mm diameter wafers) can be achieved across a wafer. Excess resist on the edges and the wafer backside will cause cross-contamination of other equipment, and it must therefore be removed. Such contamination residuals may prevent, for example, accurate contact between the wafer and the chuck of the exposure tool, resulting in reduced registration and/or loss in focus latitude. The removal is done in a step called "edge bead" removal, generally via dissolution in an organic solvent.

Table 3 Solvents for Photoresists

Name	Abbrev.	Chemical formula	MW	Density (g/cm^3)	Boiling point (°C)	Surface tension (dyne/cm)
Ethyleneglycolmonoethyletheracetate (Ethoxyethylacetate)	EGMEA	C_2H_5-O-C_2H_4-O-CO-CH_3	132.1	0.98	156	28
Butylacetate		C_4H_9-O-CO-CH_3	116.2	0.88	127	25
Xylene		CH_3-C_6H_4-CH_3	106.2	0.88	144	28
Propylenglycolmonoethyletheracetate	PGMEA	CH_3-O-C_3H_6-O-CO-CH_3	132.2	0.97	146	27
Methyl-3-methoxyproprionate	MMP	CH_3-O-C_2H_4-CO-O-CH_3	118.1	1.01	143	30
Ethylpyruvate	EP	CH_3-CO-CO-O-C_2H_5	116.1	1.04	156	31
2-Heptanone		CH_3-CO-C_5H_{11}	114.1	0.81	150	30
Ethyllactate	EL	CH_3-CHOH-CO-O-C_2H_5	118.1	1.03	154	29

4. Resist Development

The present discussion will be restricted to wet-development using aqueous alkaline solutions, which are preferred for the development of conventional positive tone resists or of chemically amplified positive or negative tone resists. Three modes of development are feasible: immersion, spray, or puddle. Immersion development is generally used in production to process a whole batch of wafers simultaneously, whereas single wafers are processed sequentially with spray and puddle development. The latter two operation modes offer better reproducibility and allow, for example, the continuous use of fresh developer.

They also generate better CD uniformity and are more economical due to their lower developer consumption.

Development is a rather complex process, with a strong impact on the overall performance of the applied resist system, influencing, e.g., resolution, photospeed, and process latitudes. Optimization of the process includes, for example, a proper choice of chemistry and concentration of developer agent and surfactant (if included, to improve wetting uniformity), determination of the optimum temperature (to be kept stable within ±0.5°C or better), and accurate positioning and adjustment of the developer nozzle (especially important in case of spray development). In order to achieve acceptable throughputs, development times are generally kept in a range of 30–90 s. After completion of the actual development step, remnants of developer are removed by a water-rinse step. Finally, the wafer is dried by spinning. Any residual water will evaporate during the subsequent postbake.

Two types of aqueous-alkaline developers are normally used in practice: (1) metal hydroxides (typically NaOH or KOH) with or without inorganic buffers, or (2) quarterny alkyl ammonium hydroxides such as tetramethyl ammonium hydroxide, TMAH, or choline (trimethyl (2-hydroxyethyl) ammonium hydroxide). Alkali hydroxide solutions are inexpensive, nonvolatile, and easy to purify, store, and dispose. However, their metal ion content, with its potential for device degradation, has become a major concern. As a consequence, the application of metal-ion-free (but more costly) quaterny ammonium hydroxide developers (predominantly TMAH) has become the standard for the fabrication of more demanding products, such as 4-Mbit and follow-on DRAM generations.

Advanced MUV photoresist materials exhibit extremely low dissolution rates in unexposed resist regions, allowing the application of more aggressive developers without loss in performance. Consequently, a gradual switch to higher developer concentrations (from 0.21–0.23 to 0.26–0.27 normality) has occurred as a means to reduce the required exposure time and to enhance the exposure tool throughput.

5. Application of Antireflective Layers

Any kind of thickness change in a thin-film stack penetrated by light will cause modifications in light interference. For photolithography applications, such a film stack might consist of resist and one or more underlying (transparent) substrate layers, like oxide or nitride. Local variations of light coupling into the resist will lead to local variations with respect to the required exposure dose to achieve a certain CD. However, as the whole chip is exposed uniformly with one particular dose, interference-induced CD variations are inevitable. A simple model by Brunner [5], derived from classical optical interference equations, allows insight into process options to suppress or at least alleviate this detrimental effect. Established was the following relationship between the exposure dose required to develop through a resist of given thickness (D), the reflectance of the air/resist interface (R_1) and the reflectance of the resist/substrate interface (R_2):

$$S = 4(R_1R_2)^{1/2}e^{-\alpha D} \tag{6}$$

The swing curve ratio (S) of the exposure dose versus resist thickness swing curve is the peak-to-valley difference in intensity divided by the average intensity required for exposure. The quantity α is the absorption coefficient of the resist. (Increasing α by incorporation of dye is one option to reduce the interference effect, as discussed in Sec. III.A.5.)

Reduction of S (correlating with improved CD control) can be achieved by lowering R_1 or R_2. Two different types of antireflective layers (ARLs) have been employed to obtain this goal:

1. Bottom antireflective layers (BARL). BARLs are either organic polymers with strongly absorbing dyes or inorganic dielectrics reducing R_2 via interference and/or absorption. They are applied prior to resist coating. The BARLs can be removed by wet processing or dry etching; in some cases, inorganic BARLs may be left as an integral part of the device.
2. Top antireflective layers (TARL), also termed TARC or ARCOR ("antireflective coating on resist") in the literature. TARLs are transparent films coated on top of the resist in order to reduce R_1 via interference. They can be removed either before or during resist development.

Both BARL and TARL techniques help to alleviate interference-induced CD variations, but only BARLs (since they are applied on the substrate) can prevent reflections from the substrate, which are capable of causing resist line notching. Despite this advantage of BARLs, TARLs have gained increased attention during the past years, because interference-dependent linewidth variations themselves have already become increasingly troublesome due to the tightened tolerances required for feature sizes below 0.5 µm. The advantage of TARLs compared with BARLs also lies in its easier integration into the overall fabrication process.

Optimum efficiency of both BARL and TARL materials operating via interference is only obtained if certain conditions with respect to refractive index (n) and ARL thickness (d) are met. The corresponding criteria for nonabsorbing materials and vertical incidence of light are shown in Eqs. (7)–(10) (the much more complicated, generally valid formula—taking also absorption into account—may be derived from the literature [6]):

$$n_{BARL} = (n_{substrate}\, n_{resist})^{1/2} \quad (7)$$

$$n_{TARL} = (n_{air}\, n_{resist})^{1/2} \quad (8)$$

$$d_{BARL} \frac{\lambda}{4} n_{BARL} \quad (9)$$

$$d_{TARL} = \frac{\lambda}{4} n_{TARL} \quad (10)$$

Examples for antireflective materials and their application areas are as follows:

Inorganic BARL Materials. Amorphous silicon (a-Si) deposited by PECVD has been used successfully as an antireflective layer over highly reflecting aluminium substrates (i.e., cold-sputtered Al films have reflectivities >80%). This material operates via destructive interference, its nominal thickness (a few nm) depending on the exposure wavelength exployed. Nonstochiometric silicon nitride compounds (SiN_x) have been mentioned in the literature as suitable ARLs over doped Si or metal silicide substrates (see Fig. 5). Note that in Fig. 5 the substrate reflectivity increases for deviations from vertical incidence of light, a situation occurring over sloped substrate regions. TiN is also used over Al and metal silicides. Inorganic BARLs are difficult to remove. In many cases they become part of the device architecture. Their applicability is limited to certain BARL/substrate combinations.

Organic BARL Materials. Organic BARLs operating via absorption, on the other hand, offer general applicability for all levels. Dry- and wet-developable materials are commercially available, but their chemistry generally is considered

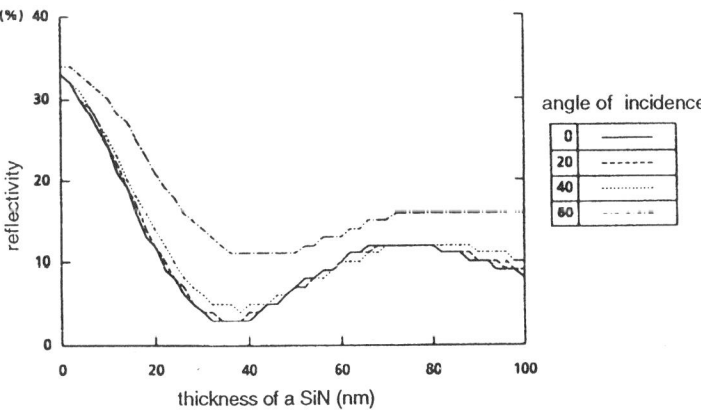

Figure 5 Reflectivity of a-SiN on molybdenum silicide into resist. (From Ref. 7.)

as proprietary. Wet-developable materials suffered in the past from a narrow baking temperature range and resist scumming after development, but these problems have been alleviated by further product optimization. Interfacial layer formation between the resist and the organic ARL, however, still often induces various degrees of pronounced resist footing, the extent of which depends on the nature of the resist solvent. Using organic BARL products optimized for O_2 plasma etching, the lateral erosion of resist profiles during the BARL etching may become an issue if enhanced overetch is required to guarantee complete BARL removal from the topography regions in which BARL thicknesses are distinctly thicker than the nominal value (a result of BARL thickness variation over severe topography).

Significant improvements in process window and/or CD control have been reported in the literature when applying organic BARLs for i-line or DUV lithography. For example, steeper resist profiles and better process latitudes have been observed in MUV lithography compared with dyed resist applications, although this is achieved at the cost of increased process complexity.

TARL Materials. The application of the TARL concept to lithography has gained increased attention with several publications in 1990 and 1991 [5,8]. A variety of materials like poly(vinylalcohol), poly(dimethylsiloxane), or perfluoroalkylpolyethers has been investigated. With regard to i-line application, only fluorinated materials have exhibited refractive indices close to the ideal value of about 1.30. Simulations have shown that TARLs remain fairly effective as long as the deviations from the ideal refractive index value is within ±0.10. Low-viscosity solutions have to be applied when spin-coating TARL materials, in order to obtain the required thin nominal thickness (e.g., 70 nm for i-line application). With respect to the thickness criterion, TARL thickness variations of ±20% over topography can still be tolerated without significant performance loss, as indicated by simulation work. This is, however, no longer true if deviations from the optimum thickness become greater than ±50%, indicating that the degree of substrate topography should be moderate in order to obtain optimum benefits from TARLs. Issues of TARL application to be further addressed are, for example, intermixing between TARL and resist or insufficient adhesion of the thin TARL films.

III. RESIST SYSTEMS IN OPTICAL LITHOGRAPHY
A. Positive Tone MUV Resists

Negative tone resists based on the insolubilization of the exposed areas via a photoinitiator-induced cross-linking reaction (e.g., cyclized poly(*cis*-isoprene) resins with bisazide cross-linkers) were widely used during the early stages of MUV photolithography. However, these materials suffered from swelling and

Photolithography 219

concomitant pattern distortion occurring during the development in organic solvents or solvent mixtures. For applications demanding production resolution below 3 μm, they have been completely replaced by aqueous-base-soluble positive tone resist systems. These materials are not affected by swelling; they offer good etch resistance and, moreover, enhanced contrast. Tenacious R&D efforts resulting in continuous material improvements, the switch from g- to i-line exposure, and concomitant enhancement of exposure tools have pushed the production resolution of conventional MUV lithography (without application of enhancement techniques) to at least 0.5 μm by 1994, a goal considered unreachable just a few years earlier. These efforts have centered around the development and optimization of one particular type of material, namely diazonaphthoquinone (DNQ)/novolak resins.

1. The Reaction Mechanism of DNQ/Novolak Resist Systems

All commercial g- and i-line positive tone resists operate via the same reaction mechanism as described in Fig. 6. They rely on a distinct discrimination of the dissolution rate between exposed and unexposed resist areas. The PAC compounds with photosensitive 1,2-diazonaphthoquinone-5-sulfonic acid ester groups (1,2-diazonaphthoquinone-4-sulfonic acid derivates had been used with

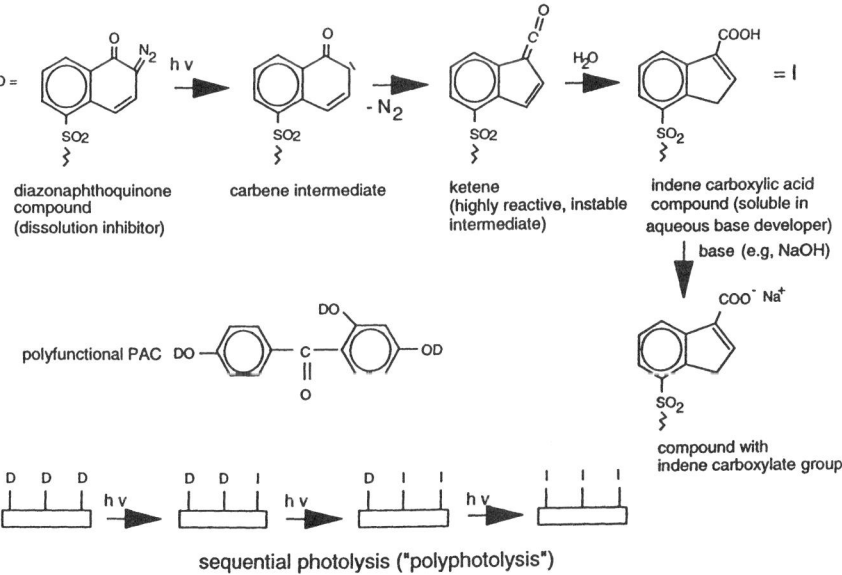

Figure 6 Chemistry of DNA/novolak positive tone resists.

some of the first-generation i-line resists) fulfill two different functions in the exposed and unexposed film regions. Being hydrophobic in the nonreacted state, they act as dissolution inhibitors in unexposed resist areas by protecting the phenolic hydroxyl groups against developer attack. Upon exposure with MUV light, however, they are converted to indenecarboxylic acid (ICA) compounds, which promote dissolution. The carbene intermediate formed by photolysis is converted to highly reactive ketene via a Wolff-type rearrangement reaction (Suess reaction). In the presence of water (the H_2O concentration in photoresists is around 0.1–0.3%) ketene reacts finally to form ICA. As an overall result, the dissolution rate of the phenolate/carboxylate-containing exposed resist regions is raised by 2–3 orders of magnitude compared with the unexposed resist areas, thus providing the high-contrast performance found with DNQ/novolak resists.

2. The Main Components of Positive Tone MUV Resists

The three main components of a positive MUV resist are a novolak resin (binder), a photoactive compound (PAC) containing diazonaphthoquinone (DNQ) chromophore groups, and an organic solvent. In addition, the resist may contain small amount of surfactant, stabilizer, or, for particular applications, a dye/dye mixture.

The Novolak Resin. Novolaks, the first wholly synthetic polymer material produced by the industry (around 1910), are generated via an acid-catalyzed condensation of phenols with formaldehyde. For resist applications, cresols are used as the phenolic component. Oxalic acid is the most frequently applied catalyst. Three different isomers of cresol exist: ortho- (o-), para- (p-), and meta- (m-) cresol. The specific isomeric structure of the cresol compounds used for novolak synthesis determines the possibilities for methylene bond formation (see Fig. 7) and their reaction rate. The polymerization rate for meta-cresol is much higher than that of ortho- or para-cresols. Only two out of three unsubstituted positions on the cresol ring are accessible to substitution. Whether methylene bridges are preferentially formed at the o- or p-position, the phenolic functionality is influenced by the acidic conditions as well as by the choice of reaction solvent.

Different secondary polymer structures are formed, depending on the nature of the methylene bridge bonding between cresol rings. O–o', p–p', or o–p' coupling is possible. The secondary structure determines the spatial configuration of phenolic OH groups within the resist. P-cresol-based novolaks, for example, form chains of high regularity, as methylene bonds are only possible on o–o' positions [9]. This resin type contains mainly intramolecular hydrogen bonds, as shown in Fig. 8a for a p-cresol tetramer. O–p' coupling of m-cresol units, on the other hand, leads to polymer structures with widely separated and outwardly directed phenolic functionalities (compare Fig. 8b). In this case, intermolecular H-bonding is distinctly favored, the stronger bonding between resin strains leading to a rise in glass transition temperature. Polyvinylphenol, popular for deep-

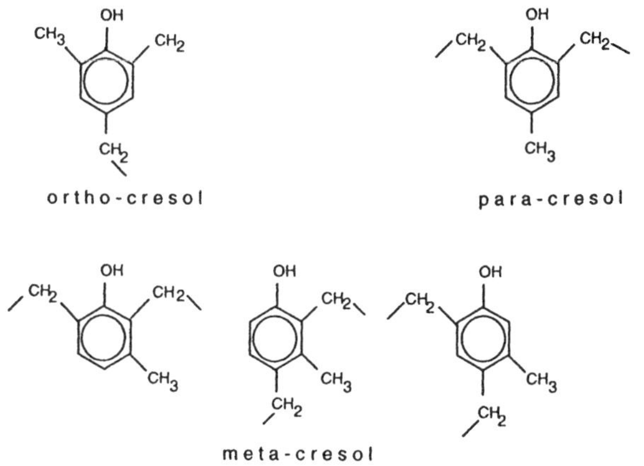

Photolithography

[Figure 7 structures: ortho-cresol, para-cresol, meta-cresol]

Figure 7 Possible methylene bond positions for cresol isomeres.

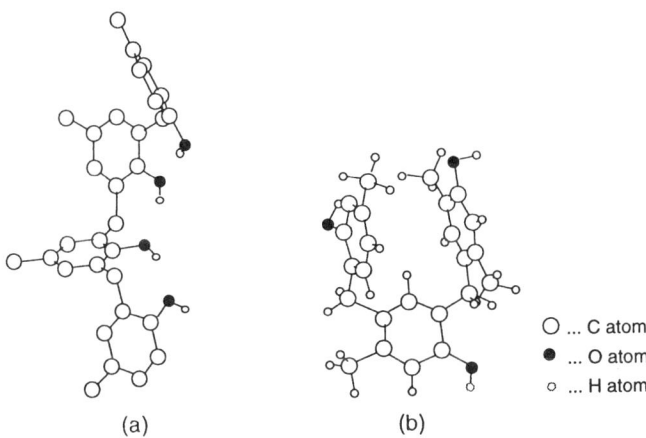

Figure 8 Influence of secondary structure of novolak on nature of H-bonding; (a) o–o'-coupled p-cresol novolak tetramer, intramolecular H-bonding favored; (b) o–p'-coupled m-cresol novolak trimer, intermolecular H-bonding favored. (Adapted from Ref. 9.)

UV lithography applications (see Sec. III.B), behaves in a similar way as o–p' coupled novolaks. Novolaks with predominantly intermolecular H-bonding exhibit much faster dissolution rates in TMAH developers compared with those with strongly intramolecular H-bonding. The nature of the hydrogen bonding has also a distinct influence on the interaction with the photoactive compound.

The Photoactive Compound. As already indicated in Fig. 6, MUV resists contain polyfunctional PACs consisting of polyhydroxy compounds esterified by 1,2-diazonaphthoquinone-5-sulfonyl groups. Benzophenone-type molecules have frequently been applied as backbone (ballast) structure. This particular kind of chemistry has been employed since the early stages of g-line resist synthesis, and the reasoning behind it was described in detail by Trefonas and Daniels [10]. Diazonaphthoquinone (DNQ) groups on a polyfunctional ballast compound undergo sequential photolysis, termed "polyphotolysis" by Trefonas, resulting in the superposition of latent images of different photoproducts with varying numbers of reacted DNQ groups. The final DNQ-free poly-ICA product, as shown in Fig. 9 for a tri-DNQ PAC, becomes dominant after the exposure dose passes a certain threshold level. Under such exposure conditions, the latent image of the final photoproduct will exhibit a distinctly improved modulation contrast compared with the modulation transfer function (MTF) of the aerial image (see Fig. 10). Optimum exploitation of the principle of polyphotolysis, leading to a highly nonlinear relationship between exposure dose and dissolution rate, can be achieved if the resist system is optimized to furnish a high dissolution selectivity toward the final photoproduct.

The PAC structure in itself has a negligible influence on photobleaching, but a major one on dissolution behavior. The inhibition efficiency of the PAC (its maximum value is limited by the PAC solubility in the novolak/solvent mixture)

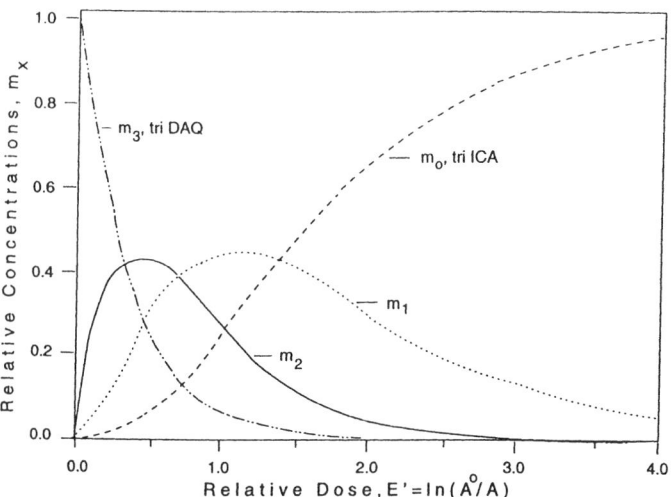

Figure 9 Polyphotolysis of tri-DNQ PAC: normalized concentrations of photoproducts (m_2 to m_0) produced by irradiation of tri-DNQ PAC (m_3) as a function of energy, E'. A^0, initial absorbance; A, absorbance after irradiation with dose E'. (From Ref. 10.)

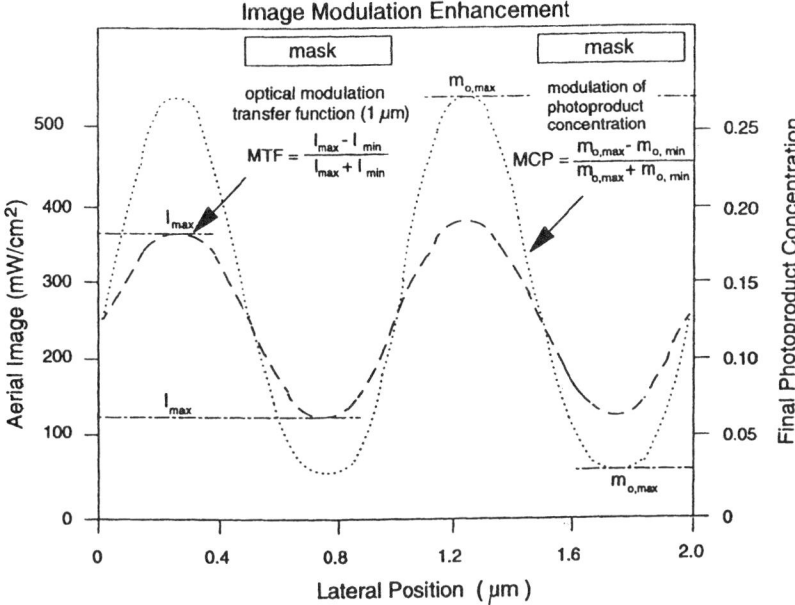

Figure 10 Modulation of concentration of Tri-ICA produced from tri-DNQ by an aerial image of modulation 0.5. (From Ref. 10.)

rises with increasing hydrophobicity. Distinctly higher hydrophobicity has been reported in the literature. For example, this occurs when substituting benzophenone-type ballast groups with similar compounds in which bulky and less polar– $(CH_3)_2C-$ or $-Ph(CH_3)CH-$ groups replace the carbonyl functionality. On the other hand, a high number of DNQs per backbone structure does not necessarily lead to enhanced inhibition. Apparently, in this case steric constraints between adjacent DNQs—the number of positions available for substitution is limited—prevent an optimum interaction between the DNQs and the novolak. As seen from Fig. 11, a benzophenone PAC substituted on three adjacent (2,3,4) positions is clearly a less effective dissolution inhibitor than the (2,4,4') compound having the three DNQ groups distributed along two aromatic rings. The steric hindrance becomes even more adverse for DNQ numbers ≥4 per ballast group.

The Solvent Component. The choice of a proper solvent is a crucial issue. Sufficient solubility of the individual solid components must be guaranteed over a reasonable shelf-life period as well as under stressed operating conditions, such as under pump pressure during coating. Moderately polar solvents or solvent mixtures (see Table 3), exhibiting generally carbonyl, ester, or ether functionalities, seem to offer the best solubility characteristics. The first generations

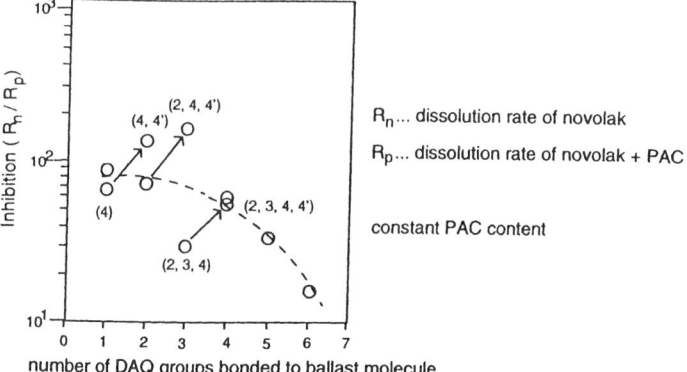

Figure 11 Inhibition versus functionality of benzophenone PACs. (From Ref. 29.)

of g-line resists typically contained a solvent mixture of the first three compounds mentioned at the top of Table 3, with EGMEA as the main component. Single-solvent systems have recently become more prevalent in order to maintain more reproducible evaporation behavior during bake steps. Moreover, rising concerns in the industry with regard to the use of EGMEA—due to its linkage to teratrogenic effects—have led to a replacement of EGMEA by so-called "safe solvents" (in reality, safer solvents) such as the last five compounds listed in Table 3. Essentially all current i-line resists and most of the g-line resists are based on safe-solvent systems.

3. Chemical Interactions

Novolak–Photoactive Compound. Different models have been presented in the literature to explain the dissolution inhibition that results from the interaction between DNQ and the matrix resin. Cross-linking via azocoupling under the influence of the base developer, or azoxy bond formation have been suggested, for example, as reasons for the reduced dissolution rate in unexposed resist areas. Alternative or complementary explanations have been given based on infrared (IR) spectroscopy results: Changes of the peak intensities at 2159 and 2118 cm^{-1} (FTIR measurements) have been observed when comparing pure DNQ compounds with DNQ/novolak mixtures. On the other hand, it has been noted that a blue shift in the O–H stretching bond frequency (region around 3400–3170 cm^{-1}) of novolaks rich in o–o' coupling vanishes upon DNQ addition. The blue shift has been interpreted as an indicator of strong intramolecular H-bonding in the resin that is then obviously disrupted in the presence of PAC, most likely as the result of an interaction between the carbonyl and diazo groups of the PAC with adjacent HO groups of the novolak. Bonding of this type should lead to an effective masking of the affected hydroxyl groups of the novolak against developer attack. The resin regions with low hydroxyl group content, as found in o–o' coupled novolaks (compare Fig. 8a), are strongly hydrophobic anyway, making access of hydrophilic moieties to the resin strain extremely difficult. Moreover, enhanced interaction between hydrophobic (naphthalene portion) parts of PAC molecules and hydrophobic novolak regions would be facilitated. All these effects may contribute to the high efficiency of dissolution inhibition observed with novolaks rich in o–o' bonding sequences.

Resist–Developer. The dissolution of photoresists in aqueous base developers is a surface-limited process involving the repetition of different reaction steps like diffusion of the base to the phenolic hydroxyl groups, ionization of the phenolic protons, stabilization of the resulting monoanions by hydrogen bonds, diffusion of cations to form cation/phenolate pairs, and so on. Once a critical stage of deprotonation and phenolate ion solvation has been reached, the polymer structure starts to unfold quickly. Lower-molecular-weight resist strains are dissolved distinctly faster, thus creating increased surface area for the developer attack on higher-molecular-weight components ("stone wall" model by Hanabata et al. [12]).

The acid–base reactions occurring at the resist surface are very fast processes, making the diffusion of developer components (OH^- ions and their countercations) the rate-limiting step. Under developing conditions with equal hydroxyl ion and cation concentrations (no salts added), the rate limitation seems to come in many cases from the rate of cation diffusion directly at the interface to the polymer. Large dissolution rate differences were observed that correlated to the rate of the unsolvated ions, indicating that the hydration shell of the cations is left behind when penetrating into the boundary zone between polymer and developer.

4. Resist Design

Some design trends, e.g., an increase in the p-novolak content or the incorporation of resin components with strong o–o' coupling, have been mentioned in the literature as means to arrive at improved performance of MUV resist, the details being proprietary knowledge of the resist vendor companies. The key for successful resist design seems to be that of finding a well-balanced mixture of resin and PAC components that exhibits optimum inhibition and dissolution characteristics. Trade-offs have to be accepted in many cases when trying to optimize one particular resist parameter to its extreme. Optimizing molecular weight distribution and resin structure in order to gain very high thermal stability, for example, cannot be obtained without performance losses in resolution and/or photospeed. Some of the trade-offs encountered are listed in Table 4. Resist materials widely accepted in the industry offer generally very good performance with respect to one or two criteria of interest (e.g., resolution capa-

Table 4 Effect of Five Aspects of Novolak Resin Design on Dissolution Rates and Resist Performance

	Dissolution rate in					
	Unexposed region, R_p	Exposed region, R_p	Ratio[a] R_p/R_o	Resolution	Sensitivity	Thermal stability
Molecular weight MW ↑	↓ +	↓ −	→	→	↓ −	+ ↑
Molecular weight distribution M_w/M_n ↑	↘↗ − +	+ ↑	+ ↗↘ −	+ ↗↘ −	+ ↑	+ ↗↘ −
Isomeric structure para/meta ratio ↑	↓ +	↓ −	+ ↑	+ ↑	↓ −	+ ↑
Methylene bond position o–o' bonding ↑	↓ +	→	+ ↑	+ ↑	→	↓ −
DNQ/novolak ratio ↑	↓ +	+ ↗↘ −	+ ↑	+ ↑	+ ↗↘ −	+ ↑

[a]Ratio R_p/R_o correlated to contrast of resist system.

↑ increase   ↗↘ increase, then decrease   − deterioration
↓ decrease   ↘↗ decrease, then increase   + improvement
→ no change

Source: Adapted from Ref. 12.

bility, profile quality, process latitudes, or photospeed) while exhibiting no severe weakness.

5. Performance Data of MUV Resists

MUV Resists Without Dye. Data of the performance range of commercially available state-of-the-art MUV resists without dye content, used in combination with advanced g- or i-line steppers, are listed in Table 5. Building on the experience gained in the area of g-line lithography, more recent resist design work has focused on the continuous improvement of i-line resists, as indicated by the corresponding data in Table 5 for the years 1989–1991 vs. 1992–1993. Most of the i-line materials surpass their best g-line competitors in terms of resolution capability, as expected from the Rayleigh criterion. The main advantage of i-

Table 5 Performance Data for Nondyed MUV Resists

	g-Line resists	i-Line resists	
	1.2–1.3 μm NA = 0.54/0.55 1988–1990	1.1–1.3 μm NA = 0.52/0.54 1989–1991	1.0–1.1 μm NA = 0.52/0.54 1992–1993
Limit of resolution	0.38–0.45 μm	0.38–0.45 μm	0.34–0.38 μm
Wall angle	87–88°C	86–89°C	87–89°C
Photospeed	150–250 mJ/cm^2	150–250 mJ/cm^2	120–200 mJ/cm^2
Dose latitude[a] for			
0.6 μm L/S	30–45%	30–45%	
0.5 μm L/S	20–25%	25–30%	25–35%
Focus latitude[a] for			
0.6 μm L/S	1.5–1.8 μm	1.8–2.2 μm	
0.5 μm L/S	0.9–1.4 μm	1.3–1.9 μm	1.7–1.9 μm
0.4 μm L/S			1.1–1.4 μm
0.35 μm L/S			0.4–0.7 μm
Thermal stability[b] for			
width 1–5 μm	120–145°C	115–135°C	125–135°C
width >100 μm	110–130°C	105–125°C	115–125°C
Dill parameter			
A	0.60–0.85 μm^{-1}	0.50–0.75 μm^{-1}	0.65–0.90 μm^{-1}
B	0.09–0.14 μm^{-1}	0.10–0.13 μm^{-1}	0.05–0.08 μm^{-1}

[a] CD tolerance is ±10% of nominal linewidth; resist height loss <10% of nominal height.
[b] CD tolerance is ±10% of nominal linewidth.

line lithography lies, however, in its focus latitude, making design rules in the range of 0.5 μm or somewhat lower feasible for standard lithography, without applying enhancement techniques. Thus, i-line lithography has become, for example, the method of choice for 0.5 μm lithography ground rules.

Table 6 shows some physical parameters of MUV resists as well as additional performance parameters important in a production environment. The specifications for particle density, metal ion content, or batch-to-batch uniformity with respect to photospeed, for example, have all been tightened over the years to accommodate the needs of the industry.

Dyed MUV Resists. The controlled increase of light absorption by MUV resists via the incorporation of a dye/dye mixture in resist materials remains the cheapest and therefore most widely employed strategy to keep line notching over highly reflective substrates, such as aluminum (reflectivity range 80–90%) or metal silicides, under control. Further beneficial effects of dye addition to a photoresist are an attenuation of the influence of local reflectivity variations on CD control, and a reduction of the standing wave effect. The applicability of

Table 6 Physical Parameters/Specifications for DNQ/Novolak Resists

Solid content	20–35% (aiming at 1–2 µm resist thickness)
Water content	0.10–0.30%
Density	1.0–1.1 g/cm^3
Viscosity	10–80 centipoise (solvent/resist thickness dependent)
Refractive index	$n_{436} = 1.66 \pm 0.02$ for g-line resists
	$n_{365} = 1.71 \pm 0.02$ for i-line resists
Particle specifications	<500/cm^3 size >0.3 µm, g-line resists, (1988)
	→100/cm^3 size >0.2 µm, i-line resists, (1993)
Metal ion specifications	≤200–100 ppb for Na, Fe (1990/91)
	≤100–50 ppb for K, Ca, Cu, Mn
Batch-to-batch uniformity	±5% (1990)
	±3% (1993)
Guaranteed shelf life	12 month (g-line resists), 6–12 months (i-line resists)

this strategy depends generally on the requirements relative to CD control, wall angle, or process window. Many dyes have been found suitable for MUV lithography applications (e.g., diaklyl chalcone compounds, strepto cyanine dyes, coumarine dyes) [13]. The maximum possible dye content is limited for several reasons: Too much dye increases the danger of particle formation from dye fallout. It may also reduce shelf life and lead to unacceptable degradations in resolution and/or profile wall angle, although most advanced dyed resist systems have improved with regard to the latter aspect compared with their predecessor materials. A reduction in photospeed by 10–50% has to be accepted for the majority of dyed resists. However, this is not a general rule. The dose requirement may be more strongly governed by changes in dissolution characteristics than the increase in absorption. Depending on the nature of the mutual interaction among the dye, PAC, and resin components, a dye addition may either increase or decrease the solubility in developer, or leave it practically unaffected. An increase in dissolution rate results, for example, upon addition of curcumin dyes. Compounds of this type undergo tautomerization, and the acidic enol form is capable of enhancing solubility in base developers. The dye concentration, as well as the applied resist thickness, may also influence whether optical or dissolution effects gain dominance with respect to dyed resist performance.

The capabilities of commercial dyed g- and i-line resists can be derived from Table 7. A comparison with the performance data of undyed resist materials (see Table 5) confirm the above-mentioned trade-offs inherent to dye addition. The Dill parameters B listed in Table 7 characterize the total absorption of a resin component and a nonactinic dye. High B values ($B \geq 0.50$ µm^{-1}) are generally appropriate in case of very high substrate reflectivity (e.g., aluminum). $B \leq 0.40$ µm^{-1} may suffice when applying resist over substrates with somewhat

Table 7 Performance Data for Dyed MUV Resists

	g-Line resists[a] NA = 0.54/0.55	i-Line resists[a] NA = 0.52/0.54
Limit of resolution	0.40–0.50 μm	0.45–0.50 μm
Wall angle	83–86°	83–85°
Photospeed	200–300 mJ/cm^2	200–300 mJ/cm^2
Dose latitude		
0.6 μm L/S	25–35%	20–30%
0.5 μm L/S		15–25%
Focus latitude		
0.6 μm L/S	1.1–1.3 μm	1.4–1.6 μm
0.5 μm L/S	→ 0	0.6–1.0 μm
Thermal stability		
linewidth 1–5 μm	115–135°C	115–135°C
linewidth >100 μm	100–125°C	100–125°C
Dill parameter B[b]	0.60–0.40 μm^{-1}	0.55–0.35 μm^{-1}

[a]Resist thickness = 1.2 ± 0.1 μm.
[b]Selected B parameter range corresponds to medium to moderately high dye concentration.

lower reflectivity, for example, for poly-Si substrates, which exhibit i-line exposure reflectivities around 60%.

B. Deep-UV Resists

1. Conservative Design Approaches

Attempts to apply conventional DNQ/novolak resists—used so successfully in MUV lithography—to 248 nm wavelength exposures have resulted in unsatisfactory resolution and insufficiently steep wall angles. This mediocre performance stems a.o. from the high nonbleachable absorbance of both DNQ photoproducts and novolaks. To make it more effective, modifications of PAC and resin components are necessary.

Bleachable aliphatic diazoketones, containing the 1,3-diacyl-2-diazochromophore, showed more suitable absorption characteristics than DNQ PACs. However, low-molecular-weight compounds of this type (e.g., Meldrum's acid or diazodimedones) exhibited an unacceptably high volatility during the prebake step, and these were eventually replaced by high-molecular-weight analogs, such as diazocyclohexanediones. The ketene intermediates, resulting from photolysis of diazoketone compounds, showed higher reactivity and therefore unfortunately a stronger tendency to undergo the lithographically detrimental side reaction of ester formation, instead of the preferred ICA formation. Diphenyl iodonium and triphenyl sulfonium salts are also suitable inhibitors for novolak resins.

However, even in the case of alternative photochemistry, which exhibited appropriate bleaching characteristics of PACs, there remained the even greater challenge to the resist designer to reduce the residual unbleachable absorbance of the resin component. Commonly used novolak formulations manufactured from o-, m-, and p-cresol mixtures exhibit absorbance values $\geq 0.5/\mu m$. The high absorbance of the binder resin attenuates the light intensity arriving at the bottom of the resist, inducing an acid concentration gradient and, consequently, sloped side walls. Novolak optimization aimed at higher transparency resulted in products with absorbance values around $0.35/\mu m$, still too high, however, to produce the required resist profile steepness. Pure p-cresol novolaks, although being advantageous with respect to transparency, suffer from low thermal stability. Polyvinyl-phenol (= poly hydroxystyrene), on the other hand, offers suitable transmission (absorbance of $0.17/\mu m$ at 248 nm) combined with high thermal stability, but this polymer is difficult to inhibit when using DNQ. It is, however, the material of choice for negative tone DUV resist applications.

Despite the various DUV resist design approaches described above, none furnished the desired advantages expected when switching from i-line to the shorter 248 nm exposure wavelength. Further innovative approaches resulted in the concept of "chemical amplification," and the majority of presently available commercial DUV resists are based on the latter design strategy.

2. Chemical Amplification Resists

General Remarks. The concept of chemical amplification (CAM) involves a two-step mechanism; CAM starts with the photogeneration of a reactive species, which then catalyzes a subsequent reaction capable of introducing distinct dissolution changes within the exposed resist areas. The second reaction step can be driven to completion wherever a certain threshold concentration of catalyst is available. This makes CAM resists less sensitive in regard to resin absorbance under the condition that the minimum light intensity, necessary for the production of the threshold catalyst concentration, can penetrate through the film.

The principle of chemical amplification has general applicability, and it has been employed for the synthesis of advanced DUV, e-beam, X-ray, and even MUV resist materials. Both positive and negative tone systems can be designed, the operation mode being dependent on the mechanism employed and the choice of the development medium, which is preferably an aqueous base solution. Due to the catalytic nature of the second CAM step, effective quantum yields as high as 100—1000 can be obtained. This reduces the dose requirements to as low as 1–5 mJ/cm^2 in some cases. Conventional MUV DNQ/novolak based materials, by comparison generally furnish quantum yields of 0.2–0.3, demanding exposure doses of 100–300 mJ/cm^2. The comparative gain in resist sensitivity of applied CAM resists was of particular interest during the early stage of DUV lithography development, because the available light sources (Hg lamps) at that time delivered low intensities in the DUV region and therefore required very long exposure times. This issue became of lesser concern with the introduction of excimer laser technology, which generated intensities up to 300 mW/cm^2. Nevertheless, increased photospeed has always been appreciated as a leverage to enhance throughput, thereby allowing a reduction of the increasingly prohibitive investment costs for innovative exposure equipment.

Several different mechanisms applicable for DUV resist synthesis have been proposed in the literature [14,15]. Most of the development work focused on materials using acid-catalyzed reaction schemes, and the majority of the commercial products available so far operate via this reaction type. Acid-catalyzed chemical amplification can affect, for example, polymer deprotection (thermolysis of polymer pendant groups), scission of polymer main chains, cross-linking reactions, or ring-opening polymerization of oxirane pendant groups. Radical-initiated CAM is also possible, e.g., cross-linking of acrylated polyols.

The first step in the acid-catalyzed CAM reaction is the production of a strong acid as a result of the photolysis of a so-called photoacid generator, PAG. A variety of compounds can serve this purpose [15]. The photolysis reactions for three types of PAGs are shown in Table 8. Inorganic onium salts, like triphenyl sulfonium hexafluoroarsenate (or antimonate), were the first PAG compounds used for the synthesis of chemically amplified resists. However, these materials exhibit poor solubility in organic solvents, and their metal ion content raised serious concerns about their applicability under manufacturing conditions. Consequently, inorganic onium salts were displaced by organic ones containing, for example, triflate ions as anions. The intramolecular photooxidation reaction of o-nitro-benzyl esters of strong organic acids can also be applied for acid-catalyzed CAM schemes. The 2-nitro or 2,4 (or 6)-dinitro benzyl esters of tosic acid have been used for this purpose. These materials offer improved solubility in organic solvents, good thermal stability, and excellent quantum efficiencies.

Table 8 Photochemical Reactions of Photoacid Generators (PAGs)

Type of PAG	PAG → photoproducts	Comment
Inorganic onium salt	$ArS_3S^+X^- \rightarrow Ar_2S + Ar\cdot + R + H^+X^-$	$X^- = AsF_6^-$, SbF_6^- Ar, aryl R, radical
Organic onium salt	$ArS_2J^+X^- \rightarrow ArJ + Ar\cdot + R + H^+ X^-$	$X^- = CF_3SO_3^-$
o-nitrobenzl tosylate	o-NO_2-C$_6$H$_4$-CH$_2$-OTs → o-NO-C$_6$H$_4$-CHO + TsOH	

Resists Operating via Acid-Catalyzed Deprotection Mechanism. Already in the initial development work stage (in the early 1980s) poly(tert.-butoxy carbonyl oxystyrene), PBOCST, was proposed as a suitable matrix resin for CAM resists functioning via a deprotection mechanism [16]. The thermally stable but acid-labile tert.-butoxycarbonyl group (t-BOC)—a protective group widely applied in peptide synthesis—protects the phenolic functionalities of the poly(hydroxy-styrene), PHOST. The reaction mechanism leading to deprotection is described by

$$-(CH_2-\underset{R}{C})- \underset{\bigcirc}{\bigcirc} \underset{O-C-O-CH_3}{\overset{CH_3}{|}} \quad \xrightarrow{H^+,\Delta} \quad -(CH_2-\underset{R}{C})- \underset{\bigcirc}{\bigcirc} \underset{OH}{} + CH_2=C\underset{CH_3}{\overset{CH_3}{\diagup}} + CO_2 \quad (11)$$

deprotection of t-BOC groups

A trace of PAG embedded in the polymer generates a strong acid upon DUV irradiation, which catalyzes the removal of the protective groups with a subsequent postexposure bake (PEB). Two conditions have to be met to make this type of catalyzed thermolysis chemistry feasible: (1) bonding between an oxygen atom and a carbon atom capable of forming a stabilized carbenium ion, and (2) availability of a hydrogen atom in the α-position at the site of the carbenium ion formation to allow the elimination of a proton. Suitable functionalities are, e.g., tertiary alkyl, tertiary carbonate, secondary allylic, or benzylic pendant groups. A proton is consumed during the carbenium ion formation, but another one is liberated during the elimination reaction, thus guaranteeing the continuation of the catalytic cycle. In practice, the catalytic chain length is limited to a few hundred due to side reactions of the photogenerated acid either with its counterion, with reactive sites in the polymer, or with contaminants. The deprotection of the hydroxyl functionality causes a large change in polarity and its solubility in the exposed resist regions as the lipophilic t-BOC protected resin is converted into a PHOST-type compound. The extremely high contrast (for some materials, >10) derives from a nonlinear dissolution response depending on exposure dose. Dissolution in base developer occurs only after the majority of the t-BOC groups ($>95\%$) has already been removed.

The photolysis step is a very fast reaction, the extent of which is determined by its quantum efficiency. The rate of deprotection, on the other hand, is a function of acid concentration, time and temperature of the postexposure bake step, and the diffusion rate of the acid into the matrix resin. The latter parameter depends on temperature as well as the nature of the PAG and resin. For PBOCST-type resist systems, the thermolysis reaction occurs only to a minor extent at room temperature, but the rate is high at temperatures around 100°C, which is well below the T_g for both PBOCST or PHOST. For many CAM resists, the photospeed is strongly dependent on the PEB temperature. An Arrhenius relationship between exposure dose and PEB temperature with an E_a of about 130 kJ/mol has been reported for a PBOCST-type resist with a nonmetallic PAG. Thus, very good PEB temperature stability over time, as well as excellent temperature homogeneity across the hot-plate, are required for many CAM resists in order to maintain CD control. Fortunately, the stringent requirements can be met by modern hot-plate equipment with temperature uniformity specifications of ±0.5°C. In general, the following ranking is valid with respect to the effect of process parameter variations on CD control: PEB temperature > exposure dose > PEB time.

More troublesome is the sensitivity of acid-catalyzed resist systems to airborne contamination. The concentration of a strong photogenerated acid, which catalyzes deprotection, generally remains at very low concentration levels (as low as 10^{-8} moles). If the acid moiety happens to be neutralized at the resist surface prior to PEB, due to the reaction with volatile base contaminants like NH_3, amines, or compounds capable of reacting to NH_3 or amines (i.e., HMDS or N-methylpyrrolidone, NMP) in the ambient air, then an insoluble surface crust may be formed, resulting either in degraded T-shaped resist profiles ("T-topping") or in complete bridging between adjacent resist structures. Detrimental effects have already been observed for contamination levels as low as a few ppb when the bake is delayed after exposure for a few minutes ("time delay effects"). Two strategies have been applied so far to cope with this problem: (1) coating the CAM resist with a protective coating (e.g., an acrylic copolymer with a thickness of 50–100 nm), and (2) installation of air purification systems (charcoal filters). Resist designers are working intensively to make CAM resists of the above-mentioned type less sensitive toward airborne contamination without trade-offs in overall performance. A reduction of the permeability of the polymer films with respect to the diffusion of base contaminants is certainly beneficial. One way to meet this goal is to decrease the free volume in the resist, which can be achieved, for example, by keeping the temperature of the prebake above the glass transition temperature of the material.

Some PBOCST-type positive tone resists have been used in pilot line or volume production of Megabit DRAM devices. Exposure doses of 10–40 mJ/cm² are generally required when tailoring the resist processing to an optimum compromise between throughput and resolution/latitude performance. IBM was the first full-scale manufacturer to use materials of this kind in patterning 1-Mbit DRAM chips by DUV lithography. Autodecomposition of polymer materials with acid-labile pendant groups is an undesirable side reaction contributing to deterioration of resist contrast and sensitivity. In the case of PBOCST, conversion to PHOST may occur at temperatures as low as 130°C.

To increase the autodecomposition stability, protecting groups forming intermediate carbenium ions less stable than the t-butyl carbenium ion have been reported in the literature. A styrene–maleimide copolymer (see Fig. 12) with a secondary alkyl carbonate pendant group [17], for example, exhibits a stability toward autodecomposition up to 230°C. The incorporation of the maleimide units results also in enhanced thermal stability (T_g after deblocking = 220°C).

Negative Tone CAM Resists Operating via a Cross-Linking Mechanism. Most commercial negative tone DUV resists rely on a photogenerated acid-catalyzed condensation reaction leading to cross-linking in the exposed resist areas. In general, the solid content of such resist materials consists of three components:

A radiation-sensitive PAG (see above)
An acid-labile, preferentially multifunctional cross-linking agent (i.e., melamine or tetramethyloxy glycoluril)
A matrix resin with reactive sites for cross-linking reactions (like PHOST or novolak)

The photogenerated acid reacts catalytically with the cross-linking agent, which then reacts further with a generally phenolic resin, primarily via O-alkylation; this is described by

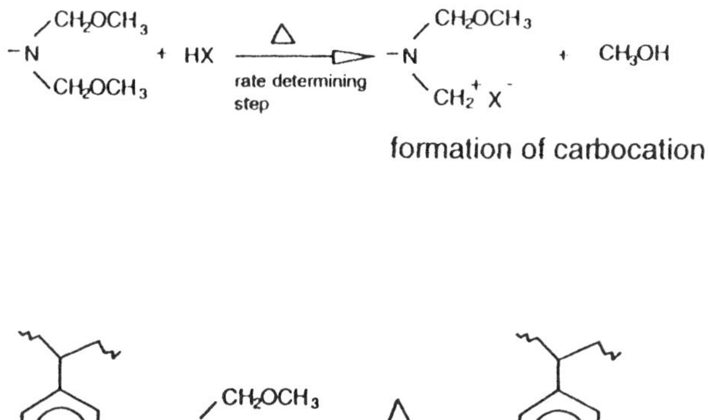

$$-N(CH_2OCH_3)_2 + HX \xrightarrow[\text{step}]{\Delta \text{ rate determining}} -N(CH_2OCH_3)(CH_2^+ X^-) + CH_3OH \quad (12a)$$

formation of carbocation

$$\text{HO-C}_6\text{H}_4\text{-R} + -N(CH_2OCH_3)(CH_2^+ X^-) \xrightarrow{\Delta} \text{R-C}_6\text{H}_4\text{-O-CH}_2\text{-N(CH}_2\text{OCH}_3) \quad (12b)$$

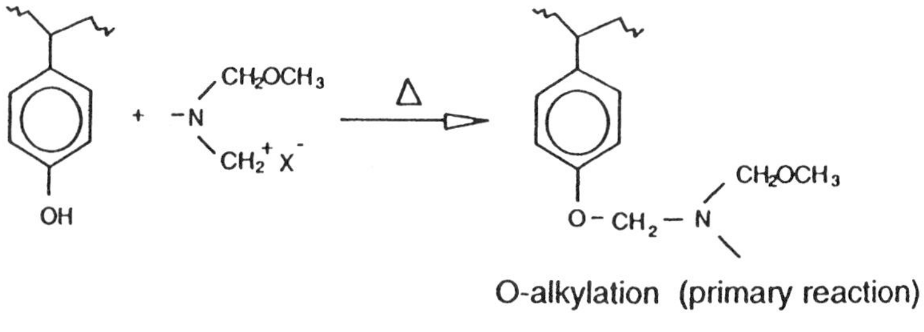

O-alkylation (primary reaction)

The resulting highly cross-linked polymer structure in the exposed resist areas exhibits distinctly lower solubility in aqueous base developers than the unreacted resist portion. The activation energy for the cross-linking reaction is relatively

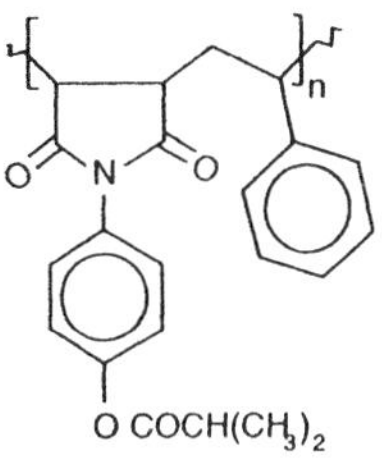

Figure 12 Styrene–maleimide copolymer with pendant alkyl carbonate groups exhibiting enhanced stability against autodecomposition and thermal flow. (From Ref. 17.)

high (40–50 mJ/mol). The formation of a carbocation is the rate-determining step, making annealing after exposure necessary to obtain feasible sensitivity values. The PEB temperature plays again a key role in determining photospeed (generally 10–40 mJ/cm^2) and linewidth.

Electrophilic aromatic substitution reactions can also be utilized for the design of a negative tone CAM resist system. The reaction scheme involved is very similar to the one described above for the condensation mechanism. In this case, the photogenerated acid reacts with a latent electrophil (generally the acetate of arylmethylol) to produce a benzylic carbocation capable of undergoing electrophilic substitution with aromatic polymers (generally PHOST or novolaks). Only C-benzylated phenols are present at the end of the reaction. The small amounts of benzyl-phenyl ether products generated during the reaction undergo an acid-catalyzed conversion to C-benzylated products.

Different resist designs based on electrophilic aromatic substitution are possible. Both latent electrophil as well as the aromatic moiety susceptible to substitution may be combined in one polymer structure:

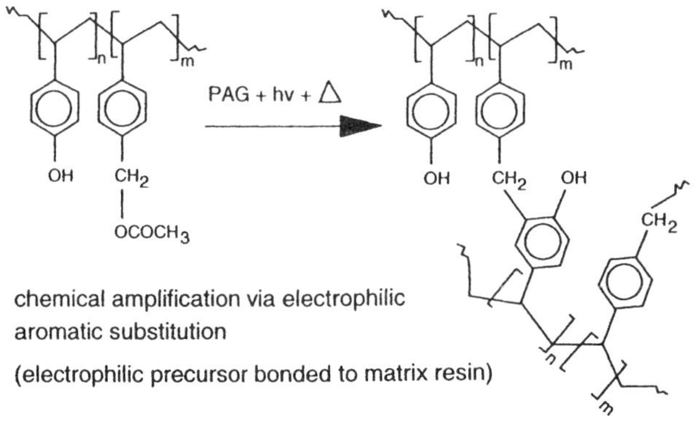

(13)

chemical amplification via electrophilic aromatic substitution

(electrophilic precursor bonded to matrix resin)

A polyfunctional electrophil precursor (like 1,4-diacetoxymethyl benzene) may also be embedded in PHOST- or novolak-type resins.

Concepts Combining CAM and Dissolution Inhibition. The photoactive compounds described in Sec. III.B.1, being more suitable to DUV exposure than conventional DNQ PACs, all suffer from a similar drawback when used as dissolution inhibitors. These PACs have quantum yields well below 1. Consequently, considerable amounts of unchanged hydrophobic products are left over after exposure, often together with hydrophobic photoproducts. A complete change of inhibitor solubility—as it occurs with conventional DNQ/novolak systems—is, however, desirable. This goal was achieved, for example, for a two-component (solvent not considered) DUV resist system consisting of a poly(hydroxy-styrene) matrix resin and a specially modified sulfonium compound (as PAC) with acid-cleavable sidegroups [18]. The PACs employed have the formula

$$(\text{t-BOC-O-Ph})_2 \; S^+PhX^-, \quad \text{whereby } X^- \text{ is } AsF_6^- \text{ or triflate} \qquad (14)$$

The acid-labile group is stable during the photochemical reaction, but it is removed completely during PEB treatment at 100°C. During the secondary reaction, unchanged hydrophobic photoinitiator as well as hydrophobic photoproducts react to form phenolic compounds. As all moieties existing within the exposed resist areas have been converted into base-soluble products, high dissolution contrast is guaranteed. Resolution capabilities down to 0.4 µm could be demonstrated for model resists when using a KrF excimer stepper with NA = 0.37 (dose requirements 20–70 mJ/cm^2).

Three-component positive tone DUV resist systems have also been designed, containing, e.g., a novolak resin, a dissolution inhibitor with acid-labile pendant groups and a photosensitive onium salt [19]. Examples for the PAC compounds employed are given in Fig. 13. Photolysis of the onium salt initiates deprotection of the dissolution inhibitor, leading to the desired drastic solubility change within the exposed areas. The deprotected inhibitor should exhibit high solubility. T-butyl ester pendant groups, capable of conversion to carboxylic acid functionalities, are therefore the most suitable among the available sidegroup options.

The inhibitor species is not photosensitive in this case, making low absorption at the actinic wavelength an advantage. Therefore, aliphatic PAC compounds rather than aromatic ones seem to be more attractive. Low volatility during softbake, good miscibility in novolak, and no detrimental influence on resist coating behavior are further demands on an ideal PAC. T-butyl cholate (see Fig. 13d) met all these requirements. This material is stable up to 250°C, has practically no absorption in the DUV region, and can moreover be easily synthesized from natural cholic acid.

Performance Data of Chemically Amplified DUV Resists. Resolution capability as well as dose and focus latitude values reported in the literature for positive and negative tone DUV resists are listed in Tables 9 and 10, respectively. A tolerable CD variation of ±10% from the nominal feature size and the fulfillment of requirements regarding profile quality/resist thickness retention were generally applied as criteria when determining process windows.

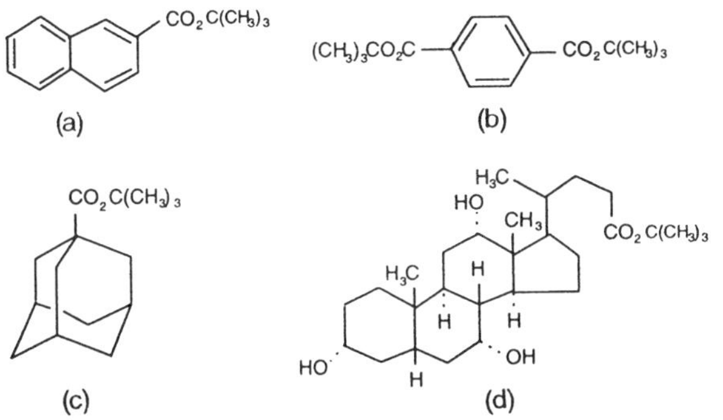

Figure 13 Dissolution inhibitors with acid-cleavable pendant groups. (Adapted from Ref. 19.)

Table 9 Resolution Capability and Process Latitudes for Chemically Amplified Positive Tone DUV Resists

	1991–1992 NA = 0.42–0.45 D = 1.0–1.1 µm	1992–1993 NA = 0.48–0.53 D = 0.84–0.90 µm	1993–1994 NA = 0.48–0.53[a] D = 0.70–0.81 µm
Resolution for L/S patterns	0.26–0.30 µm	0.25 µm	0.22–0.25 µm
Resolution for contact holes		0.30–0.35 µm	0.25–0.33 µm
Dose latitude for			
0.35 µm L/S	18–24%	26–32%	26–>40%
0.30 µm L/S		17–21%	17–31%
0.25 µm L/S		15–18%	16–18%
Focus latitude for			
0.35 µm L/S	1.1–1.8 µm	1.2–1.4 µm	1.3–1.8 µm
0.30 µm L/S			1.0–1.6 µm
0.25 µm L/S		0.6–1.0 µm	0.7–1.2 µm
0.35 µm CH			1.1–1.4 µm
0.30 µm CH			0.7–1.1 µm

[a] KrF excimer stepper and step and scan.

Table 10 Resolution Capability and Process Latitudes for Negative Tone DUV Resists (1991–1994)

	NA = 0.42–0.50[a] D = 0.7–1.0 μm
Resolution for L/S patterns	0.25–0.30 μm
Dose latitude for	
0.35 μm L/S	20–36%
0.30 μm L/S	18–40%
0.275 μm L/S	21%[b]
Focus latitude for	
0.35 μm L/S	1.0–1.6 μm
0.30 μm L/S	0.9–1.3 μm
0.25 μm L/S	0.6 μm

[a]KrF stepper and step and scan.
[b]One literature value only.

The data of Table 9 reflect performance differences resulting from continuous resist improvement, a shift toward lower resist thickness, and the use of exposure tools with higher NA. Resolution of L/S patterns with features sizes of 0.22–0.25 μm and of contact holes with 0.25–0.30 μm diameters has become feasible. Mask linearity was generally observed down to 50–100 nm above the resolution limit. Using the most advanced exposure equipment and resist materials (available in 1994), dose latitudes of ±8–9% and focus latitudes around 1.0 μm could be obtained for 0.25 μm L/S patterns. Due to the larger variety in materials employed in the innovative area of DUV resist design and the stronger dependence of photospeed on bake conditions, exposure dose requirements for different positive tone DUV resists may vary considerably. The literature suggests values between 4 and 80 mJ/cm^2, although most of the materials ranged between 15 and 35 mJ/cm^2. More recently developed resists have become less sensitive with respect to the post-exposure delay (PED) effect. With some of these products, the impact on CD and profile quality remained acceptable as long as the PED time did not exceed 1/2–2 hours. With respect to CD variation as a function of PEB temperature, the lowest values reported for 0.25–0.35 μm features ranged between 3 and 15 nm/°C.

Some literature data on negative tone DUV resists is listed in Table 10. With respect to the overall performance, negative tone resists seem to be somewhat inferior to their positive tone rivals. The best of the positive DUV resists showed, for example, noticeably better DOF values for 0.25 μm L/S patterns. Most positive and negative tone resists exhibited good and even excellent profile quality (with wall angles ranging generally between 88 and 91°). Thermal stability for larger features of positive resists could be maintained up to 120–160°C, while negative DUV resists (operating via a cross-linking mechanism) showed an advantage in that all data indicated thermal stability up to at least 130°C.

C. Multilayer Systems

Meeting the required CD tolerances continues to become increasingly difficult when entering the submicron resolution regime, especially for enhanced wafer topography and/or high substrate reflectivity. This has led to efforts to develop various alternative lithographic techniques, differing from the conventional single-layer wet-developable resist approach, to reduce topography/reflectivity-related problems. These are trilevel, bilevel, and surface imaging techniques. Although differing in process flow, all of these techniques utilize a dry etch step (dry-development) to produce a relief image with steep sidewalls. Silicon chemistry is involved in all three cases, as the etch mask for the dry-development step is formed either by a silicon-containing intermediate pattern, a patterned silicon-containing resist, or a selectively silylated area in the top region of a surface imaging resist.

1. Trilayer Systems

Trilevel techniques, which entered the field of lithography in various forms in the 1970s, involve a film stack consisting of a thick planarizing bottom resist, an inorganic intermediate layer, and an imaging top resist. Pattern formation is carried out in the thin top resist (via wet-development). The developed top resist then functions as a mask for the pattern transfer into the intermediate layer using a fluorine-containing plasma. The intermediate layer, generally spin-on-glass, in turn acts as an etch mask for the anisotropic etching of the bottom resist in O_2 plasma. The hard-baked bottom resist has a high thermal stability and exhibits adequate etch resistance during substrate etching. By incorporating a dye either in the planarizing resist or the intermediate layer, reflectivity-induced problems can be reduced effectively, leading to improved linewidth control. Nevertheless, trilayer processes have never become popular in production. The complexity of processing was certainly a major deterrent. Lateral linewidth loss during the oxygen etch step was also a problem, as was the susceptibility to pinhole or particle formation of some trilevel systems.

2. Bilayer Systems

The bilevel systems (around 1980) brought a reduction in process complexity compared with trilayer systems with equal or even better levels of performance. Bilayer systems consist of a thin imageable top resist coated over a thick planarizing bottom resist, which is highly absorptive to the exposure wavelength. The bilayer concept (see Fig. 14) relies on the presence of organometallic species in the patterned top resist that are capable of forming plasma-resistant ox-

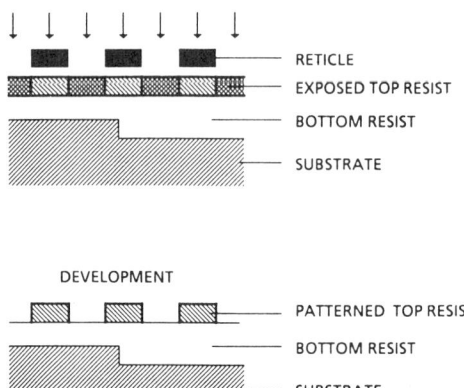

Figure 14 Processing scheme for a bilayer technique using an organosilicon top resist. (From Ref. 7.)

ides. Elements like Ge, Sn, Ti, P, or B fulfill the requirement. In practice, however, silicon, is the material of choice due to the large variety of available synthesis options. The image transfer from the top to the bottom resist is carried out using anisotropic reactive ion etching (RIE) in O_2 plasma. During the initial stage of the dry-development step, the surface region of the Si-containing top resist is converted into an oxide-type crust acting as an etch mask as the etching continues.

Two approaches are possible to introduce the required Si into the top resist: (1) the design of silicon-containing top resists, and (2) the incorporation of Si into a generally Si-free patterned resist during a silylation step prior to dry-development. One requirement for the wet-developable top resist material is that it is soluble in alkaline developers (industry standard). A variety of positive and negative tone organosilicon top resists suitable for MUV or DUV applications have been designed within the past decade [20]. Different synthesis strategies can be followed with regard to positive tone resists: the introduction of Si into a side chain of an alkali-soluble polymer such as novolak or PHOST, or the introduction of base-soluble moieties (like phenol groups) into side chains of a Si main-chain structure containing, for example, polysiloxanes, polysilesquioxanes, or polysilanes. The latter approach is more favorable for etch resistance. Negative tone top resists like polysiloxanes have been used for deep-UV applications. Examples of possible bilayer top resist formulations are given in Fig. 15.

Lateral linewidth erosion during the dry-development step as a result of insufficient etch resistance is the main problem connected with bilayer system applications. The undesirable linewidth loss depends on the silicon content and wall angle (which is generally less steep than with Si-free photoresists) of the top resist, as well as the etch conditions employed. A minimum Si content of about 10 w % is a prerequisite for any usable bilevel resist to achieve with optimized etch conditions, etch selectivities of about 1:10 between the top and bottom resists. However, linewidth loss (often 50–100 nm/structure) generally cannot be completely eliminated. This reduces the applicability of organosilicon polymer–based bilevel systems for ultimate resolution demands. Introducing a mask bias may be—in some, but not all cases—a feasible way to compensate for the lateral linewidth loss. Underexposure does not help, as it reduces the wall angle.

The second approach, that of obtaining a O_2 plasma–resistant etch mask by incorporating silicon into the top resist via gas- or liquid-phase silylation prior to etching, requires an additional process step, but also offers advantages. Si-free top resists can be applied with potentially better imaging performance, resulting at once in steeper wall angles. Moreover, cleverly designed top resist/silylating agent combinations allow a higher incorporation of Si into the resist compared with organosilicon polymer–based bilevel systems. Liquid-phase silylation of an anhydride-containing top resist with silicon-rich diamino compounds (see Fig. 16) can lead, for example, to high Si contents of >20 w %.

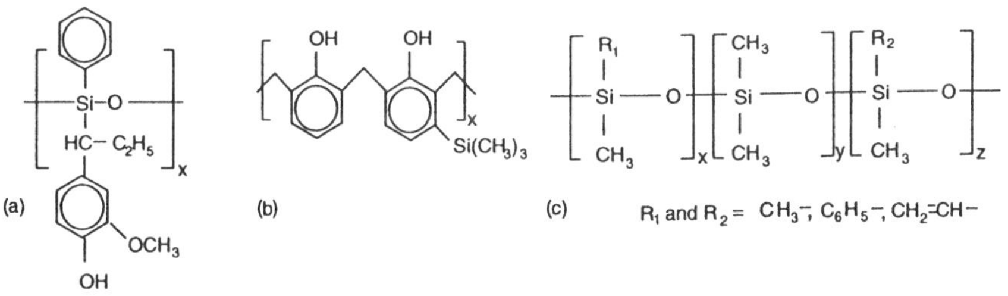

Figure 15 Examples for bilayer top resists.

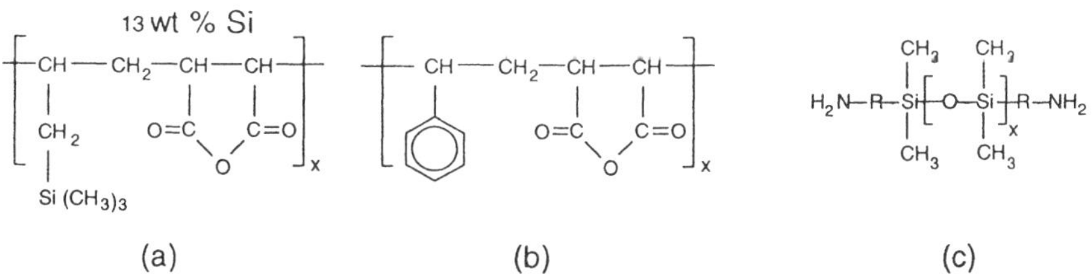

Figure 16 Liquid-phase silylation of bilayer top resist: (a) Si-containing top resist; (b) Si-free top resist; (c) silylation agent. (Adapted from Ref. 22.)

This results in improved etch resistance. No linewidth loss was observed for this particular resist system [21]. Further increased Si content can be obtained when combining a Si-containing top resist with additional silylation. A time-controllable linewidth growth during the silylation step can even be utilized to reduce space and hole widths. The "CARL" process, i.e., chemical amplification of resist lines [22], allows, for example, patterning of contact hole sizes down to 150 nm. The silylation using aqueous amine solutions requires treatment times ≤100 s (at room temperature) and can be carried out using a conventional puddle development module. Some CARL systems optimized for MUV and DUV exposure exhibiting resolution capabilities down to 0.3 μm (corresponding to extremely low k_1 factors around 0.40) have been reported in the literature.

D. Top Surface Imaging Systems

A further simplification in process flow compared with bilayer systems has come along with the development of so-called "top surface imaging" (TSI) techniques. With this type of lithographic process, the functions of imaging and masking during dry-development in O_2 plasma are taken over by a single resist layer. By designing resists with appropriate functionalities, incorporation of silicon within the near-surface region (100–300 nm) of either the exposed or unexposed resist areas can be achieved from the gas or liquid phase, furnishing positive or negative tone patterns. Depending on device topography, resist thicknesses of 1.5–2.5 μm have been applied. Short-range topography variations (but not long-range topography variations) are leveled out by resist planarization. The DOF requirements become less stringent, as the imaging process is only relevant to the top surface region. This is a distinct advantage when using exposure tools with high numerical apertures. In contrast to conventional single-resist processes, high absorption of the resist material is also not an issue with a TSI system. It may even be beneficial to increase the resist absorption by dye addition in order to achieve an effective suppression of detrimental reflections from the substrate.

Selective silylation can be achieved either by the creation or destruction of functional groups capable of reacting with a silylating agent, or by inducing a significant differentiation in diffusion rate of the silylation agent in a polymer with an inherently high number of functionalities. High-performance TSI resists are obtained if the silicon incorporation in resist as a function of exposure is as nonlinear as possible. With resist materials containing a phenolic base resin (e.g., novolak or PHOST) and DNQ PACs, the phenolic OH groups of the base resin have been identified as the main reaction partners during silylation. On the other hand, FTIR investigation showed no indications for a reaction of the indene carboxyl acid moiety contained in exposed resist regions. The reaction mechanism for a TSI process using phenolic resins and gas-phase silylation with HMDS is described by

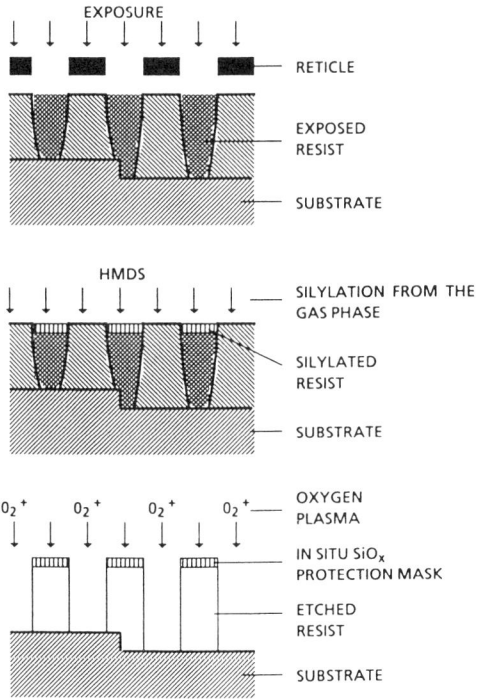

The corresponding process flow is shown in Fig. 17. A TSI process employing this type of chemistry is the DESIRE process [23], which gained much attention in the literature and has led to commercial negative tone resist materials (PLASMASK resists) suitable for g-line, i-line, and DUV exposure. For these systems, the selectivity of silylation is obtained by impeding the diffusion of the silylation agent in nonexposed areas via the formation of chemical crosslinks, which is induced by the nonreacted PAC during a presilylation bake (PSB)

Figure 17 Processing scheme for a surface imaging method using gas-phase silylation. (From Ref. 7.)

at temperatures between 160 and 190°C. At lower temperatures, hydrogen bridge formation between DNQ and resin may also contribute to the mechanism. Multifunctional PACs are required in order to make cross-linking more effective in the nonexposed areas. The width of the process window depends on resist formulation, PSB temperature and time, PAC type and concentration, molecular weight, and MW distribution. Silylation in the exposed resist regions results in almost 100% conversion of the phenolic groups of the phenolic resins, leading to Si contents to guarantee sufficient etch selectivity. Silicon contents of 9–11 w% have been reported, for example, for different PLASMASK versions, when applying gas-phase silylation with HMDS. Other compounds proposed in the literature for gas-phase silylation are 1,1,3,3-tetra-methyl disilazane (TMDS), N,N-dimethyl disilazane (TMSDMA), or dimethylsilyl dimethylamine (DMSDMA). In comparison with HMDS, these silylation agents exhibit reduced swelling during Si incorporation, better latitude, and higher sensitivity, and moreover, they allow application at lower temperatures.

Liquid-phase silylation, on the other hand, offers even further advantages. Silylation treatments of this type require much lower temperatures, ranging generally from room temperature to 40 °C. The silicon content can, however, be raised to 20–25 w %, i.e., a value more than 2 times higher than gas-phase silylation results, yielding improved dry-etch selectivity. Moreover, a nonlinear S-shaped response of Si incorporation as a function of exposure dose (see Fig. 18) furnishes improved overall contrast. Multifunctional silylating agents like 1,1,3,3,5,5-hexamethyl cyclotrisilazane (HMCTS) or bis(dimethylamino) dimethylsilane (B[DMA]DS) have been identified for liquid-phase silylation since monofunctional agents showed very poor selectivity. The latter compound was

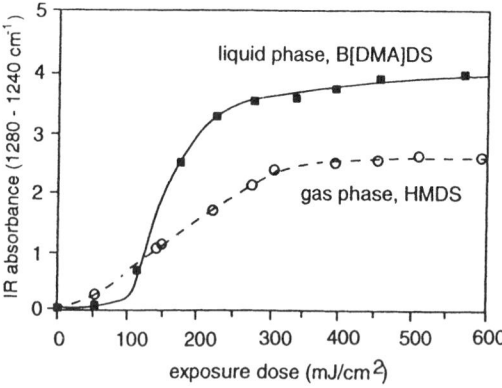

Figure 18 Response of Si incorporation into top surface of resist as a function of exposure dose. (From Ref. 24.)

applied in xylene solution. However, TSI systems have also been developed to allow more environment-friendly usage of aqueous silylation agent solutions. Bis-amino-siloxane has been employed, for example, for the liquid-phase silylation of a TSI resist system consisting of a copolymer of maleic anhydride and styrene. In this case, the silylation treatment could be carried out using conventional puddle track equipment, requiring process times of less than 2 minutes, including rinsing.

As mentioned, TSI systems can be designed for both MUV and DUV exposures. In the latter case, deep-UV induced cross-linking may become an issue, limiting the diffusion rate of the silylating agent in exposed resist areas. This has been considered in the design of surface imaging resists optimized for 248 nm wavelength exposure.

Although the increase in process complexity of a TSI technique is moderate in comparison with the application of a conventional single-layer wet-developable resist system, it does involve a more costly dry-development step, thereby introducing several additional process parameters that influence the final imaging result. Power, pressure, O_2 flow rate, and the amount of over etch determine, for example, the quality of the anisotropic etch step. Si incorporation in supposedly Si-free resist regions (due to insufficient silylation selectivity), leading to silicon-containing polymer residues (termed "grass") after dry-development, has been the major processing issue in the early stage of TSI development. Improvements in resist formulation and processing have diminished this problem significantly. "Grass" formation (nonuniform dry etch removal of material, leaving behind very "grasslike" residual) can be prevented, for example, by a two-step etch procedure whereby the uppermost surface region is removed nonselectively during the initial etch step using a fluorine-containing gas (e.g., CF_4 or $C_2F_6 + O_2$). Another possibility is to increase power and reduce flow rate during the initial stage of a pure O_2 etch step, resulting in enhanced physical sputtering. Removal of the topmost resist layer also leads to a better definition of the interface between silylated and nonsilylated areas, thereby enhancing performance. Uniformity of the magnetic field in the plasma system has also been identified as a crucial parameter. Best results have been obtained using magnetron-enhanced etch tools furnishing high etch selectivity as well as high throughput. The use of standard RIE systems requires low power to obtain good selectivity, resulting in much longer etching times. In general, TSI resist development and process optimization have made significant improvements over the years. Independent of the exposure wavelength employed, resolution capabilities corresponding to k_1 factors of 0.40–0.45 have been observed. Resolution of 0.30 µm patterns over topography has been reported, for example, when using Plasmask 200I in combination with an i-line stepper using only a moderately high numerical aperture of 0.48. With state-of-the art DUV exposure equipment, the resolution capability of 0.20 µm can be achieved.

IV. SUMMARY

Continuous use of optical lithography techniques in the industry as the mainstream option for device patterning has brought this technology to a high level of maturity with respect to all its subsystems, like pattern design, mask fabrication, exposure equipment, resist materials and processing optimization. This success is fostering continued efforts to extend the capability of optical lithography as far as possible in order to avoid a costly shift to any major new technology. Such technology options—possibly X-ray lithography, application of e-beam reduction optics, or photocathode-based electron image projectors—will have to offer distinct advantages in order to take over the current role of optical lithography.

The concomitant efforts in all areas involved have carried optical lithography successfully across the supposedly insurmountable production resolution barriers of first 1.0 µm, and later 0.5 µm. Extensive research activities are continuing to extend the feasibility of optical lithography to the 0.25 µm regime by applying so-called enhancement techniques to improve the resolution and/or process window. Modifications in the illumination system of the exposure tools and—in particular—mask technology are under intensive investigation. Theory predicts increased depth of focus of periodic structures when applying an oblique rather than the standard normal incidence of light onto the reticle. Different versions of "off-axis illumination" using annular- or quadrupole-type apertures have been tested [25]. Improvements in DOF (due to an reduction in effective NA), as verified experimentally, depend, however, on pattern type (periodic structures benefit, but not isolated features), pattern size, and, with quadrupole illumination, also on pattern orientation. These restrictions prevent a general applicability of this enhancement method, limiting it probably to a few selected device-level applications.

In mask technology, phase-shifting mask (PSM) application has drawn wide attention since the late 1980s [26,27]. Different PSM versions—for example, alternating (Levenson), attenuating (halftone), rim shifter, or chromeless phase shifters—can be applied, delivering significant gains in the image intensity slope (and hence also improved process windows). In the case of "strong" PSM techniques (alternating and chromeless PSMs) gains can also be achieved in resolution capability. Alternating PSMs offer the highest level of improvement. However, they suffer from design restrictions. Thus, there is still a long, difficult road ahead for the successful implementation of PSM technology into production. The necessary infrastructure is starting to develop, primarily focusing on i-line lithography. An optimization of shifter materials and PSM fabrication processes continues. Areas of major concerns are PSM inspection and repair, where a lot of development work is still required. Nevertheless, an introduction of the PSM technology to production can be expected, the most likely candidate for early implementation being attenuated phase shifters.

With regard to resist design, research activities will continue to focus on the development of improved CAM resists suitable for the 248 nm exposure wavelength. One particular goal will be the reduction of the sensitivity to airborne contamination. Production of devices with 0.25 µm ground rules can be expected in the second half of the 1990s. By then, the availability of DUV resist systems, which do not require the installation of air filtering systems, would be highly economical. Deep-UV lithography will also profit from a further increase of the numerical aperture of DUV exposure tools to values of 0.60 or even higher.

A considerable amount of R&D work will probably be directed toward top surface imaging techniques. They are adaptable to different wavelengths without having to worry about the absorption issue. The required increase in process complexity compared with single-layer wet-developable resist systems seems justifiable in view of the expected advantages. The combination with phase-shifting mask technology for further performance enhancement is feasible. Following this direction, very promising results have been published (in 1992) that show resolution down to 0.2 µm in case of exposure at 248 nm [28]. The major issue concerning TSI applications will be to prove that this technique can work in production with sufficiently low particle levels to guarantee the expected yields. Thus, a major corresponding effort might have to be directed toward optimization of dry-development equipment/processing.

In the meantime, another step along the well-established path of reducing the exposure wavelength for optical lithography has been initiated with R&D activities focusing on the 193 nm (ArF excimer laser) exposure wavelength. However, in this case lens and resist material issues are even more challenging than for 248 nm. The absorption of 193 nm light by air will be another problem, requiring vacuum environment during exposure. With respect to resist materials, novolak and PHOST resins are unacceptable (unless top surface imaging is applied) due to their high optical densities at 193 nm. Resins with suitable absorption characteristics, like methacrylates, suffer, on the other hand, from insufficient etch resistance. However, efforts are under way to gain improvements with the incorporation of alicyclic pendant groups into the resist materials. Some TSI options are also under development. For example, silicon-containing polymers (polysilynes, linear polysilanes, or alkylsilanes) have been reported that undergo photooxidation when irradiated with 193 nm light. HBr plasma is applied to develop these materials.

In summary, there are still a variety of options available to further push the resolution barrier for optical lithography. Of the strategies mentioned above, possibly one of them, a combination of them, or even new ones still hidden in some genius's mind will be successful, giving confidence that the capability of

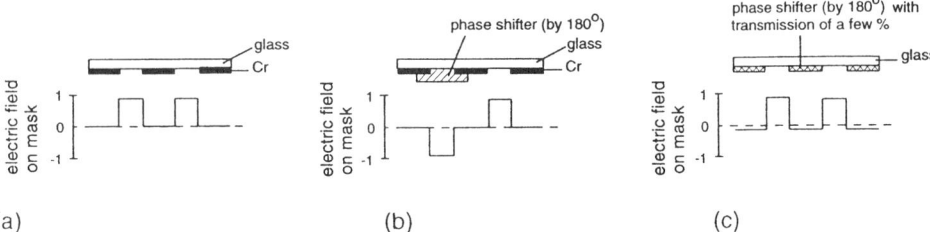

Figure 19 Cross sections of (a) chrome-on-glass (COG) mask, (b) alternating phase-shifted mask (PSM) and (c) attenuated PSM, all with corresponding electric field amplitudes.

optical lithography can be extended to a dimensional range of 0.15–0.20 µm, and possibly even further. This might make optical lithography still a candidate for the production of devices in the year 2001.

EXERCISES

1. Calculate the resolution capability for advanced i-line lithography when employing a stepper with NA = 0.55. (Assume a k_1 factor of 0.7.)
2. Resist systems have to provide a certain process window with respect to variations from the optimum focus position in order to become suitable under production conditions. Name potential contributions from resist, exposure equipment, wafer, and processing resulting in degradation of depth of focus.
3. Modify the model of Dill (Sec. II.B) describing the absorption characteristics of resist materials to render it applicable to dyed resists. Assume two cases: (a) the resists contains a nonbleaching dye; (b) the resist contains a dye that bleaches upon irradiation with actinic light. Both types of dyes have been actually employed with commercial dyed resists.
4. Describe the chemical reaction between HMDS and a silicon wafer, as occurring during wafer priming. (The chemical structure of HMDS is depicted in Eq. (15). *Hint*: There exist SiOH functionalities at the silicon surface.)
5. Calculate the optimum values for the thickness and refractive index of a top antireflective layer suitable for i-line exposure. (Assume a refractive index of 1.70 for the resist applied.)
6. The application of phase-shifting masks (PSMs) is regarded as a promising technology option to enhance conventional optical lithography. Cross sections for a conventional COG (chrome-on-glass) mask and two types of PSMs, as well as the corresponding electric fields on these masks, are de-

picted in Fig. 19. Discuss for each mask type the variations in electric field amplitude and intensity, and their impact on imaging performance.

REFERENCES

1. L. F. Thompson, C. G. Willson, and M. J. Bowden, eds., *Introduction to Microlithography* 2nd ed., ACS Professional Reference Book, 1994, pp. 1–138.
2. Y.-C. Lin, A. J. Purdes, S. A. Saller, and W. R. Hunter, *J. Appl. Phys.* 55(4):1110 (1984).
3. F. H. Dill, W. P. Hornberger, P. S. Hauge, and J. M. Shaw, *IEEE Trans. on Electron. Devices ED-22*:445 (1975).
4. H. Yanazawa, *Coll. Surfaces* 9:133 (1988).
5. T. Brunner, *Proc. SPIE 1466*:297 (1991).
6. O. S. Heavens, *Optical Properties of Thin Solid Films*, Academic Press, New York, 1955, p. 53.
7. W. Beinvogl and A. Gutmann, Submicron patterning techniques in integrated circuits, in *Low-Dimensional Structures in Semiconductors*, (A. R. Peaker and H. G. Grimmeiss, eds.), Plenum Press, New York, 1991, p. 89.
8. T. Tanaka, N. Hasegawa, H. Shiraishi, and S. Okazaki, *J. Electrochem. Soc. 137*: 3900 (1990).
9. M. K. Templeton, C. R. Szmanda, and A. Zampini, *Proc. SPIE 771*:136 (1987).
10. P. Trefonas and B. K. Daniels, *Proc. SPIE 771*:194 (1987).
11. K. Honda, B. T. Beauchemin Jr., E. A. Fitzgerald, A. T. Jeffries, S. P. Tadros, A. J. Blakeney, R. J. Hurditch, S. Tan, and S. Sakaguchi, *Proc. SPIE 1466*:141 (1991).
12. M. Hanabata, Y. Uetani, and A. Furuta, *J. Vac. Sci. Technol. B7*:640 (1989).
13. J. C. Housley, D. J. Williams, and Y. Horiuchi, *Semiconductor Intl.* 2:142 (1988).
14. J. R. Sheats, *Solid State Technology*, June 1989, p. 79.
15. T. X. Neenan, F. M. Houlihan, E. Reichmanis, J. M. Kometani, B. J. Bachman, and L. F. Thompson, *Proc. SPIE 1086*:2 (1989).
16. C. G. Willson, H. Ito, J. M. J. Frechet, and F. Houlihan, Proc. IUPAC Macrom. Symp., 28th, Amherst, Massachusetts, July 1982, p. 448.
17. W. Brunsvold, W. Conley, and D. Crockatt, *Proc. SPIE 1086*:357 (1989).
18. R. Schwalm, H. Binder, B. Dunbay, and A. Krause, *Polym. for Microelectr. Abstracts*:68 (1989).
19. M. J. O'Brien, *Polymer Engineering and Science*, Mid-July 1989, p. 846.
20. E. Reichmanis, G. Smolinsky, and C. W. Wilkins Jr., *Solid State Technology*, August 1985, p. 130.
21. M. Sebald, R. Leuschner, R. Sezi, H. Ahne, and S. Birkle, *Proc. SPIE 1262*:528 (1990).
22. M. Sebald, R. Sezi, R. Leuschner, H. Ahne, and S. Birkle, *Microelectron. Engin.* 11:531 (1990).
23. F. Coopmans and B. Roland, *Proc. SPIE*:34 (1986).
24. K.-H. Baik, L. Van den Hove, and B. Roland, *J. Vac. Sci. Technol. B9*:3399 (1991).
25. N. Shiraishi, S. Hirukawa, Y. Takeuchi, and N. Magome, *Proc. SPIE 1674*:741 (1992).

26. M. D. Levenson, N. S. Viswanathan, and R. A. Simpson, *IEEE Trans. on Electron. Devices ED-59*:1828 (1982).
27. H. Fukuda, A. Imai, and S. Okazaki, *Proc. SPIE 1564*:14 (1990).
28. O. Joubert, B. Dal'Zotto, B. Picard, A. Sahm, and S. Tedesco, *Microcircuit Engin. 17*:75 (1992).
29. K. Uenishi, Y. Kawabe, T. Kokubo, S. Slater, and A. Blakeney, *Proc. SPIE 1466*: 102 (1991).

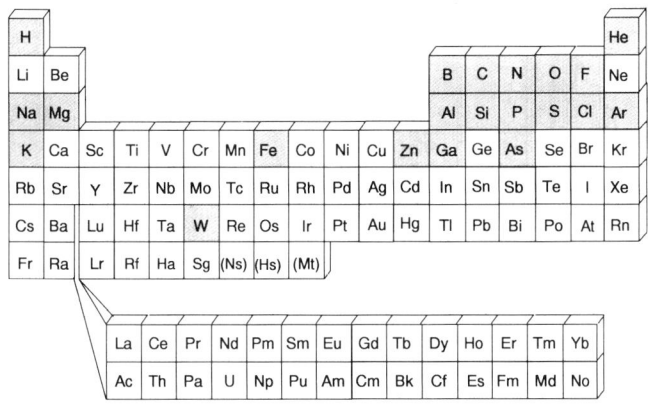

7
The Chemistry of Plasma Etching

Demetre J. Economou   *University of Houston, Houston, Texas*

I. INTRODUCTION

Patterning of thin solid films is a crucial operation in the fabrication of microcircuits and other microdevices at the heart of the "information age." Patterned layers of metals, semiconductors, and insulators are encountered in microelectronic devices, printed circuit boards, magnetic recording heads, and microelectromechanical devices. A layer may be patterned either by subtractive or additive methods with the aid of a photoresist (a polymer) mask. In the subtractive method (Fig. 1a), the portion of the layer not protected by the mask is removed by a process known as *etching*. In the additive method (Fig. 1b), the film is first deposited through a mask. The mask is then lifted off, leaving behind the desired pattern.

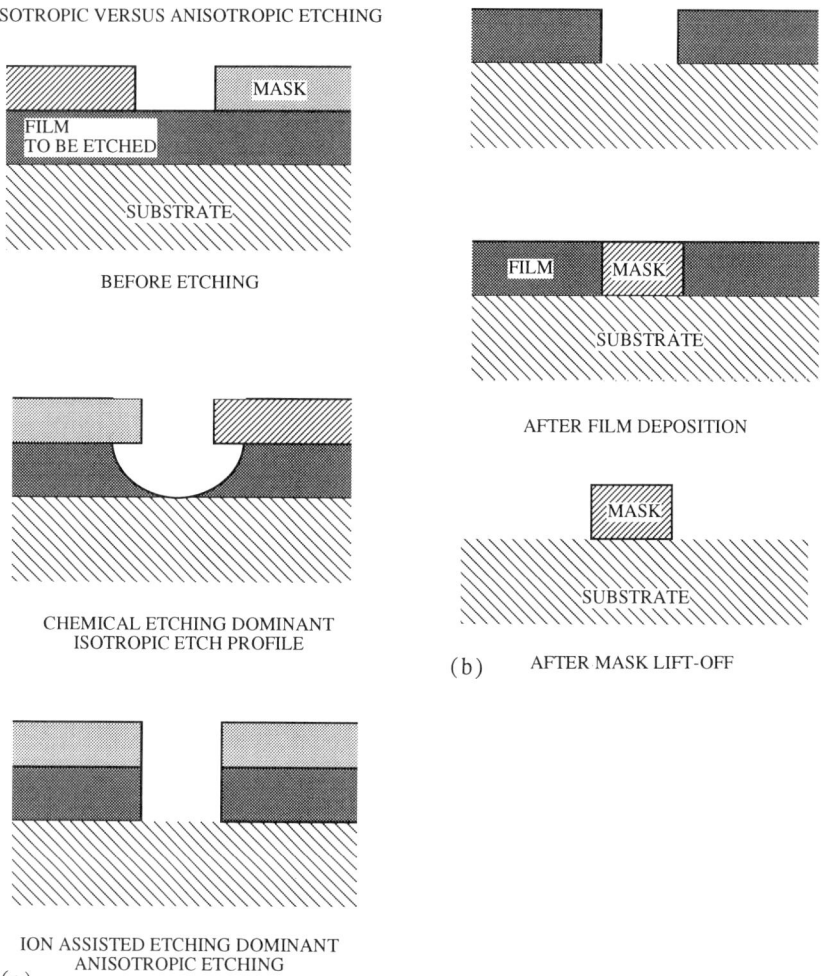

Figure 1 (a) The subtractive method of patterning a thin film; isotropic versus anisotropic etching. (b) The additive method of patterning a thin film. Method (a) is dominant in microelectronic device fabrication, where pattern dimensions are < 1 μm.

Microelectronic devices are composed of stacks of thin films of different materials (silicon, silicon dioxide, silicon nitride, metals, silicides) with complex patterns etched into these films. An example is Fig. 4 of Chapter 2, which shows the structure of a dynamic random access memory (DRAM) device. As microelectronic devices become smaller, faster, and more functional, even more stringent requirements are placed on the fabrication process.

There are two distinct etching methods: wet and dry. Wet processes make use of a liquid etchant. For example silicon dioxide can be etched in buffered hydrofluoric acid according to the reaction

$$SiO_2 + 6HF \rightarrow H_2SiF_6 + 2H_2O \tag{1}$$

Wet etching offers higher reaction rate and selectivity, but the reaction proceeds at comparable rate in all directions (except for the so-called crystallographic etching of single-crystal materials). This *chemical etching* results in mask undercut (Fig. 1a, middle), which limits the resolution of the method to a few microns. It follows that wet etching is advantageous for thicker films of not very fine dimensions. Wet etching is reviewed in a monograph by Ghandhi [1]. Dry etching makes use of gaseous plasmas [2]. Plasma etching* can provide very high resolution (better than 0.25 µ) by a combination of neutral radicals and ions bombarding the substrate.

Figure 2 is a schematic of a plasma etch process. The case of polysilicon etching in a chlorine plasma is shown as an example. The plasma is generated between a pair of parallel plate disk electrodes in a low-pressure (<1 torr) chamber (Fig. 2a). The silicon wafer to be processed rests on the lower electrode. The Cl_2 feedstock gas is attacked by plasma electrons to produce Cl radicals and Cl_2^+ ions. Radicals diffuse toward the wafer and adsorb on the surface. Ions accelerate in the plasma sheath naturally occurring over the wafer, and they bombard the wafer vertical to its surface (Fig. 2b). The combination of radical and ion bombardment produces $SiCl_4$ product, which is removed by the gas flow. Radicals do not have any directionality and can etch equally well in all directions. When the surface chemistry is dominated by the spontaneous reaction of radicals (chemical etching), mask undercut results (Fig. 1a, middle). When ion bombardment is necessary for reaction to occur, *anisotropic* (vertical) sidewall profiles can be obtained (Fig. 1a, bottom) since ions bombard only the bottom of the feature.

Wet and plasma etching methods are compared in Table 1. Plasma etching has totally replaced wet etching for patterning micron and submicron features in advanced microelectronic device fabrication. Also, the subtractive method far outperforms the additive method for creating high-resolution (submicron) patterns with controlled shapes.

---

*The term *plasma etching* is used in this chapter to refer to any process that uses a glow discharge to etch a wafer immersed in or in close proximity to a plasma. In colloquial terminology, plasma etching usually implies etching under relatively high pressure (>100s mtorr) and weak ion bombardment. The terms *reactive ion etching* and *reactive sputter etching* are used to describe etching under low pressure and intense ion bombardment. Such distinctions are not made in this chapter. Also, the terms plasma and glow discharge are used interchangeably in this chapter.

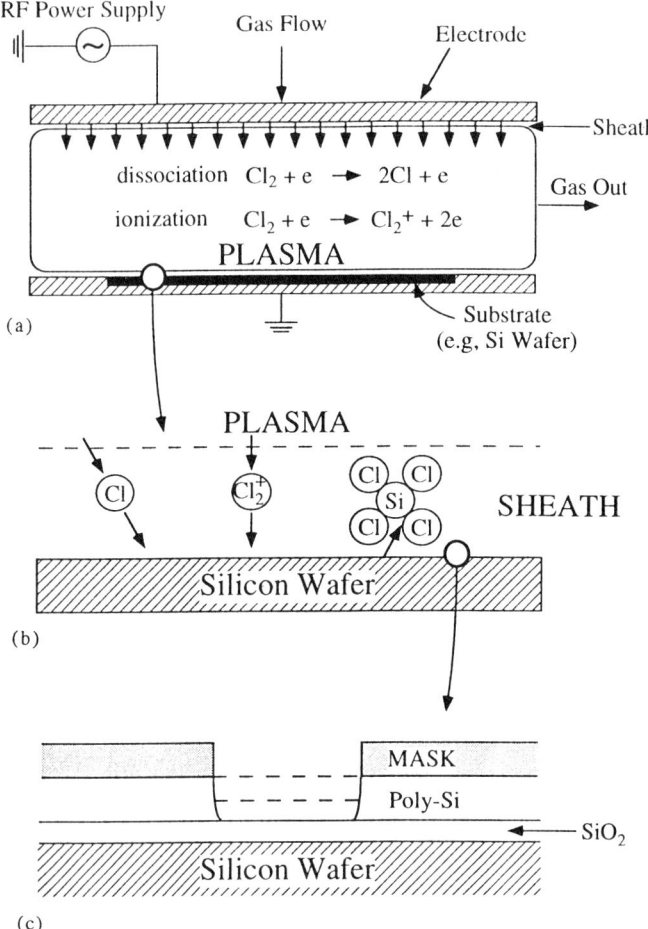

Figure 2 Operational characteristics of plasma etching. (a) Radicals and ions are generated in the plasma by electron impact of gas molecules. (b) The wafer is etched by the combined action of radicals and ions to yield a volatile product. (c) Ions accelerate in the sheath and bombard the wafer along the vertical direction, thereby inducing anisotropic etching of microscopic features. Dashed lines in (c) show the Poly-Si surface receding as a result of etching.

The goals of any plasma etch process are to achieve high *etch rate, uniformity*, and *selectivity*, controlled shape of the microscopic features etched into the wafer (degree of anisotropy), and no *radiation damage*. Manipulation of the plasma chemistry, coupled with the appropriate reactor design, is crucial for meeting these goals. There are many externally controlled variables that can influence the plasma chemistry, and in turn the process output. These are shown in Fig. 3.

High *etch rate* is desirable to increase the process throughput (wafers/hour). However, etch rate is often sacrificed to achieve better uniformity, selectivity, and anisotropy.

Uniformity refers to achieving the same etch rate across the wafer, which may be more than 20 cm in diameter. Nonuniform etching and/or nonuniformities

Table 1 Comparison of Wet and Plasma Etching

Characteristic	Plasma etching	Wet etching
Maturity	Industry standard	Industry standard
Cost	High	Low
Process control	Fairly easy	Not so easy
Etch rate	Moderate	High
Selectivity	Moderate	High
Resolution limit	<0.25 µm	>2 µm
Sidewall profile control	Good	Difficult
Waste disposal problem	Low	High

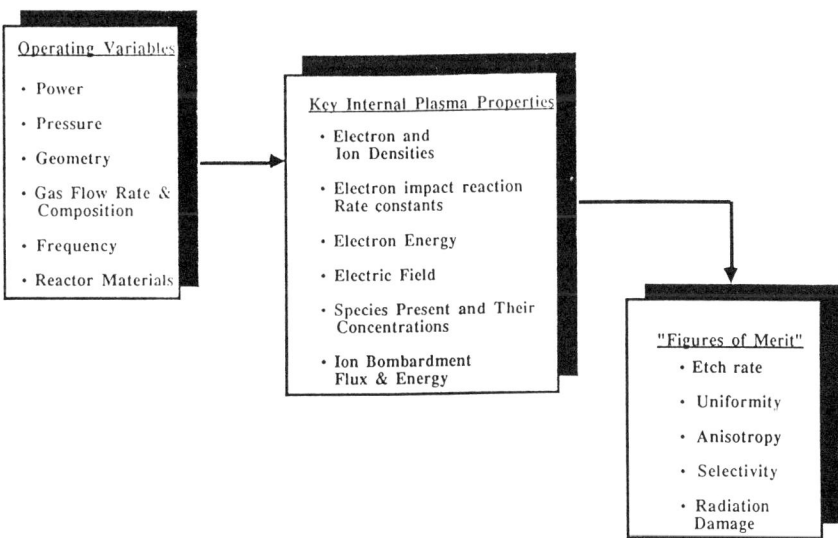

Figure 3 Representation of the parameter space in plasma etching. The key internal plasma properties are the bridge between externally controlled variables and the figures of merit (process output).

in the starting film thickness necessitate *overetch* beyond the *endpoint* of the process. Overetch must be minimized since the layer under the etched film (Fig. 1a) is then exposed to potentially harmful plasma radiation in some areas of the wafer while other areas are yet to clear. In addition, plasma uniformity is needed to avoid device damage (see below).

Selectivity refers to the relative rate of etching of one material with respect to another. Etch processes must be selective with respect to the mask and the underlying film. The mask must not be etched; otherwise, the desired pattern will be distorted. Selectivity with respect to the underlying layer is particularly important when that layer is thin or when the process uniformity is not good and overetch is required. For example, a process may call for etching polysilicon over a thin (100 Å) gate oxide. The selectivity of this process must be very high (>50:1); otherwise, a substantial thickness of the oxide may be lost.

The shape of the sidewall profile of the microscopic features etched into the wafer is of paramount importance. Most often anisotropic* sidewall profiles are required, perhaps with some taper at the bottom of the feature.

Radiation damage refers to structural damage of the crystal lattice or more importantly to electrical damage of sensitive devices caused by plasma radiation (ions, electrons, UV, and soft X-ray photons). For example, spatially nonuniform current flowing from the plasma to the wafer can lead to charging and breakdown of thin oxides; or intense ion bombardment can lead to structural damage of the top atomic layers of the etched film.

Plasma process development has been based largely on trial-and-error procedures guided by experience and intuition. A more rational selection of plasma etch chemistry and also of the appropriate plasma reactor design can be based on understanding the fundamentals of the chemical and physical processes taking place in the plasma and on the wafer surface.

In what follows, the fundamentals of plasma etching are discussed with emphasis on plasma chemistry. Discussion pertains to the kind of plasmas used in electronic materials processing. First, the general physical characteristics of plasmas are outlined. Gas-phase and surface chemistry are covered in some detail. Etching and chemical mechanisms for the most important material systems follows. The chapter concludes by discussing novel plasma etching systems.

Plasmas are also used for the low-temperature deposition of thin solid films. Reviews of plasma deposition can be found in Ref. 3.

---

*Literally speaking, anisotropic is anything not isotropic. In the plasma etching jargon, however, anisotropic etch is often taken to mean one with vertical or nearly vertical sidewalls.

II. PHYSICAL CHARACTERISTICS OF PLASMAS

A plasma is an ionized gas that contains light particles (electrons) and heavy particles (ions and neutrals). The vast disparity in the mass between the light and heavy particles is responsible for many of the unique characteristics of plasmas. There is a wide range of plasmas, depending on the charge number density, the electron temperature, and the gas temperature [2, p. 14]. Plasmas used in electronic materials processing are low-pressure *glow discharges* with operational and physical characteristics summarized in Table 2, which compares two commonly used plasma reactors. These plasmas are nonequilibrium systems (electrons have much higher temperature than ions and neutrals), and the degree of ionization is small (weakly ionized gases); there are 100 to 100,000 neutrals for every electron in the plasma.

Depending on the excitation frequency, plasmas range from direct current (dc) to radio-frequency (rf), to microwave frequency. The dc plasmas are not as common since very often the electrode (and/or wafer) is covered by an insulating material that cannot pass dc current. The rf plasmas are excited at 13.56 MHz, a frequency allowed by the Federal Communications Commission (FCC) for industrial operations, although other frequencies have also been used. Microwave plasmas are almost always excited at a frequency of 2.45 GHz, another FCC-allowed frequency.

Although a plethora of reactor configurations and methods for plasma generation exist, the parallel plate capacitively coupled reactor (also called the *diode*) shown in Fig. 2a is a typical example. A semiconductor wafer rests on one electrode and is showered with gas entering through the other electrode. The substrate electrode can be made larger to hold many wafers, but due to the tendency of increasing wafer sizes (>20 cm in diameter), single-wafer reactors

Table 2 Typical Parameters for RF Diode and High-Density Plasma (HDP) Reactors

Parameter	RF diode	HDP	Units
Pressure	10–1000	0.5–50	mtorr
Power	50–2000	100–5000	watts
Frequency	0.05–13.56	1–2450	MHz
Volume	1–10	1–50	liters
Magnetic field	0–100	0–1000	Gauss
Plasma density	10^9–10^{11}	10^{10}–10^{13}	cm^{-3}
Electron temperature	1–5	2–12	eV
Heavy particle temperature	<0.1	<0.25	eV
Ion bombardment energy	100–1000	20–250	eV
Fractional ionization	10^{-6}–10^{-3}	10^{-4}–10^{-1}	—

(processing one wafer at a time) are favored for improved uniformity and better process control. The wafer temperature is too low (near room temperature) for reaction to take place in the absence of a plasma. A voltage is now applied between the electrodes. The induced electric field causes gas ionization, and a plasma is generated. The electric field imparts forces on charged species (electrons and ions) but not on neutrals. Because the gas is only weakly ionized, charged species collide mainly with neutrals rather than with one another.

In the diode reactor, the wafer is bombarded by ions, and anisotropic etching is possible. However, in some applications anisotropy is not important; the removal (stripping) of the photoresist film from the wafer surface after etching is an example. In fact, in these applications ion bombardment may cause electrical damage to devices built onto the wafers. A reactor configuration that separates the wafer from the plasma is then used. The volume-loaded *barrel* reactor shown in Fig. 4a has the added advantage of high throughput. The plasma is generated in the annulus between the quartz barrel wall and the etch tunnel, a perforated

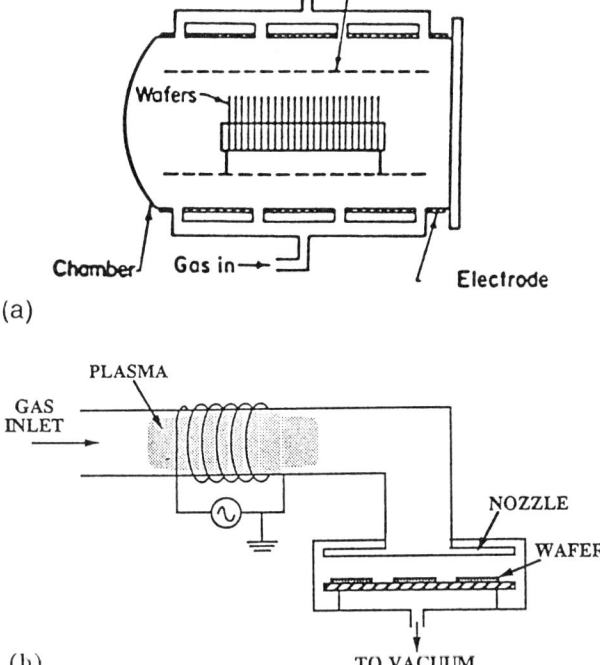

Figure 4 Plasma reactors in which etching is purely by neutral radical species (chemical etching): (a) barrel reactor; (b) downstream etching reactor.

metal cylinder. The wafers are loaded on a quartz boat that is placed inside the tunnel. Reactive radicals formed in the plasma diffuse through the perforations of the etch tunnel and then diffuse in between the wafers, where they etch the film. When the plasma is confined in the annular space outside the tunnel, the wafers are not subjected to ion bombardment. In the *downstream etching* reactor (Fig. 4b), the wafer is placed a long distance (10s to 100 cm) from the plasma. Reactive radicals formed in the plasma are transported by gas flow downstream to the wafer chamber, where etching occurs. If the plasma is far enough from the substrate, charged particles recombine before reaching the wafer. Thus, the wafer is exposed to neutral species alone.

The fact that electrons are much lighter than ions and neutrals has several consequences. First, electrons can absorb energy from the field much more efficiently compared with ions. The work done by a constant electric field E acting for time t on a charged particle with mass m and charge q is

$$W = \frac{(Eqt)^2}{2m} \tag{2}$$

Since the electron mass m_e is much less than the ion mass M_i, the work done on electrons is much higher than that done on ions. The second consequence is that electrons lose very little energy in elastic collisions with neutrals and ions. This can be understood by looking at the energy transfer function (ETF) in an elastic collision, defined as the fractional energy transferred from mass m_1 to mass m_2:

$$\text{ETF} = \frac{4 m_1 m_2 \cos^2 \theta}{(m_1 + m_2)^2} \tag{3}$$

where θ is the angle of attack. For an electron colliding with a neutral with mass $M \gg m_e$, the maximum value of ETF (corresponding to head-on collision, $\theta = 0$) is $4 m_e/M$; for electron collisions with argon, this fraction is 5.5×10^{-5}. Conversely, ions lose a substantial amount of energy in collisions with neutrals because their mass is comparable with that of neutrals.

Since electrons pick up most of the energy from the field, and they do not lose this energy efficiently in collisions with neutrals, electrons are "heated" to much higher temperatures than ions and neutrals (see Table 2). It is this nonequilibrium (different species have different temperatures) that makes plasmas suitable for electronic materials processing. Specifically, the energetic electrons can break bonds (bond energies are several eV) and initiate radical chemistry in a gas at near room temperature. One would need much higher temperatures (>500°C) to perform that chemistry by thermal excitation (e.g., thermal chemical vapor deposition) in the absence of a plasma. Thus, it is often said that glow discharge (cold) plasmas are capable of performing high-temperature chemistry at low temperatures. As device dimensions continue to shrink, low-temperature processing is becoming increasingly important in microelectronics to minimize dopant redistribution and defect generation. Low-temperature processing and anisotropy are the most important technological attributes of plasma etching.

A. Sheath Formation and Potential Distribution

Another consequence of the much smaller mass of electrons compared with ions is the formation of a so-called *sheath* over any material surface in contact with the plasma (Fig. 2b). When a *floating* surface (electrically disconnected) is first exposed to the plasma, the surface receives a much higher electron flux compared with the positive ion flux. For a Maxwellian velocity distribution of the particles, the current density due to particle j striking the surface is given by

$$J_j = \frac{qn_j \bar{c}_j}{4} = \frac{qn_j}{4}\sqrt{\frac{8kT_j}{\pi M_j}} \tag{4}$$

where n_j, T_j, M_j, and $\bar{c}_j$ are number density, temperature, mass, and mean thermal speed of particle j, and k is the Boltzmann constant. Since typically $n_e \approx n_i$ and $T_e \gg T_i$, it follows that $J_e \gg J_i$. Hence the surface charges negatively with respect to the plasma. An electric field is thus established that attracts ions and repels electrons so that, at steady state, the net current flowing to the surface is zero. The sheath thickness over a floating substrate is of the order of the *Debye length*, λ_D, where

$$\lambda_D = \sqrt{\frac{kT_e \varepsilon_0}{n_e q^2}} \tag{5}$$

is the distance over which macroscopic fields can penetrate the plasma. The sheath can be considerably thicker (10s of λ_D) when a *bias* voltage is applied to the substrate.

Formation of the sheath and the resulting sheath potential drop are of critical importance in plasma processing. The bulk of the plasma is approximately electrically neutral and cannot sustain a large electric field. The sheath electric field is much stronger (normally ~100 times stronger than that in the bulk) and is pointed such that positive ions are accelerated toward the wall (Fig. 5). Since the sheath thickness is much smaller than the lateral dimensions of the electrode, the electric field in the sheath is perpendicular to the wafer. Hence, ions acquire directional energy in the sheath and bombard the wafer normal to the surface, resulting in anisotropic etching (Fig. 1a, bottom). Ions coming from the plasma (process d) accelerate in the presheath (separated from the plasma by the dashed line in Fig. 5) and pick up energy of the order of kT_e (as required by the so-called *Bohm criterion*) before entering the sheath of thickness t. In the absence of collisions between ions and neutrals, the energy of ions bombarding the elec-

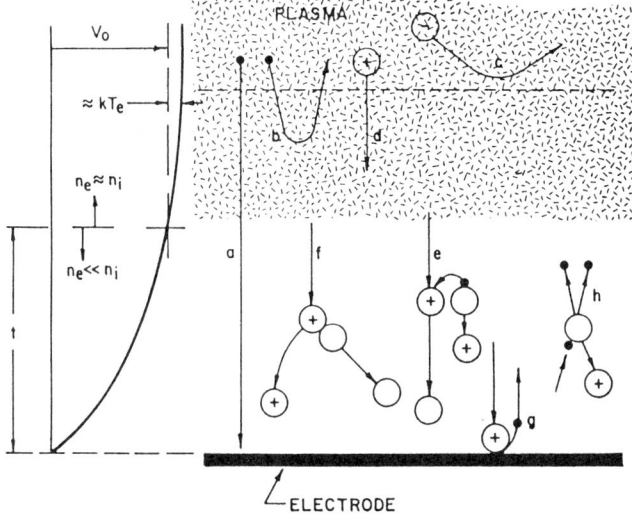

Figure 5 Representation of phenomena occurring in the sheath: (a) fast electrons can overcome the potential barrier of the sheath, thus escaping to the electrode; (b) slow electrons are reflected by the potential barrier; (c) negative ions are also reflected by the potential barrier, (d) positive ions are continuously injected into the sheath; (e) positive ions can suffer charge exchange with neutrals, or (f) elastic scattering by neutrals; (g) ion bombardment of the electrode can release secondary electrons, which (h) can induce ionization in the sheath. Schematic at left shows relevant time-average potential and other quantities: t = sheath thickness. V_0 = sheath potential, n_e, n_i = electron and ion density, respectively, T_e = electron energy in eV. (From Ref. 4, with permission.)

trode (or a wafer resting on the electrode) equals the sheath potential. However, ion collisions with neutrals in the sheath (processes e and f in Fig. 5) lower the ion bombardment energy. In ion–neutral scattering, the directionality of the ion is disrupted. Charge exchange is a glancing angle collision in which an ion turns into a (fast) neutral and a neutral turns into a (slow) ion. Fast electrons from the plasma can overcome the sheath potential escaping to the walls (process a), while slow electrons (process b) and negative ions (process c) are reflected by the sheath potential. Since the vast majority of plasma electrons cannot overcome the sheath potential, the electron density in the sheath is much lower than the positive ion density. Also, ion bombardment can cause *secondary electron emission* from the wall (process g). Secondary electrons are accelerated back into the plasma and can participate in the plasma chemistry by dissociating or ionizing the gas. Depending on their mean-free-path, [3]* secondary electrons can cause ionization within the sheath (process h), creating an *avalanche* of electrons.

The potential drop across the sheath over a floating substrate is approximately equal to $V_F = 0.5\ kT_e \times \ln(2.3\ m_e/M_i)$. For an argon plasma with $T_e = 2$ eV, one obtains $V_F = -10.4$ V; i.e., the substrate potential is 10.4 V lower than the plasma potential. (The sheath potential is the *difference* between the wall and plasma potentials.) The potential drop across a *biased* or grounded sheath can be much larger. However, for given voltage, the sheath thickness decreases as the Debye length decreases, i.e., as electron temperature decreases or electron density increases. The potential distribution in the reactor and in particular the sheath potential are of critical importance in plasma etching. The sheath potential determines the intensity of ion bombardment and in turn the etch rate, anisotropy, and selectivity. The sheath potential can be controlled by adjusting the operating variables (power, pressure, frequency; see Sec. III.C) and also the reactor design (for example area of substrate electrode compared with the counterelectrode area, see Eq. 6).

Figure 6 shows a typical potential distribution between the electrodes of a dc discharge. The cathode (left electrode) is biased and the anode is grounded. As mentioned, the bulk plasma (away from the electrodes) is electrically neutral and sustains only a small electric field (the electric field is the negative gradient of the potential, $E = -dV/dx$). However, large potential gradients (hence large fields) develop in the sheath regions near the electrodes. The sheath is much

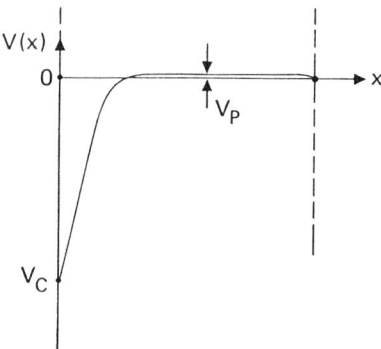

Figure 6 Potential distribution in a dc discharge sustained between two parallel plates of infinite extension. V_p is the plasma potential and V_c is the cathode potential.

---

*The mean-free-path is the average distance a molecule travels between collisions with other molecules.

thicker over the cathode, which is biased with a large negative voltage (say -1000 V). The cathode sheath contains a net positive charge and is essentially devoid of electrons. Positive ions accelerate in the cathode sheath and bombard the electrode with high-energy ejecting secondary electrons. These are essential for sustaining a dc discharge.

In capacitively coupled reactors, an alternating (ac) voltage is applied to the electrode through a capacitor (blocking capacitor), which does not allow any *net particle* (dc) current to flow through the electrode. This mode of coupling has important consequences regarding the potential distribution developed in the reactor, specifically the sheath voltage. Figure 7 shows the spatial and temporal variation of the potential in a parallel plate reactor of the type shown in Fig. 2a. The left electrode is driven by a sinusoidal radio-frequency voltage $V_{rf} = V_0 \sin(2\pi F t)$. The case shown is for a chlorine discharge with $V_0 = 150$ V and $F = 13.56$ MHz. In rf discharges, electrons oscillate back and forth between the electrodes. The potential distribution is such that the electrodes are bombarded by positive ions during the whole period of the rf cycle. Electrons "leak" to the electrode during a short time in the cycle when the sheath potential is low; in Fig. 7 this will occur at about $\tau = 0.25$ on the left electrode and at $\tau = 0.75$ on the right electrode (the two electrodes are 180° out of phase). The electron flux is such that when integrated over the rf cycle, the electron current is equal to the positive ion current, since no dc current is allowed to pass through the electrode. In rf discharges, electrons are "heated" by the oscillating electric

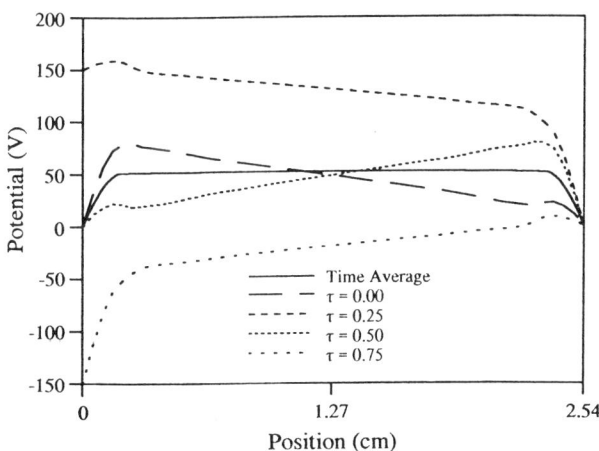

Figure 7 Spatiotemporal profiles of the potential distribution in a 13.56 MHz chlorine discharge sustained between two parallel plates of infinite extension. The left electrode is driven by a sinusoidal voltage of 150 V amplitude. Note that the plasma potential (potential in the bulk, away from walls) is more positive than either electrode potential. (From D. Lymberopoulos, Ph.D. thesis in progress, University of Houston.)

field near the plasma/sheath interface and in the bulk plasma. Secondary electron emission is not necessary to sustain rf discharges, although it is operative under certain conditions, especially at frequencies below ~1 MHz.

Figure 7 is for a symmetric reactor, i.e., one in which the two electrodes have the same area. Then the time-average potential distribution is symmetric as well. In asymmetric reactors, the time-average potential distribution looks much like the one in Fig. 6. In fact, assuming capacitive voltage division between the two sheaths, the time-average sheath potentials scale with the inverse electrode area ratio as

$$\frac{V_1}{V_2} = \left(\frac{S_2}{S_1}\right)^n \qquad (6)$$

The exponent has the theoretical value of $n = 4$, but a value of $n = 1$ has been determined experimentally [5]. In any case, Eq. (6) shows that the smaller-area electrode develops a larger sheath voltage; hence, it receives stronger ion bombardment.

In Fig. 2a, the wafer is positioned on the grounded electrode and the counterelectrode is powered. Often, the reverse configuration is used; the wafer rests on the powered electrode and the counterelectrode is grounded. In capacitively coupled reactors, it is not important which of the two electrodes is powered; what is important is the area ratio of the two electrodes (Eq. 6).

The time evolution of the electrode and plasma potential are shown schematically in Fig. 8 (a: symmetric, b: asymmetric). The solid lines show the potential difference between the driven (powered) electrode and the grounded electrode as a function of time. The time-average electrode potential for a symmetric system is zero. This is not the case for the asymmetric system (unequal-area electrodes). A negative *dc self-bias* develops between the electrodes for this system (shown as V_T in the figure). The dc self-bias is the difference between the time-average (dc) voltages developed across the sheath over the electrodes. The plasma potential is shown as the "wavy" dashed lines in Fig. 8. A striking feature of the plasma potential is that it is the most positive potential in the system (see also Figs. 6 and 7). This is again a consequence of the smaller electron mass, hence higher electron mobility.

III. PLASMA CHEMISTRY

There is a plethora of chemical species in a plasma, including electrons, positive ions, negative ions, molecules, atoms, and radicals in the ground state as well as in excited states. These species are not in thermodynamic equilibrium by any means. For example, electrons have an equivalent temperature of 10,000s K, ground state neutrals and ions are at slightly above room temperature (300–500

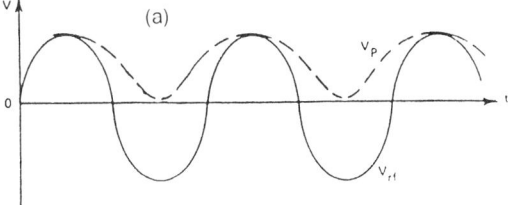

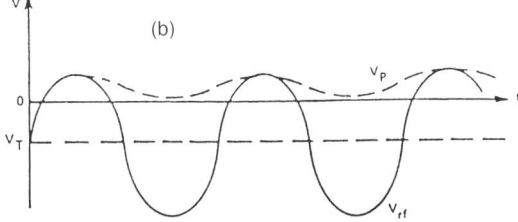

Figure 8 (a) Solid line: Time variation of the electrode potential V_{rf} of a capacitively coupled rf symmetric (equal areas of powered and grounded electrodes) discharge. There is no dc self-bias developed on the electrode (i.e., time-average potential is zero). Dashed line shows the plasma potential V_p, always the most positive potential. (b) Same as (a), but for an asymmetric (unequal areas of powered and grounded electrodes) discharge. A negative dc self-bias V_T develops at the powered electrode. (From Ref. 5, with permission.)

K), and excited neutrals and ions are at 1000s K. Plasma particles participate in a complex reaction network, for which kinetic information is scarce. One can distinguish between reactions occurring in the gas phase (volume or homogeneous chemistry) and those on surfaces (heterogeneous chemistry).

Figure 9 shows a "molecular" picture of the steps typically involved in plasma etching. Radicals generated in the plasma diffuse or are convected by gas flow to the surface, where they adsorb. The adsorbed species (adspecies) may react with the surface to form products. The products then desorb and diffuse back into the gas phase. The surface processes may be strongly influenced by energetic particle bombardment of the surface, including positive ions, electrons, and photons. Of these, positive ion bombardment is thought to be the most important. Negative ions are excluded because they are not energetic enough to overcome the sheath potential barrier (process c in Fig. 5). It must be emphasized that product *volatility* is a necessary condition for plasma etching to occur. Otherwise, the products block the surface, and the reaction stops altogether. The plasma chemistry is chosen so that the film to be etched forms volatile products while the mask and the underlying film (see Fig. 1a) do not. This allows for selective etching. However, in the presence of ion bombardment,

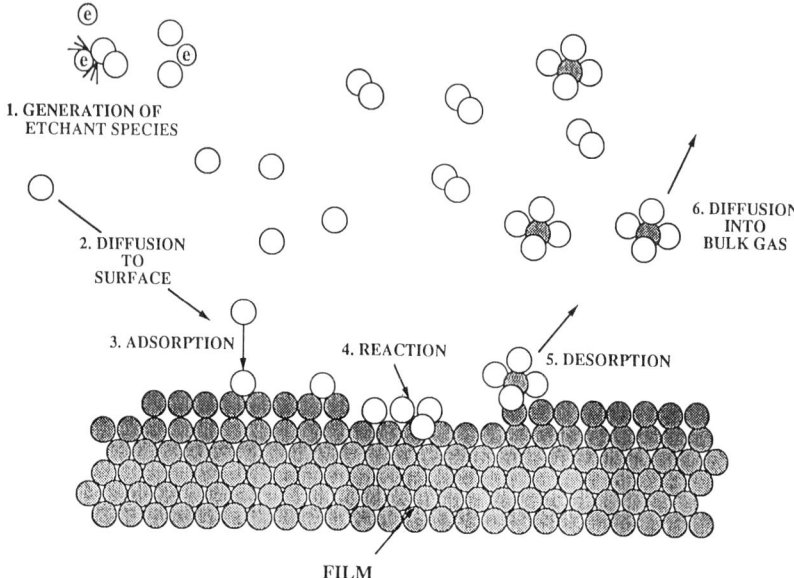

Figure 9 "Molecular" representation of primary processes occurring in plasma etching. The basic steps are etchant generation, diffusion to the surface, adsorption, reaction on the surface to form product, product desorption, and product diffusion into the flowing gas. The surface processes can be strongly affected by energetic ions bombarding the surface. (From Ref. 6, with permission.)

there is always some etching of the mask and the underlying film. The goal is to minimize that etching, thus maximizing selectivity.

A. Gas-Phase (Volume) Chemistry

Thermal dissociation reactions are not important in plasma systems because of the low gas temperature (<200°C). Dissociation of molecules is induced by energetic electron bombardment instead. In contrast, thermal dissociation is the mechanism for radical generation in thermal chemical vapor deposition (CVD) systems.

Tables 3 and 4 provide a list of the most important types of reactions taking place in glow discharges. Two rather extreme cases are shown. The argon discharge (Table 3) is a typical example of an inert gas plasma, in which negative ions do not play a role (*electropositive* discharge). Pure argon discharges are not used for etching, but they find widespread application in film deposition by *sputtering*. These discharges form an abundance of metastables by excitation of the ground state species. Metastables are high-energy species (11.6 eV for Ar)

Table 3 Representative Reactions in Argon Plasma

Excitation to metastables	$Ar + e^- \rightarrow Ar^* + e^-$	T1
Metastable ionization	$Ar^* + e^- \rightarrow Ar^+ + 2e^-$	T2
Ground state ionization	$Ar + e^- \rightarrow Ar^+ + 2e^-$	T3
Two-metastable ionization	$Ar^* + Ar^* \rightarrow Ar^+ + Ar + e^-$	T4
Metastable quenching to resonant	$Ar^* + e^- \rightarrow Ar^r + e^-$	T5
Metastable quenching to higher states	$Ar^* + e^- \rightarrow Ar^h + e^-$	T6
Metastable superelastic collisions	$Ar^* + e^- \rightarrow Ar + e^-$	T7
Radiative decay of resonants	$Ar^r \rightarrow Ar + h\nu$	T8
Excitation to higher states	$Ar + e^- \rightarrow Ar^h + e^-$	T9
Radiative decay of higher states	$Ar^h \rightarrow Ar^* + h\nu$	T10
Momentum transfer	$Ar + e^- \rightarrow Ar + e^-$	T11

Table 4 Representative Reactions in Chlorine Plasma

Molecular chlorine excitation	$Cl_2 + e^- \rightarrow Cl_2^* + e^-$	T12
Atomic chlorine excitation	$Cl + e^- \rightarrow Cl^* + e^-$	T13
Molecular chlorine radiative decay	$Cl_2^* \rightarrow Cl + h\nu$	T14
Atomic chlorine radiative decay	$Cl^* \rightarrow Cl + h\nu$	T15
Molecular chlorine ionization	$Cl_2 + e^- \rightarrow Cl_2^+ + 2e^-$	T16
Atomic chlorine ionization	$Cl + e^- \rightarrow Cl^+ + 2e^-$	T17
Dissociative attachment	$Cl_2 + e^- \rightarrow Cl^- + Cl$	T18
Dissociation of chlorine	$Cl_2 + e^- \rightarrow 2Cl + e^-$	T19
Momentum transfer to molecular chlorine	$Cl_2 + e^- \rightarrow Cl_2 + e^-$	T20
Momentum transfer to atomic chlorine	$Cl + e^- \rightarrow Cl + e^-$	T21
Positive–negative ion recombination	$Cl^- + Cl_2^+ \rightarrow 3Cl$	T22
Electron–ion recombination	$Cl_2^+ + e^- \rightarrow 2Cl$	T23
Wall recombination of atomic chlorine	$Cl + Cl + wall \rightarrow Cl_2 + wall$	T24
Volume recombination of atomic chlorine	$2Cl + Cl_2 \rightarrow 2Cl_2$	T25

that are forbidden from emitting radiation, and they have a comparatively long lifetime (10s of ms). As such, metastables can be active participants in plasma chemistry. For example, they can be ionized much more readily than ground state argon (*two-step ionization*) or they can dissociate a molecular gas into radicals. This happens in discharges that use argon as a component of the etch gas mixture. The chlorine discharge (Table 4) is an example of a molecular plasma that dissociates into radicals and also forms negative ions very readily (*electronegative* discharge). It is used for etching silicon, metals (e.g., aluminum), and compound semiconductors (e.g., GaAs).

1. Electron-Impact Reactions

Electrons are the heart of any plasma chemistry. Electrons ionize molecules and atoms (e.g., reactions T2, T3 in Table 3, and T16, T17 in Table 4), producing more electrons to counterbalance the electron losses and, thus, sustain the discharge. Also, electrons dissociate molecules to form highly reactive atoms and radicals, which are potential etchants (T19). Electron bombardment of neutrals and ions can excite bound electrons to higher energy levels (T9, T12, T13). As the excited species relax to a lower energy state, they emit photons, giving the plasma a glow characteristic of the gas (T10, T14, T15). Since there are very few electrons in the sheath region (because the sheath potential repels electrons, process b, Fig. 5) the excitation rate is very low in the sheath. Hence, the sheath appears dark (it is also called *dark space*) relative to the bulk plasma glow. In electronegative gases (such as Cl_2 or SF_6), electron attachment to molecules creates negative ions (T18).

The rate of electron impact reactions is given by

$$R_{ej} = k_j n_e n_j \tag{7}$$

where the rate coefficient

$$k_j(t) = \sqrt{\frac{2}{m_e}} \int_0^\infty \sigma_j(\varepsilon) f(\varepsilon,t) \varepsilon \, d\varepsilon \tag{8}$$

can be a function of time when the plasma is sustained by a time-varying electric field. Here n_e and n_j are the number density (particles/cm^3) of the electrons and the collision partner, respectively; $\sigma(\varepsilon)$ and $f(\varepsilon, t)$ are the collision *cross section* and the *electron energy distribution function* (EEDF) as a function of electron energy, ε.

Electron collisions can be categorized as elastic and inelastic. Elastic collisions (T11, T20, T21) do not alter the internal energy of the heavy particle and result in a small energy loss by the electrons (see Eq. 3). Inelastic collisions change the internal energy of the collision partner. Inelastic collisions (ionization, dissociation, excitation, etc.) usually have a threshold energy below which the collision cannot happen (cross section is equal to zero). In a typical example of an inelastic process, the cross section rises rapidly with electron energy above the threshold energy, reaches a maximum, and then declines at high energies. Exothermic collisions may have no threshold, meaning that electrons with very low energy can participate in the collision. A notable example is attachment of electrons to electrophilic gases to create negative ions (T18). In fact, the attachment rate may be highest for thermal electrons, i.e., electrons having temperature close to that of the gas. Another example is electron–positive ion recombination (T23), which is dissociative or radiative, i.e., results in light emission. The collision cross section for this process increases as the electron energy decreases.

A set of collision cross sections for selected electron impact reactions with molecular chlorine is shown in Fig. 10. An extensive review of electron collision processes and collision cross sections for different gases is provided in Ref. 7.

It is worth noting that the reaction rate depends on the relative velocity between the electron and the projectile. Since the electron velocity is much higher than the neutral velocity, the latter is neglected. Hence, the electron-impact reaction rate coefficients are a function of the electron temperature rather than the gas temperature. A notable exception is reactions that can happen even for low-energy electrons (e.g., attachment). At these low electron energies, the velocity of the molecules also matters. Thus, it comes as no surprise that the attachment rate is a function of gas temperature as well as electron temperature [7, p. 556].

2. Electron Energy Distribution Function

Determination of the electron energy distribution function is one of the central problems in understanding plasma chemistry. Plasma chemistry starts by electron-impact reactions, and the rate of such reactions depends on the EEDF (see Eq. 8). The EEDF is obtained by solving the Boltzmann transport equation [8], and this requires knowledge of the electron collision cross sections and the collision partners (i.e., plasma gas composition). Plasmas contain a variety of radicals and excited states for which the collision cross sections are not known. The plasma composition is also unknown. Because of these complexities, the

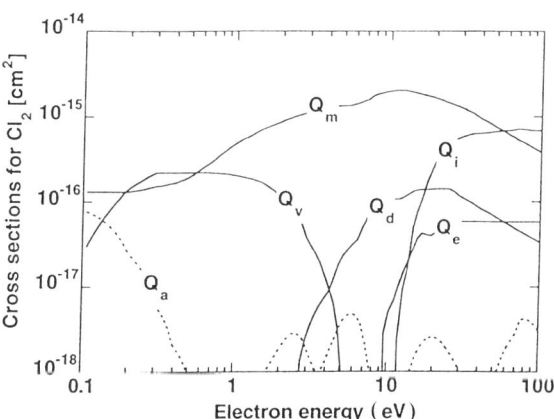

Figure 10 Electron energy dependence of collision cross section for representative electron-impact processes with molecular chlorine. Q_m: momentum transfer, Q_a: attachment, Q_v: vibrational excitation, Q_d: dissociative excitation, Q_e: electronic excitation, Q_i: ionization. Note threshold energy for endothermic processes (vibrational excitation, ionization, etc.). (Courtesy of E. Meeks and J. Shon, Sandia National Labs, Livermore, California.)

EEDF is often assumed to follow the Maxwell–Boltzmann distribution (also called *Maxwellian*):

$$f(\varepsilon) = \frac{2}{\sqrt{\pi}} \frac{\exp(-\varepsilon/kT_e)}{(kT_e)^{3/2}} \qquad (9)$$

with a mean energy

$$\langle \varepsilon \rangle = \frac{3}{2} kT_e \qquad (10)$$

where T_e is the electron temperature. It should be noted that an electron temperature is defined properly only when the EEDF is Maxwellian. For non-Maxwellian distributions, an equivalent electron "temperature" may be defined based on the mean electron energy, by using Eq. (10). A Maxwellian distribution would result if the electron–electron collisions are frequent enough for the electron gas to achieve equilibrium. For typical plasmas of interest, electron–molecule collisions are by far the most important since the degree of ionization is low ($n_e/N \ll 1$). In high-density (large n_e), low-pressure (small N) plasmas, electron–electron collisions become more important.

Figure 11 shows a plot of a Maxwellian EEDF with $T_e = 5$ eV. The collision cross section for ionization of molecular chlorine is superimposed on this plot.

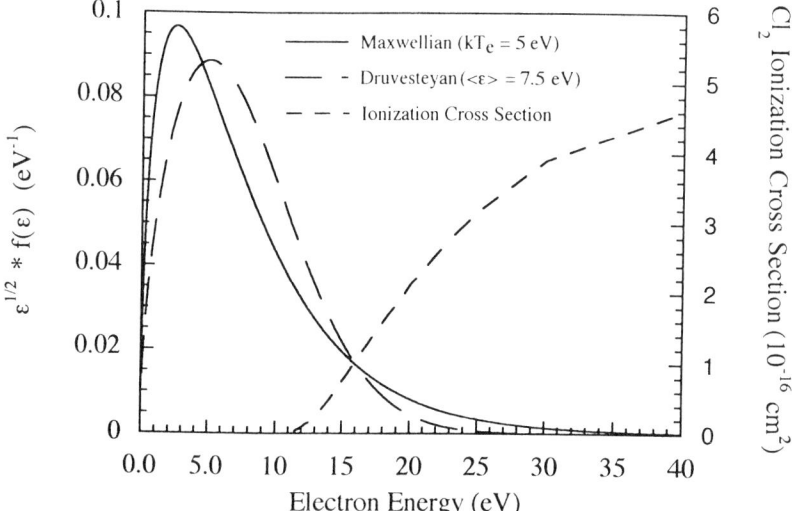

Figure 11 Maxwellian and Druyvesteyn electron energy distribution functions with the same average electron energy of 7.5 eV. The molecular chlorine ionization cross section is also shown. Only the "tail" electrons can perform ionization.

Only electrons with energy higher than the threshold of 11.5 eV can perform ionization. The Druyvesteyn distribution is given by [9]

$$f(\varepsilon) = 1.04 \langle \varepsilon \rangle^{-3/2} \exp\left(-0.55 \frac{\varepsilon^2}{\langle \varepsilon \rangle^2}\right) \tag{11}$$

A plot of a Druyvesteyn distribution with the same average energy as the Maxwellian (7.5 eV, see Eq. 10) is also shown in Fig. 11. It is evident that a smaller fraction of electrons can perform ionization if the EEDF happens to be Druyvesteyn. It is thought that this behavior better represents reality, since the high-energy electrons are depleted by inelastic collisions.

Figure 12 shows the actual $f(\varepsilon)$ in a static (dc) electric field in Cl_2 gas, calculated using the cross sections of Fig. 10 [10]. The electric field to neutral density ratio (E/N) is an important "scaling" parameter for gas discharges. E/N is a measure of the energy imparted on an electron during a mean-free-path. The higher the electric field E and/or the lower the gas density N (i.e., the longer the mean-free-path), the higher the energy imparted to an electron. Thus, as the field strength increases, the distribution extends to higher electron energies, i.e., the average energy increases. In Fig. 12, the EEDF is normalized such that

$$\int_0^\infty \sqrt{\varepsilon}\, f(\varepsilon)\, d\varepsilon = 1 \tag{12}$$

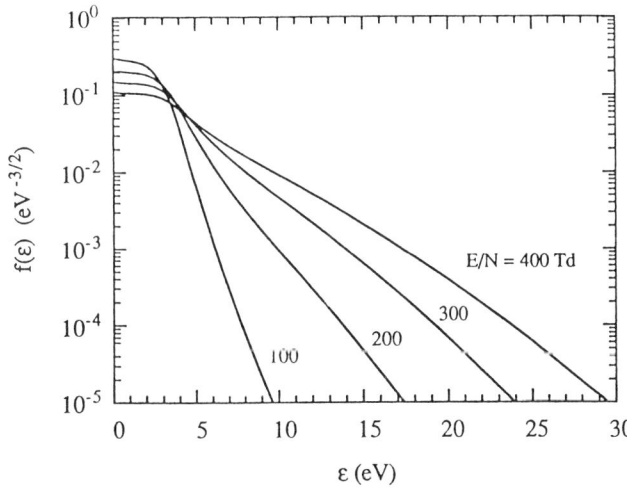

Figure 12 Electron energy distribution function (EEDF) in molecular chlorine as a function of the electric field to neutral density ratio E/N (1 Td = 10^{-17} V-cm^2). A Maxwellian EEDF would be a straight line on this plot. (From Ref. 10, with permission.)

Using this normalization, a Maxwellian distribution would be a straight line in a semilogarithmic plot of $f(\varepsilon)$ vs. ε. Evidently, the EEDFs in Cl_2 are not Maxwellian for the range of E/N shown. This is because inelastic collisions deplete the tail of the distribution. Figure 13 shows the calculated rate coefficients for electron attachment to molecular chlorine and total ionization (i.e., both Cl and Cl_2), again in a dc field. The calculation was done using Eq. (8) with the cross sections taken from Fig. 10 and the EEDF taken from Fig. 12. The rate coefficients increase with E/N. The strong dependence of the ionization rate on E/N is evident. This is characteristic of high-threshold processes (e.g., ionization). In contrast, low-threshold processes (e.g., attachment) have a much weaker dependence on the electric field. The threshold energy corresponds to the *activation energy* of conventional thermal chemical reactions.

The time response of the EEDF in an ac field [11] is determined by the relative magnitudes of the angular frequency of the applied electric field ($E = E_0 \sin(\omega t)$, $\omega = 2\pi F$) and the electron momentum (ν_m) and energy (ν_u) relaxation frequencies. The collision frequency is given by $\nu_j = \sigma_j u N$, where σ_j is the cross section for process j, and u is the electron velocity ($\varepsilon = 1/2 m u^2$). When $\omega \gg \nu_u$, the characteristic time for energy relaxation is much greater than the period of the applied field. As a consequence, the EEDF cannot respond to the time variation of the applied field and does not change appreciably over a cycle. At

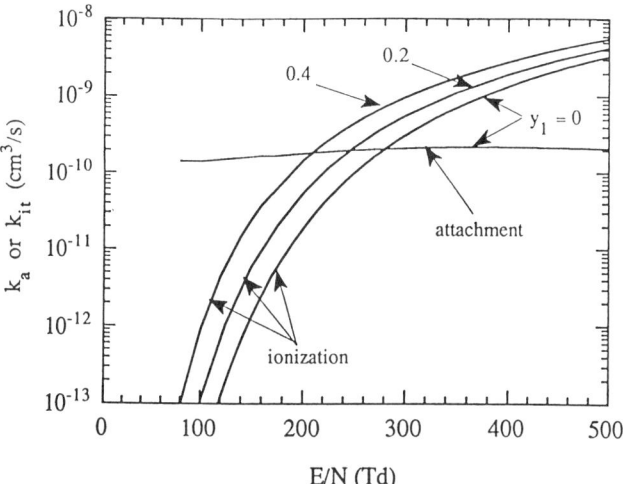

Figure 13 Rate coefficients in chlorine as a function of the electric field to neutral density ratio, E/N. k_{it} is total ionization including molecular and atomic chlorine; the atomic mole fraction is denoted by y_1. k_a is the coefficient for attachment to molecular chlorine. (From Ref. 10, with permission.)

the other extreme, $\omega \ll \nu_u$, the EEDF can follow the applied field faithfully. Under this condition, the EEDF at any time would be identical to the EEDF obtained in a static field of magnitude equal to the magnitude of the ac field at that particular moment. This is the so-called quasi-steady-state approximation. Since the collision frequency is proportional to N, the quantity ω/N is another scaling parameter of importance in describing time-dependent phenomena in gas discharges. As ω/N decreases, the EEDF and, correspondingly, the rate coefficients (Eq. 8) are modulated deeper.

Many investigators have used the effective field approximation to describe the EEDF under high-frequency excitation [9]. In this approximation, the high-frequency field is considered equivalent to an effective dc field of magnitude

$$E_{\text{eff}} = \frac{E_0/\sqrt{2}}{\sqrt{(\omega/\nu_m)^2 + 1}} \qquad (13)$$

This obviates the need for including the time dependence when solving the Boltzmann equation, since the sinusoidal field $E = E_0 \sin(\omega t)$ is replaced by a dc field.

3. What Determines the Electron Temperature, T_e?

Before the plasma chemistry can be described, one needs to know the EEDF. Even if the EEDF is assumed to be Maxwellian, the electron temperature T_e is not known a priori; it depends on the type of gas, reactor geometry, and operating conditions (e.g., gas pressure). This is in direct contrast to thermal CVD, where the (gas) temperature can be established basically by setting the wall temperature. In plasmas, the electric field and, concomitantly, the electron temperature adjust such that a self-sustained discharge is obtained: one in which the electron production and loss rates balance each other. In order to understand the factors controlling the electron temperature, a simple physical model, the so-called *ambipolar diffusion* model, will be considered. In this model, the electron density balance is described by a simple diffusion equation of the form

$$D_a \frac{d^2 n_e}{dx^2} + k_i N n_e = 0 \qquad (14)$$

written here in one spatial dimension (along x) for simplicity. Equation (14) suggests that the electrons are lost to the walls of the plasma reactor by ambipolar diffusion (first term on left-hand side), and this loss is exactly balanced by production of electrons through ionization of neutral atoms (second term on left-hand side), with number density N. Two-step ionization, which is common in noble gas discharges having an appreciable concentration of metastable atoms (see reactions T1 and T2 of Table 3), is neglected. Ambipolar refers to the diffusion of electrons under the combined effect of concentration gradients and the induced space-charge electric field [8]. The space-charge field is generated as the mobile electrons escape to the walls, leaving behind a positive charge. The field slows down the electrons and speeds up the positive ions, so that at steady state the net current to the wall is zero. k_i is the ionization rate coefficient, which depends on the EEDF (see Eq. 8), and D_a is the ambipolar diffusion coefficient, given by

$$D_a = D_+ \left(1 + \frac{T_e}{T_g}\right) \tag{15}$$

where D_+ is the diffusion coefficient (free diffusion in the absence of any field) of positive ions. Assuming spatially constant D_a and k_i, the solution to Eq. (14) for a parallel plate geometry, under the boundary condition that the electron density vanishes on the walls, is

$$n_e = n_{e0} \cos\left(\frac{\pi x}{2L}\right) \tag{16}$$

where $-L \leq x \leq L$, and n_{e0} is the electron density at the center ($x = 0$). This solution is obtained only for a specific combination of the system parameters (eigenvalue problem), namely,

$$k_i N = \frac{D_a}{(2L/\pi)^2} = \frac{D_a}{\Lambda^2} \tag{17}$$

where the *electron diffusion length* is $\Lambda = 2L/\pi$, for infinite parallel plates separated by $2L$. For a cylindrical reactor of radius R and length $2L$,

$$\frac{1}{\Lambda^2} = \left(\frac{2.406}{R}\right)^2 + \left(\frac{\pi}{2L}\right)^2 \tag{18}$$

Λ is given for different geometries by Cherrington [8].

Equation (17) is the condition that must be satisfied for a self-sustained discharge, i.e., one in which the electron production exactly balances the electron loss. Equation (17) further implies that the EEDF must be established in such a way so that the resulting ambipolar diffusivity and ionization rate coefficient (the two quantities in Eq. (17) that depend on the EEDF) satisfy that equation. If one assumes a Maxwellian EEDF, D_a and k_i may be expressed in terms of T_e. Then, Eq. (17) becomes an equation for electron temperature. The electron density in the discharge center n_{e0} is determined by the amount of power delivered to the plasma.

Most of the approximations we have made are not necessary; they were introduced in order to derive simple expressions such as Eqs. (16) and (17). For example, k_i may be space-dependent, or two-step ionization may be included.

4. Scaling Parameters

As noted in the discussion of the EEDF, E/N and ω/N are important scaling parameters in gas discharge physics and chemistry. For example, E/N determines the electron temperature and rate coefficients of electron-impact reactions. It is desirable to express E/N as well as other key internal plasma properties (e.g., electron density) as a function of easily measurable quantities such as the reactor dimensions, power delivered to the plasma, and so on. It turns out that the following two parametric groups are functions of $N\Lambda$ only [10]:

$$\left(\frac{E}{N}\right)_{\text{rms}} = F_1 \; (N\Lambda) \tag{19}$$

$$\frac{\langle n_e \rangle V_p}{P_{\text{rf}} \Lambda} = F_2 \; (N\Lambda) \tag{20}$$

These expressions were developed for an rf discharge operating at 13.56 MHz. $(E/N)_{\text{rms}}$ is the root-mean-square of the electric field to neutral density ratio (for a sinusoidal field, $E = E_0 \sin \omega t$, the rms value is equal to $E_0/\sqrt{2}$). Quantities in brackets, $\langle \rangle$, denote values time-averaged over an rf cycle. The dependence of $(E/N)_{\text{rms}}$ on $N\Lambda$ and atomic chlorine mole fraction is shown in Fig. 14 for a pure chlorine discharge. In this system, both ambipolar diffusion and attachment are important electron loss mechanisms. The electric field to neutral density ratio becomes independent of $N\Lambda$ above 2.5×10^{16} cm^{-2}, as the loss of electrons via diffusion becomes negligible compared with attachment. For high values of $N\Lambda$, the self-sustained electric field is determined solely by the dependence on E/N of the total ionization and attachment rate constants. It is found as the intersection point of the two curves as shown in Fig. 13.

The dependence of electron density on operating conditions is shown in Fig. 15, in which the group $(\langle n_e \rangle V_p)/(P_{\text{rf}} \Lambda)$ is plotted as a function of $N\Lambda$, for different y_{Cl}. V_p is the plasma volume and P_{rf} is the power deposited in the plasma. At constant power and constant Λ, the electron density first increases with pressure, goes through a maximum, and then decreases with further increase in pressure. The electron density changes only moderately for moderate variations of the atomic chlorine mole fraction.

Since the electron-impact reaction rate coefficients are functions of E/N and y_{Cl}, their time-averaged values are functions of $N\Lambda$ only, for a given atomic chlorine mole fraction. The time-averaged dissociation rate coefficient $\langle k_d \rangle$ was calculated by time-averaging Eq. (8) and is shown in Fig. 16. The dissociation rate coefficient is relatively independent of operating conditions for high $N\Lambda$.

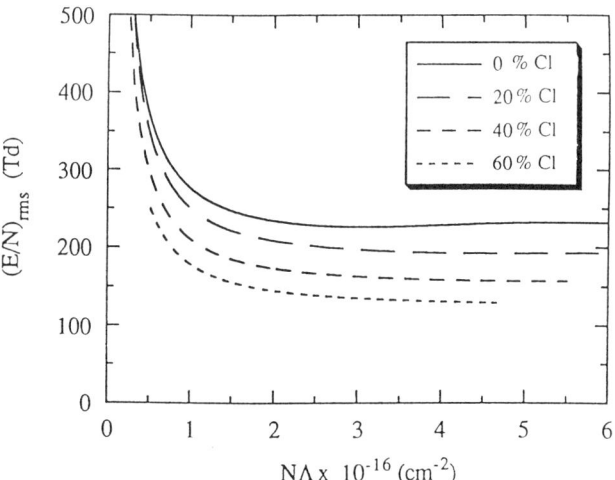

Figure 14 The electric field to neutral density ratio, E/N, as a function of $N\Lambda$ in a self-sustained chlorine discharge containing different atomic chlorine mole fractions. (From Ref. 10, with permission.)

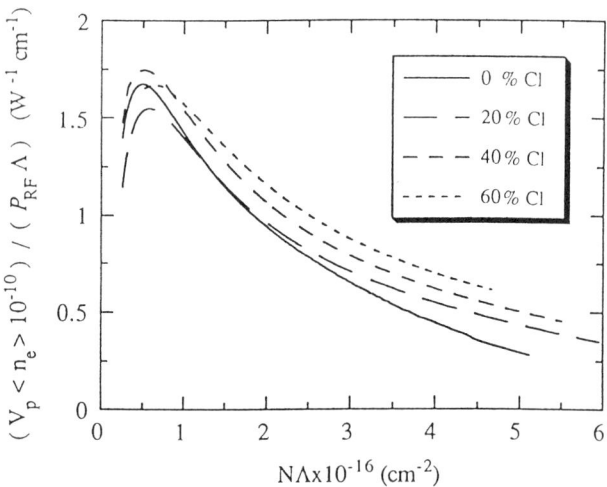

Figure 15 The parameter $V_p \langle n_e \rangle /(\Lambda P_{rf})$ (V_p is the plasma volume) as a function of $N\Lambda$ in a self-sustained chlorine discharge containing different atomic chlorine mole fractions. (From Ref. 10, with permission.)

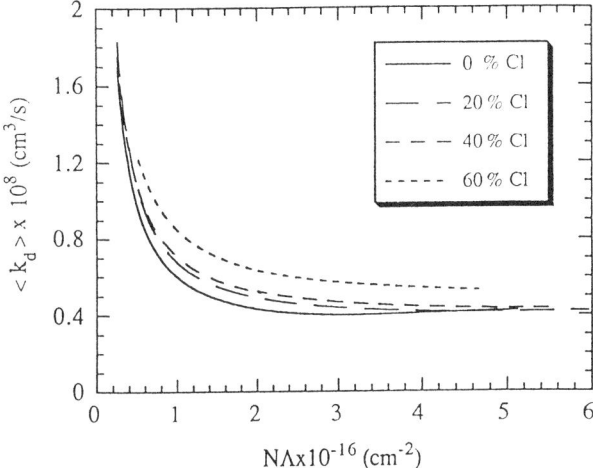

Figure 16 The molecular chlorine dissociation rate coefficient as a function of $N\Lambda$ in a self-sustained chlorine discharge containing different atomic chlorine mole fractions. (From Ref. 10, with permission.)

However, $\langle k_d \rangle$ increases drastically at low $N\Lambda$ as the electron diffusion loss and hence $(E/N)_{\mathrm{rms}}$ required to sustain the plasma both increase at low $N\Lambda$.

Figures 14–16 may be used to determine the electron density and the dissociation rate coefficient in a chlorine discharge, for given reactor geometry (L, R), power input to the bulk plasma P_{rf}, and total number density N (particles/cm^3). The latter is found by knowing the gas pressure and temperature, $N = P/kT_g$. Some iteration may be required since the atomic chlorine mole fraction is not known a priori. Consider, for example, Figs. 15 and 16. One may first estimate the dissociation rate coefficient and the electron density by assuming $y_1 = 0$. One may then use the material balance, Eq. (44) following, to determine the atomic chlorine mole fraction, and use Figs. 15 and 16 again to obtain a better estimate of the rate coefficient.

It should be noted that Figs. 14–16 are for pure chlorine and are applicable to the bulk plasma of a relatively high-pressure (>100s of mtorr) diode reactor. Equations (19) and (20) are more general. Also, it should be pointed out that the power that enters the left-hand side of Eq. (20) is the power absorbed by the electron gas. In practice, one measures the total power delivered to the plasma; there is always a fraction of this total power dissipated by ions in the sheath. The sheath power may be estimated if the ion flux and sheath potential are known or can be estimated.

5. Heavy-Particle Reactions

Common types of chemical reactions among heavy particles are also shown in Table 4. An important reaction is radical recombination in the gas phase according to

$$A + B \xrightarrow{k_1} (AB)^* \tag{21}$$

$$(AB)^* \xrightarrow{k_2} A + B \tag{22}$$

$$(AB)^* + M \xrightarrow{k_3} AB + M \tag{23}$$

The overall reaction is given by

$$A + B + M \xrightarrow{k_4} AB + M \tag{24}$$

Upon recombination, an amount of energy equal to the bond energy is released. In order for the resulting complex $(AB)^*$ to be stabilized, this excess energy must be carried away; otherwise, the complex will brake apart into radicals A and B again (Eq. 23). The excess energy can be carried away by a third body M, usually an atom or molecule (see reaction T25 in Table 4). Energy transfer from $(AB)^*$ to M must occur in a time shorter than the activated complex lifetime, which may be less than 1 ps. This means that all three species (A, B, and M) must participate in the collision simultaneously. The rate of these termolecular reactions is given by

$$R_{vr} = \frac{k_1 k_4 n_A n_B n_M}{k_2 + k_4 n_M} \tag{25}$$

where R_{vr} is the volume (gas-phase) recombination rate. The species number densities can be related to the system pressure through the corresponding mole fraction y, i.e., $n_j = Py/kT_g$. For a given mole fraction, the number density is proportional to pressure. At low enough pressures such that $k_4 n_M \ll k_2$, Eq. (25) becomes

$$R_{vr} = \frac{k_1 k_4}{k_2} n_A n_B n_M = k_r n_A n_B n_M \tag{26}$$

whereas, at the other extreme of high pressure $k_4 n_M \gg k_2$, and Eq. (25) becomes

$$R_{vr} = k_1 n_A n_B \tag{27}$$

In typical plasma etching systems, these reactions are in the low-pressure limit or in the transition regime. In the former case, Eq. (26) shows that recombination reactions are strongly dependent on pressure (cubic dependence, assuming that the gas composition does not change drastically with pressure). Usually, these reactions become insignificant in plasma systems at pressures below ~100 mtorr.

The rate coefficient has an Arrhenius dependence on gas temperature,

$$k_r(T_g) = A(T_g) \exp\left(-\frac{E_A}{RT_g}\right) \qquad (28)$$

where the preexponential factor $A(T_g)$ is a weak function of temperature, often masked by the temperature dependence of the exponential factor. Thermochemical kinetics is discussed by Benson [12].

Another important kind of heavy-particle reaction consists of those between ions and molecules or atoms. Typical examples are charge exchange, and ion–molecule recombination,

$$A^+ + B \rightarrow A + B^+ \qquad (29)$$

$$A^+ + B \rightarrow AB^+ \qquad (30)$$

Charge exchange between a molecule and the daughter ion (symmetric charge exchange) is particularly efficient. Charge exchange transforms a fast ion and a slow neutral into a fast neutral and a slow ion. When this process happens in the sheath (process e in Fig. 5), the energy by which ions bombard the wafer surface is reduced. Ion-insertion reactions are responsible for generating molecular ions in atomic gas plasmas. Finally, ion–ion neutralization reactions such as

$$A^- + BC^+ \rightarrow AB + C \qquad (31)$$

are important as negative ion sinks in the gas phase, since negative ion loss to the walls is negligible (process c in Fig. 5). These reactions are often dissociative (see reaction T22 in Table 4). Data on reactions involving ions can be found in the book by McDaniel [13].

It should be noted that, although most of the species in a plasma are ground state neutrals vibrationally or electronically excited species may play an important role in plasma chemistry. The problem is complicated by the fact that the "temperature" of these excited states (say the vibrational temperature) is unknown. In fact, even the (translational) temperature of the ground state species is seldom known; it is difficult to estimate, and accurate measurements are cumbersome to make.

B. Surface Chemistry

1. Sputtering, Chemical Etching, and Surface Recombination

One can distinguish between several types of processes that can occur on the surface of a solid exposed to a plasma (Fig. 17). The surface is bombarded by neutral radicals and molecules, positive ions, electrons, and photons. Of these, neutral species and positive ion bombardment are generally considered most important. *Sputtering* (Fig. 17a) refers to the process of ejection of surface atoms induced by ions transferring momentum to the surface. Sputtering can be thought of as atomic-scale sandblasting. This is a physical rather than a chemical mechanism, and as such it has very low selectivity. For low-energy ions (<1000 eV)

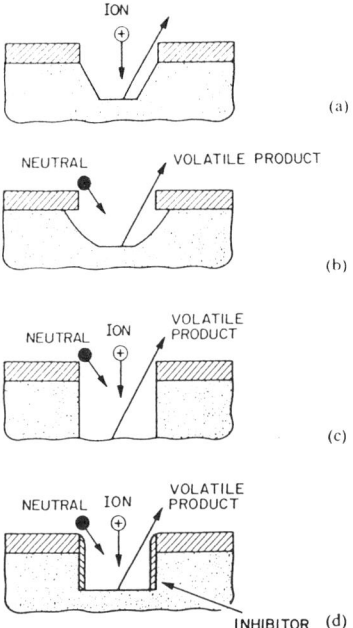

Figure 17 The four basic mechanisms of plasma etching: (a) sputtering; (b) chemical etching by the action of neutrals only; (c) ion-induced etching; reaction occurs only on surfaces undergoing ion bombardment; and (d) ion-enhanced etching; neutrals could etch even in the absence of ion bombardment; the latter increases the etch rate. In order to avoid etching of the sidewalls by neutrals in case (d), the plasma chemistry is selected so as to "coat" the sidewalls by an extremely thin layer, which inhibits sidewall etching. Ion bombardment clears the inhibitor film, allowing etching to proceed anisotropically. (From Ref. 11, with permission.)

of interest to plasma processing, the sputtering rate may be obtained by an expression of the form

$$ER_s = Y_s \frac{I_i}{\rho} = \alpha(\varepsilon_i^x - \varepsilon_{th}^x) \frac{I_i}{\rho} \tag{32}$$

where $x \approx 0.5$, Y_s is the sputtering yield (atoms of substrate material removed per incident ion), I_i and ε_i are the ion flux and energy, respectively, and ρ is the density of the substrate material. α is a constant characteristic of the material ($\alpha = 0.0337$ and 0.0139 eV$^{-0.5}$ for Si and SiO$_2$, respectively), and ε_{th} is the sputtering threshold energy (≈ 20 and 18 eV for Si and SiO$_2$, respectively). The latter is the minimum ion energy required for sputtering to occur at all. The sputtering yield for some solid/ion combinations is shown in Ref. 5, p. 374. Plasma etching systems should be designed to minimize sputtering by avoiding high-energy ion bombardment of the electrodes and the reactor walls.

In *chemical etching* (Fig. 17b), neutrals react with the surface spontaneously to yield product. An example is etching of Si with F atoms. Rapid surface fluorination leads to SiF$_4$, which desorbs into the gas:

$$Si + F \rightarrow SiF \tag{33}$$

$$SiF + F \rightarrow SiF_2 \tag{34}$$

$$SiF_2 + F \rightarrow SiF_3 \tag{35}$$

$$SiF_3 + F \rightarrow SiF_4 \tag{36}$$

Ion bombardment is not necessary for this reaction to occur. Generally, chemical etching has very high selectivity. The chemical etching reaction rate is often proportional to the etchant (e.g., F atoms) concentration (see also Eqs. 46–48 following). The rate can also be written as the product of the etchant flux to the surface times the reaction probability (see Eq. 41 following).

Instead of etching, radicals can recombine on the surface (reaction T24 in Table 4). This may be an important loss mechanism of potential etchant species. Recombination can occur between two radicals adsorbed on the surface at adjacent sites, (Langmuir–Hinshelwood mechanism [14]). The product of recombination desorbs into the gas:

$$A(g) + S \rightarrow A.S \tag{37}$$

$$B(g) + S \rightarrow B.S \tag{38}$$

$$A.S + B.S \rightarrow AB(g) + 2S \tag{39}$$

Here S is a surface site and $A.S$, $B.S$ denote species adsorbed on such surface sites. Another mechanism is for a gas-phase radical to abstract another radical adsorbed on the surface (Rideal mechanism [14]). The latter case can be shown schematically as

$$A(g) + B.S \rightarrow AB(g) + S \tag{40}$$

The wall recombination reaction rate is often taken first order with respect to gas-phase concentration of the respective radical. It is also written as the product of the thermal flux of the radical and a recombination probability (γ):

$$R_{sr,j} = k_{sr} n_j = \gamma I_j = \frac{\gamma}{4} n_j \sqrt{\frac{8kT_j}{\pi M_j}} \tag{41}$$

It follows that the first-order surface recombination rate coefficient is related to the reaction probability by

$$k_{sr} = \frac{\gamma}{4} \sqrt{\frac{8kT_j}{\pi M_j}} \tag{42}$$

Recombination probabilities differ markedly depending on type of species and the material of the surface. For example, F atoms recombine on quartz with a probability of $\sim 10^{-4}$; only one in 10,000 collisions results in atom recombination. On the other hand, Cl atoms recombine on stainless steel with a probability of 0.1. The recombination probability also depends on the surface temperature, but data are very scanty. Also, the value of γ may change drastically depending on the condition of the surface. For example, a clean metal and the same metal covered with its native oxide (e.g., Al and Al_2O_3) can have very different γ. When etching and wall recombination occur simultaneously, the reaction probability is a composite of the two reactions.

2. Ion-Assisted Chemistry

Ion bombardment can influence one or more of the surface reaction steps in plasma etching. For example, ion bombardment disrupts the crystal lattice, generating "active" sites (dangling bonds) where neutrals can adsorb (step 2 in Fig. 9), or ions promote the reaction between adsorbed species (step 3) to generate new species that either desorb spontaneously or are easier to remove by sputtering compared with the substrate material (Fig. 17c). In plasma systems where a thin polymeric film deposits on surfaces, ions can clear the bottom of the microscopic features by sputtering away such deposits, thereby exposing "fresh" surface to the incoming etchant flux (Fig. 17d). In *ion-induced etching*, reaction between a species with the surface occurs only in the presence of ion bombardment. An example is etching of undoped silicon with chlorine. Figure

18a shows the result of a well-defined experiment that demonstrates the principle [15]. A piece of silicon wafer at ambient temperature can be bombarded under ultrahigh vacuum (UHV) conditions by a thermal beam of Cl_2 molecules and an energetic beam (~1000 eV) of argon ions. No etching occurs when the surface is bombarded by chlorine alone. Dissociative adsorption of chlorine on the silicon surface does occur, but surface reaction leading to the volatile $SiCl_4$ compound cannot occur under these conditions. When the surface is bombarded by the ion beam alone, a small etch rate is observed, corresponding to sputtering of the surface. When the chlorine beam is also turned on, so that the surface is bombarded by molecules and energetic ions simultaneously, the etch rate is much higher than the sputtering rate. The ion beam has now induced a reaction of chlorine with the surface, leading to $SiCl_x$ product formation.

In *ion-enhanced etching*, ion bombardment accelerates the reaction of the neutral with the surface. Without ion bombardment that reaction would still occur, but at a reduced rate. An example is shown in Fig. 18b. This is an experiment analogous to that of Fig. 18a. When the silicon surface is exposed to XeF_2 alone, a small etch rate (chemical etching) is observed. Apparently, fluorination of the silicon surface leading to volatile SiF_4 (Eqs. 33–36) does not require ion bombardment. Also, when the surface is exposed to argon ions alone, a small sputtering rate is observed. However, simultaneous exposure to the neutral and ion beams leads to an etch rate much higher than the sum of the chemical etching and sputtering rates. In fact, the etch yield can exceed 25, meaning that 25 Si atoms are removed per incoming ion.

In general, the etch rate can be expressed as a combination of three components: physical sputtering, chemical (spontaneous) etching, and ion-assisted etching:

$$ER_{tot} = ER_s + ER_c + ER_i \qquad (43)$$

For ion-induced etching situations, the spontaneous etching component $ER_c = 0$. The components due to sputtering and chemical etching are given by Eqs. (32) and (41), respectively (in Eq. (41), γ should now be the etching reaction probability). The kinetics of ion-assisted etching are unknown, although phenomenological descriptions in terms of adsorption/reaction/desorption (sputtering) concepts have been proposed. However, these descriptions require several "fitting" parameters to describe the reaction rate in terms of neutral and ion fluxes and ion energy. There is some evidence that an expression similar to that given by Eq. (32) can be used to describe the ion-assisted etch rate, with different values for α (higher) and ε_{th} (lower) compared with physical sputtering.

The experiments shown in Fig. 18, although conducted in UHV (outside of a plasma), clearly demonstrate the synergism between neutrals and energetic ions in promoting the surface reaction in plasma etching. It should be stressed that neutrals and not ions are the primary etchant in most systems of interest.

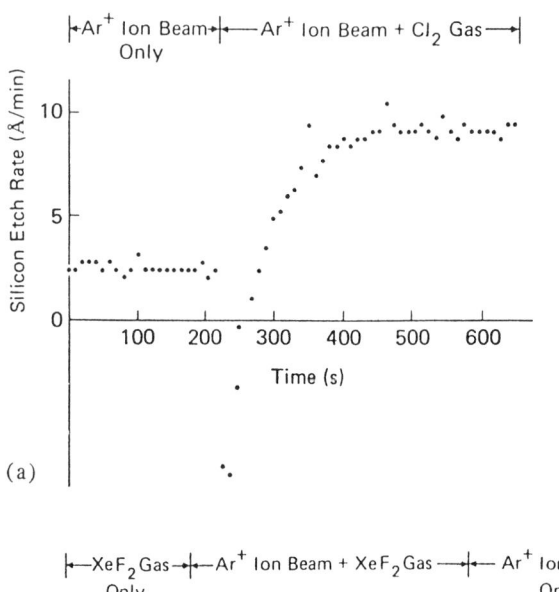

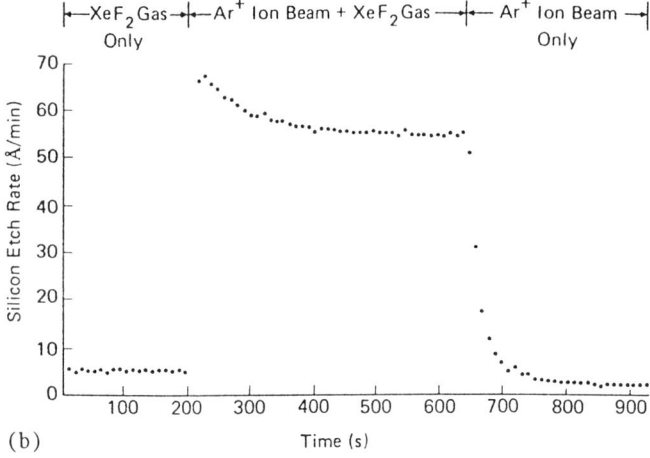

Figure 18 (a) Demonstration of ion-induced etching. Chlorine gas alone does not etch silicon. Ion bombardment alone yields a low sputtering rate. Simultaneous exposure to ions and neutrals results in an enhanced etch rate (much higher than the sputtering rate). (b) Demonstration of ion-enhanced etching: XeF_2 can etch silicon (chemical etching) even in the absence of ion bombardment. Ion bombardment alone yields a low sputtering rate. Simultaneous exposure to ions and neutrals results in greatly enhanced etch rate (much higher than the sum of chemical etching and sputtering). (From Ref. 15, with permission.)

Ions may also react directly with the surface, but the ion flux is too low to support the observed etch rates. For example a Cl_2^+ ion flux of 1 mA/cm^2 corresponds to 6.24×10^{15} particles/cm^{2}s. Even if all impinging chlorine reacted to yield SiCl$_4$ product, the etch rate would be 374 Å/min, much lower than observed etch rates. Thus, the primary function of the ions is to aid the reaction of neutrals with the surface. However, direct ion etching may be dominant at higher ion fluxes, especially when the impinging flux of (reactive) ions (e.g., Cl$^+$) is similar in magnitude to the neutral flux (see Sec. V).

At the molecular level, ion bombardment induces mixing of the adsorbed gas (e.g., F) with the crystal by the so-called *collision cascade*. The depth of the mixed layer depends on the ion energy, and typically it extends several atomic layers into the crystal lattice. Energy deposited by the ions in the mixed layer favors the formation of weakly bound species (SiF$_x$, $x \leq 4$, with binding energy <1 eV) that are either sputtered away or desorb spontaneously into the gas phase. Molecular dynamics simulations, which follow the motion and reaction of atoms in a finite volume of the crystal, have shown that simultaneous bombardment of silicon by chlorine radicals and low-energy (<200 eV) argon ions promotes attack of Si–Si bonds by Cl to form SiCl$_x$ reaction products.

All the aforementioned processes (shown in Fig. 17) can occur simultaneously when a surface is exposed to a plasma. The relative importance of one mechanism over the others depends on the material system (film and gas), the ratio of neutral to ion flux, and the ion energy. In general, conditions are selected such that ion-assisted chemistry dominates.

In some cases, ion bombardment has no effect on etching. A notable example is etching of aluminum in a Cl$_2$/BCl$_3$ plasma [11]. The etch rate is independent of the wafer bias voltage, which controls the energy of ions bombarding the wafer. In still another situation, ion bombardment can retard etching, as in the case of copper etching by chlorine.

3. Anisotropy Mechanisms

Anisotropic etching refers to the situation where reaction proceeds only along the bottom surface of a microscopic feature, or more precisely, only along surfaces that are not perpendicular to the macroscopic wafer surface. Anisotropic etching is based on the premise that energetic ion bombardment is naturally directed along the normal to the wafer surface. Of course, the finite ion temperature and/or ion collisions in the sheath (processes e and f in Fig. 5) can result in a spread of ion velocities around the surface normal. Plasma etching processes are designed to minimize this angular distribution of ion flux. Also, the ion energy must be optimized such that it is high enough to cause the desired surface reaction, yet low enough not to cause undesired sputtering or radiation damage. For many ion-assisted etch processes, ion bombardment energies in the range of 50–300 eV may be most beneficial.

There are basically two mechanisms to achieve anisotropy (Figs. 17c and d). When gasification of the solid by neutral species is not spontaneous, etching can occur only on surfaces exposed to ion bombardment. Thus, anisotropy can be readily achieved. An example is etching of undoped silicon with chlorine plasma. However, when the chemical reaction between neutrals and the surface is spontaneous, a "wall passivation" mechanism is necessary to achieve anisotropy. An example is etching of polysilicon in a F-atom-containing plasma. Wall passivation means deposition of thin polymeric films or "inhibitors" that block etching of the surface by the neutrals. The passivation film does not form on surfaces exposed to ion bombardment (ions sputter away any such films), allowing etching to proceed unimpeded on these surfaces. The deposition of passivation films is controlled by judicious selection of the plasma chemistry (see Sec. IV.A.1 for specific examples). Often, the photoresist (polymer) used as mask can contribute to the formation of passivation films.

C. Effect of Externally Controlled Variables on Process Output

The relation between the externally controlled variables and the key plasma properties, and in turn the process output (see Fig. 3), is very complex. Hence, only general observations can be made. In discussing the effect of one variable, the other variables are assumed to be kept constant.

1. Effect of Pressure

The reactive species (atoms and radicals) density first increases with reactor pressure since there are more parent gas molecules to be dissociated as pressure increases. As pressure increases, however, the electron energy (hence radical production rate coefficient, k_d in Fig. 16) declines, and volume recombination reactions come into play (note cubic power dependence on pressure, Eq. 26), causing the radical density to decline. The pressure that maximizes the species density depends on the system.

The electron (Fig. 15) and ion density depend on pressure similarly to the neutral species, but the maxima in their respective densities occur at different values of pressure. Ion bombardment energy decreases with pressure as ions suffer more collisions in the sheath. This in turn leads to better selectivity but worse anisotropy. The etch rate also goes through a maximum as pressure increases. Etch uniformity generally improves at lower pressures as the species diffusivity increases, smoothing out concentration gradients.

2. Effect of Power

Increasing power generally results in higher number density of radicals, electrons, and ions. This in turn results in higher etch rate. The sheath thickness decreases (smaller Debye length, see Eq. 5) reducing the number of ion–neutral collisions in the sheath, in turn yielding improved ion flux directionality. In addition, the sheath voltage increases with power, resulting in enhanced ion bombardment energy. This may lead to better anisotropy, but selectivity can worsen.

3. Effect of Gas Flow Rate

Flow rate affects the plasma chemistry mainly through its effect on the species residence time in the plasma. At very low flow rates, the feedstock gas has plenty of time to be dissociated and the etch rate is limited by reactant supply—i.e., most of the feedstock gas entering the system is consumed to form etch product. For example, etching a 20 cm diameter wafer covered with polysilicon at a rate of 0.5 µm/min requires 58.4 sccm of Cl_2 (assuming $SiCl_4$ as the etch product). At very high gas flow rates, the gas residence time is so short that the chance for gas dissociation is very small. Hence, the etch rate declines with flow rate. Figure 19 shows the etch rate of a polymer in an oxygen plasma. At constant power, the etch rate reaches a maximum with flow rate. Also, the etch rate is an increasing function of power.

Convective gas flow does not have a direct impact on charged species density, because the charged species velocity due to the electric field (drift velocity) is much larger than the gas convective velocity. However, in complex gas plasmas,

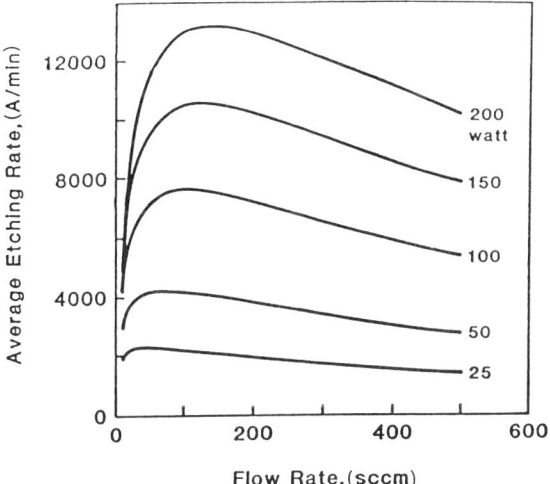

Figure 19 Typical response of the variation of the average (over the wafer radius) etch rate of photoresist in an oxygen plasma as a function of the feed gas flow rate. At low flow rates, etching is limited by reactant supply. At high flow rates, convective losses reduce the concentration of the etchant species (atomic oxygen). Etch rate is also seen to increase monotonically with power. (From Ref. 16, with permission.)

where a plethora of reactions occurs, the effect of flow may be more subtle; depending on the relative importance of reaction time compared with residence time, some reactions will be favored over other reactions, altering the plasma gas composition. This in turn can affect the EEDF and indirectly the charged particle densities, ion flux, and so on.

4. Effect of Frequency

Plasma systems are excited at frequencies ranging from dc to microwave (2.45 GHz); rf excitation at 13.56 MHz is most common. The effect of frequency is difficult to quantify. Frequency can affect the plasma chemistry by affecting the electron energy distribution function. It is not clear what frequency is best for maximum species production; that certainly depends on the threshold energy and the shape of the collision cross section for the reaction of interest. In capacitively coupled systems (Fig. 2a), the ion bombardment energy increases with decreasing frequency (Fig. 20). This is because the sheath potential increases at lower frequency. Hence, a larger fraction of the applied power is dissipated for accelerating ions in the sheath rather than for producing radicals in the bulk plasma. Higher ion energy generally implies better anisotropy, but selectivity or radiation damage become worse.

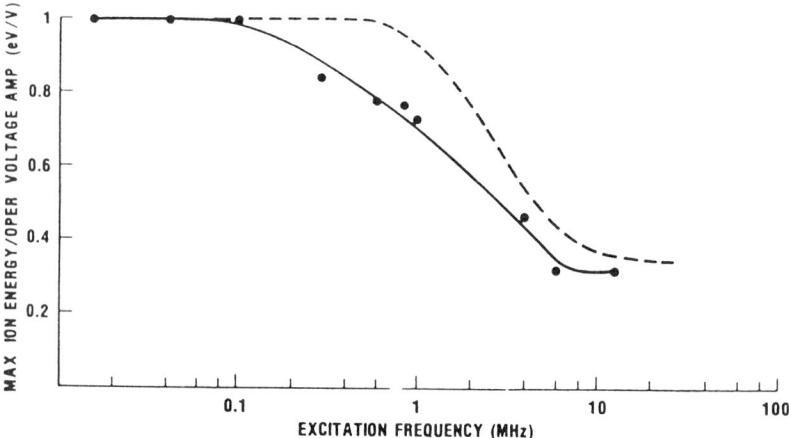

Figure 20 Dependence of the maximum ion bombardment energy on the excitation frequency in a capacitively coupled chlorine discharge at a pressure of 0.3 torr. At low frequencies, ions traverse the sheath in a short time compared with the period of the field. The maximum ion energy is then equal to the peak of the sheath voltage. At high frequencies, the ion transit time is many rf cycles; ions respond only to the time-average sheath voltage. The transition happens at the ion plasma frequency, which for the conditions here is about 1 MHz. (From Ref. 17, with permission.)

5. Loading

Loading refers to a situation where the etch rate depends on the etchable area exposed to the plasma. The etchable area can be increased, for instance, by using more (or larger) wafers. The origin of loading is rather simple. For given plasma conditions (i.e., given production rate of radicals), increasing the etchable area increases the loss rate of radicals, with concomitant decrease in the radical concentration. Since etch rate is often an increasing function of concentration, more etchable surface means lower etch rate. The loading effect can be described quantitatively by again taking the chlorine plasma as an example system. If Q_0 is the feed rate of molecular chlorine (in standard cm^3/min, sccm), then the component balance for atomic chlorine can be written as (see Table 4 for reactions involving atomic chlorine)

$$\frac{8.96 \times 10^{17} Q_0}{V_p} \frac{y_{Cl}}{2 - y_{Cl}} = (k_a + 2k_d)N(1 - y_{Cl})n_e$$

$$- 2k_{vr}N^3 y^2_{Cl} - (k_{sr} + k_c f)Ny_{Cl} \frac{S}{V_p} \quad (44)$$

where y_{Cl} is the atomic chlorine mole fraction (the remainder, $1 - y_{Cl}$, is molecular chlorine), V_p is the plasma volume, $k_a + 2k_d$ is the rate coefficient for production of atomic chlorine, and k_{vr} is the rate coefficient for volume recombination of atomic chlorine.* The left-hand side of Eq. (44) is the rate at which atomic chlorine leaves the plasma volume by convective flow. The first term on the right-hand side represents the rate of production of Cl by dissociative attachment and dissociation. The second term on the right-hand side represents the loss rate of Cl by volume recombination. The last term describes surface recombination (k_{sr}) and chemical etching (k_c) occurring on a fraction f ($0 \leq f \leq 1$) of the surface; S/V_p is the surface-to-volume ratio.

When the EEDF follows the temporal variations of the field (e.g., rf case), the electron-impact reaction rate coefficients are modulated at twice the excitation frequency. Thus, the atomic chlorine production by electron-impact dissociation is time-dependent. However, volume and wall recombination losses of Cl occur on a time scale much longer than the period of the applied field, resulting in a Cl density that is not modulated in time. Therefore, k_a and k_d in Eq. (44) can be time-averaged over the period of the applied field.

Neglecting volume recombination (e.g., <100 mtorr) and assuming that the Cl atom mole fraction is small, i.e., $y_{Cl} \ll 1$, Eq. (44) becomes

---

*For simplicity it has been assumed that both Cl and Cl$_2$ can act as a third body (see Eq. 24) with the same efficiency for recombination (same k_{vr}). In practice, Cl$_2$ should have a higher efficiency compared to Cl.

$$y_{Cl} = \frac{(k_a + 2k_d)n_e}{\frac{8.96 \times 10^{17} Q_0}{V_p} + \frac{S}{V_p}(k_{sr} + fk_c)} \tag{45}$$

One observes that the Cl mole fraction (or concentration, for constant N) decreases as the etchable surface (fraction f) increases. From Eq. (45), it is also seen that in order to have loading, the etchable material must be a significant sink for the radicals. Suppose, for instance, that wall recombination is much more effective than etching ($k_{sr} \gg k_c$). Then doubling, say, the etchable surface fraction is not going to make a difference in y_{Cl} since the wall recombination loss dominates. Similarly for high feed flow rates Q_0, convective losses will dominate the etching reaction loss, and loading will be insignificant.

Figure 21 shows the etch rate distribution along the wafer radius in a parallel plate reactor of the type shown in Fig. 2a. The etch rate decreases as the wafer diameter increases because of loading. The radical density profile can also be affected by loading; this in turn can affect the etch uniformity. Figure 22 shows the atomic oxygen concentration profiles along the electrode radius in an oxygen plasma. These profiles were measured by optical emission actinometry (see Sec. VII.A.1). The electrode is covered with a reactive film of 3.75 cm radius. The rest of the electrode is relatively inert. Oxygen atoms are consumed by the film,

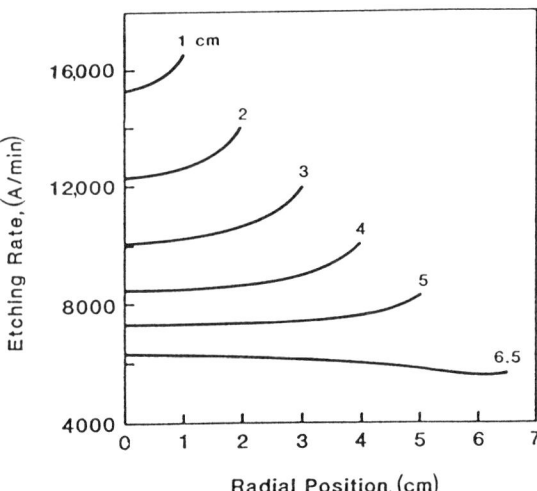

Figure 21 Etch rate of photoresist in an oxygen plasma as a function of radius along the wafer, for different wafer sizes. The loading effect is clearly seen; a smaller wafer etches faster than a larger one. Etch nonuniformities that would lead to a "bull's-eye" clearing pattern (from outside inwards) are also evident, especially for the smaller wafers. (From Ref. 16, with permission.)

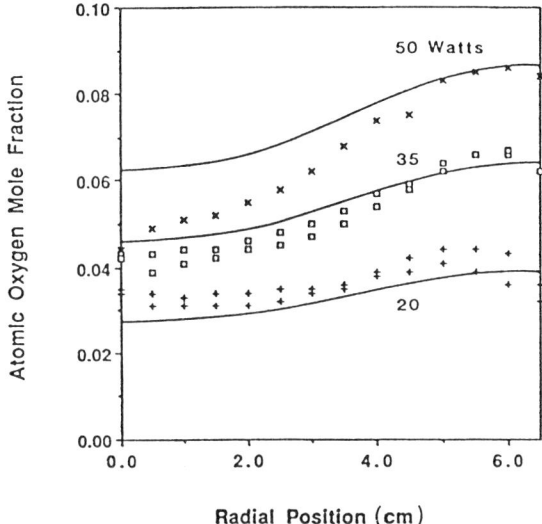

Figure 22 Atomic oxygen concentration, measured by optical emission actinometry, as a function of radius in a 13.56 MHz oxygen discharge sustained in a diode reactor. The electrode is covered with a reactive film up to a radius of 3.75 cm. This film acts as a sink for atomic oxygen, resulting in significant radial concentration gradients. Such gradients are responsible for etch nonuniformity. Solid lines show the result of mathematical model predictions. (From Ref. 18, with permission.)

and their concentration dips in that region. In contrast, the atom concentration builds up over the surrounding inert region. The gradient in the radical density thus established is most important around the periphery of the film. This gradient would lead to a so-called "bull's-eye" clearing pattern, whereby the etch rate decreases from the periphery to the center of the wafer. This "local" loading can be enhanced at higher power input to the plasma.

Loading can also occur at the microscopic feature scale (microloading). It has been observed, for example, that isolated features etch faster than dense patterns. This is due to local reactant depletion over the dense pattern caused by greater consumption of the reactant (greater etchable surface area exposed by the dense pattern). If reactant transport is not adequately fast to alleviate concentration gradients in the reactor, then loading can manifest itself.

6. Aspect-Ratio-Dependent Etching

Aspect-ratio-dependent etching (ARDE) refers to a situation commonly observed in etching of high-aspect-ratio (depth:width) features. It has been found that the etch rate decreases as the trench aspect ratio increases (also known as reactive ion etching, or RIE, lag). Figure 23 shows a typical situation. A combination

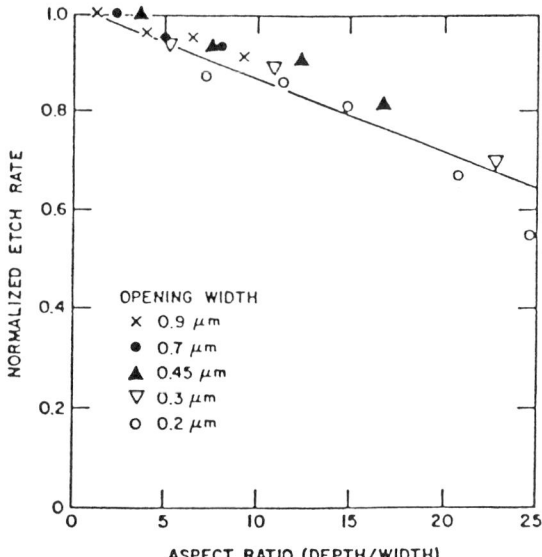

Figure 23 Aspect-ratio-dependent etching of silicon in a CCl_2F_2/O_2 gas mixture. The etch rate is a function of the aspect ratio of the feature (depth/width). (From Ref. 19, 1985 with permission.)

of effects can result in ARDE. As the feature becomes deeper, for example, ions may impact the sidewall instead of the feature bottom. This can be due to the angular distribution of the incoming ion flux, or the decollimation of ions due to electric fields developed by nonuniform charging of the mask or the feature sidewalls. As the flux of ions impinging on the feature bottom decreases, the etch rate follows suit. Also, transport limitations may reduce the flux of neutrals reaching the bottom of the feature for high-aspect-ratio features. One should not confuse ARDE with microloading. The latter can occur even for the same aspect ratio features. Finally, in a less common situation, the etch rate increases as the feature deepens (inverse RIE lag).

IV. PLASMA ETCHING SYSTEMS

Table 5 provides common examples of etching gases and the materials they etch. A brief discussion of several material systems is given below [11].

A. Silicon, Oxide, and Nitride

Silicon, silicon dioxide (oxide), and silicon nitride (nitride) are among the most important materials in microelectronics. Etching of these materials has tradi-

Table 5 Etching Gases and the Materials They Etch

Source gas	Additive gas	Etchant species	Material	Mechanism	Selective over
CF_4	O_2	F	Si	Chemical	SiO_2/III–Vs
SF_6	O_2	F	Si	Chemical	SiO_2/III–Vs
SF_6	He	F	Si	Ion-enhanced (low-temp.)	Resist/III–Vs
NF_3	None	F	Si	Chemical	Resist/III–Vs
C_2F_6	O_2	F	Si	Chemical	SiO_2
CF_4	H_2	CF_x	SiO_2/Si_3N_4	Ion-induced	Si
NF_3	Cl_2	F	Si_3N_4	Chemical	Resist
C_2F_6	CHF_3	CF_x	SiO_2	Ion-induced	Si
Cl_2	Ar	Cl	Undoped Si	Ion-induced	SiO_2
Cl_2	C_2F_6	Cl	n-type Si	SP[a]	SiO_2/Resist
Cl_2	BCl_3	Cl_2/Cl	Al	SP[a]	SiO_2/Resist
Cl_2	CCl_4	Cl_2/Cl	Al	SP[a]	SiO_2/Resist
Cl_2	O_2	Cl/O	Mo, $MoSi_2$, Cr	Ion-induced	SiO_2
Cl_2	None	Cl	III–Vs	Chemical	SiO_2/Resist
CCl_4	O_2	Cl	III–Vs	SP[a]	SiO_2
CH_4	H_2	CH_x, H	III–Vs	Ion-enhanced Chemical	Resist
O_2	H_2O	O, OH	Resist	Chemical	Si, SiO_2

[a]SP = sidewall passivation

tionally been performed in fluorine- and/or chlorine-containing plasmas. Fluorine atoms and CF_x radicals generated by electron-impact dissociation in a fluorocarbon plasma initiate the reaction sequence. Si can be etched by F atoms via a chemical mechanism (without ion bombardment) shown by Eqs. (33)–(36). Equation (36) is considered to be the rate-limiting step. The etch rate of single-crystal Si (100) has been measured to be [11]

$$R_{Si} = 2.86 \times 10^{-12}\, n_F\, \sqrt{T}\, \exp \frac{-1250}{T} \qquad (46)$$

F atoms also etch SiO_2 in the absence of ion bombardment with a rate

$$R_{SiO_2} = 0.614 \times 10^{-12}\, n_F\, \sqrt{T}\, \exp \frac{-1890}{T} \qquad (47)$$

In these expressions, the etch rate is in Å/min, T is the surface temperature in K, and n_F is in F atoms/cm^3.

The reaction of F atoms with Si and especially SiO_2 can be enhanced considerably by ion bombardment. In-plasma studies have shown that fluorocarbon etching of silicon leads to the formation of a rather thick (several atomic layers) fluorosilyl layer on the silicon surface containing variable amounts of fluorine [2]. The fluorine incorporated in this layer decreases as a function of depth.

Silicon can also be etched in chlorine-containing plasmas. Oxide is not etched in these plasmas, and so high selectivity can be obtained. Cl_2 plasmas etch undoped or p-type polysilicon and single-crystal silicon by an ion-induced mechanism; n-type single-crystal Si and polysilicon etches chemically by Cl atoms with a rate

$$R_{Si} = 4.04 \times 10^{-18} n_{Cl} N_D^{\gamma} \sqrt{T} \exp \frac{-E_a}{kT} \tag{48}$$

where N_D is the dopant density. The activation energy E_a and the exponent γ depend on the type of the substrate (polysilicon or single-crystal silicon [6]).

A doping effect has been observed in silicon etching: n-type silicon etches faster than undoped silicon, while p-type etches slower. This effect has been attributed to electrophilic atoms (such as halogens) having higher affinity to the n-doped surface, since the *chemical potential* of electrons increases with n-doping.

1. Selectivity

In the absence of ion bombardment (e.g., downstream etching, Fig. 4b), the selectivity of etching silicon over oxide is given by the ratio of Eqs. (46) and (47):

$$\frac{R_{Si}}{R_{SiO_2}} = 4.66 \exp \frac{640}{T} \tag{49}$$

In the plasma, the selectivity is determined by the intensity of the ion bombardment and the density of free fluorine atoms compared with CF_x radicals, particularly unsaturated radicals such as CF_2. Unsaturates can lead to the formation of thin polymer films that block etching. The formation of such films is more difficult on the oxide surface because the available oxygen oxidizes the carbon deposited on the surface by the radicals. In fact, the addition of oxidation (oxygen) or reducing (hydrogen) agents to CF_4 has been used to control polymer formation and thereby affect the etch rate and selectivity. Figure 24 shows the effect of oxygen addition. At low % oxygen, the formation of extra F is favored via reactions of the form

$$O/O_2 + CF_x \rightarrow COF_2, CO, CO_2 + F/F_2 \tag{50}$$

This enhances the etch rate of both silicon and oxide. However, at high % oxygen, two factors play a role: (1) the etchant is diluted, and (2) the Si surface is oxidized, and oxide etches much slower than Si. Hence, the Si etch rate drops at high oxygen additions. The optimum oxygen addition is ~10–20% for silicon. Oxide has some oxygen already, and so the optimum is shifted to higher values,

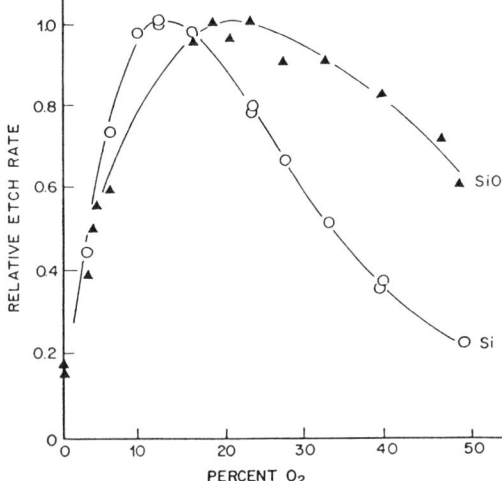

Figure 24 Silicon and silicon dioxide etch rate as a function of oxygen additions to a CF_4 discharge. (From Ref. 20, with permission.)

~20–30%. Similar results have been obtained when etching silicon in other F-containing gases, such as SF_6 and NF_3, with added oxygen. Hydrogen addition to CF_4 has a dramatic effect (Fig. 25). Hydrogen ties up F atoms in reactions such as

$$H, H_2 + F, CF_x \rightarrow HF, H, CF_{x-1} \qquad (51)$$

thereby increasing the concentration of unsaturates. Hence, the Si etch rate drops as hydrogen is added. At higher hydrogen additions, polymer formation blocks etching of silicon. However, oxygen from silicon dioxide can in part compensate for the effect of hydrogen and prevent polymer formation, allowing etching of the oxide to proceed. Thus, very high selectivity of etching oxide over silicon can be achieved.

It becomes evident from the foregoing discussion that polymer-forming radicals are crucial for selective etching in these systems. Practical systems frequently operate under conditions that form polymers on some surfaces and not on others. This may be a sensitive point of operation, in the sense that a slight variation of conditions may lead to gross polymerization in the system [15].

Oxygen or hydrogen addition basically affects the free fluorine to "active" carbon ratio F/C in the discharge. The term *active carbon* means carbon in plasma species that enters into etching or polymerization reactions at the wafer surface. Examples of such species are CF_x ($x < 4$), the corresponding ions CF_x^+, and so on. Carbon in relatively inert compounds (e.g., CO_2, CF_4, etc.) is

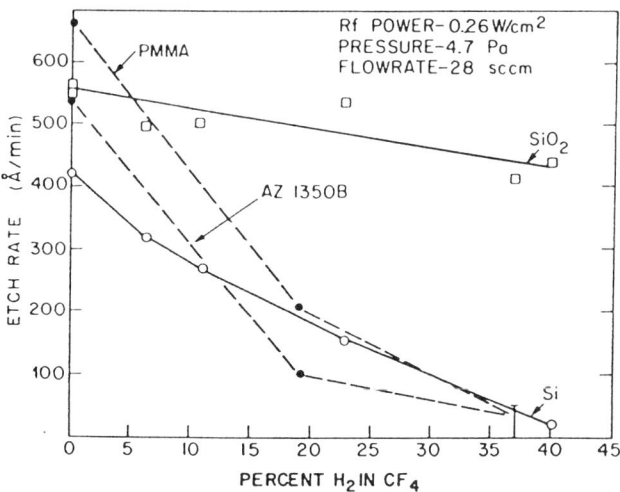

Figure 25 Etch rate of silicon, oxide, and two kinds of photoresist (AZ 1350B and PMMA) as a function of hydrogen addition to a CF_4 discharge. Etch rate stops on non-oxide materials due to heavy polymer build up at high % H_2. Oxide continues to etch due to polymer removal by the available oxygen. This yields high selectivity of etching oxide over silicon. (From Ref. 21, with permission.)

not considered active. The F/C ratio is important for controlling etch rate, anisotropy, and selectivity. For example, low values of F/C favor polymer deposition that can improve anisotropy and selectivity of etching silicon over oxide. However, the etch rate decreases with decreasing F/C. The concept of F/C ratio (or more generally, the halogen to carbon ratio) has been used to describe different fluorocarbon plasmas. For example, adding CHF_3 or C_2F_6 to a CF_4 plasma has a similar effect to adding hydrogen; the F/C ratio decreases. The polymer mask often plays an important role as well, since mask erosion contributes hydrocarbon fragments in the plasma, which can become active participants in the gas and especially the surface chemistry.

For given chemistry, selectivity is a sensitive function of ion bombardment energy. Figure 26 shows the etch rate of silicon and oxide in a chlorine plasma as a function of power. Chlorine chemistry is inherently selective with respect to oxide in the absence of ion bombardment. However, the selectivity worsens as power increases. Higher power means higher ion bombardment energy, and sputtering of the oxide is inevitable. The situation becomes worst when a photoresist is used as a masking material. Increased power leads to increased photoresist errosion; and the hydrocarbon fragments are effective oxide etchants under ion bombardment.

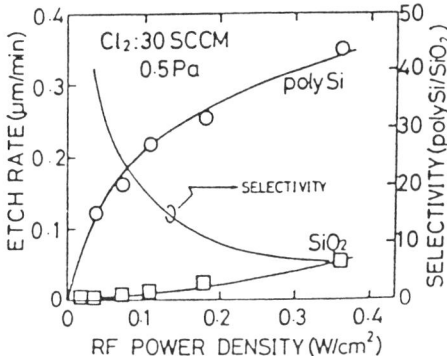

Figure 26 Etch rate of silicon and oxide and the oxide to silicon selectivity as a function of power density. As power increases, so does the ion bombardment energy, degrading the etch selectivity. (From Ref. 22, with permission.)

Silicon nitride reactivity in fluorocarbon plasmas is in general in between that of silicon and silicon dioxide. Also, the reactivity depends on how the film was deposited. Plasma CVD films etch faster than thermal CVD films. Silicon nitride can be etched selectively over silicon in fluorine-deficient plasmas (CF_4 + CHF_3). Selectivity over oxide can be achieved in chemical etching (without ion bombardment) in NF_3/Cl_2 mixtures, as shown in Fig. 27. The etch was carried out at 200 mtorr using a downstream etching reactor excited at 2.45 GHz microwave frequency. By adding chlorine to the NF_3 gas, very high selectivity can be achieved. However, the etch is isotropic. In the presence of ion bombardment, selectivity is compromised. A two-step process may then be used to obtain both anisotropic and selective nitride etch over oxide: first, an ion-assisted anisotropic etch to remove most of the nitride; then, a chemical selective etch to remove the remaining nitride down to the oxide underlayer.

2. Sidewall Profiles (Anisotropy)

Plasma chemistry strongly affects the shape of the sidewalls of microscopic features (degree of anisotropy) for given material and reactor operating conditions. A variety of sidewall profiles can be obtained, as shown in Fig. 28, showing silicon etching in an rf plasma reactor. The columns correspond to different feedstock gas composition and the rows to different etching times (100% corresponds to endpoint, 200% corresponds to 100% overetch, etc.). Etching with CF_3Cl or CF_4/O_2 results in mask undercut by F attack of the sidewalls (middle and rightmost columns). The latter plasma releases more F, and the undercut is correspondingly higher. Addition of enough C_2F_6 forms sidewall polymeric deposits that minimize or prevent mask undercut (first two columns). Ion bombardment clears these deposits from the bottom of the features, allowing etching

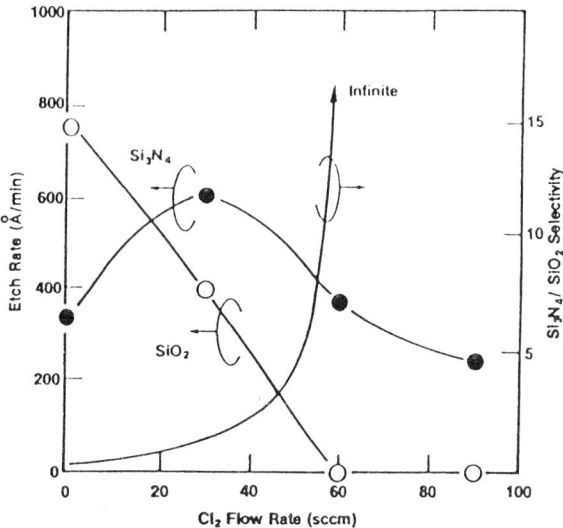

Figure 27 Etch rate of oxide and nitride and the nitride to oxide selectivity as a function of Cl_2 flow rate at 30 sccm of NF_3 in a downstream etching reactor. (From Ref. 23, with permission.)

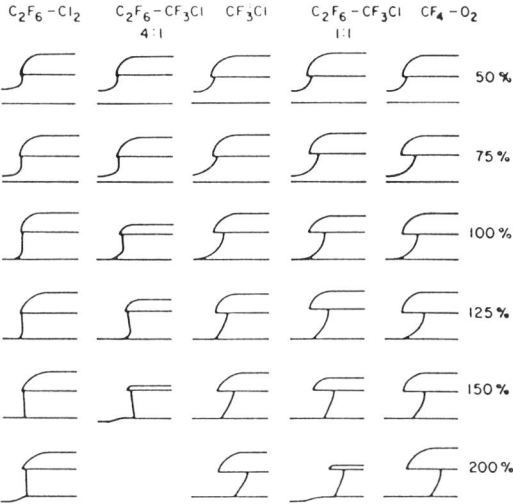

Figure 28 Sidewall profiles of microscopic features etched into silicon using different gases. (From Ref. 24, with permission.)

to proceed anisotropically. The undercut (when present) increases with overetch; in fact, overetch can accelerate undercut by the loading effect, as the surface area of the film decreases drastically when the film clears from horizontal surfaces at endpoint. In plasmas that etch the film spontaneously (e.g., Cl_2/n-type silicon), lowering the substrate temperature quenches the spontaneous reaction and minimizes mask undercut. The reaction can still proceed on ion-bombarded surfaces since ion-assisted chemistry is largely independent of temperature.

When polymer formation of the sidewalls is important, the effect of temperature can be more subtle. Figure 29 shows wall profiles obtained in oxide etching as the substrate temperature is varied between 30 and 130°C. The etched hole profile can be tailored by affecting the polymer deposition/etching balance as temperature changes. Profiles such as those obtained at 30°C are desirable for "contact" holes because they are easy to refill with metal (see Chapter 5).

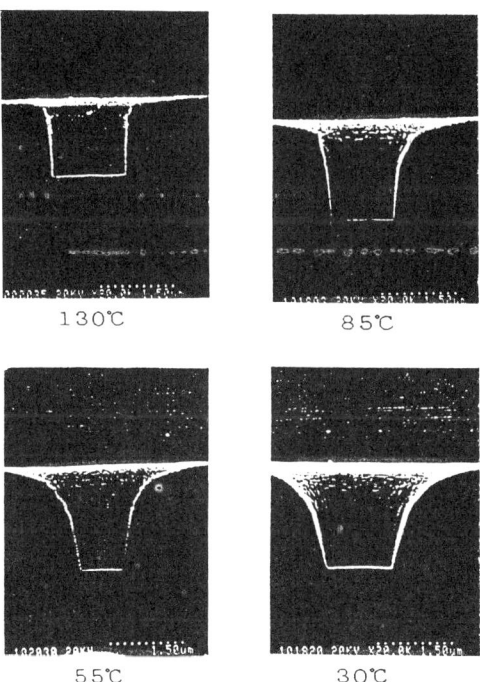

Figure 29 Sidewall profiles of microscopic features etched into oxide at different temperatures. Lower temperature promotes the deposition of thin polymeric films on the surface, affecting the sidewall profiles. (From Ref. 25, with permission.)

B. Metals

1. Aluminum

Aluminum is one of the most important metals in microelectronics. It is used as an interconnect due to its high electrical conductivity. Fluorine-containing plasmas form nonvolatile aluminum fluorides, which block further surface reaction and are therefore not useful for etching Al. In contrast, aluminum can be etched rapidly in chlorine-containing plasmas. However, the aluminum surface is covered by a native oxide (Al_2O_3), which must be removed before etching can commence. BCl_3 (or $SiCl_4$) is added to the Cl_2 gas not only to help in oxide removal but also to prevent its re-formation by scavenging any water vapor in the system. Ion bombardment helps keep the bottom of the feature "clean" so that etching can occur. A clean aluminum surface reacts with molecular or atomic chlorine even in the absence of a plasma, according to

$$Al + Cl/Cl_2 \rightarrow (AlCl_3)_2 \tag{52}$$

Since etching is chemical, the etch uniformity is strongly affected by the gas distribution. CCl_4 or $CHCl_3$ may be added to achieve anisotropy. Their fragments "coat" the feature sidewall with a film, which reacts with Cl atoms or simply blocks attack of the underlying aluminum by chlorine. This is basically the same mechanism used to obtain anisotropy in the etching of heavily doped n-type polysilicon. Most often, polymer fragments originating from the photoresist mask form sidewall-protecting layers. Figure 30 shows the effect on anisotropy of adding chlorine to $CHCl_3$. The $CHCl_3$ concentration has to be high enough to achieve anisotropy. When $CHCl_3$ is diluted by adding too much chlorine, the etch becomes isotropic.

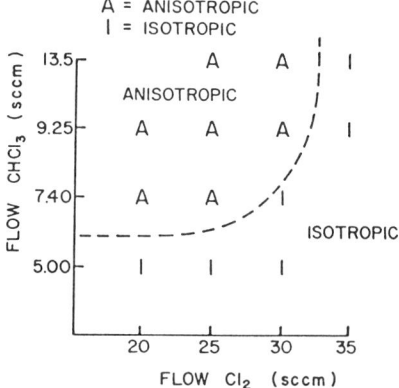

Figure 30 Anisotropic versus isotropic etching of aluminum for varying flow rates of $CHCl_3$ and Cl_2 feed gases. (From Ref. 26, with permission.)

Aluminum is often used as an alloy with a few atomic % Si and/or Cu to reduce electromigration. This alloy is difficult to etch because copper does not form readily volatile compounds. Heating of the wafer to some 200°C and intense ion bombardment to sputter off involatile products are then necessary to effect etching. Residual chlorine remaining on the wafer after aluminum etching can be deleterious because it can promote corrosion especially in Al–Cu alloys. Hence, wafer rinsing with deionized (DI) water is common. Dry passivation by exposure to a fluorocarbon plasma has also been reported to protect aluminum by converting aluminum chloride to fluoride, which can then be removed by a nitric acid wet clean.

2. Tungsten

Tungsten is a low-resistivity metal that can replace doped polysilicon conductors. Tungsten can be etched in fluorine-containing plasmas, such as CF_4, SF_6, NF_3, by forming volatile fluorides. Sometimes oxygen is added to increase the fluorine density in the plasma. Anisotropic etching can be obtained even with weak ion bombardment (10s of eV). When present, mask undercut can be suppressed by lowering the substrate temperature. Etching of the sidewalls is quenched, but the microscopic feature bottoms continue to etch under ion bombardment. Tungsten can also be etched anisotropically in Cl_2 plasmas, forming tungsten chloride.

C. Compound Semiconductors

Compound semiconductors are important for optoelectronic device fabrication (e.g., semiconductor lasers), and specialized high-speed devices. GaAs is etched rapidly (>1 µm/min) in Cl_2 plasmas. Chlorocarbons (e.g., CCl_4) will also etch GaAs, but at lower rates; in addition, polymer deposits may form. Thus, chlorocarbons are not favored as GaAs etchants. The reaction is limited by product ($GaCl_3$) desorption. The situation is similar for InP. The etch rate can be increased by heating the surface to volatilize the reaction-limiting product $InCl_3$. BCl_3 additions to Cl_2 are used to control sidewall profiles through sidewall passivation. Selectivity of etching GaAs over AlGaAs is readily achieved by small additions of oxygen to the feed gas that forms stable aluminum oxides (GaAs does not form a stable oxide). Some workers have used hydrocarbon-based plasmas (e.g., methane/hydrogen) to etch these materials. The reaction may proceed via organometallic (for Ga and In) and hydride (for As and P) compound formation. However, etch rates are rather low (<1000 Å/min). Compound semiconductor etching has been reviewed in Ref. 27.

D. Polymers

The most common polymer etching (also called ashing) is that of photoresist. Ashing is a necessary step after all etching operations that use photoresist as the mask (Fig. 1). In fact, photoresist ashing marked the first commercial application of plasma etching in microelectronics, back in 1969 [11]. Ashing is performed in barrel or downstream (see Fig. 4) reactors using oxygen plasmas to generate O atoms. The reaction rate is enhanced by heating the wafer. The reaction proceeds via O atoms attacking the carbon backbone and initiating chain scission, eventually forming volatile carbon oxides and water:

$$C_xH_y + O \rightarrow CO + CO_2 + H_2O \qquad (53)$$

Addition of small amounts of fluorocarbons to the plasma has been shown to greatly enhance the reaction rate. It is believed that hydrogen abstraction from the polymer by F atoms facilitates attack by the O atoms that ''burn'' the polymer. Ion bombardment enhances the rate of polymer etching; this is used in the so-called trilevel resist process to pattern thick polymer layers anisotropically. Polymer etching has been reviewed in Ref. 28.

E. Gas Additives

As already discussed in the foregoing sections additive gases are often used with the main feed gas (the one supplying the etchant species) to improve selectivity or anisotropy, or to achieve other goals. Table 6 provides a list of common additives and their function.

Oxidants (oxygen or oxygen-containing species, e.g., N_2O) are used to suppress polymer formation by oxidizing unsaturated halocarbons, which could otherwise lead to polymer deposits. In doing so, oxidants free up etchant species as seen in Eq. (50); this leads to a higher etch rate. Oxygen is also used to control the angle (taper) of the etched feature by controlled photoresist erosion.

Table 6 Effect of Additives in Plasma Etching

Additive	Function	Example system (additive/main gas:material)
Oxidative agent	Increase etchant density	O_2/CF_4: Si
	Suppress polymer	O_2/CCl_4: GaAs
	Control sidewall angle	O_2/CHF_3–CF_4: SiO_2
Reducing agent	Improve selectivity	H_2/CF_4: SiO_2
Sidewall protector	Improve anisotropy	O_2/HBr: Si
		C_2F_6/Cl_2: Si
Native oxide etchant	Initiate etching	BCl_3/Cl_2: Al
		CF_4/Cl_2: Si
Wall passivant	Increase etchant density	H_2O/O_2: polymer
Inert gases	Stabilize plasma	Ar/CHF_3–CF_4: SiO_2
	Improve heat transfer	He/Cl_2: Ta

Reducing agents are used to tie up etchant and increase the density of unsaturates, which then deposit polymers on nonoxide surfaces, blocking etching. Some examples were referred to in the previous discussion. Another example is oxygen addition to HBr in polysilicon etching. Oxygen oxidizes the sidewall, which then is not attacked by the etchant (Br atoms). Oxygen is sputtered away from the bottom of the feature, allowing etching to proceed anisotropically. Additives that can remove the native oxide so that etching can commence are used in situations where the film to be etched forms a stable oxide. Examples are aluminum and silicon. Aluminum oxide in particular is very stable and can re-form in improperly designed systems. Oxide or water scavengers are then used as additives to prevent reoxidation. BCl_3 in Al etching serves as both a sidewall protector and water scavenger. Wall passivants alter the reactivity of the reactor walls toward surface recombination of the etchant. For instance, the atomic oxygen concentration in an oxygen plasma can be increased when small amounts of water or nitrogen are added to the plasma. This increase has been attributed to conditioning of the reactor walls, which decreases the wall recombination probability γ (see Eq. 41). Finally, inert gases are often added to (1) improve the plasma stability, especially in strongly electronegative gases (e.g., Cl_2), (2) improve heat transfer between the wafer and the electrode, especially at low pressures, and (3) dilute the feed gas to reduce, for example, the chance for polymer and particulate formation.

V. HIGH-DENSITY PLASMAS

The drive for delineating finer features over larger diameter wafers without radiation damage has resulted in the development of a new class of plasma reactors, called high-density plasmas (HDP [29]). These reactors operate at low gas pressures (<50 mtorr) to improve uniformity and perhaps reduce contamination, and high plasma density (>10^{11} cm^{-3}) to deliver a high flux of ions to the wafer surface, thereby maintaining a high throughput (Table 2). Also, low pressure helps anisotropy since the ion mean-free-path is greater than the sheath dimensions, making ion flow to the wafer highly directional. A distinct characteristic of these sources is that the plasma is generated independently of the voltage applied to the substrate electrode. In capacitively coupled systems (Fig. 2), in order to increase the power delivered to the reactor one needs to increase the voltage applied between the electrodes. However, most of the voltage drop in these systems occurs in the sheath next to the electrodes (Fig. 7); hence, most of the power is consumed in accelerating ions in the sheath rather than in creating radicals in the bulk plasma. In addition, high ion bombardment can result in radiation damage and sputtering of the electrodes, which in turn lead to device contamination.

Because of differences in the operating conditions and how the plasma is generated, the chemistry in high-density plasma sources may be different compared with the relatively high-pressure (>100 mtorr) systems (Fig. 2). Some differences are

1. Surface chemical reactions of radicals are more important than gas-phase reactions as the operating pressure is lowered. For example, volume recombination is usually negligible below ~100 mtorr. Therefore, it is imperative to understand surface chemistry, especially the ion–surface interactions, more so in low-pressure HDP systems.
2. Because volume recombination is not efficient, and the wall materials are chosen to minimize wall recombination, the feedstock gas may be highly dissociated. Hence, radical chemistry dominates over the chemistry of the molecules. For example, in high-pressure oxygen plasmas, metastable oxygen molecules and ozone are thought to contribute to the chemistry of the system. In HDP at high enough power, atomic oxygen is expected to be the dominant species.
3. If the gas is highly dissociated in atoms, and since negative ions usually form by dissociative electron attachment with molecules, the negative ion density in HDP is expected to be much lower than in the high-pressure systems. For example, in a high-pressure chlorine discharge, negative ions outnumber electrons by a few hundred times. In a HDP chlorine discharge, the negative ion density is comparable with the electron density.
4. Because of the high ion flux delivered to the wafer surface, reactive ions may be the primary reactants (the term *reactive ion etching*, RIE, which is a misnomer when used for traditional plasma tools, may not be so in HDP sources!)

VI. ATOMIC LAYER ETCHING

Control of device dimensions down to the atomic level is important for fabrication of abrupt heterostructure interfaces and extremely thin layers for optoelectronics, quantum devices, and nanostructures [30]. This has led to the evolution of atomic layer processing, which can offer atomic layer control over deposition and etching for a variety of materials including semiconductors, metals, superconductors, and insulators. Impressive advances have been made in the development of atomic layer deposition. For example, metallorganic chemical vapor deposition (MOCVD) and molecular beam epitaxy (MBE) have been used to grow very thin layers of compound semiconductors and their alloys. Also, atomic layer epitaxy (ALE) has been developed as a technique with atomic resolution to deposit heterointerfaces that are smoother than those obtained by conventional MBE. Atomic layer epitaxy is a self-regulatory process that, in its

The Chemistry of Plasma Etching

simplest form, produces one atomic layer of a compound per operational cycle. The cycle may be repeated to deposit the desired number of layers.

In order to fabricate nanometer-scale devices, etching with atomic layer resolution must also be developed. Dry etching techniques such as plasma etching are not capable of atomic layer resolution, because they are too rapid to be able to control the process with monolayer accuracy. In addition, these techniques depend on the action of ion bombardment, which disrupts the crystal lattice, introducing damage that cannot be tolerated in extremely thin layers.

To overcome these limitations, atomic layer etching (ALET) has been developed, as shown schematically in Fig. 31. One complete cycle of ALET consists of the following steps:

1. Exposure of a clean surface to a gas and chemisorption of the gas onto the surface (e.g., Cl atoms chemisorbed onto Si).
2. Evacuation of the chamber to remove excess gas.

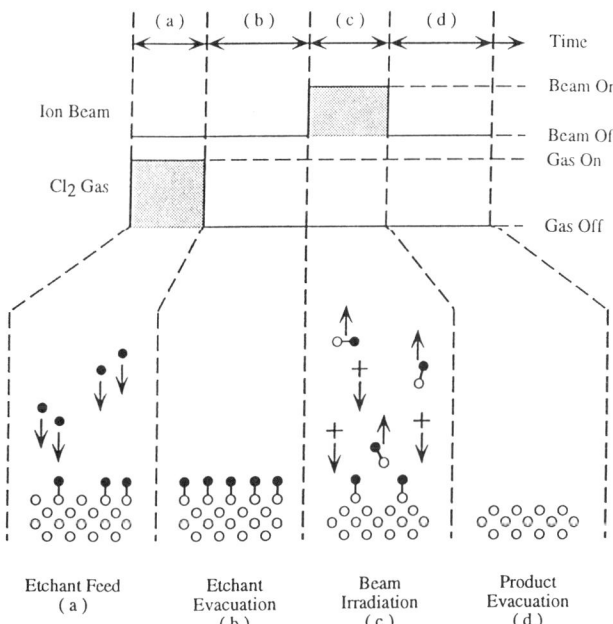

Figure 31 Steps in atomic layer etching: (a) gas feed and chemisorption to form a monolayer; (b) evacuation of excess gas; (c) exposure of the surface to energetic particle (e.g., ions) bombardment to induce etching; and (d) evacuation of product. (From S. Athavale, Ph.D. thesis in progress, University of Houston.)

3. Exposure of the surface to an energetic beam (e.g., Ar ion beam) to induce a chemical reaction between the adsorbed species and the surface atoms. In this step, a monolayer of the solid is removed as product.
4. Evacuation of the chamber to remove products.

Steps 1–4 complete one cycle, resulting in etching of one atomic layer of the film. The cycle can be repeated to remove as many atomic layers as required.

A critical requirement for ALET is that the process must be self-limiting, i.e., in step 3 further reaction stops as soon as all adsorbates have reacted. Also, the gas in step 1 must not be able to etch the substrate chemically (i.e., without ion bombardment).

Figure 32a shows the etch rate of GaAs per ALET cycle as a function of the gas feeding time (step 1). The etch rate saturates at 1 monolayer/cycle for feeding times greater than 5 s. Apparently, a minimum "dose" is required to fully chemisorb one monolayer of Cl on top of the GaAs surface. Similarly, the argon ion irradiation time (step 3) must be greater than 6 s to complete the reaction across the whole substrate surface (Fig. 32b). If the gas feeding time or the ion irradiation time are too short, etching at less than a monolayer per cycle can occur, as shown in the molecular sketch of Fig. 33. In this figure, the frames marked as "self-limited" give the desired result. If the gas exposure time is too long, multiple layers of gas may adsorb (frame d). The ion irradiation time (actually, ion dose) must then be increased to first remove the excess gas layers before inducing the surface reaction to remove a monolayer of the substrate. When either the gas feed or the ion irradiation time are too short, monolayer etching is not complete, and the surface becomes "rough" (frame b).

VII. PLASMA DIAGNOSTICS

Plasmas are very complex chemical systems where a plethora of reactions and processes occurs simultaneously. Due to the many parameters affecting the plasma chemistry (see Fig. 3) and the interdependence of these parameters, it is essential to develop techniques for monitoring key plasma properties and the process output. These *plasma diagnostics* help one to better understand and manipulate the plasma to meet specific goals. Plasma diagnostics are also used for control of etching and deposition systems. Plasma diagnostics can be categorized into gas-phase and surface diagnostics. They should preferably be selective, nonintrusive, sensitive, capable of spatial and temporal resolution, when necessary, capable of real-time monitoring, cheap, and as simple as possible. Optical techniques are particularly attractive because they combine many of these features. A brief discussion of some important plasma diagnostics is given in the following. A detailed description of plasma diagnostics is given in Refs. 33 and 34.

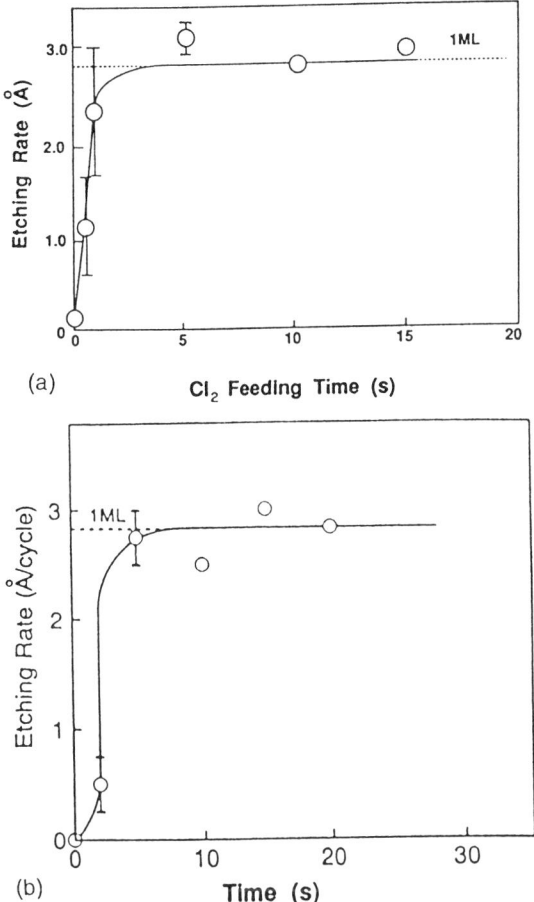

Figure 32 (a) Etch rate of GaAs as a function of Cl_2 feed time during step (a) of Fig. 31. A feed time of >5 s is required to achieve a complete monolayer of Cl adsorbed on the surface. (b) Etch rate of GaAs as a function of argon ion irradiation time during step (c) of Fig. 31. An irradiation time of >6 s is needed to fully react the adsorbed Cl into volatile product. (From Ref. 31, with permission.)

A. Gas-Phase Diagnostics

1. Optical Emission Spectroscopy

Optical emission spectroscopy (OES) is used extensively in plasma processing because it is a simple, nonintrusive technique. It is based on the light emitted by plasma species after these species have been excited by electron impact

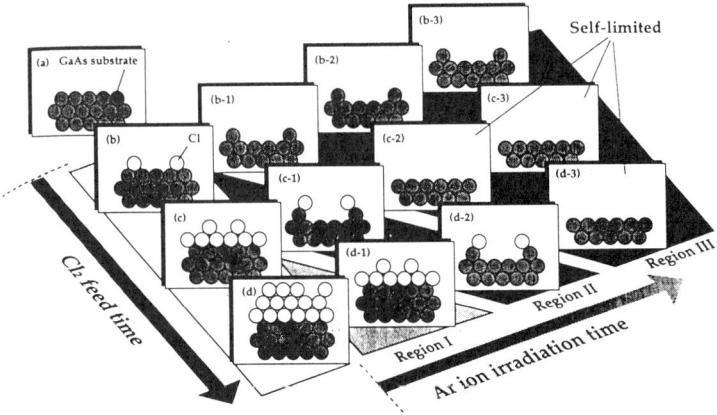

Figure 33 Model to explain molecular processes during atomic layer etching (see text for discussion). (From Ref. 32, with permission.)

(reactions T14, T15 in Table 4). Plasma emission is collected and dispersed into its wavelengths by a monochromator. The light intensity as a function of wavelength is measured with a photomultiplier tube. Photodiode array detectors and charge-coupled device (CCD) cameras have also been used to provide rapid acquisition of spatially and spectrally resolved emission. The OES spectrum is used to identify plasma species that emit in the wavelength range monitored. An example of a spectrum obtained in a 95% chlorine 5% argon (mole %) plasma is shown in Fig. 34. Cl atom and Ar emission lines are clearly seen. Common species monitored by OES and their corresponding wavelength are given in Ref. 2, p. 226.

Qualitative characterization of the plasma is the most common use of OES. Semiquantitative results can be obtained by applying a technique known as *actinometry*. In this technique, the intensity of light emitted by a species may be related to the ground state concentration of that species. In an oxygen plasma, for instance, excited oxygen atoms (O^*) are produced by electron-impact reaction given as

$$O + e \xrightarrow{k_e} O^* + e \tag{54}$$

O^* can decay by spontaneous emission,

$$O^* \xrightarrow{k_s} O + h\nu \tag{55}$$

or by quenching upon collision with other species,

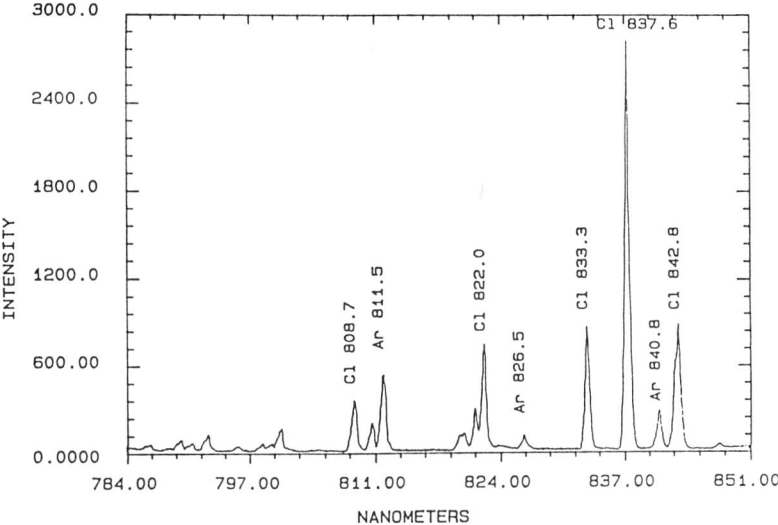

Figure 34 Optical emission spectrum of a chlorine plasma with 5 vol % added argon. Important peaks are identified. (From Ref. 35.)

$$O^* + M \xrightarrow{k_q} O + M \tag{56}$$

If quenching can be neglected, the spontaneous emission intensity, I_O is proportional to the ground state concentration of oxygen atoms (c_1),

$$I_O \sim k_e n_e c_1 \tag{57}$$

The preceding analysis assumes electron-impact excitation of atomic oxygen as the dominant mechanism for producing O^*. A reaction of the type

$$O_2 + e \rightarrow O^* + O + e \tag{58}$$

would invalidate actinometry if the corresponding O^* atoms were to emit at the wavelength of interest. In Eq. (57), the so-called excitation efficiency $\eta_e = k_e \eta_e$ varies with the reactor operating conditions. To account for this variation, Coburn [15] introduced a small amount of an inert gas (actinometer) having an excitation threshold and a cross section similar to those of the species of interest. This way, although the individual values of the excitation efficiency for the actinometer and the species of interest change with operating conditions, their ratio remains almost constant. Hence, by writing an equation similar to Eq. (57) for the actinometer (e.g., Ar), and taking their ratio, one obtains

$$\frac{I_O}{I_{Ar}} = q \frac{c_1}{c_{Ar}} \tag{59}$$

where q is a proportionality constant and c_{Ar} is the known concentration of Ar. By measuring I_O and I_{Ar} at the appropriate wavelengths, the relative change in O concentration (c_1) can be obtained. Actinometry does not provide the absolute concentration of the species of interest; the proportionality constant q in Eq. (59) is not known a priori. Monitoring the radical concentration profiles in real time can be used for process control. For example, the radical concentration must be uniform across the wafer to achieve etch uniformity especially in chemical etching systems (e.g., aluminum).

Endpoint Detection. Optical emission spectroscopy is also used widely for endpoint detection. Consider, for example, etching of silicon over silicon dioxide in an F-atom plasma (e.g., CF_4). An OES system monitoring a prominent F line may show a response similar to that of Fig. 35. When the plasma is turned on, the F atom density (and signal intensity) increases sharply. Etching of silicon does not commence immediately; the native oxide (~30 Å thick) must first clear. Since silicon consumes F atoms much more efficiently than oxide (see Eqs. 46 and 47), the F atom density in the plasma decreases as the native oxide clears and silicon starts etching. When the silicon film itself starts clearing and the oxide underlayer is exposed, the F density rises again. This signals the endpoint. It must be realized that for OES to work for endpoint detection, there must be a loading effect present, i.e., the film being etched must be a significant sink

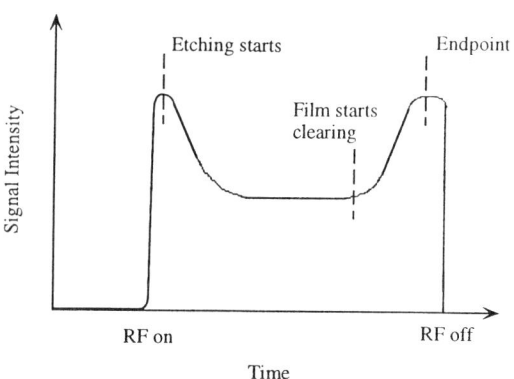

Figure 35 Use of optical emission spectroscopy for endpoint detection (schematic). During etching, the film consumes the reactant and the emission intensity is at a low level. When the film starts clearing, the reactant concentration starts building up and the emission intensity increases. The intensity stays at a constant level when the film has fully cleared. A gentle slope of the curve is often indicative of etch nonuniformity. If the starting film thickness and the etch rate are both uniform, the transition will be sharp.

for the species of interest. Otherwise, the change in plasma emission at endpoint may be too small to be detected. Endpoint detection can be used to monitor the etch uniformity as well. If the films clears gradually because of etching non-uniformities, the signal rise will also be gradual. Uniform film clearing will result in a sharp rise of the OES signal at endpoint. Emission from the reaction product can also be used for endpoint detection. When the film clears, the product intensity will correspondingly decrease, and eventually disappear.

Other endpoint detection techniques include laser interferometry, mass spectrometry, ellipsometry, and plasma impedance. When applicable, OES is the method of choice for many plasma systems.

2. Laser-Induced Fluorescence

Laser-induced fluorescence (LIF) uses a tunable laser to perform selective excitation of plasma species to a higher energy level. Upon deexcitation, the species emits fluorescence, the intensity of which is a measure of the species density. Laser-induced fluorescence is more powerful than optical emission spectroscopy because it can monitor directly the ground state species, which are the most abundant and often the most important etchant species. In contrast, OES monitors the excited state species density and indirectly (when a set of rather restrictive conditions are satisfied, see above) the ground state density. Laser-induced fluorescence can offer high selectivity, spatial (~ 1 mm) and temporal (~ 10 ns) resolution, and it can be made quantitative. However, LIF can be applied only to a limited number of species (having fluorescing excited states that can be reached by an optically allowed transition), and it is more complex and expensive than OES. Laser-induced fluorescence has been used to monitor metastable atoms in noble gas discharges, light atoms (O, Cl, H, etc.), which are thought to be important in a number of plasma systems, metals (e.g., Al, Fe), and a variety of radicals (BCl, SiF_2, CF_2, etc.).

3. Absorption

In this technique, the attenuation of light intensity as the probe beam passes through the plasma is related to the density of the absorbing species (Beer's law). Quantitative analysis is possible. Multiple passes are usually required to obtain an acceptable signal-to-noise ratio, and this limits the spatial resolution of the technique.

4. Mass Spectrometry

In mass spectrometry, the sample gas is ionized by electron impact in the instrument's ionization chamber. The resulting ions are separated in the mass filter according to their mass-to-charge ratio. The mass spectrum provides a fingerprint of the sample gas. Quantitative analysis is difficult to perform and is seldom done in plasma processing. Generally speaking, mass spectrometry has been used in two configurations: effluent gas analysis and molecular beam. In the former case, the mass spectrometer is attached to the exhaust port of the plasma reactor. In this configuration, only stable gaseous species can be detected. Radicals and ions normally have ample chance to recombine on the walls of the apparatus before reaching the mass spectrometer. Effluent gas analysis can provide valuable information about the process. For example, the SiF_3^+ ion peak originating from the reaction product (SiF_4) can be used as a monitor of the etch rate of silicon in a fluorine plasma. In molecular beam mass spectrometry, a sample is drawn directly from the plasma into the instrument as a molecular beam. The goal is to minimize or totally eliminate collisions, which can alter the sample composition. Thus, radicals as well as positive and negative ions can be detected. When detecting plasma ions, the ionizer of the spectrometer is turned off. Most often, the spectrometer is housed in a separate chamber that has a line-of-sight communication with the plasma through a pinhole [10]. This configuration is nonintrusive. Neutrals and ions effusing through this pinhole are analyzed by the mass spectrometer. Molecular beam mass spectrometry is intrusive when the head of the spectrometer is inserted into the plasma. Also, a sheath forms between the plasma and the spectrometer, which can alter the ion composition.

Nonchemical techniques to measure the plasma density include microwave interferometry (electron density) and Langmuir probes (ion density and electron temperature) [33].

B. Surface Diagnostics

Surface diagnostics are very important for understanding how the plasma interacts with the surface and how this interaction modifies the surface chemically and physically. There is a host of surface diagnostics utilizing ultrahigh vacuum (UHV) techniques [34]. These include Auger electron spectroscopy (AES) to obtain the elemental composition of the top surface layers, X-ray photoelectron spectroscopy (XPS) to obtain, in addition, chemical bonding information, and many others. These techniques are best when applied in situ, in the sense that the sample is transferred from the processing chamber to the analysis chamber under vacuum. Using these techniques, a wealth of information has been gained regarding the surface chemistry. For example, the fluorination of a silicon or oxide surface, when the wafer is etched in fluorocarbon plasmas of varying composition, can be studied by in situ XPS [2]; or the carbon deposits on the silicon surface (when the F/C ratio is too low) that provide selectivity of etching oxide versus silicon can be studied by AES [15].

1. Attenuated Total Internal Reflection

Plasma reactors operate at pressures much too high for the UHV techniques to be applied in real time during wafer processing. A technique to monitor the wafer surface in real time is attenuated total internal reflection (ATIR) [6]. This is basically an infrared absorption technique that samples species adsorbed on the wafer surface. Multiple "passes" are necessary to improve the signal-to-noise ratio, which requires a special design of the substrate. This technique has been used to monitor oxide removal off of silicon surfaces, and to study surface passivation of GaAs and wafer cleaning.

2. Ellipsometry

Some information on the plasma surface interaction can be obtained in real time by ellipsometry. In ellipsometry, the surface is irradiated with linearly polarized light. After reflection, the light becomes elliptically polarized and provides information about the optical constants of the substrate and its overlayers. The polarized light can be decomposed into a p component, lying parallel to the plane of incidence, and an s component, lying perpendicular to the plane of incidence. Ellipsometry measures the change of the amplitude ratio and the phase difference of the p and s components of the reflected light relative to the incident light. Ellipsometry, along with a "model" of the disordered layer formed by ion bombardment, has been used to study fluorination of the silicon surface in plasma etching. Ellipsometry can also be used for film thickness measurements with very high resolution (few Å).

3. Other Surface Diagnostics

Other surface diagnostic techniques are laser interferometry and quartz crystal microbalance (QCM). These techniques do not provide information about the chemical state of the surface. Rather, they are used to monitor the etch or deposition rate in real time. Interferometry (and ellipsometry) can also be used for endpoint detection. Quartz crystal microbalance measures the change in the frequency of vibration of a quartz crystal as the mass of the crystal changes by etching or depositing a film on the crystal. The technique can detect submonolayer surface coverage changes. It was used to obtain the etch rate results shown in Fig. 18. The "negative" etch rate at $\sim$220 s in Fig. 18a corresponds to adsorption of chlorine on the "fresh" silicon surface, after the chlorine gas is first turned on at 200 s.

VIII. SUMMARY

The importance of plasma etching will continue to increase in the future, as device dimensions continue to shrink to $\leq$0.25 μm and low-temperature processing (<300°C) becomes even more critical. Plasmas can provide the anisotropy needed for high-resolution patterning. Further, plasmas can generate reactive radicals at near room temperature, and this is a strong driving force for their use in a variety of thin-film etching, deposition, and cleaning applications.

More complex patterns with finer dimensions will require integrated processing and process automation with sophisticated smart sensors for real-time process control. Environmental awareness will require more careful selection of plasma feedstock gases.

Selection of the etch gases and of reactor design and operating conditions to meet the demands of the microelectronics manufacturing of the next century will be challenging. Ideally, the etch process must be fast, uniform, and selective with controlled wall profiles and without particle contamination or radiation damage. A better fundamental understanding of the plasma physics and chemistry is required to meet the challenge. Plasma modeling and simulation coupled with well-defined experiments to validate the models has emerged as a promising tool in the quest for improved and novel plasma processes to meet the evolving demands of the industry.

Understanding at the molecular level will become increasingly important. Surface processes including ion-assisted kinetics, polymerization, deposit nucleation and growth, and adhesion must be better understood to engineer materials and microstructures with tailored properties (e.g., superlattices). Atomic resolution techniques such as scanning tunneling and atomic force microscopy will find increasing applications. Atomic layer processing will be developed further to control film deposition and etching with monolayer accuracy; it is ushering the way into the world of nanoelectronics.

EXERCISES

1. (a) Calculate the velocity of an electron having an energy of 3 eV. This electron is accelerated by a constant electric field of 10 V/cm (the field is parallel to the electron velocity) for 0.1 cm. Calculate the new electron energy.
   (b) The electron resulting from the acceleration of part (a) suffers an elastic collision with an argon atom and scatters by 60° with respect to its original direction. What fraction of the electron energy was transferred to the atom?
   (c) Repeat parts (a) and (b) for an argon ion having 0.1 eV energy instead of an electron. What do you observe?
2. (a) Explain how a sheath forms over a material surface in contact with a plasma.
   (b) What is the importance of the sheath in plasma etching?
   (c) What is floating potential? Calculate the floating potential for an argon plasma having an electron temperature of 5 eV.
   (d) Why is the plasma potential the most positive potential in the system?

The Chemistry of Plasma Etching

(e) What is Debye length? What is the importance of the Debye length? Calculate the Debye length for a plasma having electron density and temperature of 5×10^{11} cm^{-3} and 5 eV, respectively.

3. Calculate the flux of Cl atoms striking a wafer immersed in a chlorine plasma when the gas temperature is 500 K, the total pressure is 10 mtorr, and the mole fraction of atomic chlorine is 0.1.

4. (a) Calculate the rate of ionization of molecular chlorine as a function of mean electron energy using a Maxwellian electron energy distribution function (EEDF), and a Druyvesteyn EEDF. Plot the results for an energy range of 2–12 eV.

   (b) Which of the two distributions of part (a) would result in a higher self-sustained electric field (E/N) in a chlorine plasma? Why?

   (c) What fraction of electrons is capable of performing ionization for a Maxwellian EEDF with an average energy of 3 eV?

5. A plasma is sustained between two parallel plates of infinite extension separated by a distance L. Electrons are produced by ionization of atoms according to

$$A + e \rightarrow A^+ + 2e$$

where A^+ is a positive ion. The electron production rate is given by $R_e = k_i n_e$, where k_i is the ionization rate coefficient. Electrons are lost to the plates by ambipolar diffusion.

   (a) Determine the spatial distribution of electrons at steady state operation.
   (b) What must the electron temperature be to assure steady state?
   (c) What is the magnitude of the ambipolar field as a function of position between the plates?

Data: Electron diffusivity $D_e = 10^6$ cm^2/s, ion diffusivity $D_+ = 100$ cm^2/s, ion temperature $T_+ = 0.040$ eV, $L = 2$ cm, rate coefficient $k_i = 2 \times 10^9 \exp(-\phi/kT_e)$ s^{-1}, where $\phi = 16$ eV, and T_e is the electron temperature.

6. A piece of single-crystal silicon is bombarded simultaneously by a noble gas ion beam (Ar$^+$, Ne$^+$, or He$^+$) and a neutral beam of XeF$_2$ under ultrahigh vacuum conditions. The measured etch yield of silicon (in Si atoms removed as SiF$_4$ per incident ion) is given in Fig. 36 as a function of the neutral beam flow rate. Explain the results of Fig. 36.

7. Single-wafer parallel-plate plasma reactors (Fig. 37) are widely used for etching in the microelectronics industry. Gases are introduced uniformly through portholes at the upper (showerhead) electrode and are pumped radially outward. In a chlorine discharge, Cl atoms are generated via the reaction

$$Cl_2 + \text{fast electron} \xrightarrow{k_d} 2Cl + \text{slow electron}$$

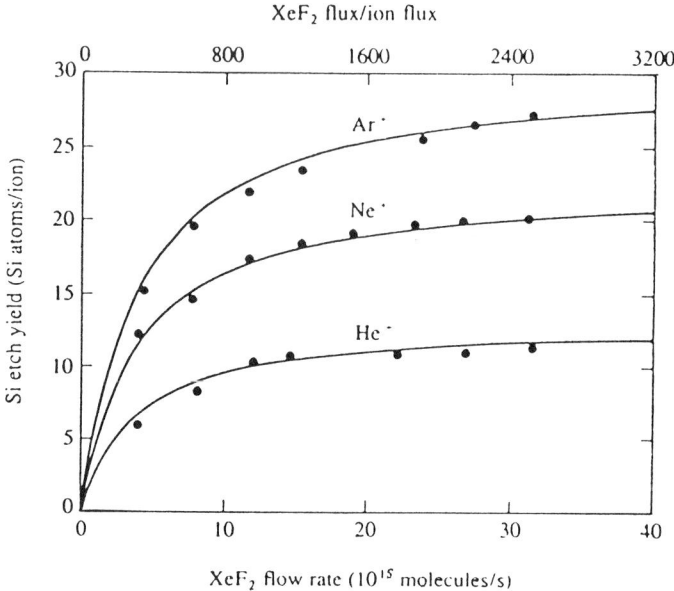

Figure 36 Si etch yield vs. XeF$_2$ flow rate. (From Ref. 36, with permission.)

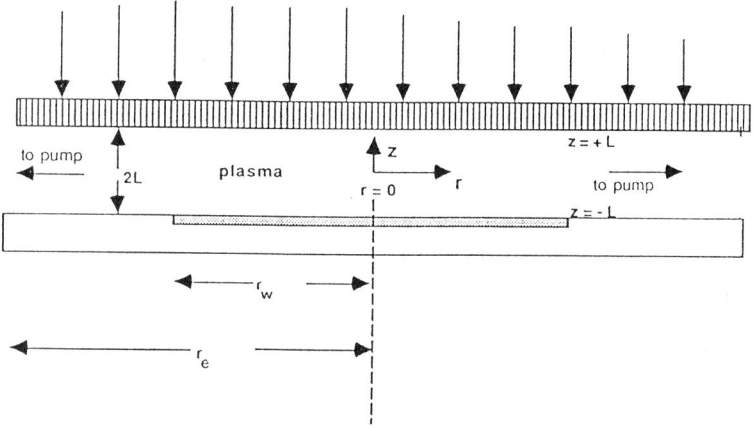

Figure 37 Reactor schematic for problem 7.

The main gas-phase elimination reaction of Cl is volume recombination:

$$\text{Cl} + \text{Cl} + M \xrightarrow{k_{vr}} \text{Cl}_2 + M, \quad \text{where } k_{vr} = 2.96 \times 10^{-32} \text{ cm}^6 \text{ s}^{-1}$$

where M is a third body (Cl_2 in this case).

Wall recombination of Cl atoms also takes place on the nonwafer surfaces with a rate

$$R_{sr} = \frac{n_{Cl}\bar{c}_{Cl}\gamma}{4} \frac{S}{V}$$

where n_{Cl} is the atomic chlorine density, $\bar{c}_{Cl}$ is the mean thermal speed of chlorine atoms, and S/V is the surface-to-volume ratio; γ is the reaction probability, which depends on the wall material.

(a) First consider the case of an "empty" reactor (no wafer). Using Figs. 15 and 16, calculate the atomic chlorine concentration in the reactor for electrode spacings of 1, 2, and 3 cm, for a well-mixed reactor. Assume a constant pressure of 0.5 torr and a gas temperature of 350 K. Also, the power delivered to the plasma is 100 W and the pure chlorine feed flow rate is 50 sccm. Do the calculation for stainless steel ($\gamma = 0.1$) and anodized aluminium ($\gamma = 0.007$) electrodes. Comment on the results.

(b) A 4-in. silicon wafer is now placed on the lower electrode. The wafer is covered with a P-doped polysilicon film (dopant density 10^{19} cm^{-3}). Plot the etch rate versus spacing for anodized aluminum and stainless steel electrodes. Which electrode material would you select? Why?

(c) Do you expect uniform etching in case (b)? Why or why not?

(d) In case etching is nonuniform, suggest two methods to improve uniformity. Explain your choices.

(e) Suggest a method for endpoint detection in this system.

8. The data shown in Fig. 38 were obtained in a downstream reactor shown in Fig. 4b of the text. Reactive radicals are generated by an upstream oxygen plasma. The radicals are then flown to the etching chamber, where a polymer-coated wafer is etched. In contrast to expectation, the etch rate increases by increasing the exposed wafer area (an *inverse* loading effect). A reasonable hypothesis for this behavior is that atomic oxygen reacts with the polymer P according to

$$\text{O} + \text{P} \rightarrow \text{X}$$

to give product X, which itself reacts with the polymer:

$$\text{X} + \text{P} \rightarrow \text{R}$$

to give a (nonreactive) product R. Based on this hypothesis, develop a well-

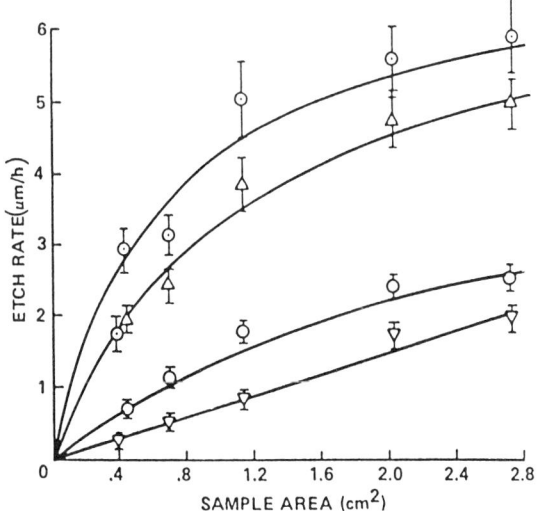

Figure 38 Etch rate of a function of sample area. (From Ref. 37, with permission.)

mixed reactor model for the etching chamber to explain the results of Fig. 38.

9. (a) An etch process calls for selective and anisotropic etching of a SiO_2 film on top of a polysilicon film. Suggest a reactor configuration and dry etch chemistry to carry out the task.
   (b) In plasma etching systems designed to etch anisotropically, it is often necessary to water cool the electrodes on which the wafer sits; i.e., it helps to keep the wafers cool. Explain why this might help the performance of these systems.
   (c) Suggest a reactor configuration and etch chemistry to etch anisotropically and selectively silicon nitride over silicon dioxide.
10. What is the selectivity of etching silicon over oxide in a fluorine-containing downstream plasma at a substrate temperature of 200°C? Plot the selectivity versus the temperature.
11. What are the main requirements for optical emission actinometry to work? Which wavelength of the O-atom emission would you use to monitor the O-atom concentration by actinometry? Why? (See Reference 18.)

ACKNOWLEDGMENTS

A portion of the work described herein has been supported financially by the National Science Foundation, The Welch Foundation, and the State of Texas through the Texas Advanced Technology Program. The author is grateful to the above agencies for their support.

NOMENCLATURE

$\bar{c}_j$	mean thermal speed of species j, cm/s
D_a	ambipolar diffusion coefficient, cm^2/s
D_+	diffusion coefficient of positive ions, cm^2/s
E	electric field, V/cm
E_0	amplitude of electric field, V/cm
E_{eff}	effective electric field, V/cm
E_a	activation energy, J/mol
F	plasma excitation frequency, s^{-1}
$f(\varepsilon,t)$	electron energy distribution function, eV$^{-1.5}$
J_j	current density due to species j, A/cm^2
I_j	flux due to species j, 1/cm^2-s
k	Boltzmann constant, J K^{-1}
k_j	rate coefficient for process j, cm^3/s, cm^6/s, or cm/s
k_a	attachment rate coefficient, cm^3 s^{-1}
k_d	dissociation rate coefficient, cm^3 s^{-1}
k_i	ionization rate coefficient, cm^3 s^{-1}
k_{sr}	surface reaction rate coefficient, cm s^{-1}
k_{vr}	rate coefficient for volume recombination, cm^6 s^{-1}
L	half of interelectrode spacing, cm
m_j, M_j	mass of species j, g
n_j	number density of species j, cm^{-3}
$\langle n_e \rangle$	time-average electron density, cm^{-3}
N	total gas number density, cm^{-3}
P_{rf}	rf power, W
P	pressure, torr
Q_0	feed gas flow rate, standard cm^3 per s (sccm)
q	elementary charge, 1.602×10^{-19} C
R	reactor radius, cm
R_j	reaction rate in gas phase or on surfaces (units vary)
S_j	area of surface j, cm^2
t	time, s
T_j	temperature of species j, K or eV
V_T	dc self-bias of substrate, V
V_F	floating potential, V
$V_{1,2}$	sheath potential over electrode 1, 2, V
V_p	plasma volume, cm^3
x	Cartesian coordinate, cm

y	species mole fraction
Y_s	sputtering yield

Greek

γ	surface reaction probability
ε	electron energy, eV
$\langle\varepsilon\rangle$	mean electron energy, eV
ε_+	energy of ions striking wall, eV
ε_0	permittivity of free space, F/cm
Λ	electron diffusion length, cm
λ_D	Debye length, cm
ν_m	electron momentum exchange collision frequency, s^{-1}
ρ	mass density, g/cm³
$\sigma_j(\varepsilon)$	collision cross section for process j, cm²
ω	excitation frequency, rad s^{-1}

Subscripts

e	electrons
c	chemical
Cl	chlorine
i	ions
+	positive ions
r	recombination
rf	radio frequency
sr	surface recombination
vr	volume recombination
th	threshold

REFERENCES

1. S. K. Ghandhi, *VLSI Fabrication Principles: Silicon and Gallium Arsenide*, Wiley, New York, 1989.
2. S. M. Rossnagel, J. J. Cuomo, and W. D. Westwood, eds., *Handbook of Plasma Processing Technology*, Noyes Publications, Park Ridge, New Jersey, 1990.
3. J. L. Vossen and W. Kern, eds., *Thin Film Processes II*, Academic Press, Orlando, Florida, 1991.
4. D. L. Smith, "High Pressure Etching" in *Plasma Processing for VLSI* (N. G. Einspruch and D. M. Brown, eds.), Academic Press, Orlando 1984.
5. B. Chapman, *Glow Discharge Processes*, Wiley, New York, 1980.
6. J. A. Mucha, D. W. Hess, and E. S. Aydil, Plasma etching, *Introduction to Microlithography*, 2nd ed. (L. F. Thompson, C. G. Willson, and M. J. Bowden, eds.), ACS Professional Reference book, American Chemical Society, Washington, DC 1994.

7. L. G. Christophorou, ed., *Electron Molecule Interactions and Their Applications*, Vols. 1 and 2, Academic Press, Orlando, Florida, 1984.
8. B. E. Cherrington, *Gaseous Electronics and Gas Lasers*, Pergamon Press, Oxford, 1979.
9. A. T. Bell, Fundamentals of plasma chemistry, *Techniques and Applications of Plasma Chemistry* (J. R. Hollahan and A. T. Bell, eds.) Wiley, New York, 1974.
10. E. S. Aydil and D. J. Economou, theoretical and experimental investigations of chlorine RF glow discharges, Part I: Theoretical, *J. Electrochem. Soc. 139*:1396 (1992); Part II: Experimental, *J. Electrochem. Soc. 139*:1406 (1992).
11. D. M. Manos and D. L. Flamm, eds., *Plasma Etching: An Introduction*, Academic Press, Orlando, Florida, 1989.
12. S. W. Benson, *Thermochemical Kinetics*, 2nd ed., Wiley, New York, 1976.
13. E. W. McDaniel, *Collision Phenomena in Ionized Gases*, Wiley, New York, 1964.
14. C. G. Hill, *An Introduction to Chemical Engineering Kinetics and Reactor Design*, Wiley, New York, 1977.
15. J. W. Coburn, *Plasma Etching and Reactive Ion Etching*, AVS Monograph Series, N. Rey Whetten, ser. ed., American Institute of Physics, New York, 1982.
16. S.-K. Park and D. J. Economou, *J. Electrochem. Soc. 137*:2624 (1990).
17. V. Donnelly, D. Flamm, and R. Bruce, *J. Appl. Phys. 58*:2135 (1985).
18. D. J. Economou, S.-K. Park, and G. Williams, *J. Electrochem. Soc. 136*:188 (1989).
19. D. Chin, S. H. Dhong, and G. J. Long, *J. Electrochem. Soc. 132*:1705 (1985).
20. C. Mogab, A. Adams, and D. Flamm, *J. Appl. Phys. 49*:3796 (1979).
21. L. M. Ephrath, *J. Electrochem. Soc. 126*:1419 (1979).
22. S. Noda, S. Nishikawa, and S. Ohno, *Jpn. J. Appl. Phys. 28*:2362 (1989).
23. N. Hayasaka, H. Okono, and Y. Horiike, *Solid State Technol. 31*:127 (1988).
24. A. C. Adams and C. D. Capio, *J. Electrochem. Soc. 128*:366 (1981).
25. T. Toyosato, T. Tamaki, and T. Tsukada, Proceedings of the 8th Plasma Processing Symposium (G. S. Mathad and D. W. Hess, eds.), The Electrochemical Society, Pennington, NJ, vol. 90-14, 1990, p. 723.
26. R. H. Bruce and G. P. Malafsky, *J. Electrochem. Soc. 130*:1369 (1983).
27. S. J. Pearton, F. Ren, T. R. Fullowan, A. Katz, W. S. Hobson, U. K. Chakrabarti, and C. R. Abernathy, *Mater. Chem. Phys. 32*:215 (1992).
28. M. A. Hartney, D. W. Hess, and D. S. Soane, *J. Vac. Sci. Technol. B 7*:1 (1989).
29. M. A. Lieberman and R. A. Gottscho, Design of high density plasma sources for materials processing, *Physics of Thin Films* (M. Francombe and J. Vossen, eds.), Academic Press, Orlando, Florida, 1993.
30. T. F. Kuech, P. D. Dapkus, and Y. Aoyagi, *Atomic Layer Growth and Processing*, Materials Research Society Symp. Proceedings, Vol. 222, MRS, Pittsburgh, Pennsylvania, 1991.
31. Y. Aoyagi, K. Shinmura, K. Kawasaki, I. Nakamoto, K. Gamo, and S. Namba, *Thin Solid Films 225*:120 (1993).
32. T. Meguro, M. Ishii, K. Kodama, Y. Yamamoto, K. Gamo, and Y. Aoyagi, *Thin Solid Films 225*:136 (1993).
33. O. Auciello and D. L. Flamm, eds., *Plasma Diagnostics, Vol. 1: Discharge Parameters and Chemistry*, Academic Press, Orlando, Florida, 1989.

34. O. Auciello and D. L. Flamm, eds. *Plasma Diagnostics, Vol. 2: Surface Analysis and Interactions*, Academic Press, Orlando, Florida 1989.
35. E. Aydil, *Theoretical and Experimental Investigation of Chlorine RF Glow Discharges and Polysilicon Etching.* Ph.D. thesis, University of Houston, 1993.
36. U. Gerlach-Meyer, *Surf. Sci. 103*:524 (1981).
37. N. R. Lerner and T. Wydeven, *J. Electrochem. Soc. 136*:1426 (1989).

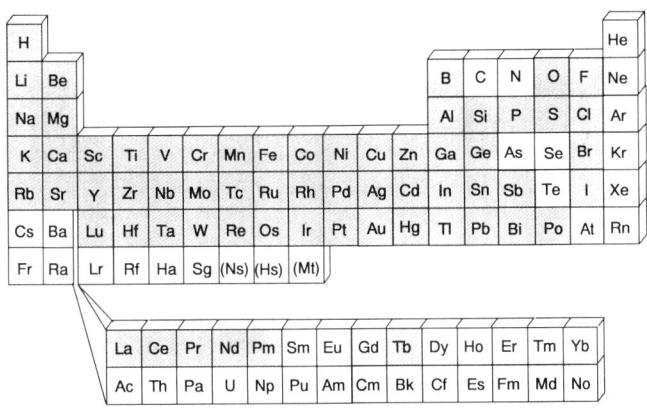

8
Electrochemistry of Corrosion: Principles and Protection

Vlasta Brusic *IBM Corporation, Yorktown Heights, New York*

I. INTRODUCTION

For most of this century, electrochemistry fought for an independent place in the realm of science; nothing contributed more to its victory than its many applications to the electronics industry. If discussion is limited only to the behavior of metals, there are four distinct areas where electrochemical reactions play a very unique role: plating, etching, electroetching, and corrosion.

It is impossible to imagine the existence of integrated circuits without plated-through holes. In this case, electroless Cu plating of the catalytically activated nonconducting board surface accomplished what no other metal deposition process could. Electroless Ni plating remains irreplaceable as a selective metallization of the circuitry on the fired ceramic substrate of multilayered modules.

Plating and, in particular, electroless CoP played a vital role in moving up the learning curve with regard to the properties of magnetic media. The media rely on the high magnetic moment of Co, but they need an alloying element to produce a microstructure with high coercivity, and an electrochemical deposition process offers a variety of possibilities. Electroplating through a mask, combined with the dry deposition of a conductive underlayer and the lithographic formulation of a photoresist pattern, was developed in the 1960s for the plating of permalloy in thin-film magnetic heads. A thorough understanding of the process and process controls resulted in the development of smart plating cells and the production of submicron metallic lines with greater than 10:1 aspect ratio, with thickness variation of less than 2% over an area of more than 20×20 mm^2. The use of jet solution delivery and laser-enhanced processes has resulted in highly controlled maskless plating as well as electroetching. Etching and electroetching are precision material removal processes that are used in micromachining and microfabrication.

The subject of this chapter is corrosion, a ubiquitous process that constantly threatens to undermine many aspects of device fabrication and limit the useful operational life of these devices. Corrosion is defined as the destructive attack of a metal by chemical or electrochemical reaction with its environment [1]. Among controlled electrochemical reactions, corrosion seems to be a calamity. Corrosion behavior varies from metal to metal and for each metal changes with seemingly small alterations in environment. Yet corrosion is a science, albeit a complex one. It requires full knowledge of electrochemistry, chemistry, metallurgy, the physical chemistry of interfaces, hydrogen embrittlement, and fracture mechanics, and there is no one person who is an expert in all these areas.

Principles that govern corrosion reactions have been well understood for more than a half century, and they apply equally to the corrosion of a large structure, such as the Golden Gate bridge, as they do to a miniature device. A device is connected to the outside world through many interconnections. An example of a packaging hierarchy that applies to silicon-based integrated circuits in the computer industry is illustrated in Fig. 1. On the chip, the size of the device, the circuit lines, and the spacing between the lines are about 1 µm. The diameter of the chip/module connection is close to 100 µm, while the wires leading to an external device and the spacing between them have the size of the order of a millimeter.

There are several reasons that the corrosion remains a concern to the electronics industry. Electronic materials are mostly used in thin-film form. They are selected for their specific properties, such as electronic switching, conduction, and magnetism, which might not be compatible with their thermodynamic tendency to corrode. Structures might be built with galvanically interactive metal stacks. Processing conditions, from device fabrication and patterning of interconnections, to fluxing and soldering, are by necessity aggressive; thus, both

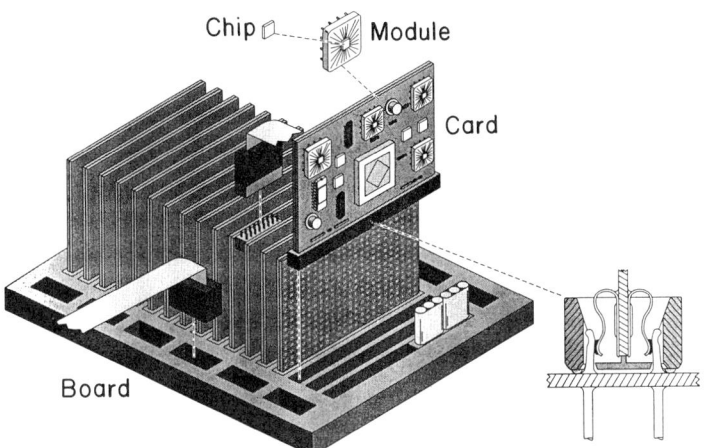

Figure 1 Schematic view of chip packaging with modules, boards, and cables leading to external devices. (Reprinted from Ref. 2, by courtesy of McGraw-Hill Co.)

processing and accidental remnants of processing reactants could enhance corrosion reactions. Furthermore, devices are typically exposed to an applied potential, thus, to conditions that could accelerate dissolution reactions. Yet, because of their small dimensions, the thin films are very sensitive to corrosion loss.

The fundamentals of the corrosion process will be briefly discussed in Secs. II–IV. An understanding of the nature of the corrosion reactions provides the basis for a selection of the methods of corrosion testing, in particular those applicable to thin-film structures (Sec. V). In Sec. VI, several established corrosion mechanisms and means of corrosion protection will be described, with examples taken from the electronics industry.

II. THERMODYNAMICS OF CORROSION

A. Nature of Corrosion Reactions

Destructive effects of corrosion can be found everywhere. From the easily immobile Tin Man, rust on cars, and the growth of unsightly corrosion products to open chip connections, the general concern is with the loss of metals and metallic properties that, in a variety of ways, interfere with product performance.

The nature of corrosion reactions is similar to that occurring in batteries. In a simple flashlight battery, the essential parts are a Zn casing (an anode), a carbon rod (a cathode), an electrolyte (NH_4Cl and $ZnCl_2$), and an external circuit in which electrons can travel from the anode to the cathode. The reaction involved can be understood by considering the dissolution of zinc in hydrochloric acid:

$$Zn + 2HCl \rightarrow ZnCl_2 + H_2 \tag{1}$$

In a corrosion process, metals invariably get oxidized, and the process can be described in terms of two separate reactions:

$$Zn \rightarrow Zn^{2+} + 2e^- \tag{2a}$$

$$2H^+ + 2e^- \rightarrow H_2 \tag{2b}$$

In the first reaction, metallic ions are formed that are combined afterward with available anions to form oxides, chlorides, sulfates, and similar compounds. In the second reaction, an available reactant has to consume electrons left after the exodus of the metallic ion. Without this reaction, which has to proceed at the same time and at the same rate as metallic oxidation, corrosion would not occur. Water and oxygen provide both the environment and the reactants for corrosion.

In a battery, however, the natural tendency of Zn to dissolve has been utilized as a source of energy. Thus, chemical energy has been converted into electrical energy. In a corrosion reaction, the two reactions occur inseparably, tenths of nanometers away from each other, and both material and energy are wasted.

B. Driving Force for the Reactions

Corrosion is a spontaneous process. In other words, corrosion proceeds under conditions in which the reaction product has a lower free energy than the reactants. One of the reactants is a metal atom on the surface; in fact, it is an ion that is stabilized by the surrounding electron "gas" and occupies its own energy well (Fig. 2). In vacuo, metals are stable, because the formation of free metallic ions (unstable gas) is a process that requires an expenditure of a fairly large amount of energy, about 6 eV. However, if the metal is immersed in a polar solvent such as water, there is an alternative deep energy well in which the metal ion is solvated by water molecules or is complexed with other ligands. Because water molecules are small and mobile, the two energy wells are very close together, and an ion only needs to acquire thermal energy of about 0.5 eV to surmount the energy barrier and go from the metal lattice into aqueous solution.

An ion is a charged particle. In general, every charged particle in the system is characterized by its electrical energy, equal to the product of its electric charge, ze^-, and the potential, Φ, at the point in space where it is located. Thus, the total energy possessed by a charged chemical entity is the sum of electrical energy and G, or chemical free energy:

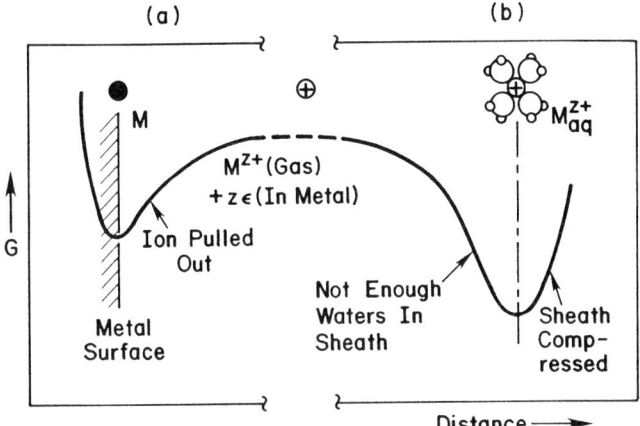

Figure 2 The chemical free energy of a family of metal ions (a) pulled out of the metal surface and (b) subsequently solvated. The curves are known as Morse curves.

$$\overline{G} = G + ze^- \Phi \qquad (3)$$

where $\overline{G}$ is electrochemical free energy. Under equilibrium conditions, the electrochemical free energy of the metal is equal to the energy of the metal ion; the net change in chemical free energy, $^+\Delta^0 G$, is an exact balance of the electrical work the ion has to do in traversing the electrode/electrolyte interface:

$$^0\Delta^+\Phi = \frac{^+\Delta^0 G}{ze} \qquad (4)$$

where $^0\Delta^+\Phi$ represents the potential difference across the metal/electrolyte interface [1]. When $^+\Delta^0 G_M$ is expressed in cal/mol, e is replaced by the Faraday F expressed as 23,060 cal/volt/g-equivalent.

As the metal dissolves, a separation of charges at the metal–solution interface occurs (Fig. 3). The metal (or an electrode) has a net negative charge, and the aquo-cations, at a new position adjacent to the metal, have a positive charge. The charges tend to line up opposite to each other in a manner reminiscent of a capacitor, forming the so-called electrical double layer. The structure of the double layer can be very complex. Aside from the plane with aquo-cations (outer Helmholtz plane), there might be specifically adsorbed anions (such as chlorides at the inner Helmholtz plane) and a diffuse region (the Gouy–Chapman layer) where a slight excess of charge is spread over some distance (possibly 1 μm long) before reaching an electroneutral bulk of the solution. The potential distribution across the interface follows the charge distribution, with a significant part of the overall potential difference occurring over a few tenths of a nm

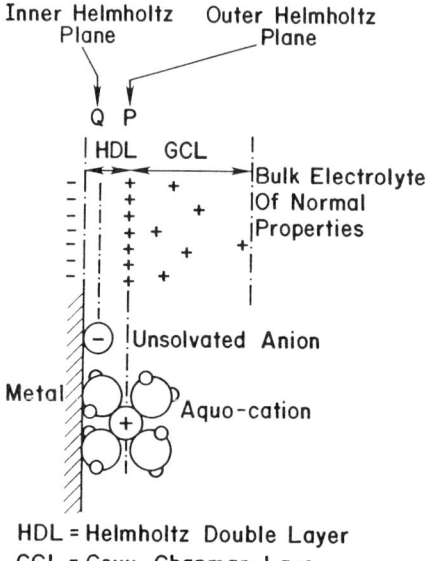

Figure 3 The structure of the electrical double layer.

(Helmholtz double layer). The buildup of the adverse electric field discourages the further exodus of ions. Usually in less than 1 s, a dynamic equilibrium is established with the number of dissolving ions perfectly balanced by a similar number of ions being deposited onto the metal.

The equilibrium potential difference of Eq. (4) is a characteristic of the specific metal electrode, but it can be measured only on a relative scale with another arbitrarily chosen electrode system. Use of a platinized Pt electrode in a standard acid solution, 1.2 N HCl, bubbled with purified hydrogen at 1 atm pressure, results in the reaction

$$H_2(g) \rightarrow 2H^+_{aq} + 2e^- \quad (5)$$

This reaction produces a separation of charges and consequently a potential difference $^0\Delta^+\Phi_H$ between adsorbed hydrogen gas on the metal and hydrogen ions, assumed, by convention, to be zero. Thus, the electromotive force (e.m.f.) or the cell potential between the corroding metal and standard hydrogen electrode becomes

$$E_M = {}^0\Delta^+\Phi_M + {}^+\Delta^0\Phi_H \quad (6)$$

E_M, measured against an arbitrary zero, is also known as the half-cell potential or single-electrode potential. Just as the chemical free energy depends on the effective concentrations of the products and reactants, so does the electrode potential:

$$E_M = E_M^0 + \frac{RT}{zF} \ln[M^{z+}] \tag{7}$$

where R is the gas constant (2 cal/g-equivalent/°C) and F is Faraday's constant (also given as 96,500 Coulombs/g-equivalent). Equation (7) is known as the Nernst equation.

Some of the standard oxidation–reduction potentials (in solutions where the activity of metal ions is 1) are listed in Table 1 [3]. They can be used to predict whether or not a metal will corrode in a given environment. This is accomplished by following the general rule: In any electrochemical reaction, the most negative half-cell tends to be oxidized and the most positive half-cell tends to be reduced. Table 1 indicates that Zn would corrode in acidic media, while copper should be stable in deaerated acids. In contrast, if the electrode potential is held below the Nernst potential, metals are "immune" to corrosion and do not dissolve.

Metals and alloys of utmost industrial importance, such as iron–chromium alloys or aluminum and its alloys, have high corrosion resistance because of the

Table 1 Standard Electrochemical Potentials

Metal	Potential, V, nhe	
$K = K^+ + e-$	−2.92	Active
$Al = Al^{3+} + 3e-$	−1.66	
$Zn = Zn^{2+} + 2e-$	−0.76	
$Cr = Cr^{3+} + 3e-$	−0.71	
$Fe = Fe^{2+} + 2e-$	−0.44	
$Co = Co^{2+} + 2e-$	−0.27	
$Ni = Ni^{2+} + 2e-$	−0.23	
$Sn = Sn^{2+} + 2e-$	−0.14	
$Pb = Pb^{2+} + 2e-$	−0.126	
$2H^+ + 2e- = H_2$	0.000	Reference
$Cu = Cu^{2+} + 2e-$	0.34	
$O_2 + 2H_2O + 4e- = 4OH^-$	0.401	
$Ag = Ag^+ + e-$	0.799	
$Pd = Pd^{2+} + 2e-$	0.83	
$Pt = Pt^{2+} + 2e-$	1.2	
$O_2 + 4H^+ + 4e- = 2H_2O$	1.23	
$Au = Au^{3+} + 3e-$	1.42	Noble

formation of a passivating oxide (rather than aquo-cations and easily soluble salts). The potential range of thermodynamic stability of such oxides is also tabulated in so-called Pourbaix diagrams [3]. An example of such a diagram is given in Fig. 4, for Fe^{2+} ion concentration of 10^{-6} M. Below the potential of -0.560 V (with respect to normal hydrogen electrode) iron is thermodynamically stable, or immune to corrosion. Line A, at -0.560 V, describes the reaction

$$Fe \rightarrow Fe^{2+} + 2e^- \tag{8}$$

As the pH increases, the solubility product is reached for the reaction

$$Fe^{+2} + 2OH^- \rightarrow Fe(OH)_2 \tag{9}$$

and $Fe(OH)_2$ will precipitate along the line B. In the pH range where $Fe(OH)_2$ is stable, Fe should passivate with the oxidation reaction according to line C:

$$Fe + 2H_2O \rightarrow Fe(OH)_2 + 2e^- + 2H^+ \tag{10}$$

and with $Fe(OH)_3$ forming at yet higher potentials.

A comparison of the potentials for Fe dissolution (A) with potentials of hydrogen ion reduction (a) and oxygen reduction (b) shows that Fe can be oxidized with both of these reactions. The only metal that does not dissolve or form oxide when exposed to air-saturated water is gold, as gold is truly thermodynamically stable, or "noble."

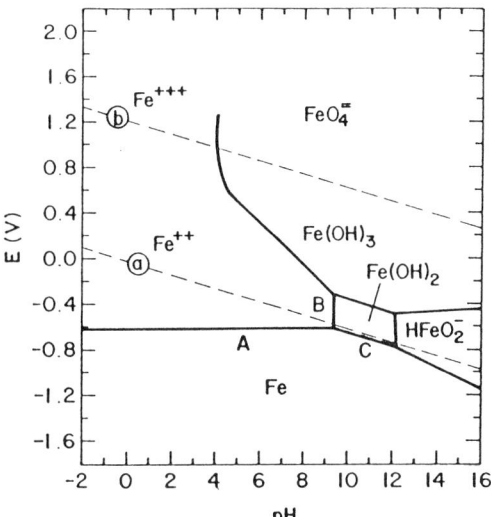

Figure 4 Pourbaix diagram for Fe.

Thermodynamic arguments can be used to (1) predict whether or not corrosion can occur, (2) estimate the composition of the corrosion products, and (3) predict environmental changes that will prevent or reduce the corrosion attack. The challenge, however, lies in the determination of the actual corrosion loss, which is a complex result of the kinetics of all the participating reactions.

III. REACTION KINETICS

A. Metal Dissolution

As in any chemical reaction, the rate of metal dissolution is proportional to the probability that some number of reacting entities (e.g., atoms deemed to constitute chemically active sites) will surmount the activation energy of the reaction. However, in a corrosion process, the reaction is accompanied by a flow of charge and can be measured as a current. The energy of activation contains a potential term, and it can be changed in either direction by a change of potential across the interface (Fig. 5):

$$i = i_0 \exp \frac{zF\beta d(\Delta\Phi)}{RT} \tag{11}$$

where i_0 is the exchange current density or reaction rate at the equilibrium potential, zF is the charge of a mole-ion, β is a symmetry factor, F is the Faraday constant, $d(\Delta\Phi)$ is the overpotential or the potential difference between the ap-

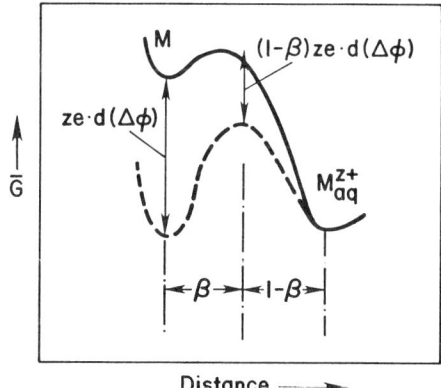

Figure 5 The electrochemical free energy profile for a metal electrode anodically polarized by $d(\Delta\Phi)$. β is the symmetry factor. Dashed line indicates the electrochemical free energy at equilibrium.

plied and equilibrium potential, R is the gas constant, and T is temperature. The exchange current density is a sensitive function of metal and environment:

$$i_0 = zFf\alpha \left(\frac{N_S}{N_0} C_{\text{lig}}^n \exp \frac{-\Delta \overline{G}^*}{RT} \right) \tag{12}$$

where f is the frequency of the activated state, $\alpha N_s/N_0$ is a fraction of active surface atoms, C_{lig} is the effective concentration of solvating ligands (even water), and n is the number of such ligands bonded to the activated ion [1].

Dissolution rates at equilibrium potentials, i_0's, are small and equal to the rates of deposition; thus, they are not responsible for substantial metal loss that can occur in corrosion. However, according to Eq. (11), if β is 0.5 and $z = 2$, there is a linear dependence between $d(\Delta\Phi)$ and $\log i$ with a slope of 2.3 RT/F, or 0.059 V/decade. ($d(\Delta\Phi)/d \log i$ slope is also known as Tafel slope.) Therefore, even a relatively small change of potential, e.g., 0.3 V, could change the rate of dissolution by six orders of magnitude. This sensitive interdependence of the current (rate) and potential provides an explanation of why corrosion can be catastrophic, offers means of experimentally evaluating corrosion phenomena, and emphasizes an essential difference between chemical and electrochemical kinetics. In addition to the factors that normally govern the rate of chemical reactions, such as the concentration of reactants, temperature, and so on, the rate of electrochemical reactions can be easily manipulated by potential, i.e., by changes of the energy barrier.

B. Electrochemical Elements of Corrosion Rate

A metal spontaneously corrodes when, in a given environment, a metal/electrolyte interface is formed with a potential difference that diminishes the activation energy for dissolution. The corrosion potential is a consequence of the kinetics of all possible reactions and not an entity found in thermodynamic tables. Figure 6 shows the corrosion potential and rate for Fe in dilute acid. There is dissolution of Fe (anodic current, increasing with an increase of potential) and reduction of hydrogen ions (cathodic current, increasing with a decrease of potential). The equilibrium potential of hydrogen oxidation/reduction is much higher than the standard potential of Fe dissolution. As a good conductor, Fe will not be at either of the two reaction potentials, but it will reach some value in between the two. As there can be no net accumulation of electrical charge during electrochemical reactions, the new, mixed, potential will be the one at which

$$i_{\text{net}} = i_{\text{ox}} - i_{\text{red}} = 0 \tag{13}$$

Thus, corrosion reactions normally proceed at potentials above the equilibrium value and at a substantially higher rate than measured as i_0 for metal dissolution.

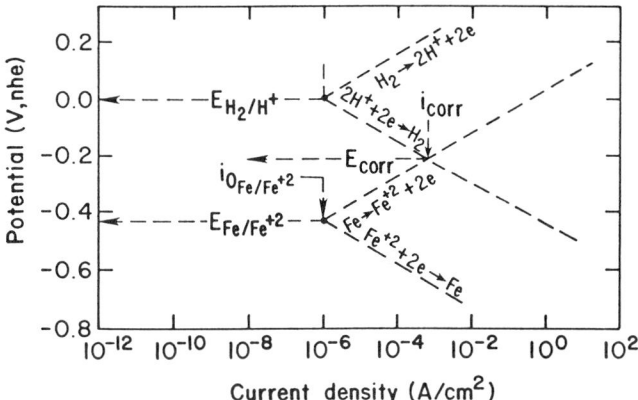

Figure 6 A schematic representation of the corrosion potential and rate in terms of the partial anodic and cathodic reactions for Fe in acidic solution.

However, the rate of the metal dissolution at the corrosion potential has to be equal to the rate of the reduction reactions. A dependence of the overall process on the ambient should not be surprising, as the environment is responsible for half of the reactants.

The spatial distribution of the contributing reactions can be random, resulting in uniform corrosion loss. Or it could be governed by galvanic contacts to other metals, geometry, morphology, crystallography, or chemistry, giving rise to a variety of localized corrosion attacks. In thin-film structures, the presence of defects such as notches, scratches, seams between metals and dielectrics, grain boundaries, intermetallic inclusions, and intermetallic contacts all present possibilities for enhanced localized corrosion and challenges to experimental evaluation.

IV. SYSTEMATIZATION OF CORROSION FIELD

The corrosion field can be systematized according to corrosion environments, corrosion mechanisms, corrosion types, metals, field of application, and methods of protection.

A. Localized Distribution of Corrosion Reactions

In a localized attack, anodic and cathodic sites may sometimes be distinguished by the naked eye. Differentiation between general and (many aspects of) local corrosion is based on a practical viewpoint; proper identification of the given corrosion form may help in an evaluation of the severity of corrosion damage as well as in the design of protection schemes. The localized distribution of corrosion reactions can produce intergranular corrosion, crevice corrosion, galvanic attack, filiform corrosion, pitting, stress–corrosion cracking, hydrogen damage, and dealloying. All of these have been observed in devices and products of the electronics industry.

B. Effects of Environment

Corrosion of devices in moist environments (water, aqueous solutions, humid atmosphere) proceeds by electrochemical mechanisms. Other relevant reactions are corrosion (oxidation) in dry gases, both at room and at elevated temperature, proceeding via chemical mechanisms, and corrosion in water-free chlorinated hydrocarbons and alcohols, also via chemical mechanisms. For example, integrated circuit (IC) board fabrication calls for multiple passes through a soldering process, in which thermal oxidation of the Cu conductor has become a serious problem. Corrosion in water-free solvent is rare and mostly traceable to impurities. However, the behavior of Al in chlorinated hydrocarbons is a well-documented example of such a corrosion loss [4].

V. CORROSION TESTING

In general, the purpose of corrosion testing could be characterization of new materials, evaluation and selection of alternative materials, or definition of the corrosion mechanism so that one can enhance or control the corrosion resistance of the material or corrosiveness of the environment. In more specific terms, corrosion testing in processing provides guidance for process specification and control in manufacturing without accidental yield losses, while corrosion testing after fabrication characterizes the reliability and lifetime of the product.

When a suspect corrosion environment is an aqueous solution, corrosion testing by electrochemical methods should lead to the direct evaluation of the corrosion rate, as well as the corrosion mechanism. When the environment of interest is an atmosphere with variable levels of humidity and pollutants, materials can be tested in that same environment (over a long time, such as years) or in gaseous atmospheric chambers under conditions that accelerate the aggressiveness of the environment (over weeks or days). In either case, knowing the complexity of corrosion reactions, tests should be carefully planned, materials should be well characterized before the test, and testing conditions should be similar to the conditions in which the product is expected to survive without corrosion loss.

A. Electrochemical Studies

A variety of available direct current (dc) techniques utilize the plentiful information on the behavior of individual reactions that current–potential curves, such as shown in Fig. 6, provide in addition to the direct evaluation of the corrosion rate. The metal to be studied is placed in an electrochemical cell that normally also contains an auxiliary electrode (of noble metal), a reference electrode, and an electrolyte. The most common approach uses a potentiostat, a form of a differential amplifier that is connected to the electrodes as shown schematically in Fig. 7. Potentiostats, corrosion cells, and computer software for corrosion tests are commercially available. Sample holders and cell can be redesigned for a specific purpose. One such design, particularly suitable for tests of thin-film samples, is shown schematically in Fig. 8. The cell consists of a masked sample, Pt mesh, and a reference electrode that are separated by filter paper. Only a droplet of an electrolyte with a volume of 0.01 ml can be used. The cell can be applied to studies in electrolytes that are normally difficult to handle, such as HF or buffered HF. Also, because of the closeness of the electrodes, the cell is suitable for tests in water and other low-conductivity electrolytes without interference of the high ohmic drop. The use of water as an electrolyte mimics the conditions that apply in processing (such as product rinsing) or in exposure to a humid atmosphere.

Electrochemical measurements done on thin films are performed under essentially the same conditions as on bulk materials. The corrosion potential measurement is simple and nondisturbing. However, in order to measure the corrosion rate at the corrosion potential (where the net current is zero), the corroding electrode has to be disturbed by the application of either a potential or current. The corrosion rate can be determined in several ways.

The corrosion rate can be obtained from the extrapolation of the cathodic reaction rate to the corrosion potential. The cathodic current is measured as the applied potential varies from the open circuit potential to some low cathodic value. With the application of an externally applied potential, the rates of the oxidation and reduction at an oxide-free surface change as shown schematically in Figs. 6 and 9. Cathodic and anodic currents are usually plotted on the same logarithmic scale, although one is negative and the other positive. When the difference between the applied and corrosion potential becomes more than about

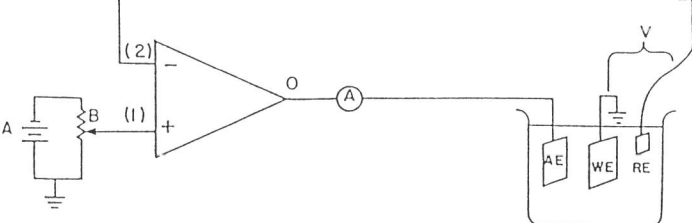

Figure 7 Schematic of the potentiostat.

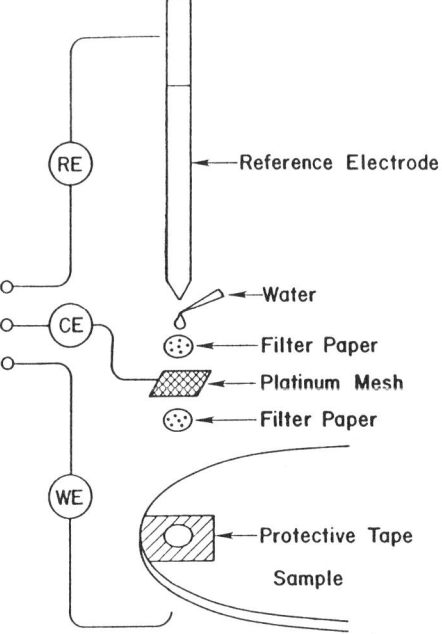

Figure 8 Schematic of the electrochemical cell for corrosion evaluation in a droplet of water or other electrolytes.

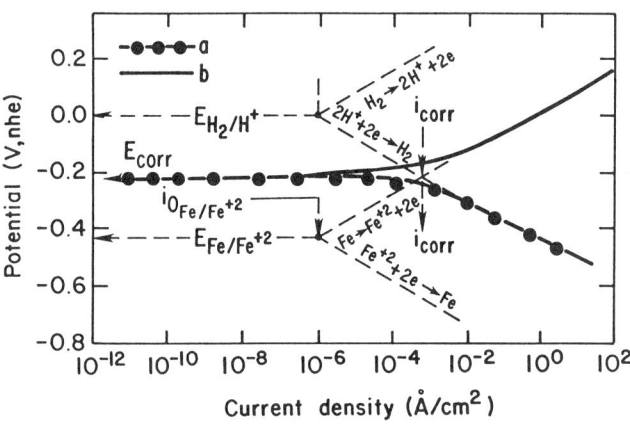

Figure 9 Determination of the corrosion potential and corrosion rate by (a) the evaluation of the cathodic rate and (b) by potentiodynamic polarization with measurements of both cathodic and anodic net currents.

50 mV in the negative direction, the rate of H_3O^+ reduction becomes much faster than the rate of Fe oxidation, and the measured current is approximately equal to i_{red}. (The opposite is true with the increase of potential.) A plot of the potential and the logarithm of current (the cathodic Tafel line) is then extrapolated to the corrosion potential (Fig. 9). The value of this extrapolated line corresponds to the cathodic rate at the corrosion potential, which is by definition also equal to the corrosion rate (Eq. 13). Other widely used dc techniques are potentiodynamic polarization and linear polarization, both based on an understanding of the electrode kinetics outlined by Eq. (11). In the linear polarization technique, the electrode potential is scanned only 10–20 mV on both sides of the corrosion potential, and the corrosion rate is then calculated from the experimental potential–current slope. Potentiodynamic polarization curves are obtained by recording the current upon the application of a potential sweep from some value below the corrosion potential (e.g., 0.25 V cathodic to the corrosion potential) to a value much more anodic. The result could be similar to curve b in Fig. 9. These and other techniques not only provide the corrosion rates, but they also reveal the kinetics and mechanisms of both the cathodic and anodic processes, the ability of metal to passivate, and an assessment of many different corrosion forms, from pitting to possible galvanic exposure. Recommended books for further reading are Refs. 1 and 3–9.

B. Atmospheric Corrosion Tests

The fabricated product is expected to operate reliably for at least 5 to 10 years in an environment, the corrosiveness of which is primarily dependent on the presence of water and oxygen. Devices and chip interconnections are normally separated from the environment by passivation layers (oxides, nitrides, polyimides, etc.). Aside from providing functional coupling to the board, i.e., the second-level package, modules (Fig. 1) provide an additional protection to the chip. In hermetic packages, which are sealed with metal or glass, the effective device isolation provides a thermodynamic solution to chip corrosion: the corrosive environment is not permitted to exist. However, even such packages can corrode if the sealed atmosphere has a high amount of moisture and contaminants, if the package develops a leak, or if the passivation layer provides only partial coverage. Polymer sealants normally adsorb water; thus, polymeric packages are not hermetic. Some of the enclosures, such as magnetic disk files, are not sealed at all, and the corrosion rates of materials in such files are the complex results of an ever-changing environment.

The accidental presence of contaminants, originating from processing, handling, or defective sealing, greatly affects device reliability. The contaminants affect the corrosion reaction by affecting the critical humidity at which water condenses at the surface, by increasing the conductivity of the resulting electrolyte, and by providing additional reactants for corrosion reactions.

Because of the circuit density, the application of a bias can lead to power dissipation and localized heating that could cause a drop in the localized humidity and the corrosion rate, if the bias is the only stress variable. In cases when water adsorption is possible, applied voltage typically increases the potential drop across the metal/adsorbed water layer interface and drives the corrosion reactions to an orders of magnitude faster rate (see Eq. 11).

1. Role of Water in Atmospheric Corrosion

The role of water in atmospheric corrosion is similar to its role in electrolytic dissolution. In both cases, water is a complexant for metallic ions and a solvent for many reactive species [10]. Both corrosion processes are electrochemical in nature, with similar corrosion mechanisms but different kinetics. As shown in Eq. (12), the metallic dissolution rate depends on water to form aquo-cations, and under some atmospheric conditions, the amount of water for ion complexation and hydrogen reduction is limited. In contrast, oxygen reduction, which depends on the solubility of oxygen in a given electrolyte and its diffusion to the reaction site, is kinetically faster under a thin water layer. Thus, under environmental test conditions, the overall corrosion rate is expected to be limited by the rate of anodic reaction.

Corrosion experts predicted that the corrosion rate should vary from a low value at very low humidities to a high value with total immersion in an electrolyte, passing through a maximum at 100% relative humidity with visible condensation. With a thickening of the condensed layer, the corrosion rate should decrease because the thicker film of moisture impairs the arrival of oxygen to the surface [6]. This projection has been experimentally confirmed. A Kelvin probe was used to measure the corrosion potential under thin electrolyte layers, and a pressure transducer was employed to register oxygen consumption, i.e., the corrosion rate of iron as the surface went from wet to dry. Figure 10 shows the variation of the corrosion potential, corrosion rate, and the electrolyte thickness with time of drying [11]. The results indicate that the corrosion rate is indeed fastest during the time of the wet-to-dry transition, just before the last water is lost and the corrosion potential increases to a new, dry and passive value.

At relative humidities (RH) below 30%, the average thickness of water is about one monolayer, and corrosion is normally not observed. Metals might be consumed by a reaction similar to thermal oxidation. At intermediate RH, $\approx$40%, thin layers of a corrosion product are detected, with the appearance of surface roughening. This may be associated with the fact that adsorption shows clustering even on homogeneous surfaces. The preferred areas for attack may be sites where water adsorption is favored, where the oxide is weak, or where there are preferred sites for adsorption of pollutants. Once corrosion is initiated, deliquescent corrosion products may aggravate local attack. At high RH, $\approx$65%,

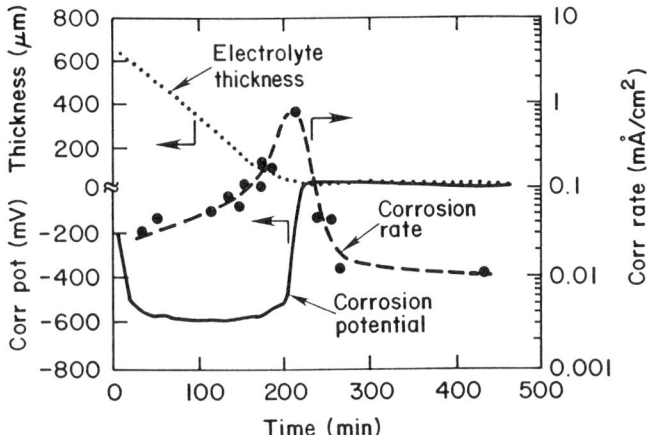

Figure 10 Variation of the corrosion potential, corrosion rate, and electrolyte thickness during transition from wet to dry surface. (Reprinted from Ref. 11, by courtesy of the Electrochemical Society.)

for which water is more than three molecules thick, water approaches the behavior of bulk solutions. This is particularly true for yet thicker films, formed because of dirt or deliquescent corrosion products, in which case corrosion is fully related to that in bulk electrolytes and can be studied in a similar fashion [10].

2. Factors Affecting Water Adsorption on Metals

The amount of water adsorbed on a metallic surface depends on the relative humidity as already described. In this case, the correlation between RH and the number of adsorbed water molecules is obtained on a relatively clean and smooth surface, at which the condensation is expected to occur at 100% RH. Real surfaces, however, are subject to localized condensation (and a much higher corrosion rate) at crevices, pores in overlayers, points of contact with asperities, and so on. The relationship determining capillary condensation is $P_1 = P_0 \exp(-2\sigma v/RTr)$, where P_1 and P_0 are the pressures of saturated vapor above a concave meniscus of radius r and above the flat surface, respectively; σ is the surface tension and v the molecular volume of a liquid, i.e., water. If the surface is hydrophilic and porous, condensation can occur at very low humidities. At 25°C, a pore with 69 nm radius will retain water at 99% RH, but in a 1 nm pore, condensation occurs at ≈40% RH [6]. Impurities and their chemistry play a significant role in the overall process of water uptake. Critical relative humidity (CRH) is defined as the RH that is in equilibrium with a saturated solution of a given compound, and it denotes the value of RH above which condensation is to be expected. Some compounds that form well-hydrated salts produce a noticeable increase of water uptake at very low humidity levels, e.g., 20% in the case of $CuCl_2 \times 2H_2O$ [12]. Thus, uncontrolled impurity levels can give rise to an unknown amount of adsorbed water and very different corrosion rates.

3. Accelerated Corrosion Tests

Accelerated tests are performed under more stressful conditions than the electronic parts will experience during actual operation. Temperature and humidity—with or without added pollutants and bias—are the normal stresses in these tests. Most commonly, tests are conducted in gaseous environmental chambers. The results, obtained in some reasonable time, are used to predict lifetime and the reliability of the product under operating conditions. A growing but competitive market demands that a new product is more reliable than the one being replaced.

Samples could be metallic films or IC structures and packages at various levels of completion. Integrated circuit test structures are made under conditions closely resembling those used in general wafer processing, but their geometry is selected to enhance the sensitivity of the measured response to the stress test while preserving the linewidth and spacing of the actual structures. Some of the test geometries are illustrated in Fig. 11. They contain parallel line tracks, interdigitated structures, or triple tracks, with the lines and separations of 100 μm

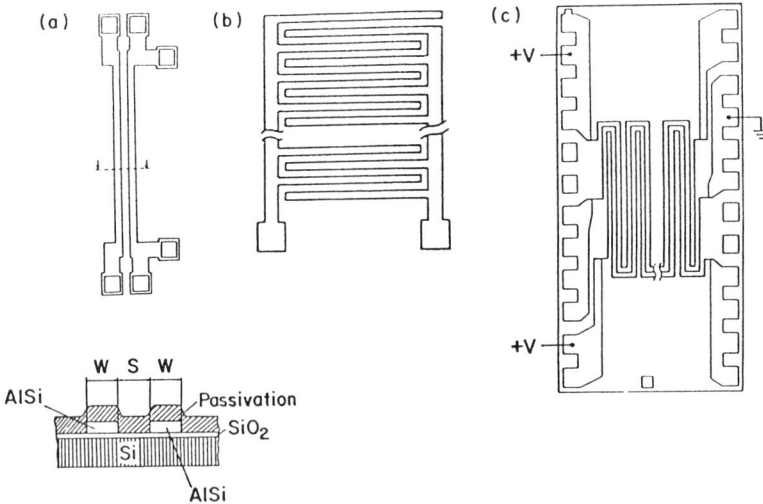

Figure 11 Examples of the test structures for temperature/humidity/bias tests: (a) parallel line tracks; (b) interdigitated structures; (c) triple tracks.

down to 1 μm dimensions. Metallic lines, consisting of Al, Au, Al–Cu, or Al–Cu–Si, might be encapsulated or not.

Temperature–humidity (TH) tests can be used to detect and map sites of accidental, localized contamination. If the character of corrosion defects can be assigned to contamination caused by a particular process step, defects can be eliminated. Contaminants may be purposefully added in a test, when the product is expected to operate in an environment where such pollutants are inevitably present. Mixtures of gases that are typically found in an office environment have been standardized in the environmental test specification:

Cl_2 3 ppb
SO_2 350 ppb
NO_2 610 ppb
H_2S 40 ppb

to be used in 70 (± 2) %RH at 30 (± 0.5)°C at a velocity of 0.5 (± 0.1) m/s [13]. Each of these gases is normally present in much smaller concentrations, with the probability that the upper limit—chosen for the test—will be detected once in a thousand days. The water layer saturated with these gases has an acidic pH of about 3.7.

Ionic contamination can be provided by adsorption of the ubiquitous submicron particles, shown to contain ammonia and surfate, and tests have been designed to monitor their effects [14].

The progress of corrosion is monitored by weight gain, defect count and mapping, and numerous surface-sensitive techniques such as ellipsometry, Auger spectroscopy, X-ray photoelectron spectroscopy (XPS), or a change in surface reflectivity often coupled with simultaneous functional tests, such as resistance measurements. A very appropriate test uses measurements of the leakage currents with applied potential between the two lines and its change with time in the stress test.

C. Life Projections

The life of a batch of electronic packages is typically described by the "bathtub" curve. The curve gives failure rate as a function of *log time*. Initially the failure rate might be large, possibly caused by gross assembly defects, and it decreases as the learning improves. As the population of early defects is depleted, the failure rate becomes relatively low and independent of time, with failure being caused by random defects. Finally, at the later stages, the failure rate increases because of wear-out of the devices. Wear-out failures are usually characterized by a rate rising to a maximum and diminishing as the population decreases and tend to conform to a log-normal distributions. Reliability is assessed according to the expected regime: early fails are unpredictable, and weeded out in early, in-house tests under operating conditions. Parts suspected of random defects are submitted to operating endurance tests used to define some mean-time-to-failure (MTF) per component. Wear-out reliability is time-dependent, and determined by accelerating the applicable failure mechanisms in a predictable way. Most of the models for life prediction rely on the determination of the MTF by testing a large sample size at different temperatures and relative humidities, harsher than operating conditions. The MTF can be determined, for example, from the measurement of line resistance change with the firmly set limits of the end-of-life value. The determined value of MTF under accelerated conditions and the knowledge of the MTF variation with T and RH allows an estimate of the MTF under normal conditions. The ratio of the two times is known as the acceleration factor, AF. Most failure mechanisms show the Arrhenius-type time dependence:

$$\text{Degradation rate} = \alpha \exp \frac{-E_a}{kT} \tag{14}$$

with a corresponding acceleration factor of

$$\text{AF} = \exp \frac{E_a}{k(1/T_0 - 1/T_T)} \tag{15}$$

A more general relation between MTF and RH is found to have the form (15):

$$\text{MTF} = A \exp \left\{ \frac{E_a}{kT} + B \text{ RH} \right\} \tag{16}$$

where E_a is experimentally obtained activation energy and A and B are numerical fitting constants.

VI. CORROSION MECHANISMS OF ELECTRONIC MATERIAL SYSTEMS

To minimize corrosion loss, several approaches have been used, conveniently divided into techniques that change either the thermodynamic tendency of the material to react with the environment, or the kinetics of the possible reactions.

The thermodynamics of the expected reactions can be influenced by alloying with metals that tend to form well-passivating oxides, or by use of metallic overlayers, such as gold or electroless Ni. Use of an external source to bring the metal into the potential range of "immunity" or use of a sacrificial metal that would corrode preferentially and offer galvanic protection is known in the construction industry, but their application to electronic devices is not very practical.

In kinetic methods, the corrosion rate is influenced by the removal and separation of reactants. The aggressive nature of the environment can be nullified and cathodic reactants removed if the relative humidity is kept at ≤30%. The exchange current density of the reactions can be minimized by use of inhibitors that are particularly effective if a metal is exposed to electrolyte. Metals could be purposefully oxidized by anodic, thermal, or plasma oxidation, or by use of conversion coating to form a surface film that provides a barrier to environmental reactants. One of the most widely used techniques is application of polymer coatings, glass layers, deposited oxide films, oils, and so on, all of which prevent water and oxygen diffusion but also act to separate anodic and cathodic sites on the metal and thereby reduce the corrosion rate.

The overall solution to corrosion might come from kinetic methods or engineering solutions. In any case, a knowledge of corrosion mechanisms is essential so that protection schemes can be properly devised. Identification of the particular corrosion form can be used for the same purpose.

Several examples of corrosion and protection of electronic materials can be given in processing, storage, and operation.

A. Effects of Design on n + Si/PtSi Galvanic Corrosion

1. Galvanic Attack of Emitters in Contact with PtSi

Fabrication of devices on logic chips is occasionally accompanied with a failure traceable to the PtSi/Si interface. Under some processing conditions, the Si emitter in contact with PtSi (used for making Schottky barriers and ohmic contacts in integrated circuits) is preferentially dissolved. Dissolution can produce large ''mouse holes'' under the PtSi and subsequently cause a malfunction of the device (Fig. 12). The failure has been diagnosed to be due to galvanic attack by PtSi when the wafers are exposed to buffered HF (BHF). The attack is exacerbated if PtSi is accidentally oxidized or if the area ratio of PtSi/Si is increased [16]. Electrochemical data obtained on blanket films of n-Si, p-Si, and PtSi not only identify a galvanic attack, but they provide quantitative values for galvanic corrosion loss and how that can change with dilution. Results suggest that there are two mechanisms of Si dissolution, one applicable in concentrated BHF (Fig. 13), another in dilute solutions (Fig. 14). Similarly, as reported for concentrated and dilute HF [17], the dissolution of Si is suggested to be

$$Si + 2HF + \lambda + \to SiF_2 + 2H^+(2 - \lambda) - \tag{17}$$

with the unstable SiF_2 changing further into a stable tetravalent form by a disproportionation reaction and, in addition, by hydrogen evolution. Si dissolution proceeds with consumption of two holes (+) at low potentials and four-hole participation at high anodic potentials with formation of Si^{4+} compounds. At

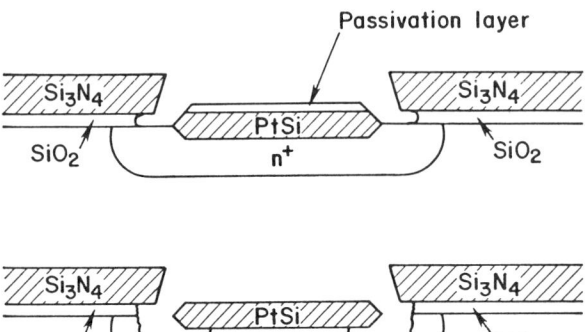
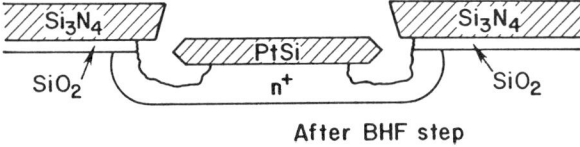

Figure 12 Schematic view of PtSi-emitter contact with galvanic attack.

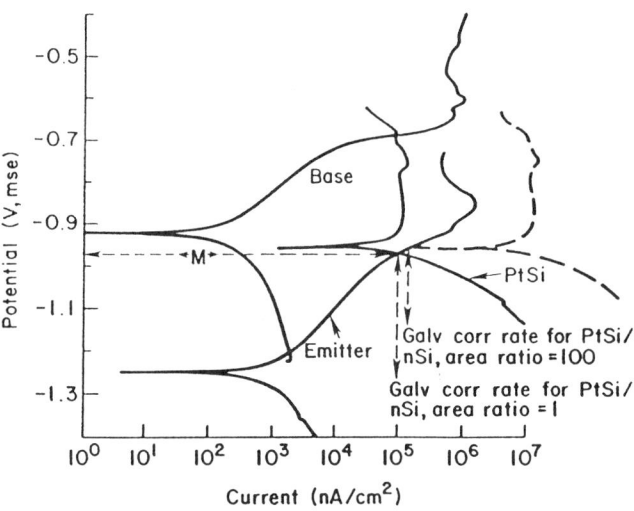

Figure 13 Potentiodynamic curves on emitter, base, and PtSi in buffered HF.

high potentials, water is plausible, and in dilute solution, it is the preferred reactant with silicon. This could correlate well with the appearance of two slopes in the potential–log i curves in concentrated BHF. The initial anodic slope is high, indicating that a significant part of the overall potential drop across the semiconductor/electrolyte interface is not across the space charge region of the semiconductor (as this slope is expected to be 0.060 V/decade), but has to be

Electrochemistry of Corrosion

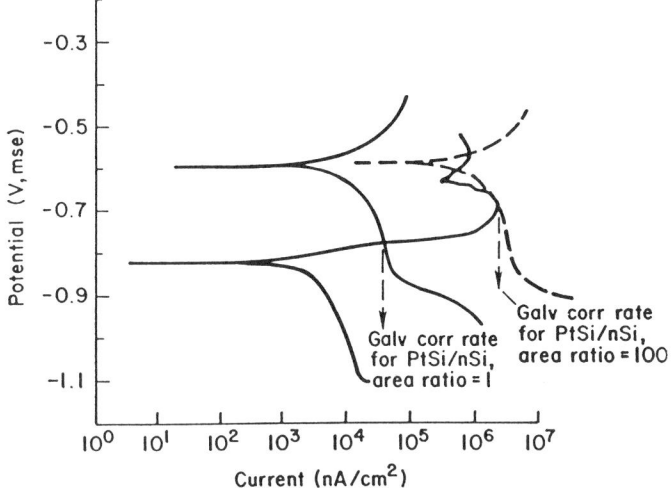

Figure 14 Potentiodynamic polarization curves on emitter, base, and PtSi in a mixture of dilute BHF (with BHF/H_2O of 1:9).

across the solution side of the interface. Heavily doped Si acts as an electrode without a space charge region. Thus, in a corrosive environment a heavily doped surface is much more vulnerable to corrosion than is an intrinsically pure semiconductor, because such a surface reacts readily to any change of surface potential.

Galvanic attack is possible only if the following conditions are met: there must be at least two electrochemically different metals, they have to be in electrical contact, and they have to be exposed to the same electrolyte. PtSi/Si contact fulfills these conditions. The presence of PtSi greatly changes the effective potential across a Si/electrolyte interface. As shown in Fig. 13, the emitter dissolves with an average corrosion current of 1×10^{-6} A/cm^2. According to Faraday's law, one gram-equivalent of any material is chemically altered for each 96,500 coulombs or one Faraday, F, of electricity passed through the electrolyte; thus, the rate, R, in cm/s can be calculated as $R = (i \times M)/nF\rho$, with M being the molecular weight, n the number of electrons in the reaction, and ρ the density. The current then corresponds to a Si dissolution rate of 0.36 nm/min assuming that primary dissolution leads to the formation of bivalent Si, and 0.18 nm/min upon formation of tetravalent Si. PtSi, placed in BHF, loses its oxide and proceeds to dissolve. Judging from the shape and magnitude of the potentiodynamic polarization curves, the primary oxidation reaction is preferential dissolution of Si (that is faster on PtSi than from Si alone). The observed corrosion potential on PtSi is about -0.9 V (against mercurous sulfate electrode,

MSE, which itself has potential of 0.656 V against normal hydrogen scale). In this potential range, Si dissolves as fast as hydrogen can reduce. Hydrogen reduction is very fast on PtSi because of the catalytic effects of Pt. This material takes the role of a cathode, and an emitter in contact with PtSi rapidly dissolves at a new mixed potential. Even for the 1:1 area ratio, the mixed potential is close to one measured on PtSi, and Si should dissolve with a rate of 1×10^{-4} A/cm^2. On product wafers, the area ratio of PtSi is close to 100 (dotted line) and, undoubtedly, the potential of the galvanic couple will be equal to the potential of PtSi.

2. Effects of Dilution

In dilute BHF, both Si and PtSi tend to be spontaneously oxidized. The dissolution mechanism for Si is changed, showing a much smaller slope of $dV/d \log i$; thus, any contact capable of increasing the potential of Si only 0.06 V above its normal corrosion potential causes an increase of Si dissolution by an order of magnitude; hence, the galvanic contact with PtSi could cause catastrophically fast dissolution rates. Thus, dilute BHF, or a rinsing operation, is worse for corrosion than BHF alone, but one without the other is unthinkable.

3. Elimination of the Problem

One way of eliminating the problem would be to keep the potential of PtSi at the potential of the emitter, or below, during the entire process. Use of oxygen scavengers, such as sodium sulfite, can minimize oxidation, limit the potential increase of PtSi, and limit the attack. However, the problem can be fully eliminated only by means of a change in design. Use of a mask that fully covers the PtSi and prevents its exposure to the electrolyte nullifies one of the three conditions needed for the galvanic attack to occur, thus eliminating the attack.

B. Corrosion of Al–Cu in Buffered HF

Very-large-scale integrated (VLSI) devices are most commonly fabricated with Al–Cu alloy lines in sputter-deposited SiO_x. Etching of the glass dielectric in BHF brings the Al–Cu into contact with that etchant. Electrochemical evaluation of Al–Cu in BHF, in comparison with Si in BHF, brings out some similarities, but also some interesting differences. Figure 15 shows that Al–Cu is passive in BHF, as the current remains small even with a significant increase of the potential. The small anodic dissolution is most likely caused by the poorly soluble AlF_3, the presence of which makes any galvanic contact with PtSi or p-Si less effective. The corrosion of Al–Cu is counterbalanced by H$^+$-reduction that is significantly reduced by an increase of pH (or an increase of NH_4F/HF ratio) or by additions of glycerol (Fig. 16). Glycerol not only reduces the cathodic reaction but apparently promotes Al-passivation. Dilution with water results in less effective surface passivation and more pronounced galvanic effects by PtSi.

Electrochemistry of Corrosion

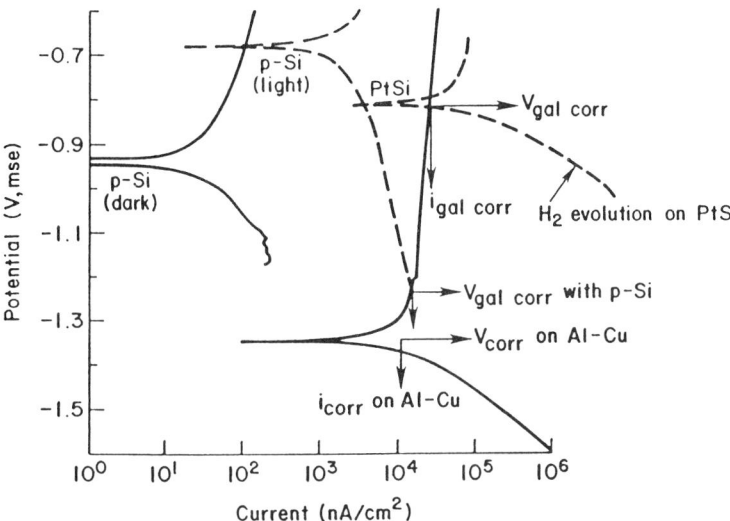

Figure 15 Potentiodynamic polarization curves on Al–Cu, PtSi and p-Si in BHF. Al–Cu is passive, and galvanic attacks by more noble materials are negligible.

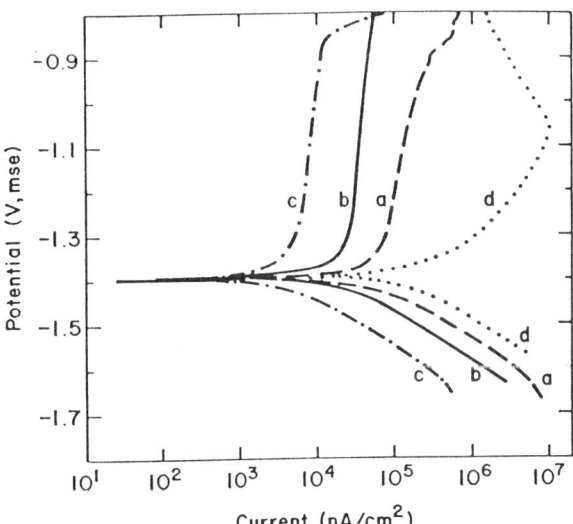

Figure 16 Potentiodynamic polarization curves on Al–Cu in (a) 5:1 BHF, (b) 5:1 BHF with 12% glycerol, (c) 5:1 BHF with 40% glycerol, and (d) 5:1 BHF with 90% water.

Both departure from the passive behavior and galvanic effects are smaller as the NH_4/HF ratio increases. Thus, the unwanted corrosion should be minimized by using 40:1 BHF with glycerol (with the etching of the dielectric proceeding at a slower rate).

C. Removal of Corrosives: AlCu and Reactive Ion Etching Process

When Al–Cu lines are patterned by reactive ion etching (RIE), use of a Cl-containing plasma leaves the wafer surface and photoresist mask contaminated with chlorine and chlorides. If they are not quickly removed, Al-chlorides can hydrolyze and form voluminous, unprotective corrosion products and prevent the formation of normally passivating Al-oxide. The reaction sequence can be described by

$$AlCl_3 + 3H_2O \rightarrow Al(OH)_3 + 3HCl \tag{18}$$

$$Al(OH)_3 + 3HCl + 3H_2O \rightarrow AlCl_3 \cdot 6H_2O \tag{19}$$

$$2(AlCl_3 \cdot 6H_2O) \rightarrow Al_2O_3 + 9H_2O + 6HCl \tag{20}$$

The chloride ions, freed by the hydrolysis of the original chloride compounds, are quite capable of attacking the poorly passivated aluminum to continue a cycle of continuous, accelerated corrosion:

$$Al + 4Cl \rightarrow AlCl_4^- + 3e^- \tag{21}$$

$$AlCl_4^- + 3H_2O \rightarrow Al(OH)_3 + 3H^+ + 4Cl^- \tag{22}$$

The problem is aggravated by the fact that etching of Al–Cu alloys results in a surface decorated with Cu-rich residues, which are difficult to remove and which enhance localized corrosion of the aluminum matrix.

Electrochemical data, obtained on samples prepared by the partial etching of the photoresist-masked blanket wafers and exposed to a variety of cleaning conditions, provide an excellent measure of the effectiveness of cleaning steps and the acquired corrosion resistance of the pattered lines. Electrochemical tests can be conducted in a droplet of triple distilled water added to a small area (e.g., 0.32 cm^2, delineated by a mask), minutes after the RIE processing and cleaning. The residual chlorides that did not already react with ambient humidity will react with added water and provide a proportionally aggressive environment. The ionic content can be determined separately with a separate application of a water droplet, followed by ion chromatography and inductively coupled plasma (ICP) analyses. The results show an excellent correlation between the corrosion rate and halide content, and they indicate that the immediate application of a water rinse step provides a cleaner and more corrosion-resistant

Electrochemistry of Corrosion

surface than the use of a CF_4/O_2 plasma (Fig. 17). The most effective halide cleaning sequence seems to be an immediate water rinse, followed by photoresist stripping (in O_2 plasma), with or without a chromic–phosphoric acid etch. (Electrochemical data alone show a somewhat higher dissolution current density after the application of the acid etch, which could be a result of an aggressive dissolution and surface roughening in that process.) Post-RIE cleaning steps are of most importance in processing, with more permanent protection expected to be obtained by the application of sputtered SiO_2 or a similar dielectric layer.

D. Intergranular Corrosion of Conductive Metals

The principal material for the back end of the line (BEOL) interconnections on chips is AlCu (with 0.5 to 2% Cu), with Cu being considered as a replacement because of its good conductivity. Both Al and Cu films, and to a lesser extent their alloys, form pronounced grains when heated. Grain boundaries normally contain high-index planes with larger interatomic distances and weaker interatomic bonds than expected on the face of the grains. When exposed to a cor-

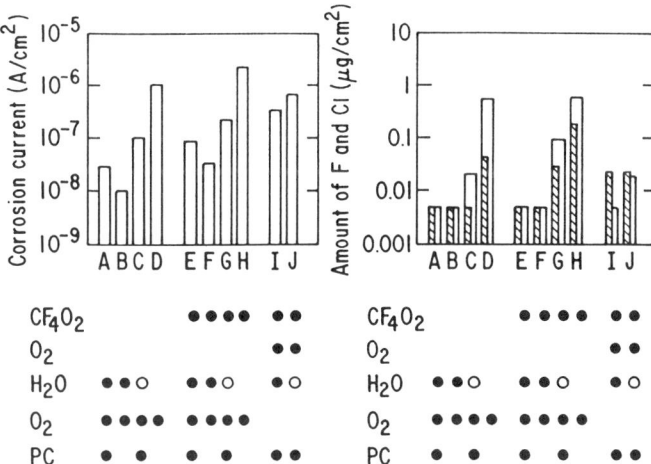

Figure 17 Corrosion rate and halide content on AlCu surface exposed to RIE and different post-RIE cleaning steps. The amount of the chloride is given by the clear bar graph, and of the fluoride by the shaded graph. Full circles represent the process steps in various cleaning alternatives (A–J): CF_4–O_2 plasma was applied immediately after the chlorine plasma etch or not at all, O_2 ashing was used for a removal of the photoresist, before or after the water rinse. H_2O rinse was applied immediately (solid circle) or with a 30 min delay (empty circle) and PC denotes chromic–phosphoric etch. (Reprinted from Ref. 18, by courtesy of the NACE Int.)

rosive environment or to an oxidizing atmosphere, the grain boundaries react faster than the grain surface. Figure 18 shows the surface of annealed copper before and after thermal oxidation. The oxide thickness in grain boundaries is significantly higher than everywhere else on the surface. In the case of Al–Cu and Al–Cu–Si, the initial stages of corrosion in solutions containing chlorides can be seen to occur at grain boundaries and triple points [19]. However, annealing at elevated temperature seems to improve the corrosion resistance of Al alloys. As expected, aluminum grains are growing, Cu is redistributing and going into solid solution with the Al matrix, and, unless the atmosphere is indeed inert, there is the possibility of thermal oxidation and an improvement of native oxide. The strength of this oxide is of prevailing importance for the corrosion resistance of Al alloys [19]. In the case of Cu, one cannot expect such protection. For example, as the lines in a Cu BEOL structure decrease in width and height, and their geometry becomes similar to a bamboo structure, a significant increase of the corrosion rate at grain boundaries should enhance the vulnerability of the lines to corrosion. Thermal oxidation of Cu is fast, and oxides are not very protective; thus, annealing of Cu in an oxidizing atmosphere is not recommended. Alloying of Cu with Al reduces a tendency to form grain boundaries and eventually (for alloys with >13 at. % Al) leads to passivation by Al-oxide. However, the conductivity of Cu–Al alloys decreases with Al content to an unacceptable level. Thus, for Cu, a more permanent protection against corrosion is expected to come from protective overcoats.

E. Use of Inhibitors for Protection of Cu

In contrast to Al (and Al–Cu alloys), Cu is not well protected by its native oxide. According to thermodynamics, Cu cannot be dissolved upon reduction by hydrogen ions; in water containing oxygen, however, its dissolution rate is

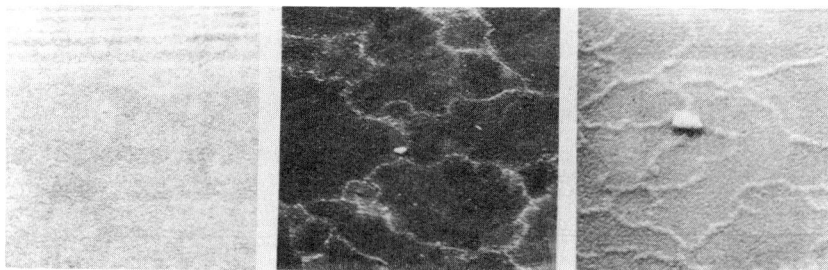

Figure 18 TEM replica of as-deposited thin-film Cu (left), annealed in He at 250°C (middle), and oxidized in air at 250°C (right).

about three orders of magnitude higher than measured on Al (covered by the native oxide). The corrosion resistance of Cu has been much improved with the use of corrosion inhibitors, notably benzotriazole (BTA). At alkaline and neutral pH levels, the growth of the Cu–BTA film is logarithmic. In slightly acidic electrolytes the film is formed by a parabolic growth law, while at pH 2 the growth becomes less reproducible and is best described by a dissolution–precipitation mechanism. The protectiveness of the film reflects its structure and growth. The most protective film is the thinnest and the most polymerized Cu(I)–BTA film, formed on a cleaned, preoxidized Cu surface in a neutral 10^{-2} M BTA solution. Although only 2 nm thick, once formed, this film suppresses the corrosion rate of Cu in water by a factor close to 1000× and offers a durable protection to Cu, even in the absence of further supply of the inhibitor [20].

Two widespread uses of Cu in the electronics industry, Cu tubes for cooling water in mainframe computers and Cu on PC boards, rely on the protectiveness of the Cu–BTA film. Addition of BTA to a closed water system depleted by oxygen has reduced Cu corrosion in that application to low, acceptable levels. Use of Cu–BTA films on PC boards, however, give very useful but limited protection. This film is protective against corrosion caused by the presence of water and oxygen. However, it has limited resistance against a Cl-containing atmosphere and almost no resistance against thermal oxidation [21]. Also, the shape of the potentiodynamic polarization curves obtained on Cu covered with the Cu–BTA film and exposed to water (without BTA) indicates that the Cu–BTA film cannot be very resistive under the applied field in operation.

F. Corrosion and Protection of Magnetic Media

Magnetic storage media provide an alternative approach to that of chip technology for handling large amounts of information. The information is stored on a magnetic disk as a string of recognizable magnetization patterns. Writing on the disk is accomplished by a magnetic head, the pole tips of which are "flying" in close proximity to the disk. Once magnetized in a programmed fashion, the pole tips invoke magnetic patterns on the disk that can be read by another head. The storage density increases with a decrease of the thickness of the magnetic film on disks, of pole tip dimensions and pole tip gap, and of the head–disk distance (flying height). The quest for smaller and more powerful computers has resulted in replacement of the particulate disk and ferrite heads by higher capacity thin-film disks and heads. A magnetic material with excellent corrosion resistance, the rust itself, was replaced with corrosion-sensitive metals, in the form of thin films, that are sandwiched in a galvanically incompatible structure, fabricated in harsh environments and expected to perform in a partially controlled ambient. The schematic representation of a disk housing is given in Fig. 19. One or more disks rotate in close proximity to the head in an unsealed file

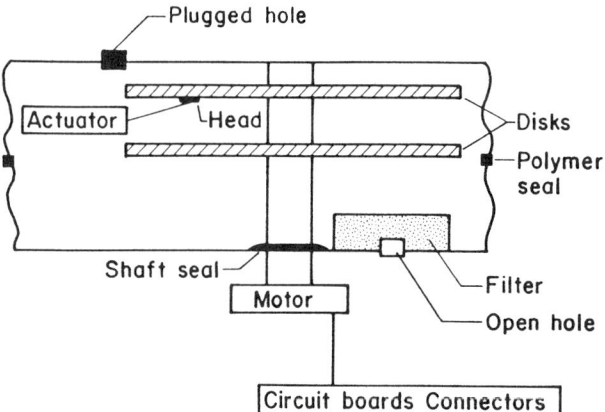

Figure 19 Schematic view of a disk housing.

that might have a filter at the "breathing" hole for aerosol particles. An air exchange is apparently independent of the file operating mode, and it is dominated by diffusion through the small hole [22]. For a hole with a diameter of 7 mm and depth of 2 mm, the leak rate was measured to be about 100 cm^3 (NTP) h^{-1}. If a gaseous atmospheric pollutant is present in the outside air in amounts of 1–100 µg m^{-3}, it would diffuse through the hole at a rate of only 1–100 µg per year [22].

1. Thin-Film Heads

Permalloy (80% Ni–20% Fe) is the most common material in thin-film inductive as well as magnetoresistive heads. The surface of the alloy has a thin layer of air-formed Fe_2O_3 over more protective Ni-oxide that assures the corrosion resistance of the alloy. The corrosion rate of permalloy varies greatly, depending on deposition method, deposition conditions, and surface treatments. Films deposited by plating corrode faster than vacuum-deposited films, which can be further passivated by thermal oxidation. The sensitivity of plated films is attributed to incorporated impurities, notably sulfur, that prevent the formation of a well-passivating oxide film [23].

Considerable effort was spent to find a material that is more corrosion resistant than permalloy. Evaluation of Fe–Cr and Fe–Cr–B alloys provides good insight into the role of Cr and B in corrosion. Electrochemical measurements were conducted in dilute sulfuric acid, pH 3, i.e., in an acidic environment similar to one provided by the mixture of gaseous pollutants in the environmental chamber. Figure 20 shows the variation of the corrosion potential with time in sulfuric acid. Fe and Fe–Cr alloys with ≤8.6 at. % Cr show a significant

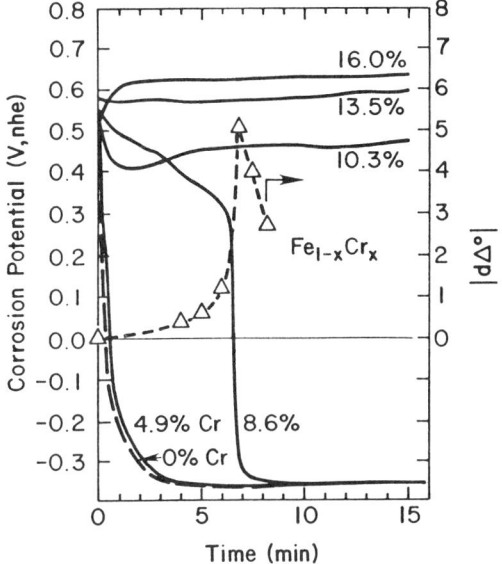

Figure 20 Variation of corrosion potential with time in dilute sulfuric acid measured on Fe and Fe–Cr alloys. Dotted line shows an ellipsometric determination of the spontaneous dissolution of the native oxide on Fe–Cr alloy with 8.6% Cr.

decrease of the corrosion potential with time. At this pH level, Fe oxides are not stable (Fig. 4), and in situ applied ellipsometry indicates that a decrease of potential is accompanied by removal of the native oxide, as noted by the dotted line for the alloy with 8.6% Cr. Alloys with ≥10.3 at. % Cr have primarily Cr-oxide at the surface, hence the high, noble, corrosion potential. Once the corrosion potential showed a stable value, cathodic polarization curves were utilized to determine the corrosion rate (Fig. 21). For Fe and low-Cr alloys, the corrosion rate is about 1.7×10^{-5} A/cm^2, or close to 2 nm/min, occurring with the simultaneous H-evolution reaction. Corrosion of Fe–Cr alloys with ≥10.3 at. % Cr occurs from an oxide-covered surface with oxygen reduction reaction and at a rate three orders of magnitude smaller than measured on Fe. The corrosion rate is a stepwise function of the Cr content (Fig. 22). The figure also shows that boron, which forms glassy, amorphous alloys, does not significantly improve the corrosion resistance. Observe that the data from electrochemical tests are comparable with the data from an environmental chamber given as a weight gain (lower lines on figure). In both cases, significant corrosion resistance has been observed only if the alloys contained ≥10.3 at. % Cr. The corrosion rate

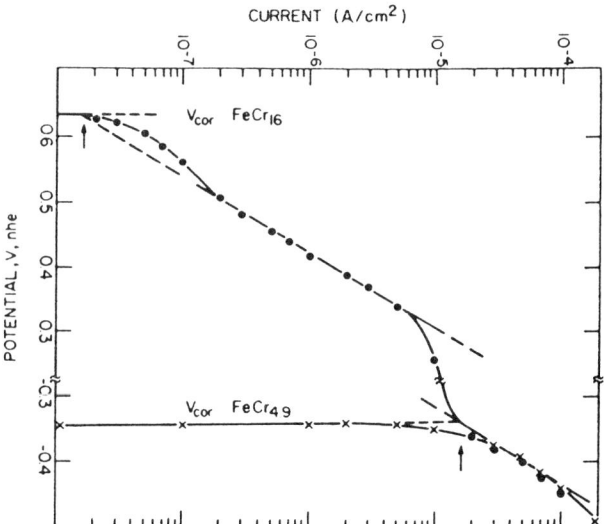

Figure 21 Evaluation of the corrosion rate on some Fe–Cr alloys from the cathodic current–potential curves.

of these alloys is similar to the corrosion of sputter-deposited permalloy and about 10× smaller than measured on the plated films.

In spite of considerable concerns for permalloy's behavior in processing and data showing that additions of Cr [24] or Rh [25] increase the corrosion resistance of permalloy, plated permalloy, with no further alloying, remains in use and is often used as an unofficial standard for comparison of the corrosion behavior of other magnetic alloys, Co-based alloys with high magnetic moment, of magnetoresistive layers with FeMn and FeTb, that are, by comparison, yet more corrosive. In order to reduce corrosion, it was suggested that some of the most sensitive materials should be used recessed from the surface that is exposed to environmental conditions and wear, with processing adjusted during fabrication and gases scavenged in the head–disk assembly [26].

2. Thin-Film Disks

A schematic representation of the thin-film disk structure is given in Fig. 23 (top), with an AlMg substrate, covered with a thick NiP layer, an optional undercoat (Cr, 25–100 nm), Co alloy as a magnetic layer (10–100 nm), a carbon overcoat (10–50 nm), and a lubricant (1–3 nm) for wear resistance. A more realistic structure for textured disks is given in Fig. 23 (bottom), with notations of several different reactions. Carbon is a chemically stable material, but electrochemically active. If it were to cover the underlying layers perfectly, the

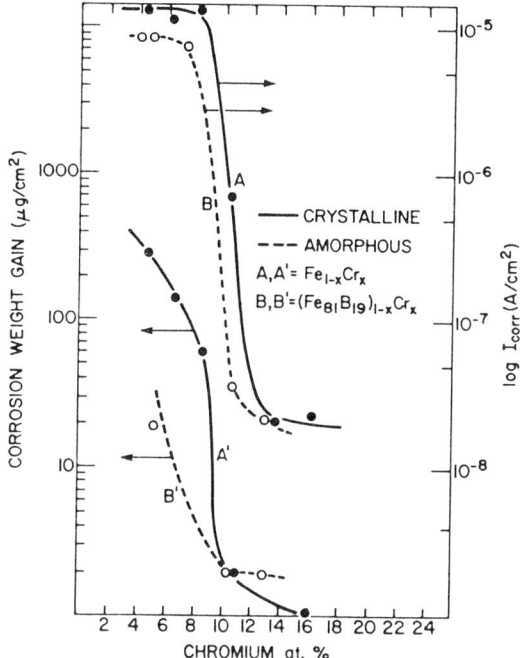

Figure 22 A comparison of the corrosion rate on Fe–Cr alloys measured by electrochemical techniques in dilute sulfuric acid and by weight gain after an exposure to the environmental corrosion chamber with acidic pollutants.

corrosion activity of the disk would be negligible. The roughness of the textured disk and the thinness of the overcoat preclude the existence of a fully covered surface and provide areas of incomplete coverage.

Corrosion Mechanism. Electrochemical data obtained on individual layers are given in Fig. 24. They allow the following extrapolations: if the magnetic alloy is electroless CoP covered with dc sputtered carbon, the overcoat would act as a galvanic aggressor and cause enhanced localized corrosion at the sites of incomplete coverage, with CoP being a primary reactant (galvanically protecting NiP, if NiP is exposed). However, if the magnetic alloy is CoCrPt, galvanic attack by the carbon should be considerably less pronounced. CoCrPt alloy was developed with corrosion protection in mind. Cr is added to provide a passive oxide, and Pt is added to keep Cr in the passive state [27]. However, in this case, NiP is the least "noble" and if NiP is not completely covered, it will react preferentially in a strong galvanic attack. This is confirmed experimentally on tests in an environmental corrosion chamber kept at room temperature, with

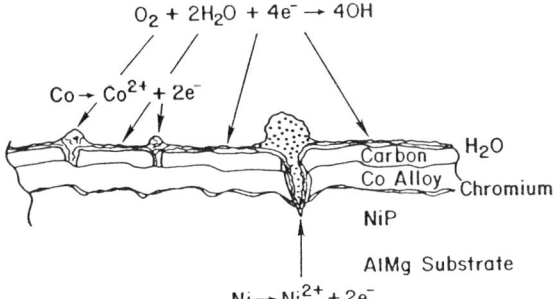

Figure 23 Schematic and realistic structure of a thin-film disk.

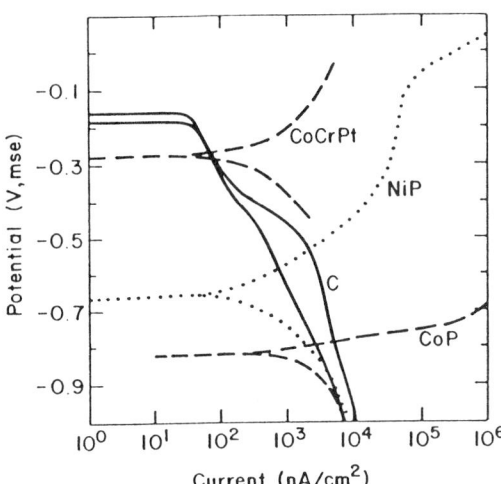

Figure 24 Potentiodynamic polarization curves on CoP, CoCrPt, NiP, and C.

70% RH and 10 ppb of chlorine, where the localized corrosion reactions can be used as markers for the sites of an incomplete coverage (Fig. 25). Both CoP and CoCrPt disks show an enhanced localized corrosion, with $CoCl_2$ as the corrosion product in one case and $NiCl_2$ in the other. Thus, although alloying of Co has resulted in an alloy with much higher corrosion resistance, the overall corrosion defect density (on a textured disk) is not much different. A significant reduction of the corrosion rate should come from the geometry and processes that use smooth substrate, and overcoats that provide the full coverage of the magnetic layer and have a high ohmic resistance.

Lifetime Projection. Predictions of useful lifetimes for magnetic storage devices come from the tests in which the progress of corrosion is measured by an entity proven to interfere with the functional life of the disk. As seen in Fig. 26, a typical corrosion defect with $CoCl_2$ can be several μm high, more than an order of magnitude higher than the head–disk distance. Thus, the accumulated corrosion product interferes with flying of the head; a head crash, or an end of a lifetime, is normally observed before any magnetic errors are recorded. The corrosion rate, monitored by the evaluation of CoCr corrosion products over the carbon surface as a function of time, for disks exposed to different temperatures and relative humidities follows the relationship $\Delta C = a \exp\{\Delta RH/22\} \exp\{-\Delta G^*/RT\} t^{1/2}$, where ΔC is the concentration of Co detected over carbon,

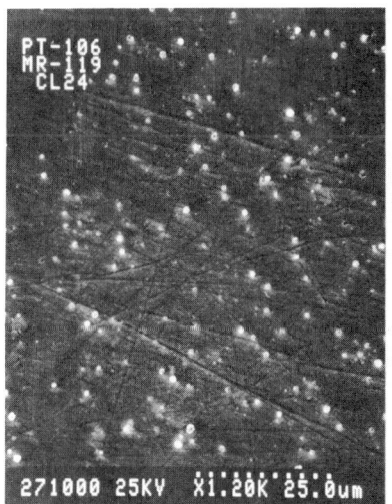
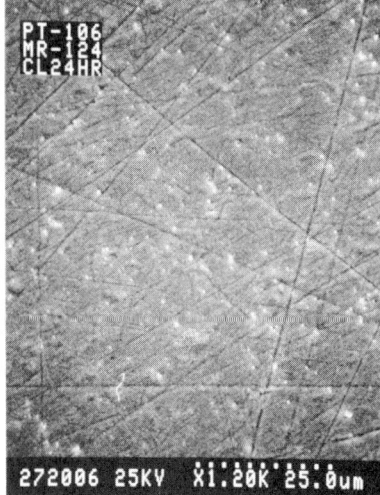

Figure 25 Localized corrosion, on NiP/CoP/C disk with $CoCl_2$ (left) and on NiP/Cr/CoCrPt/C with $NiCl_2$ (right).

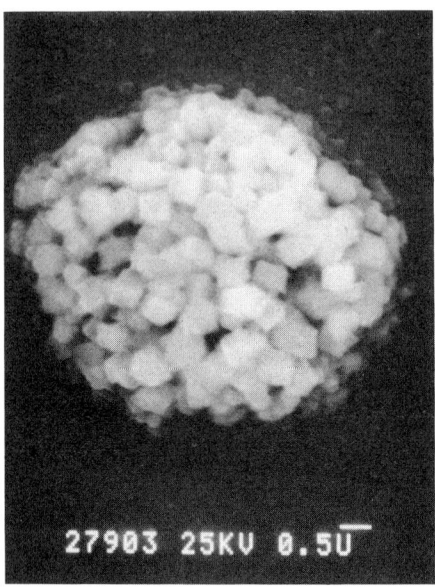

Figure 26 Magnified view of the corrosion defect on CoP/C disk, identified to be $CoCl_2$ (magnification 10,000×).

increasing with time t, relative humidity RH, and temperature T, and ΔG^* is the thermal activation energy. With the reported activation energy of about 0.4 eV [28], the calculated life at 30°C and 40% RH is 150 times longer than the life at 90°C and 90% RH; or, if a disk survives two weeks at 90°C/90% RH it should survive 5.7 years under operating conditions.

Some of the key parameters for the calculation, e.g., ΔG^*, are observed to be very sensitive functions of the film preparation condition, varying for CoCr alloy from 0.07 to 0.3 eV [29]. For films with a small activation energy, the tests at elevated temperature provide barely any acceleration for the reactions, and a considerable diminishing of applicable acceleration factors. As there are no reports on corrosion losses in actual disk structures, the magnetic films in the product most likely have a high activation energy.

G. Corrosion Mechanisms of Conductors With and Without an Applied Bias

1. Copper in PC Circuits

One of the failure modes in printed circuit boards is caused by conductive filaments formed between Cu conductors. The failure is caused by the coupled anodic corrosion and the cathodic electrodeposition processes, which lead to dendrite formation and, eventually, a short circuit in the device. Figure 27 shows a typical dendritic structure formed when two neighboring lines under a bias are exposed to a humid atmosphere. Under dry conditions, two lines at different voltages will not dissolve, because the *IR* drop between them is large. With the adsorbed water, the potential difference is only in part consumed by *IR* drop, and in part used up to form a potential drop at a metal line/substrate interface. At the anode this partial potential difference increases the rate of metal dissolution. Metal ions diffuse toward the cathode. At the cathode, they form a metal deposit.

The mechanism of dendrite formation has been studied by electrochemical techniques with test structures resembling the line geometry applicable to packaging, i.e., with the linewidth and space between the lines on the order of 100 μm. Data indicate that the tendency for dendritic growth increases with a decrease of pH and an increase of Cu^{2+} and overpotential. In the absence of sufficient Cu^{2+}, the principal cathodic reactions become hydrogen evolution and oxygen reduction, which increase the local pH and cause precipitation of dissolved metal oxides and hydroxides [30]. The required minimum overpotential for the dendrite formation is consistent with classical studies on dendrite growth. The minimum bias is of the order of 0.5 V, which is easily exceeded in a practical device. An evaluation of the mean-time-to-failure (MTF) with a variation of RH, temperature, bias, and substrate material resulted in the empirical relation

$$\text{MTF} = a(\text{RH})^b \exp \frac{\Delta G'^*}{RT} + \frac{dL^2}{V} \tag{23}$$

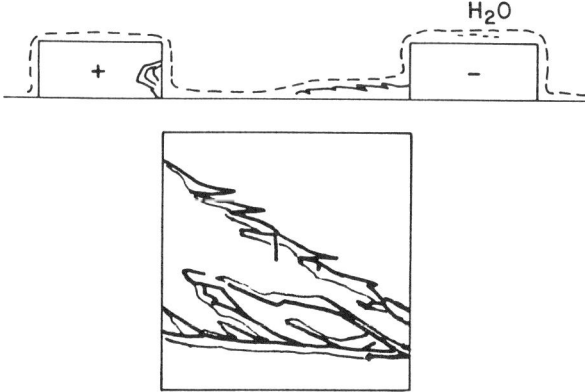

Figure 27 Dendritic growth of Cu at the cathode reaching the anode.

where a, b, d and ΔG^* depend on material, L is the space between the lines, and V is the bias [30].

In many applications, Cu is covered with eutectic PbSn solder, Ag, or Au. However, all of these metals, including gold, have a tendency to form dendrites. Thus, improved reliability can only be assured by control of the environment (dry and clean) or with overcoated structures.

2. Corrosion of Al–Cu in Integrated Circuit Structures

Under dry conditions, leakage currents are small and independent of contamination. The electrochemical failure rate thus becomes vanishingly small, as would be predicted by thermodynamics. Under very clean conditions, AlCu is passive and does not corrode even at high relative humidities and without encapsulation [31]. In the case when both contamination and humidity are present, AlCu lines exhibit failures that depend on the nature of the contaminant. A variety of corrosion mechanisms that are evaluated in laboratory experiments (mostly by electrochemical methods) correlate well with the actual device failures. Lifetime projections, however, obtained from the TH and THB tests and empirical and semiempirical models, vary greatly from model to model, indicating the need for a more comprehensive relationship between component lifetime and stress tests.

Failure Mechanisms. In a device, AlCu metallization is covered by "passivation" layers, such as sputtered SiO_2 or P-doped glass. Contact areas are uncovered and connected with more noble metals, such as Au wire, with the possibility of a galvanic attack. When defects in passivation layers, adsorbed water, contaminants, and bias provide an environment for aluminum dissolution, corrosion can occur either on an anode or a cathode. In the presence of halides, the failure occurs at the anode, with the localized breakdown of passivity resulting in pitting. In the presence of impurities that increase surface conductivity but are not deleterious to the aluminum oxide, such as found with P-doped glass, the corrosion occurs at the cathode. Cathodic corrosion is caused by the hydrogen evolution, followed by a local increase of the hydroxyl ion concentration, and a dissolution of aluminum leading to the formation of an aluminate, similar to that observed in the etching of Al in alkaline solutions. The corrosion proceeds by the following reactions:

$$H_2O + e^- \rightarrow OH^- + \frac{1}{2} H_2 \qquad (24)$$

$$OH^- + Al + H_2O \rightarrow AlO_2^- + \frac{3}{2} H_2 \qquad (25)$$

In a low-temperature region, the second reaction is rate-determining and the corrosion rate increases with temperature with the energy of activation of 0.47

Electrochemistry of Corrosion 361

eV, equal to that observed for the chemical dissolution of Al in NaOH. In the high-temperature region, the first reaction is rate-determining and the corrosion is proportional to the current density, independent of temperature [32]. (Although the corrosion occurs at the cathode, it is important to note that the local potential difference across cathode/electrolyte interface is still anodic to the equilibrium potential for Al dissolution.) The cathodic corrosion is observed much more often than the anodic corrosion because in many instances an anodic bias leads to an enhanced Al-oxide growth and passivation [33]. (Si should behave similarly if exposed to similar conditions.)

Life Projections. The activation energy is found to be in the range 0.4–1.2 eV [15], which greatly affects the MTFs and the projected AFs (Eqs. 15 and 16). Device lifetimes, calculated for normal operating conditions of 30°C/70% RH, using different models and an assumption that they all show MTF of 1000 h on a standard stress test of 85°C/85% RH, could vary from 0.8 to 18.7 years [15]. The variation is in part caused by the sensitivity of the test to a variation in material and process conditions [33]. The models are largely empirical in nature, and the discrepancy among them emphasizes the need for developing a better understanding of the interactions among the device technology, the use conditions, and the test environment, which would allow for improvements in reliability modeling.

3. Circuit Miniaturization and Corrosion Loss

As dimensions of devices decrease to submicron sizes, the sensitivity of the structure to contamination and corrosion loss increases further. A line model used to calculate the time to failure depending on the leakage current and the cross-sectional area of the line is shown in Fig. 28. The lower part of the figure shows the time needed to dissolve the width of the line assuming that the measured leakage current between the two lines originates solely from a Faradaic dissolution of one defective site [34]. Calculations show that a leakage current of 10^{-12} A measured on dimensions applicable to printed wiring boards would not destroy the metal in more than 100 years. However, the same current would cause the failure of a 0.1 μm line in tens of seconds.

This indicates that submicron structures will have reasonable lifetimes only if the leakage currents are not permitted to be higher than 10^{-18} A, which means that fabrication would have to be done in a nearly particle-free environment and with passivation layers that would keep low humidity levels. Submicron airborne contaminants are shown to be ionic in nature [14], and their presence on the surface can immediately interfere with water adsorption and corrosion chemistry.

A potential replacement for AlCu in VLSI structures by the more conductive Cu [35] increases corrosion concerns. Potentiodynamic polarization curves for Al–Cu and Cu in aerated water show not only a difference in the respective corrosion rates of several orders of magnitude, but a significant difference in

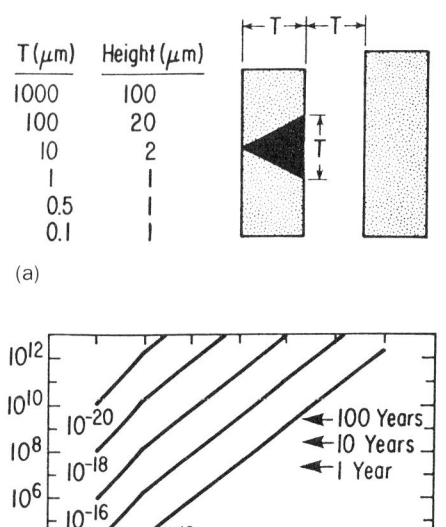

Figure 28 Calculated time to failure in seconds as a function of leakage current and line dimension (b), if the dissolution propagates from a single defect site across the width of the line according to (a). (Reprinted from Ref. 34, by courtesy of the Electrochemical Society.)

dissolution rates at anodic potentials (Fig. 29). Copper lacks a self-passivating capability that is shown by aluminum. Aluminum can be tested without benefit of passivating overcoats, but not copper. The overall corrosion protection of Cu relevant to reliability should be achieved by application of protective overcoats. However, it should be emphasized that an applied bias and defects in the Cu overcoats would lead to fast dissolution of the Cu even in very clean, nondry environments.

VII. SUMMARY

Corrosion is a constant reminder of the temporary nature of the metallic state. Metals spontaneously react in varying degrees with various environments, and

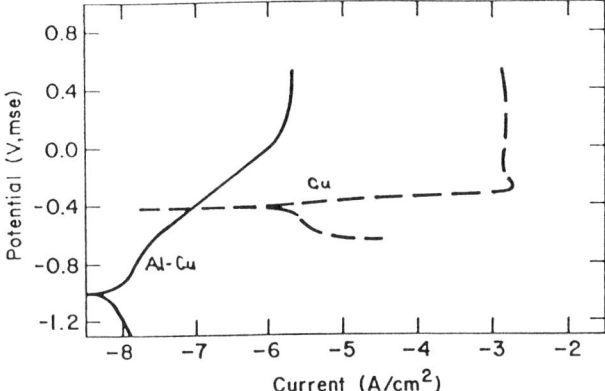

Figure 29 Potentiodynamic polarization curves on Al–Cu (solid line) and Cu (dashed line) measured in a droplet of distilled water.

they form oxidation products similar to the ores from which they were originally extracted with an investment of much energy. The electrons liberated in oxidation must be consumed in reduction of available reactants, with the two reactions occurring simultaneously and at the same rate. Each of the reactions obeys the laws of electrochemical kinetics, and the rate of each reaction can be significantly increased or decreased by the potential across the metal/environment interface. Distribution of the participating electrochemical reactions can be uniform or not, giving rise to many forms of localized corrosion.

Electronic devices are fabricated with thin films, which require a balance between some specific property and corrosion resistance. They are often used with the application of an electrical potential. Thus, both the nature and the application of the materials make a device more prone to corrode and more sensitive to corrosion loss.

Through the years, the scientific community has developed many approaches to corrosion prevention. Regarding thin-film structures, all possible approaches should be diligently used, with respect to the demands on product quality and restrictions of their dimensions.

Corrosion can be minimized by the proper choice of materials and with control of environment. Materials, if possible, should be noble metals, metals that passivate (alone or alloyed), and material combinations that are galvanically compatible. Reduction of the aggressive nature of the environment can be accomplished by controlling the relative humidity at below 30%, application of surface oxidation, use of inhibitors, or by encapsulation with inorganic or organic layers.

In many ways, thin-film deposition techniques can be effective in controlling and optimizing thin-film properties, either alone or in multilayer structures. A good example is the development of CoCrPt alloy for the thin-film disk and the use of overcoats in chip fabrication. Corrosion is a natural process; nature is capricious, and none of us can exactly predict what it will do. However, corrosion can be minimized through better dissemination and wider application of well-understood principles.

EXERCISES

1. Given the standard free energies of Na_{aq}^+, Fe_{aq}^{+2}, and Cr_{aq}^{+3} as -62.6, -20.3, and -49.0 kcal/mol, respectively, calculate the standard electrochemical potential for Na/Na$^+$, Fe/Fe^{2+} and Cr/Cr^{3+}.
2. Given the standard potential of Al/Al_{aq}^{+3} as -1.66, calculate the single potential for 10^{-3} M and 10^{-6} M solutions of aluminum ions.
3. If the anodic Tafel slope for Fe dissolution is 0.06 V/decade, calculate the anodic current on iron in a solution of unit ferrous ion activity at the potential of -0.32 V. Assume that i_0 is 10^{-8} A/cm^2.
4. Standard potentials for Fe, Zn, Sn, and Cu are given as -0.44, -0.76, -0.14, and $+0.34$ V. Ionic activities in seawater at 25°C for all metallic ions are $\approx 10^{-6}$ M. i_0 for Fe, Zn, Sn and Cu in seawater are 10^{-12}, 10^{-9}, 10^{-8}, and 10^{-8}. Anodic Tafel slope for Fe dissolution in Cl$^-$-containing solution is 0.059 V/decade; for Zn 0.028; for Sn 0.015; and for Cu 0.02 V/decade. The exchange current densities for H$_2$ evolution are 10^{-6} A/cm^2 on Fe, 10^{-11} A/cm^2 on Zn, 10^{-10} A/cm^2 on Sn, and 10^{-7} A/cm^2 on Cu, with the cathodic slope of -0.12 per decade.

Calculate the overall corrosion current when 1 cm^2 of iron is immersed in neutral deaerated seawater at 25°C and connected to (*a*) itself, (*b*) 100 cm^2 of Zn, (*c*) 100 cm^2 of Sn, and (*d*) 100 cm^2 of Cu. What is the rate of attack in A/cm^2 on the iron in each case? Use v-log i graphs in evaluation.

REFERENCES

1. J. M. West, *Electrodeposition and Corrosion Processes*, D. Van Nostrand Co., London, 1965.
2. D. Seraphim, R. Lasky, and C.-Y. Li, eds., *Principles of Electronic Packaging*, McGraw-Hill.
3. M. Pourbaix, *Atlas of Electrochemical Equilibria in Aqueous Solutions*, 1st English ed., Pergamon Press, New York, 1966.

4. H. H. Uhlig, *Corrosion and Corrosion Control*, Wiley and Sons, New York, 1963.
5. J. M. West, *Basic Corrosion and Oxidation*, Ellis Horwood Ltd., New York, 1980.
6. N. D. Tomashov, *Theory of Corrosion and Protection of Metals* (B. H. Tyrell, I. Geld, and H. Preiser, trans. and eds.), McMillan and Co., New York, 1966.
7. M. G. Fontana and N. D. Greene, *Corrosion Engineering*, McGraw-Hill, New York, 1967.
8. G. Wranglen, *An Introduction to Corrosion and Protection of Metals*, Chapman and Hall, New York, 1985.
9. D. A. Jones, *Principles and Prevention of Corrosion*, Macmillan, New York, 1991.
10. P. B. P. Phipps and D. W. Rice, in *Corrosion Chemistry* (G. R. Brubaker and P. B. P. Phipps, eds.), American Chemical Society, Washington, DC, 1979, pp 235–261.
11. M. Strarmann, The investigation of the corrosion properties of metal surfaces covered by condensed or adsorbed electrolyte layers, The 1st Int. Symp. on Corrosion of Electronic Materials and Devices (D. Sinclair and R. Frankenthal, eds.), Electrochemical Society, Pennington, New Jersey, 1991.
12. M. Zamanzadeh, Y.-S. Liu, P. Wynblatt, and G. W. Warren, *Corrosion 45*:643 (1989).
13. IBM Engineering Specification C-H 109700-010.
14. R. E. Lobnig, R. P. Frankenthal, D. J. Siconolfi, and J. D. Sinclair, in The 2nd Int. Symp. on Corrosion of Electronic Materials and Devices (D. Sinclair and R. B. Comizzoli, eds.), Electrochemical Society, Pennington, New Jersey, 1993.
15. L. T. Nguyen, SPE 46th ANTEC, Atlanta, (Georgia, 1988.)
16. J. J. Gajda and H. Wildman, 20th Ann. Proc. Reliab. Phys., 1982, p. 106.
17. R. Memming and G. Schwandt, *Surf. Sci. 4*:109 (1966).
18. V. Brusic, G. S. Frankel, C.-K. Hu, M. M. Plechaty, and B. M. Rush, *Corrosion 47*:35 (1991).
19. S. K. Fan and J. W. McPherson, *IEEE Ann. Proc. Reliab. Phys.*
20. V. Brusic, M. A. Frisch, B. N. Eldridge, F. P. Novak, F. B. Kaufman, B. M. Rush, and G. S. Frankel, *JECS 138*:2253 (1991).
21. V. Brusic, M. A. Frisch, B. N. Eldridge, F. B. Kaufman, T. A. Petersen, A. Schrott, and G. S. Frankel, Growth kinetics, polymerization and protection of Cu–X–BTA films, International Symposium on Control of Copper and Copper Alloys Oxidation, July 6–8, 1992, Edition de la Revue de Metallurgy, Paris, 1993.
22. L. Volpe, Proc. Symp. Corrosion Effects of Acid Deposition and Corrosion of Electronic Materials, 168th Meeting of the Electrochemical Society, Las Vegas, 1985, pp. 379–386.
23. G. S. Frankel, V. Brusic, R. G. Schad, and J. W. Chang, *Corrosion Sci. 35*:63 (1993).
24. D. W. Rice, J. C. Suits, and S. J. Lewis, *J. Appl. Phys. 47*:1158 (1976).
25. D. W. Rice and J. C. Suits, *J. Appl. Phys. 50*:5899 (1979).
26. J. K. Howard, U.S. Patent No. 4,789,598 (December 6, 1988).
27. V. J. Novotny, A. Itnyre, A. Homola, and L. Franco, *IEEE Trans. Magn. MAG-23*:3645 (1987).
28. K. Tagami and H. Hayashida, *IEEE Trans. Magn. MAG-23*:3648 (1987).

29. G. W. Warren, P. Wynblatt, and M. Zamanzadeh, *J. Electron. Mater. 18*:339 (1989).
30. T. L. Welsher, J. P. Mitchell, and D. J. Lando, 18th Ann. Proc. Reliab. Phys., 1980, p. 235.
31. M. Iannuzzi, *IEEE Trans. Comp., Hybr., Mfg. Techn. CHMT-4*:181 (1983).
32. E. P. G. T. van de Ven and H. Koelmans, *JECS 123*:143 (1976).
33. R. B. Comizzoli, Materials developments in microelectronic packaging; performance and reliability, Proc. 4th Electronic Mat. and Proc. Cong. (P. J. Singh, ed.), ASM Int., Materials Park, Ohio, 1991.
34. J. D. Sinclair, *JECS 135*:89C (1988).
35. Y. Shacham-Diamand, *J. Micromech. Micreoeng. 1*:66 (1991).

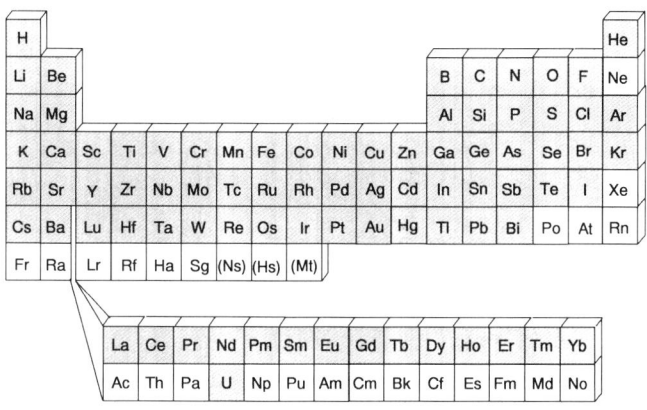

9
Micro and Nano Analyses of Materials

Homi Fatemi   *Advanced Micro Devices, Sunnyvale, California*

I. INTRODUCTION

To remain a viable competitor in the electronic product marketplace, every manufacturer of electronic material devices is under constant pressure to produce products with increased performance and functionality, and at lower costs. In part, this is achieved by shrinking the device dimensions and increasing the number and types of devices per unit area. As discussed in Chapter 2, fabrication of advanced integrated circuits requires hundreds of process steps, each of which must be highly optimized in terms of process parameter control and operated in a contamination-free environment. These production challenges are highly dependent on the use of ultrapure precursor materials and supportive chemicals, whose high purities must be deliverable to the point of use—the chemical process reactor. Strong inverse correlations exist between the presence of unwanted contamination (either as chemical impurities or as particulates during the production process) and the final product yields. Such electrical or physical "defects" can lead to unacceptable leakage currents, insulator breakdown, degraded optical or magnetic properties, or a host of other poor device characteristics. Any one of such defects can be the cause of actual device failures and, as discussed in Chapter 2 and Chapter 8, may contribute to device reliability issues. Even very low levels of atomically dispersed contaminant metals, present on or in a wafer, may aggregate during high-temperature processing steps to create atomic scale defects. This demand for higher purity of materials and processes will continue. For example, the integrated circuits of the mid-1990s require the metal contamination levels to be less than 10^{10} atoms/cm^2 (or less than 10^{-5} of a monolayer). By the year 2000, it is anticipated that these levels must be further reduced to less than 10^8 atoms/cm^2.

Thus, the ongoing electronic material product density trends mandate a continuous effort and focus on reducing contamination and defects, starting with the various development phases and continuing through to the final manufacturing process. In the development mode, each process step must be exhaustively characterized for any defect identification and defect generators, as well as for verification of their successful elimination. In the manufacturing mode, the continuous vigilance for, and the efforts to further reduce, any remaining defect generators are a way of life. This applies to materials, chemicals, processes, and process tools in addition to the human operators, even though the processes are already exercised in exceedingly clean environments. Due to this high degree of process discipline and constant vigilance, any still existing contaminants or defects will actually be very few, and they will typically be very small.

To seek out such remaining defects or contaminants requires sophisticated analytical tools stressed to their resolution limits. In the above example of metal contamination, the detectability limits for metals in the mid-1990s was approximately 5×10^9 atoms/cm^2—limits that are inadequate for the needs in the year 2000. New tools and techniques with improved detection sensitivities will be required for future progress. In addition to the resolution issue, the rate of the technology progress is also greatly dependent on the ability to repeatedly locate an isolated defect or contamination area either on the surface, at interfaces, or within the body of a material, and then analyze their morphology and chemical composition distribution. Hence, the chemical analysis, including its various forms of material characterization, has become an equal partner with the electronic material product technology, process developer, and manufacturer.

Despite possible future limitations, the existing arsenal of analytical techniques and tools can provide incredibly detailed information concerning material topography, structure, and chemistry of the first 2–5 nm of a sample's surface or of the impurity content within material films and structures. To assure unambiguous identification, various tools and techniques must often be employed to complement one another, since any single tool pressed to its analysis limits may not yield sufficiently conclusive results. For example, the investigation of point and crystallographic structural defects generally involves the use of optical spectroscopy, scanning electron microscopy (SEM), x-ray diffraction, or transmission electron microscopy (TEM) methods.

The schematic in Fig. 1 of a typical complementary metal oxide semiconductor (CMOS) device illustrates the various areas that are of potential interest and some of the various analytical tools that are used to generate the desired data. Secondary ion mass spectroscopy (SIMS) techniques are routinely employed to profile free-carrier and impurity distributions, especially in ion-implanted and annealed regions of a semiconductor. Focused ion beam (FIB) has become a powerful tool to assist in circuit mode modifications and in failure analyses. Microanalysis of very small samples is done with energy dispersive x-rays (EDX), a technique also very appealing for monitoring of the chemical nature of a surface after various cleaning procedures. Chemical information on conductive films can be generated with Auger electron spectroscopy (AES).

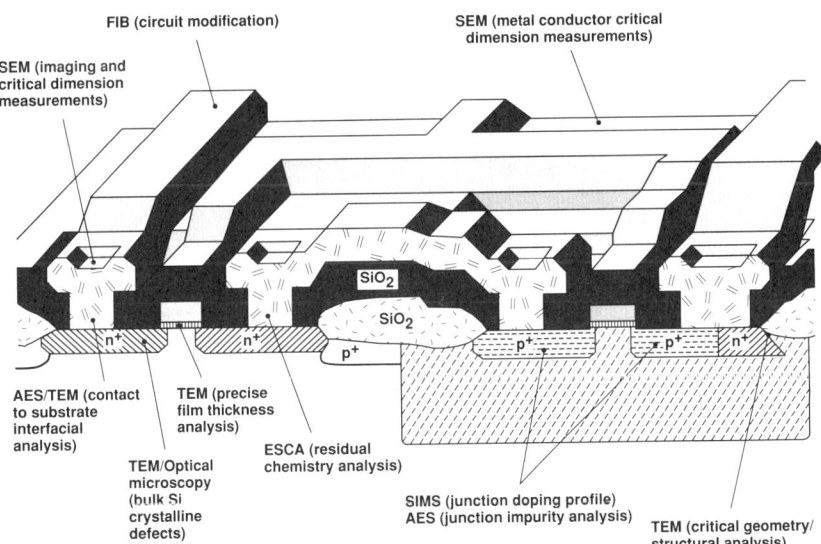

Figure 1 Schematic of a typical CMOS device cross-section and examples of analytical techniques utilized for the acquisition of chemical and physical microanalytical details.

Many modern analytical tools depend on some form of optics. The relationships of the basic electron optics, ion optics, and x-ray optics technologies to various analysis methods are shown in Fig. 2. Also, most modern methods of microanalysis involve the bombardment of a specimen by either electrons, photons, ions, or nuclear particles to generate a signal that is subsequently captured for detection. Some typical examples are depicted in Fig. 3 along with some of the commonly used acronyms for the various techniques: ESCA (electron spectroscopy for chemical analysis), EPMA (electron probe microanalysis), AES and SAM (Auger electron spectroscopy and scanning Auger microscopy), ES-TEM (electron spectroscopy in transmission electron microscopy), ISS (ion-scattering spectrometry), nuclear backscattering, SIMS (secondary ion mass spectrometry), nuclear activation analysis (also called neutron activation analysis), and so on. Other techniques, not depicted in Fig. 3, exist. For example, SCANIIR (surface composition by analysis of neutral and ion impact radiation), LPMS (laser probe

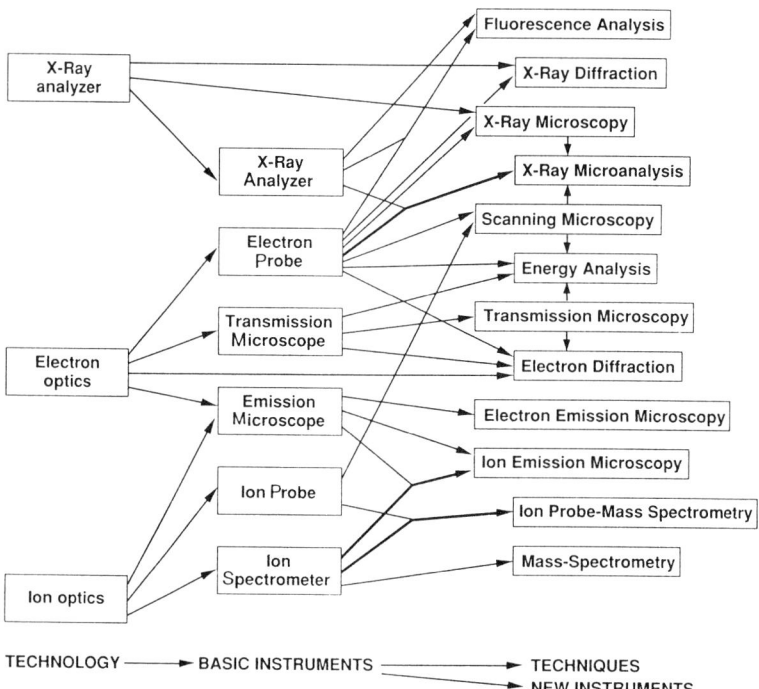

Figure 2 Basic relationships between various optic probes and types of analytical instruments.

Micro and Nano Analyses of Materials 371

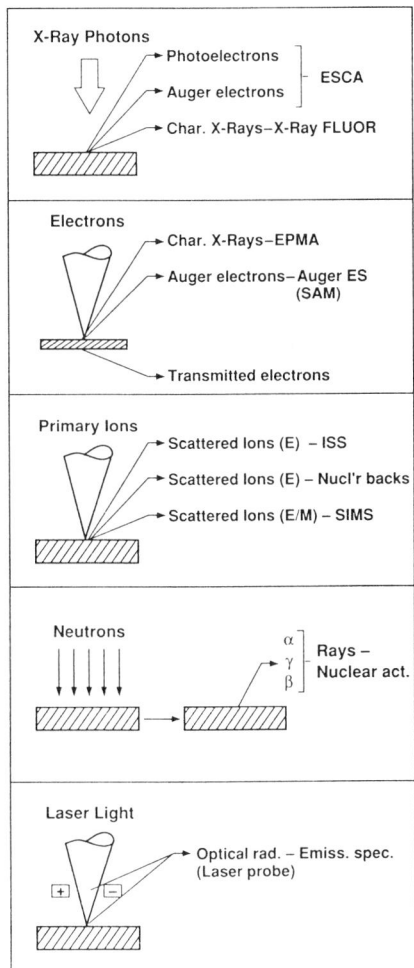

Figure 3 Examples of several microanalytical techniques as related to different material bombardment sources.

mass spectroscopy), FDMS (flow discharge mass spectroscopy), and CLS (cathodoluminescence spectroscopy) have been omitted, since their applications are generally restricted to very special cases. Their omission should not be interpreted to imply they are unimportant. Furthermore, new techniques are constantly being explored, for example high-resolution near-field optical microscopy (NFOM).

The following sections will focus on a number of major analytical tools as they apply to high-resolution analysis of material defects, impurities, and crystallographic information, along with their capabilities and limitations in solving development and manufacturing processes and material problems. The chapter concludes with methods of how these tools can assist in the identification of device failure modes. Although the present discussion tends to focus on applications with examples of semiconductor devices, and thus is closely related to the subject matter of the preceding chapters, these methods and techniques apply equally well to many of the analysis needs with respect to the material topics discussed in the remaining chapters of this book. Although specific material issues may be different, the information generated tends to be of equal importance to the developer or manufacturer involved with those material systems. The various analytical techniques fall into three general categories:

1. Optical methods—optical, SEM
2. Structural methods—TEM, FIB
3. Chemical methods—EDX, AES, XPS, SIMS

These analytical techniques and their respective detection limits, depth resolutions, and typical applications are summarized in Table 1. Expanded details on all of the techniques can be found in many excellent references.

II. OPTICAL METHODS

As indicated, much current work in electronic device and circuit development is focused on creating smaller devices with individual elements in the submicron and even Angstrom range. Verification of these dimensions is of primary importance, as is the need to understand the interspatial relationship between device features, various geometrical configurations (i.e., angles, surface contouring), and the physical integrity of material structures (i.e., absence or presence of film delamination, cracks, voids, etc.). Many of these details can be gathered with optical analysis techniques, in part because of the wide-ranging capabilities of commercially available instrumentation. At the very basic level, optical techniques involve the simple, direct visual observation of a sample with white-light at resolutions down to 0.2 μm. In essence, these optical tools yield surface analysis information, much of it nondestructively. An optical microscope can yield objective magnifications of up to about 1000×, while with a scanning electron microscope, magnifications well beyond 100,000× are obtainable along with scanning resolutions as low as 6 Å (0.6 nm). This latter capability yields images with the ''look'' of a traditional optical microscope image, but with a much greater depth of field. Although special SEM tools do allow the nonde-

Table 1 Summary of Analytical Tool Capabilities

Analytical technique	Typical applications	Signal detected	Elements detected	Detection limits	Depth resolution	Image mapping	Lateral resolution probe size
Optical	Imaging	Visible light, IR, UV	—	—	0.5–1 μm	Yes	0.2–0.5 μm
SEM	Imaging & elemental microanalysis	Secondary backscattering electrons and x-rays	B-U (EDX mode)	0.1–1 at %	1.5 μm	Yes	7 Å
u-FTIR	Identification of polymers, organic films, liquids	IR absorption	H_2-U	0.1–100 ppm	—	No	5 μm
TEM	Imaging, elemental microanalysis	Secondary backscattering transmitted electrons and x-rays	B-U	> 0.5%	—	Yes	1.8–50 Å
FIB	Ion milling, imaging, deposition	Ions, secondary electrons	—	—	0.25 μm	Yes	250 Å
EDX	Elemental analyses	Characteristic x-rays	B-U	0.05–1 at %	0.05–5 μm	Yes	0.3–3 μm
Auger (AES)	Surface analysis and high-resolution depth profiling	Auger electrons from near surface atoms	Li-U	0.1–1 at %	< 20 Å	Yes	150–1000 Å
XPS/ESCA	Surface analysis of organic and inorganic molecules	Photoelectrons	Li-U	0.1–1 at %	10–100 Å	Yes	75 μm–2 mm

Table 1 Continued

Analytical technique	Typical applications	Signal detected	Elements detected	Detection limits	Depth resolution	Image mapping	Lateral resolution probe size
SIMS	Dopant and impurity depth profiling	Secondary ions	H_2-U	10^{12}–10^{16} at/cm^3	50–300 Å	Yes	30 μm

structive viewing of full 200-mm-diameter wafers, a more typical SEM analysis relies normally on a small wafer section. Such sampling allows viewing not only from the top, but also in cross-section, thereby yielding both vertical and horizontal dimensional and structural data. The depth of field of a SEM is 500 times that of an optical microscope.

A. Optical Microscopy

The optical microscope is typically the first tool applied to any physical investigation of a sample. It is quick, and it is nondestructive. It is the only tool that permits the evaluation of specimen properties such as color, relative opacity, and relative refractive index. These properties are closely associated with the normal visual process and can provide a wealth of information about a specimen that would otherwise be missed if only electron microscopy techniques were employed. The optical method also provides considerable information regarding the surface structure of a sample [1]. For example, a discrete device or a fully integrated circuit chip may be examined, usually without any preparation, to: (1) ascertain various construction elements, (2) detect defects or anomalies that might have been introduced during manufacturing or are the cause of observed device failures, or (3) monitor the effect of other analytical procedures carried out at various preparation steps of a particular sample.

1. The Basic Optical Microscope

Of all the various analysis tools, the optical microscope is the least sophisticated, and it is possibly the one most familiar to the reader. High-resolution microscopes rely on high-intensity (up to 500-watt xenon source) illumination, which is typically incident to the sample. For most analysis applications, a planochromatic objective lens is usually adequate. This lens has an intermediate color correction with a well-corrected flat field, necessary for optimum photomicrography resolutions. The bright-light illumination mode provides information about relative reflectivities of the various areas on a sample's surface. When the brightness of such areas is independent of the light's wavelength, it is referred to as brightness contrast. If the reflectivity varies, it is referred to as color contrast. The latter mode is often quite useful in elucidating more detailed structural surface information.

2. Special Features—Phase Microscopy

With the appropriate microscope optics, phase relationships of reflected rays from different points on the sample's surface may be observed. These are due to slight differences in the local surface heights existing on the sample. The simplest phase microscopic technique relies on oblique illumination, in which the incident light reaches the sample at an oblique angle from one direction. Such illumination accentuates any surface irregularities on the sample, much as the late-afternoon sun accentuates the landscape. The more oblique the light, the greater the phase illumination. If this illumination is arranged to come from all directions around the viewing axis, at an angle so that no direct illumination is reflected back into the viewing objective lens, the directional effects are minimized and the surface contrast maximized. This technique, referred to as darkfield illumination, is quite useful to study surface particle densities or surface stains.

In a slightly different interference contrast mode (the Linnick mode), the sample and a reference surface are arranged to be optically coincident and very nearly parallel. An interference contrast illuminator may be adjusted by tilting or displacing the reference surface. This can yield an image in which colors delineate areas of different elevations. Fringes are produced with increases in the angle between the sample and reference surface, allowing for quantitative measurement of sample elevation and depressions with resolutions to as low as 20 Å. This technique requires considerable care in the creation of the fringes and is sample-vibration-sensitive. An alternative mode (the Normanski interference system) automatically produces fringes and is vibration-insensitive. However, the fringe patterns are more difficult to interpret for complex structures.

Another quantitative surface measuring method is based on a multiple-beam interferometer. It relies on a partially reflecting mirror being placed in physical contact with the surface of the sample during observation, limiting this technique to samples of planar configuration. If highly monochromatic light is used for illumination, very sharp, narrow fringes are produced, making measurements very precise. The method permits measurements of surface asperities as small as 5 Å, or about 2 atomic layers in thickness.

In materials science, too many things look alike; however, their structures may be quite different internally and, if crystalline, quite unique. Ordinary white light cannot be used to study such materials, principally because the light vibrates in all directions and consists of a range of wavelengths, resulting in a composite of information, which is analytically useless. However, material structure studies can be done with a polarized light microscope. If a polarizer is placed in the light's path before the sample, light is made to vibrate in only one direction, isolating those properties of a material that are aligned in specific orientations.

3. Applications

The characteristics of materials that may be determined with light microscopes include morphology, size, transparency or opacity, color (reflected and transmitted), refractive indices, dispersion of refractive indices, dispersion staining colors, crystal system, birefringence, sign of elongation, fluorescence (ultraviolet, visible, and infrared), melting point, eutectics, degree of crystallinity, and microhardness.

Surface studies and spatial structure relationships of cross-sectional device structures are often executed with optical microscopy when extreme resolutions are not required. Many of these samples normally consist of material composites, each with different chemical and physical properties. Utilizing material-specific etchants, some degree of surface relief can be created between these different materials, making them visible in the microscope. In situations where such differentiations cannot be achieved (e.g., multiple metal lines), a mechanical polishing operation can create a similar degree of delineation due to material hardness differences. The harder components in a sample will stand out slightly in relief versus softer components.

B. Scanning Electron Microscopy (SEM)

The scanning electron microscope is based on scanning a finely focused electron beam across the surface of a specimen. The latter reflects the beam into two directions (X and Y). These reflection signals are collected, and their intensities are displayed on a cathode-ray-tube screen by brightness modulation. As already indicated, the method allows specimen magnifications to more than $100,000\times$, while maintaining a large depth of focus. The ease of sample scanning of an SEM over large distances is quite appealing, in that a large sample-viewing area is first surveyed at generally low magnifications to seek out particular areas of interest, followed by high magnification of those specific areas for subsequent detailed investigations. Hence, specific surface irregularities, for example, known to be present or noted at low magnifications can be identified and further investigated at significantly higher magnifications. Such studies can highlight unexpected geometrical configurations, unique shapes of particulates, or the degree of film-contouring efficiency or deficiency. The SEM is also extensively employed for the generation of dimensional and spatial relationship details of structure elements. This tool is a primary monitoring and analysis instrument in support of many manufacturing process and product issues.

1. Fundamental Electron-Material Interactions

The SEM tool relies on the generation of electrons that are accelerated through an electric field to acquire sufficient kinetic energy. These energized electrons are then directed onto the material to be investigated. The electron interaction with the material results in a number of different energy dissipation modes, as depicted in Fig. 4. The particular type of released energy depends on the energized electron interaction with the various orbital electrons of the material. If the ejected orbital electron is weakly bound, it emerges with only a few eVs of energy. These are termed secondary electrons. (Strictly speaking, any electron

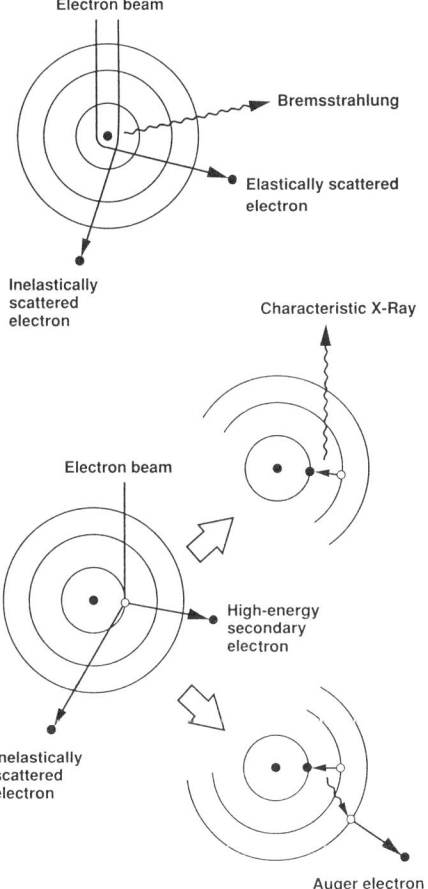

Figure 4 Primary electron beam interaction models with orbital electrons and various types of possible secondary radiation.

ejected from an atom in the sample is a secondary electron—whatever its energy. To the electron microscopist, however, secondary electrons are those with energies below about 50 eV.)

Secondary electrons generated sufficiently deep within the material will be reabsorbed by that material before they can reach the surface, whereas those generated near the surface can escape and therefore are detectable. As shown in Fig. 5, secondary electrons created at topographic peak areas in a material will have a greater chance to escape than those generated in topographic troughs. Furthermore, since the greatest density of secondary electrons is created by the primary beam of energized electrons before they spread into the material to undergo other possible interactions, they have a high spatial resolution relative to any other possible energy signals. The ability to capture both the topographic sensitivity and the spatial resolution, forms the basis for high-resolution microscopy of material surfaces as measured by an SEM tool. As long as a material surface exhibits some degree of surface irregularities (these may actually be intentionally created with etchants), it will generate a SEM micrographic image.

The ultimate spatial resolution of an SEM image is proportional to the tool's ability to generate an electron current density. The development of the field emission gun by Crewe in 1968 greatly advanced the resolution of an SEM [2]. The gun creates extremely high electron current densities by forcibly emitting electrons from a needle-pointed metal tip under an intense electric field and ultra-high-vacuum conditions.

2. Special Features—Backscattering Electrons

In the case in which the primary electrons interact with the nucleus of a material atom (as opposed to interaction with orbital electrons), the electron itself may be scattered in any direction with little energy loss. Those that are emitted back out of the sample become detectable. These backscattered electrons (BSE) are much more energetic (>50 eV) than the secondary electrons and consequently may escape from greater sample depth. Although these electrons do not carry as much information about a sample's topography or spatial resolution, their

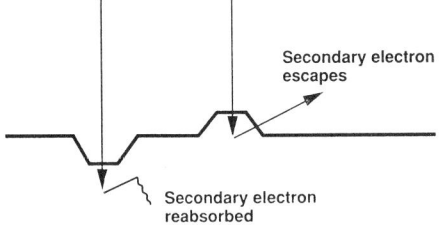

Figure 5 Topographic influence on low-energy secondary electrons.

energy is influenced by the mean atomic number within the interaction volume of the material. The higher the atomic number of an atom, the greater its positive charge within the nucleus, and the more likely that its interactions with the electrons will produce BSEs. These therefore carry some material composition information.

3. SEM Analyses

Examples of the ability to make high-fidelity SEM analyses of surfaces, cross-sectional views, or contaminants appear within this chapter as well as in other chapters of this book (i.e., Chapter 2, Fig. 2; Chapter 3, Fig. 17; Chapter 7, Fig. 28; Chapter 8, Fig. 18, Fig. 26). Indeed, SEM photomicroscopy tends to be the preferred means to obtain any initial high-resolution data of a particular sample. It is quite useful for most dimensional and structural shape information, including feature-to-feature comparisons to evaluate consistencies or abnormalities—as represented by the tungsten-filled via holes of Fig. 6. This form of SEM analysis relies on the most commonly used secondary electron (SE) imaging mode. The data suggest a process concern in that the left via exhibits incomplete tungsten filling due to the presence of residues left behind from a previous cleaning process.

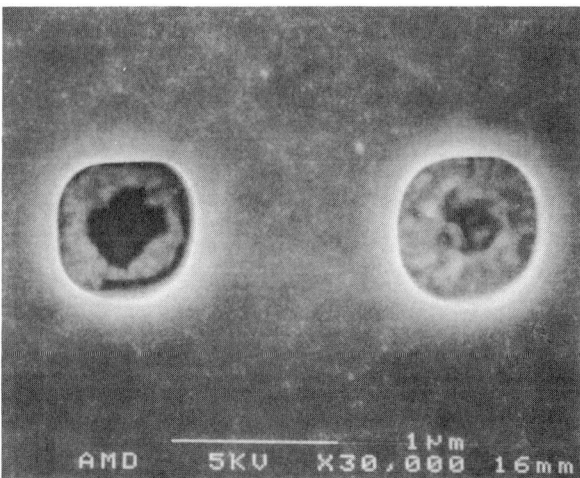

Figure 6 SEM surface image of tungsten-filled contact holes (vias). The right via is properly filled, whereas the left one is only marginally filled due to a residual photoresist film around the periphery and bottom, which was not fully removed during a preceding surface cleaning process step.

Microanalysts routinely extend the SEM's ability with the use of the backscattering electron (BSE) imaging mode to gain additional specific sample information. While SE imaging shows topography, the BSE images depict contrast due to differences of the electron density in materials. Such data may indicate material differences where none are expected (i.e., contaminants) and may therefore encourage additional in situ SEM or other analyses. The in situ SEM investigation is frequently supported by an attached energy dispersive x-ray spectrometer (EDX, see below), to allow the collection of x-ray spectra for elements present, or elemental mapping of the sample surface. Such combined SEM/EDX analysis is illustrated in Fig. 7. The circular area represents the pad area on a Ni substrate surface, where an attached PbSn solder ball was removed. The SE-SEM image (Fig. 7a) outlines the topographic pad area, whereas the BSE mode (Fig. 7b) identifies actual material contrasts across the pad. With the use of an EDX spectroscopy analyzer, the pad area was fingerprinted for several chemical elements. The Sn content in the PbSn solder is of the order of 3%,

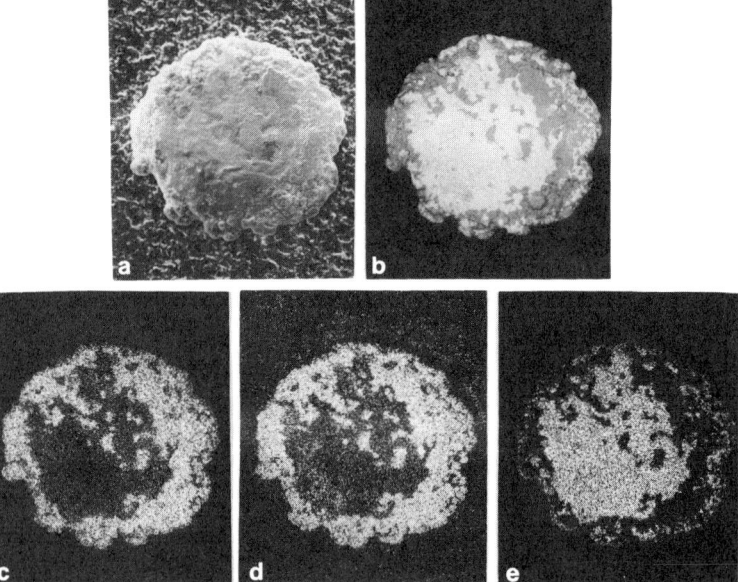

Figure 7 Surface SEM images of a solder ball pad area after the removal of the Pb/Sn solder ball. (a) Normal SEM (SE) image; (b) backscattered SEM (BSE) mode; (c, d, e) energy-dispersive x-ray images of Ni, Sn, and Pb, respectively. Magnification 400×. (Courtesy of D. L. Falcon and P. J. Brofman, IBM Microelectronics Division, Hopewell Junction, NY.)

and it interacts readily with the Ni substrates to form a Ni/Sn alloy. Ni and Sn predominate in the periphery of the solder pad area (Fig. 7c,d), whereas the Pb (Fig. 7e) is more centered and corresponds to the lighter center area of the BSE image. Additional analysis maps would indicate the location of other elements present in the sample. By such combined modes of analysis, much can be learned about the nature, content, and distribution of chemical elements in a particular sample area. The data can be of great assistance in process development, manufacturing process control, or sample failure analyses issues.

III. STRUCTURAL METHODS

Although the use of modern SEM microscopy can generate impressive physical and structural details, many development, manufacturing, or failure analysis activities require even greater resolution capabilities. These needs are dictated by the increasingly more complex material compositions and interstructural relationships of advanced electronic material products. The ability to generate unambiguous information regarding structural fidelity or defects of often very small material features or to clarify interfacial material structures is often key to a proper understanding of material shortcomings or causes of functional device failure. Such level of material detail can be realized with the use of transmission electron microscopy, an analysis tool capable of producing atomic scale resolutions. Complementing the TEM data with information derived from other techniques can result in not only structural, but also chemical, details of a particular material sample.

Another high-resolution tool is the focused-ion beam (FIB) system. The FIB technique is not so much an analytical tool as it is a means to prepare specific sample sections for analysis in conjunction with other high-resolution tools. It also allows for the ability to selectively remove or add material layers, the latter resulting in modifications (or repair) of defective device products or other material structures.

A. Transmission Electron Microscopy (TEM)

The high spatial resolution information generated with the TEM most often encompasses material morphology (i.e., porosity, voids, etc.) or crystallographic structure. With the addition of special attachments, the chemical composition of a material may also be determined. Although many of these analysis methods are key to elucidate a variety of material characteristics or problems in the various device technologies, the TEM is most extensively utilized in the generation of high spatial resolution images of the morphology or shapes of pattern features and material crystal grains. Two different tools are available: the standard TEM and the scanning TEM, or STEM. Both produce the primary image as a result of transmitted electrons (secondary for TEM, primary for STEM) passing through a thin material specimen. In a conventional TEM, electrons from a source are focused by a condenser lens, passed through the sample, and imaged onto a fluorescent screen, photographic film, or image converter plate. In a STEM, the electron beam is focused to a small diameter and rastered across the sample. Examination by either a TEM or STEM requires the material under study to be sufficiently thin so that the electron beam is not fully absorbed within the sample, but is adequately transmitted. Denser materials (i.e., metals) dictate the need for even thinner samples. Some latitude of sample thickness can be accommodated by varying the acceleration energies of the electron beam.

The outstanding feature of the TEM is its superior resolution. In fact, TEM analysis provides the highest magnification and resolution imaging capability of a material structure due to the very small wavelength of its high-energy electrons. The accelerating voltage (V_c) dictates the electron's energy and therefore the wavelength, as per Eq. (1):

$$\lambda = \frac{h}{[2mV_c e\,(1\,+\,eV_c/2mc^2)]^{1/2}} \tag{1}$$

where h is Planck's constant, m the mass of an electron, e its charge, and c the velocity of light. At 100 keV, λ is 3.7×10^{-3} nm, implying a resolution well below the size of atoms. However, this potential resolution is not fully realizable because of tool limitations and magnetic lens aberrations. The best observed point-to-point resolutions are near 0.2 nm for 300–400-kV instruments. Such resolution capabilities are favored for atomic scale interface and defect studies.

1. Basic Microscope Elements

A TEM microscope consists of an electron source and an assembly of magnetic lenses arranged in a vertical column. The electrons are produced by thermionic emission from either a W or LaB_6 filament. With the introduction of improved vacuum technologies (10^{-5} Pa), LaB_6 cathodes have become the preferred electron source, since these can generate higher emissions and exhibit better filament lifetimes. The higher emission efficiencies are essential for dedicated STEM applications, since these tools focus the scanning electron beam onto a small spot (0.5 µm diameter) of the sample, as opposed to a spot size of about 4 µm in regular TEMs. In medium-voltage microscopes, the electrons are accelerated with energies between 20 and 400 keV. High-voltage instruments accelerate them up to 1–1.25 MeV to achieve better penetration through thicker samples.

The emitted electrons are focused by a condenser lens system to produce the desired sample illumination. Those electrons that pass through the sample are focused by objective lenses to form the diffraction patterns. Additional optics transfer the diffraction patterns onto a viewing fluorescent screen, where they can be captured onto a photographic plate or monitored on a TV system.

2. Sample Considerations

A sample to be investigated with a TEM may be viewed in one of two possible orientations: (1) a horizontal or plan view, or (2) a cross-sectional view. Each requires a different preparation technique, although both involve sample thinning of the area of interest to near or submicron dimensions. The plan view sample allows the investigation of the horizontal surface region, whereas the cross-sectional view captures a cross-section through the various vertical structures or layers perpendicular to a sample's surface.

Sample Thickness. The maximum sample thickness that will allow the generation of useful quality images depends on the specific type of electron-specimen interaction and the analysis mode. Fig. 8 indicates the practical limiting thickness as a function of the accelerating electron voltage. The criterion for the limiting thickness is based on the disappearance of fringe contrasts from twin and stacking fault boundaries, using the (111) crystallographic reflections. However, for day-to-day sample analyses, the actual film thickness's are always below the limiting values. For example at a 200-keV accelerating voltage, the preferred sample should only be approximately 0.8 μm thick. If more detailed information is desired, the thickness may need to be reduced further. Still further adjustments to these thicknesses may be required to accommodate the different electron interaction efficiencies with specific types of materials (i.e., metals, insulators, crystallinity, or porosity). The useful thickness of crystalline material is of the order of 50–300 μm for 100 keV, about 1 μm for 1-MeV systems [3]. To resolve dimensions in the range of 0.5 μm with high-resolution electron microscopes (HREM), the practical sample thickness is restricted to only a few tens of nanometers. For any thicker samples, the phase-contrast image deteriorates and is swamped out by other system effects.

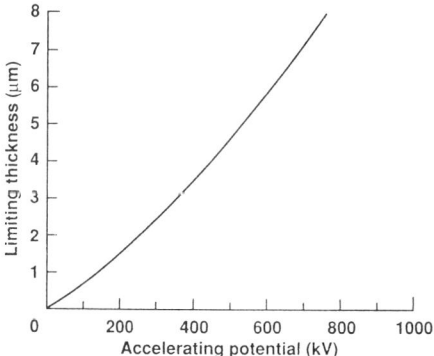

Figure 8 Limiting thickness for electron beam transmission through silicon as a function of the TEM's accelerating voltage.

Sample Preparation. The standard techniques for material specimen preparation involve either chemical thinning, electrolytical polishing, ion thinning, or some combination of these techniques. Special methods can be employed for both horizontal or cross-sectional sections to enhance certain sample features.

Plan view section samples. Fig. 9 outlines the procedure for preparing a plan view (horizontal) silicon section after the sample had been first lapped to a thickness of less than 100 µm, and after a 3.0-mm disk (with the area of interest approximately in the center) is created. That disk is mounted face down on a clear sapphire plate covered with a thin layer of wax to assure adherence. An additional thicker layer of wax covers the edges and part of the back surface, leaving a small central opening (Fig. 9a). The sample is then etched or milled to leave a shallow depression (Fig. 9b), after which the backside wax is dissolved and reapplied, this time to leave a larger area open. The etching of the sample is reinitiated and continued until a small hole forms in the initially thinned center region (Fig. 9c). This two-step etching procedure is necessary to avoid "trenching" or anomalous fast etching at the edge of the wax mask to prevent the formation of an annular hole and to enhance tapered thinning. The transparency of the sapphire aids in determining the end-point of the thinning

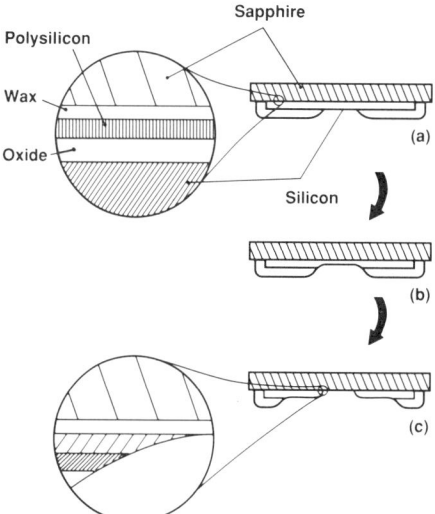

Figure 9 Schematics of plan view (horizontal) TEM sample preparation method. (a) Sample disk; (b) shallow depression formation; (c) hole formation after additional etch.

process by visual inspection. The etching sequence leaves a tapered thin material foil within the center region (see expanded view of Fig. 9c), which is adequately thin for a TEM investigation. The taper is sufficiently gentle so that the sample thickness remains transparent to 100-keV electrons over a region of about 50–100 µm away from the etched hole.

The preferred method for preparing silicon samples is by chemical thinning as opposed to the considerably slower ion milling. Specifically, a CH_3COOH:HF:HNO_3 etchant solution creates very desirable large-area thin films. In situations where a specimen sample layer would etch significantly different than silicon (such as SiO_2, aluminum, or silicides) ion milling (Ar) is exercised to achieve more equivalent material removal rates. The ion milling beam would strike the sample at an angle of 15° or less to the surface, while the sample rotates slowly about an axis normal to the sample surface.

Cross-sections. Much useful device structure information can be generated from cross-sectional (or vertical) samples to investigate structural relationships of device elements, many of which are extremely small. Most samples will therefore consist of a substrate and a stack of films or composites, as well as possibly localized different materials within the substrate (i.e., trench capacitors).

Sample preparation is somewhat more complex in comparison to horizontal sections, although the available methods have become routine and quite reliable for preparing very specific sample areas of interest. The most refined sample preparation technique utilizes polishing methods in conjunction with a tripod polishing jig [4]. A specimen sample of about 4 mm by 5 mm is cleaved out of the substrate and mounted with a thermal setting wax onto the side of a glass insert, which itself is locked into the tripod holder (see Fig. 10). This insert represents the third leg of the tripod. The other two legs can be height-adjusted with micrometer controls. The attached sample extends slightly beyond the insert edge so that it will be polished when the tripod jig is placed on the polishing wheel. Polishing progress can be monitored under a microscope and is continued until the desired area is reached. The sample is demounted and remounted onto the top of the glass insert with the initially polished surface face down. It is then placed on the polishing wheel for further thinning to realize an ultimate thickness of about 1 µm. Since the glass insert allows light transmission, the thinning sample will begin transmitting light—red in the case of about 10 µm of silicon, shifting to orange at about 7.0 µm, and becoming clear at 1.0 µm. When the polishing is completed, a 3-mm masking washer with a 1.5-2-mm narrow slotted opening is glued over the polished section and the sample is ion-milled to result in a rather uniformly thin sample foil acceptable for TEM investigations.

The advantage of the tripod polisher is that a relatively long sample can be cross-sectionally polished, thereby exposing many device structures. If the de-

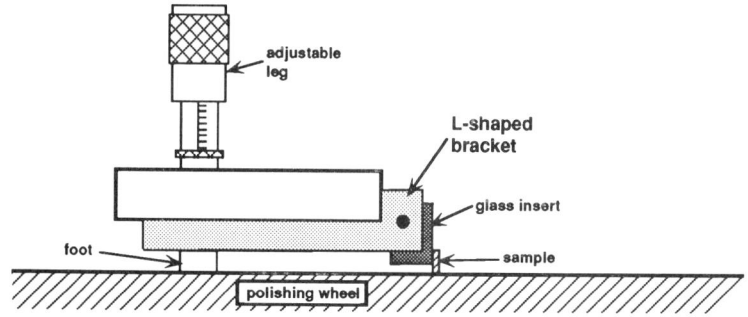

Side view

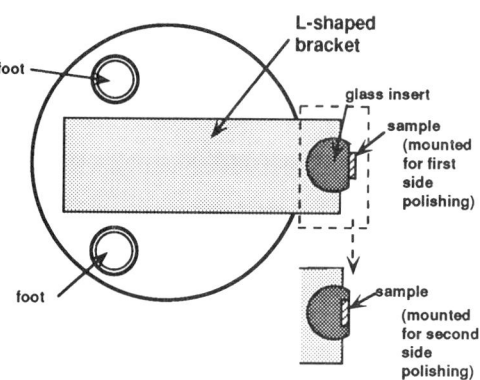

Bottom view

Figure 10 Tripod polisher jig with L-shaped bracket utilized to prepare cross-sectional TEM samples. (Courtesy of L. M. Gignac, IBM Microelectronics Division, Hopewell Junction, NY.)

vice density is high, potentially many equivalent structures can be investigated with the possibility to compare "good" device structures with "bad" ones and to understand their structural differences or the factors contributing to the inferior devices. With proper adjustments of the tripod jig's feet, wedge-shaped samples can be realized in which the films and the substrates could have different thicknesses. This can be useful in enhancing the electron transparency of more opaque samples (i.e., heavy metals). Because of its high control over polishing, the tripod technique makes it readily possible to pinpoint particular device structures that are to be investigated. The technique has greatly minimized the "hit-and-miss" statistics of older techniques and significantly reduced sample preparation times.

3. Application Studies

Due to the high-resolution capability and the refined sample preparation techniques, TEM analyses are extensively used in the process development phases to assist in fine-tuning processes and structure elements, as well as in manufacturing support applications. Most analyses are focused on characterizing internal microstructures of thin film materials, on determining the physical morphology of device structures, and on assessing the interfacial integrity between layers [5].

Fig. 11 shows an electron diffraction pattern (created with a highly focused beam) of a tungsten film. The pattern is comprised of concentric rings around

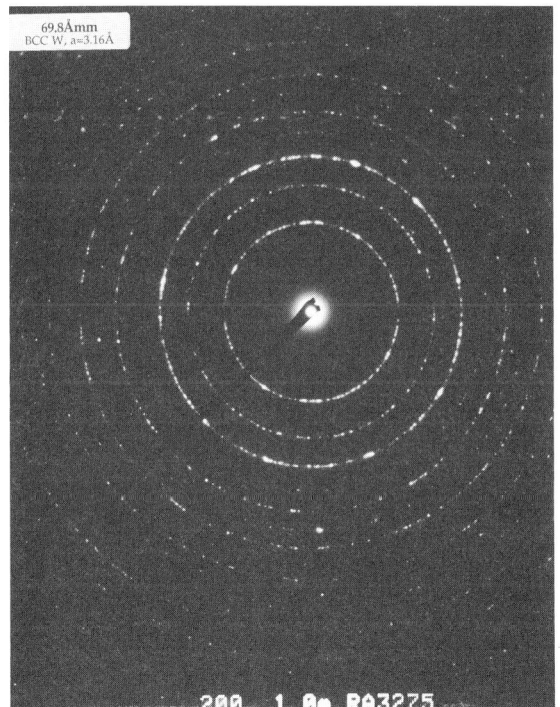

Figure 11 X-ray diffraction pattern of a tungsten film.

a single spot from undiffracted electrons. The radius of a ring is directly related to the atomic planes within the tungsten film. If the crystalline structure of a material, and the material itself, is unknown, the diffraction pattern can be indexed to determine the lattice spacings d by:

$$\lambda L = r d_{hkl,} \tag{2}$$

where λ is the wavelength of the electrons, L the camera length, and r the radius of a ring from the h, k, or l plane. Thus the d-spacing of the various planes can be derived and matched with known x-ray powder diffraction files for conclusive material identification.

A cross-sectional TEM example of a MOS logic transistor is shown in Fig. 12. Because amorphous and crystalline materials transmit electrons differently, high-contrast images can be obtained without the need for selective delineation etching techniques, as required for SEM sample preparation. The polycrystalline gate electrode texture differentiates clearly from the silicon crystal substrates, and both of them from the SiO_2 gate oxide. The device also exhibits SiO_2 spacer (or sidewall) regions, titanium silicide on top of the polycrystalline silicon gate and in the source and drain regions, as well as an overcoating of a protective SiO_2 film. The silicide serves to reduce the electrical contact resistance between the silicon regions and the contacting metal (usually W and not present in this photomicrograph) on top of these areas (see schematic in Chapter 1, Fig. 5).

A 30 times higher magnification of the gate electrode region in Fig. 12 is shown in Fig. 13. This lattice imaging TEM mode outlines the (111) orientation

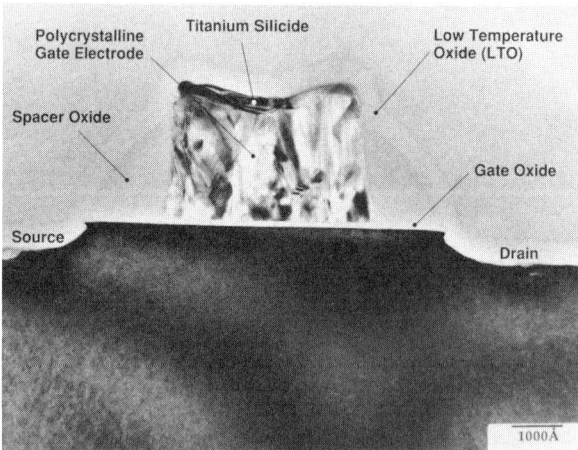

Figure 12 TEM cross-sectional view of a MOS transistor structure.

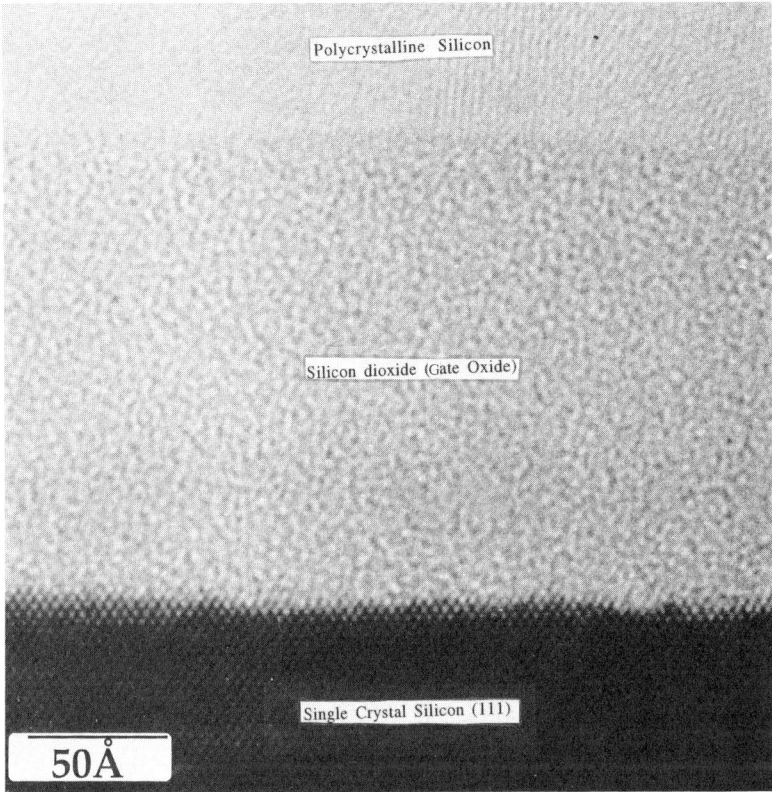

Figure 13 High-resolution lattice imaging TEM cross-sectional view of the gate oxide region of Fig. 12. Magnification approximately 7,500,000×.

of the silicon crystal at the atomic level, as well as the different crystalline orientations of the polycrystalline material above the amorphous SiO_2 gate oxide layer. The lattice spacing in the silicon substrate is approximately 2.7 Å in the vertical direction and 3.1 Å along the 45° lines of the (111) planes. The lattice image can therefore provide a very accurate "in situ" magnification calibration to establish the exact thickness of the SiO_2 gate oxide film, in this case 160 Å. The thickness of the gate oxide dictates the overall operation of this type of device and must be accurately monitored in a manufacturing environment.

Another example of a cross-sectional TEM is presented in Fig. 14. Here the polycrystalline gate and the source and drain areas are dwarfed by the metal interconnect superstructure (compare to schematic in Chapter 1, Fig. 9). The

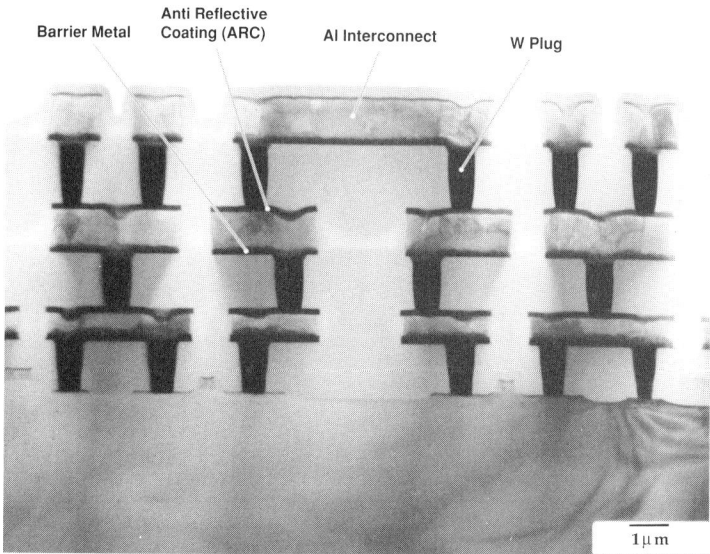

Figure 14 Cross-sectional TEM of MOS device structures with multilayered metal interconnection superstructure.

vertical W plugs serve to interconnect the various Al layers, and these to the device regions of the device.

B. Focused-Ion Beam (FIB)

More than four million transistors, resistors, capacitors, and connecting wires can be contained on a modern device chip, all within a single square centimeter of silicon, and all of which must function properly to have the chip be fully functional. Some degree of device redundancy is often designed into these high-density chips when practical to allow for the possibility of some postprocessing electrical "rerouting" around defective device elements. This is accomplished with the use of on-chip electrical fuses, which can be activated to "blow" out a set of devices. The electrical signals will then bypass these areas and flow through a redundant set of equivalent devices. This design method can only be utilized in memory device chips containing many repetitive device cells. Logic device designs must be completely defect-free and cannot make use of such a redundancy design scheme. Any attempt to exercise traditional physical methods to either repair or modify any device element in either type of device, in case of a device fault, is essentially impossible. However, the FIB system offers the opportunity to selectively modify integrated circuits by cutting or adding micron-thick wires [6].

The FIB tool generates energetic beams with the use of a liquid-metal ion source (LMIS), usually gallium. The beam consists of a high density (up to 10^6 amps/cm^2) of charged particles. Coupled with an ion optical system, the particle beam is steerable and can be focused to create atom energies from one to several hundred keVs. This beam energy flexibility permits a variety of ultraprecise tasks:

1. Micromachining—the fabrication of micron-size objects as if done by a miniaturized milling machine
2. Maskless implantation of dopants to modify semiconductor material electrical characteristics
3. Analysis of the elemental composition of surfaces
4. Nanolithography
5. Repair of optical and x-ray lithographic masks

In addition, FIB milling and cross-sectional sample imaging can provide new means of evaluating structure-to-structure overlay accuracies, oxide thickness and quality, resist profiles, aluminum grain size, and cross-sectional views of micron-size particles present on integrated circuit surface. The tool has also enabled major breakthroughs in the ability to section through submicron contacts, remove material around particles for more accurate analysis, allow precise sectioning of defect locations, and modify circuit design.

1. Basic Tool Functions

The FIB operates in a high-vacuum environment (less than 5×10^{-7} torr) to avoid particle beam interference with atmospheric gas molecules. A strong electric field applied to the LMIS (located at the top of a focusing column) extracts positively charged ions and, with a series of electrostatic lenses and octapole reflectors in the column, focuses them into a beam. With appropriate beam energies, the ions can be scanned onto specified areas of a specimen and initiate a material milling action. If a gas injection source is added to the system, appropriate metal species can be introduced from organometallic precursors [i.e., $W(CO)_6$, PtF_6], which can then be selectively deposited in the form of micron-size wires, perhaps to interconnect existing separated metal lines, or to circumvent a defective area causing a metal line discontinuity. In combination with selective milling, such defects can be first removed, followed by metal depositions. By opening vias through passivation and interlayer dielectric films, underlying conductors can be exposed and potentially interconnected, or E-beam device probing is made possible in regions not normally possible. High-resolution mask repairs with the use of carbon deposits are also performed with the FIB technique.

2. FIB Applications

The FIB tool is most often used in combination with other high-resolution instruments [7]. Fig. 15 is a cross-sectional SEM image of a localized area of a memory device exhibiting bit line failures. Sequential FIB milling was performed along the failed bit line by vertically milling away sufficiently large areas across the line, followed by SEM analysis of the cross-section of each area. The SEM image in Fig. 15 exhibits microvoids underneath the low-temperature-oxide (LTO) film, an indication of poor interfacial integrity between the film and the substrate. Such void formation is often the result of surface contaminants prior to film depositions. To pinpoint the void location by regular mechanical cross-sectioning techniques (i.e., polishing) would be quite difficult because of the small defect size and low density along the bit line, the structure's fragility, and the degree of inprecision of the polishing technique. On the other hand, the FIB method permits micrometer precision in forming a series of such cuts into the specimen, each of which can be imaged with an SEM.

The FIB method is also quite useful in failure analysis activities. Cutting through interconnection lines permits the electrical isolation of failure regions and the electrical characterization of specific structures. The SEM micrograph in Fig. 16 relates to an integrated optical disk controller device with a measured operational timing problem in the register circuit. By first disconnecting gate regions (i.e., the "cuts" in Fig. 16) via ion milling and reconnecting them with

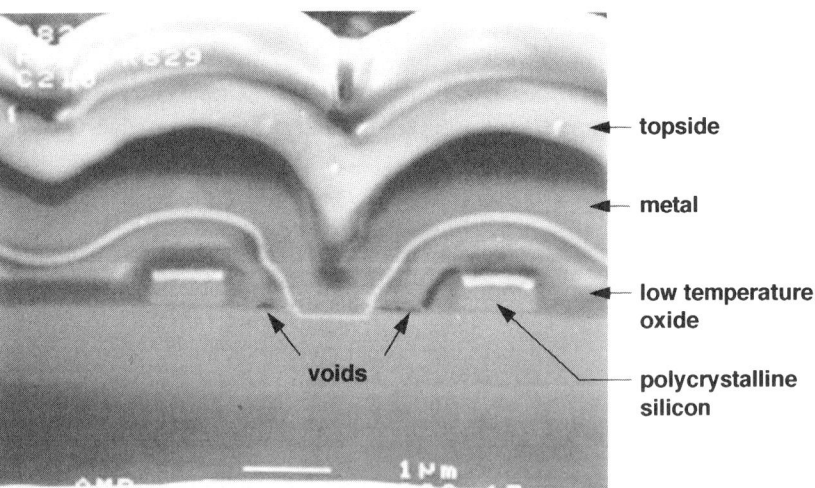

Figure 15 SEM image of a FIB-generated cross-section exhibiting voids in the low-temperature oxide film-substrate interface region.

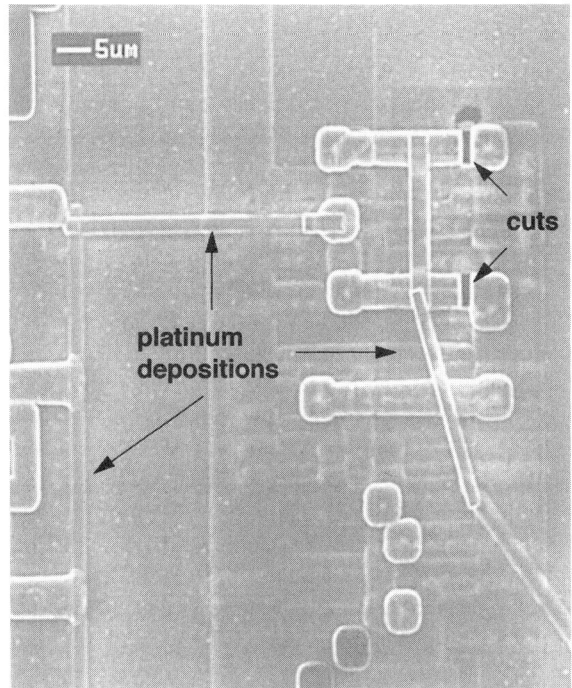

Figure 16 FIB-modified integrated circuit device using the in situ "cut and deposition" technique.

FIB platinum lines selectively deposited to another gate approximately 1 mm away (not shown in photomicrograph), the timing problem was corrected.

IV. CHEMICAL METHODS

The previously discussed optical and structural analysis methods are primarily geared to generate morphological, structural, or dimensional information. In general, they do not yield any extensive elemental analysis data of material surfaces. Yet, in many instances, the effective characterization of the chemical constitution of materials is key to successful electronic product process development, or to retrospective analyses of device failures. It is important to remember that in addition to establishing "contamination-free" rooms, tools, process methods, and materials to minimize physical particulate contamination, the absence of nonparticulate chemical contamination (i.e., organic films or ionics) is of equal importance. Chemical contaminants can greatly alter interface properties or can be the cause for initiating material corrosions. Conceptionally, the most effective means to eliminate such contaminations is to first identify their nature and composition, and then their source of origin. A wafer surface is constantly being modified by a sequential series of process steps, and hence, a variety of surface analytical techniques are required for various microscopic elemental environments. Most surface analysis techniques are complementary in terms of the information they can provide. Energy dispersive x-ray spectroscopy (EDX or EDS) is employed for analyses of all elements with atomic numbers above boron. Its depth resolution ranges from 1 to 2 µm. Auger electron spectroscopy (AES) has a lateral resolution of >150 Å and a depth resolution of a few nanometers. It can also be employed for material depth profiling. It exhibits elemental sensitivity for all elements except hydrogen. X-ray photoelectron spectroscopy (XPS) stands out for its surface sensitivity and chemical bond information. In comparison to AES, it causes less material radiation damage and less surface charging in the presence of insulator materials. XPS has been drawn in more and more for thin film thickness measurements and dry etching residue characterizations. To generate elemental and chemical composition information for specific material regions, the secondary ion mass spectroscopy (SIMS) technique is quite useful. In particular, it has become the preferred tool for impurity profiling requirements, primarily due to its unparalleled sensitivity (<ppm), high dynamic range, and superior depth resolution (<1 nm). SIMS dopant profiles of B, As, and P in silicon have become routine. The technique permits the measurement of hydrogen, oxygen, and nitrogen at trace levels (1–10 ppm).

A. Energy-Dispersive X-Ray Microanalysis (EDX)

Taken literally, microanalysis is the study of "very small" samples independent of the available technique. Historically, however, the term has had a much narrower definition. As shown in Fig. 4, electrons impinging with appropriate energies on a sample can cause x-ray emissions. The energies and their relative abundance will depend on the sample's composition. The elemental content analysis of the x-rays emitted from a microvolume (roughly one to several hundred cubic micrometers) is commonly referred to as microanalysis. To further narrow the topic, EDX concerns itself only with energy-dispersive analyses in which the emitted x-rays are sorted electronically, rather than by means of diffraction patterns. Sometimes, it is the only means of analyzing microscopic samples. EDX is sensitive to low concentrations; that is, the minimum detection limits can be as low as 0.05 atomic % in the very best situations, but more typically are about 1 atomic % of the elements' presence in a sample. The method's dynamic range covers elemental compositions up to 100%, and it maintains a relative precision of 1–5% throughout that range. Furthermore, sample preparation requirements are minimal, and the technique is almost nondestructive to the sample [8].

1. X-Ray Generation

The x-ray emission process is as follows. First, an accelerated electron, generated for example by a SEM and impinged on a material, can eject an electron from an inner shell of a material's atom. The resultant vacancy is then backfilled by an electron from a higher energy shell of that same atom. By "dropping" to a lower-energy state, this vacancy-filling electron must give up some of its energy, which is in the form of electromagnetic radiation. Since all electromagnetic radiation can be classified on the basis of its wavelength and, at the same time, can be thought of as packets of energy (called photons), the wavelength and the energy dispersive techniques can be related to the same phenomenon. This equivalence is represented in Planck's equation:

$$\lambda = \frac{hc}{E} \qquad (3)$$

where λ is the wavelength of the radiation, h Planck's constant, c the speed of light, and E the energy of the radiation. Substituting the appropriate values and rearrangement yields

$$E = \frac{12.4}{\lambda} \qquad (4)$$

The energy is measured in kiloelectron volts and the wavelength in angstroms. The emitted radiation energy is exactly equal to the energy difference between the two respective electronic levels involved in the radiation process. Since this energy difference is fairly large for inner shells, the emitted radiation appears in the form of x-rays.

However, there are many energy levels and, therefore, many potential vacancy-filling combinations within every atom. As a consequence, even a sample of pure iron will emit at several energies. The radiation energy spectra will therefore uniquely identify a material, hence the name: characteristic x-rays.

The nomenclature for an x-ray emission is defined by the shell in which the initial vacancy is created and the shell from which an electron drops to fill that vacancy. For instance, if the initial vacancy occurs in the K shell and the vacancy-filled electron drops from the adjacent shell (the L shell), a Kα x-ray is emitted. If the electron drops from the M shell—two shells away—the emitted x-ray is a Kβ x-ray. Because of the complexity of electronic structure, the nomenclature becomes more complex when the initial vacancy occurs in higher-energy shells. Microanalysts are generally concerned with K-, L-, and M-series x-rays. Therefore, the common reference to KLM lines.

In mixed samples, every element present will emit its own characteristic x-rays. Furthermore, under given analysis conditions, the number of x-rays emitted by each element is generally directly proportional to the element's concentration. Since the x-rays travel much greater distances through a material than electrons and escape from depth at which the impinging electrons have been more widely spread, x-ray signals have poorer spatial resolutions in comparison to secondary and backscattering electron signals.

2. EDX Qualitative Analyses

Converting the emitted x-rays into meaningful data is relegated to the instruments' electronics, which in the end produces a digital spectrum. Specifically, the x-ray creates a charge pulse that a detector (usually silicon) converts into a voltage pulse. The amplitude of the pulse reflects the energy of the x-ray. After an additional conversion into a digital signal, it becomes recorded as a "count" in an analyzer. The total sum of the counts at various energy levels over a particular time period is represented by an x-ray spectrum, such as the one shown in Fig. 17. In this case, the dominant features include a broad low-level background and several major spectral peaks. Those labeled "Copper" and "Zinc"—the elemental composition of brass—are the respective Kα x-rays. These are accompanied by the two-lower intensity peaks at higher energies and are due to the respective Kβ emissions.

3. Data Display Options

Frequently, the analyst is familiar with the sample's composition, but is more interested in the absence or presence of particular elements in specific areas or over a general region. Such data are shown in Fig. 18, which is a silicon substrate covered with a network of aluminum lines. The SEM photomicrograph

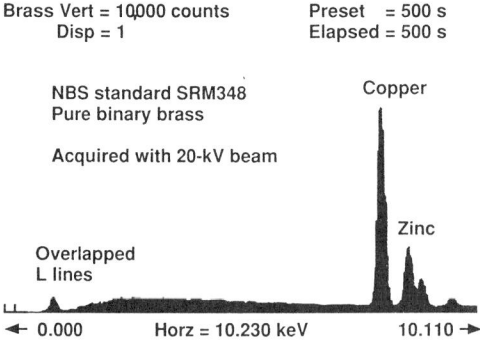

Figure 17 EDX spectrum of a brass sample acquired with a 20-keV primary electron beam. Sampling time 500 s.

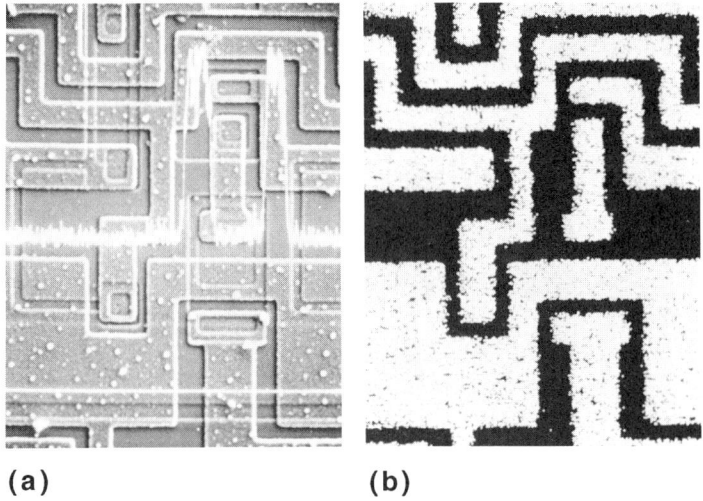

Figure 18 Silicon line EDX profile (a) and aluminum dot map (b) of an aluminum metallized integrated circuit.

(SE mode, see above) indicates the general surface structure of the sample. The light dots are due to hillocks on the surface of the aluminum layer. The superimposed signal line with its two peaks is generated by specifically monitoring the x-ray signal of silicon as the primary e-beam is scanned across the sample in the lower portion of the photomicrograph (indicated by the straight white line). Thus, as the scan passes over the speckled aluminum surface, the Si signal remains low; however, it increases over the two "clean" areas, which represent exposed silicon surfaces. With the x-ray monitoring set to read the Al signal, and mapping the whole area of the very same sample, the data in Fig. 18b are generated to outline the aluminum areas. The dark speckling in the Al regions results from the Al signal recording in each position. With longer monitoring times, these regions would become a uniform white.

B. Auger Electron Spectroscopy (AES)

As illustrated in Fig. 4, the energetic primary electron impinging onto a material can result in a number of secondary energy emissions, Auger electrons being one of them. The emission of Auger is closely coupled to photon emissions (i.e., x-rays); in fact, the combined probability of x-ray and Auger emission is unity. Since the energy range of the Auger electrons is between 50 and 2000 eV, their escape from a material is limited to the top few nanometers. Hence, detectable Auger spectra truly represent data from a material surface. The sensitivity of detecting Auger electrons spans the full range of elements from the periodic table (except H_2), as well as from very small sample volumes, making the technique very useful for elemental surface mapping. Film thickness as low as 10% of a monolayer can be detected.

1. Basic Principle

The alternative to the photon energy emission in the form of x-rays (due to vacant orbital filling by higher orbital electrons) is the initial radiationless transfer of that energy to another electron in the same orbit. This energized electron can be subsequently emitted as the so-called Auger electron. The Auger electron energy is characterized by

$$E_K = E_{L_1} - E_{L_2} - \phi \tag{5}$$

where E_i are the respective energies of an electron in the K, L_1, or L_2 orbits of a material, and ϕ a work function of the material. The Auger transition in this case is termed KLL. Certain transitions are more favored than others for specific elements. The best detectable sensitivity of elements with atomic numbers < 15 are the KLL transitions; the LMM transitions are favored for those ranging between 15 and 40; and for those above 40, the Auger signals are best monitored by the MNN transitions. Some shift in the energy signal can also arise from the influence of the chemical environment on the binding energies of the core and valence electrons, resulting in a "chemical shift" in the Auger spectra. These shifts can be utilized in assessing the chemical state of a material [9], although other techniques (e.g., XPS) may be more useful in that regard, provided sample size is not a concern.

2. Analysis Methods

The AES data are typically generated in one of two formats: (1) Auger signals intensity as a function of Auger energies, or (2) specific Auger peak monitoring as a function of material depth. The former is useful in identifying all potential elements present in the sample, and an evaluation of individual peak intensities will yield quantitative data, provided appropriate calibrations are included to accommodate a host of sample "matrix effects."

Fig. 19 is a typical energy-based Auger profile, in this case of a titanium silicide film. The spectrum is the derivative of the initial energy signals, which creates a sensitive means to pinpoint the Auger energy location. It also separates out most of the undesirable secondary electron signal background. This particular sample indicates some level of Ar (due to surface sputter cleaning) and O_2 contamination.

The Auger spectra of TiN are complicated by the superpositioning of the TiN peak at 385 eV over the N_2 peak. However, as the inserted spectra in Fig.

20 indicate, TiN has a second peak at 420 eV, which overlaps a similar one for pure Ti. Since the relationship between the insensity ratio of I_{385}/I_{420} is proportional to the N/Ti content of TiN, such calibration curves are useful when the nitrogen content in day-to-day TiN samples must be established. Depth profiling of such samples as a function of the intensity peak ratio can readily identify

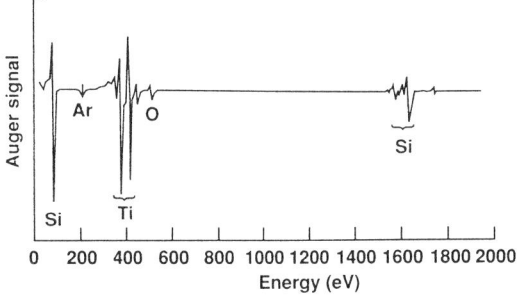

Figure 19 Auger line spectrum of a titanium silicide sample.

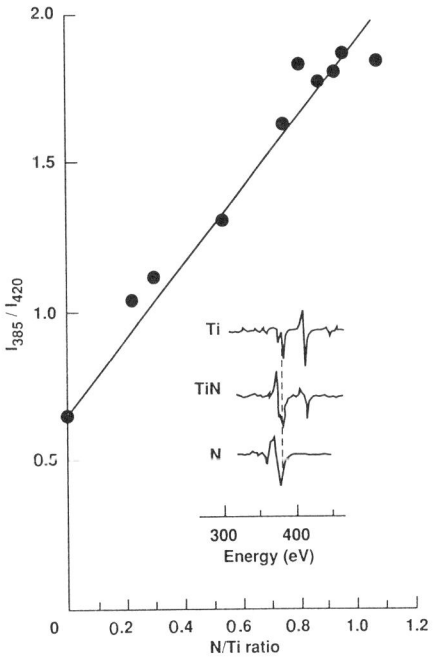

Figure 20 Auger calibration curve for titanium nitride compositions.

potential compositional variations of sputtered TiN within the film. It may also yield valuable information regarding the composition at the interface regions.

3. Material and Process Considerations

Since the Auger spectra is induced by energetic electron impingement of the sample's surface, precautions must be taken to assure that the incident electrons do not modify the surface of the material and thereby create "false" signals. This can occur via beam concentration or excessive beam energy. For instance, a primary beam current of 2 nanoamp focused into a 100-nm spot size will deposit a current density of about 20 amp/cm^2. Insulators cannot withstand such intense bombardment. As a consequence, such surfaces may undergo material decomposition, desorption, segregation, diffusion, and even melting of the very surface to be analyzed [9,10]. Defocusing of the beam can alleviate these effects, but with the penalty of lower spatial resolution. Electron-beam-induced effects are particularly dominant for fluorine-containing polymeric films. Fig. 21 indicates a very rapid fluorine desorption for a highly focused beam. The irradiated spotsize must be expanded to a 500 × 500 μm^2 area for a beam current of 250 nanoamp to minimize the fluorine desorption. Similar observations are experienced with phosphorous desorption from BPSG glasses. Even thermally grown SiO_2 is not stable under certain intense radiations.

Some degree of beam concentration may nevertheless be desirable for such insulating materials to provide a possible conductive path for any charge buildup to underlying materials exhibiting better electrical conductance. This approach is not routinely practical, and reliable spectra can often only be acquired after much time-consuming adjustment of analysis parameters [10]. In the case of conductive materials, AES analyses are relatively routine.

Beam artifacts can also be present in depth profiling. The technique involves controlled material removal by ion sputtering (normally Ar$^+$ ions), followed by

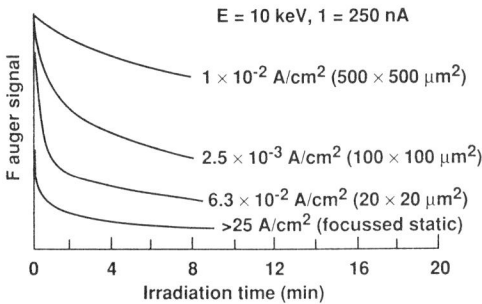

Figure 21 AES analysis of fluorine desorption as stimulated by the electron beam density.

elemental monitoring of preselected Auger peaks [11]. To maintain the Auger sensitivity with depth, the Ar+ ion sputtering is performed over an area several mm^2 in size, while the surface analysis is done only in the central portion of the large crater. Again, the data may still be sensitized by the condition of the primary beam energy. For example, in the case of a thin silicon nitride/silicon dioxide film sandwich on top of a silicon substrate, the monitored nitrogen Auger peak is steady during the sputtering away of the silicon nitride layer (as expected), yet it does not fully disappear during the SiO$_2$ film sputtering phase (Fig. 22a). In fact, as the silicon dioxide/silicon substrate is reached, its concentration tends to actually increase (arrow). The observations can be traced to effects of the primary electron beam, since the nitrogen presence in the silicon dioxide layer and the apparent accumulation at the silicon substrate surface will vary with the beam current density (Fig. 22b). The nitrogen is completely absent if the Si$_3$N$_4$ film is first fully removed by sputtering in the absence of any electron radiation (Fig. 22c). Thus, analysis conditions must be carefully defined to avoid such false readings. This is most reliably done by generating corroborating data with other analysis tools.

To assure accurate analyses in depth-profiling situations, an understanding of the material-sputtering sensitivity is of critical importance. Respective elements in compound materials may not be sputtered off during the ion bombardment with equal probabilities ("preferential sputtering"), and hence, the surface can be depleted of the species with a higher sputtering yield, leaving the surface with a disproportionate concentration of the less sputtered species. Analyses involving surface sputtering are therefore best done with the aid of reference sample analyzed under identical conditions.

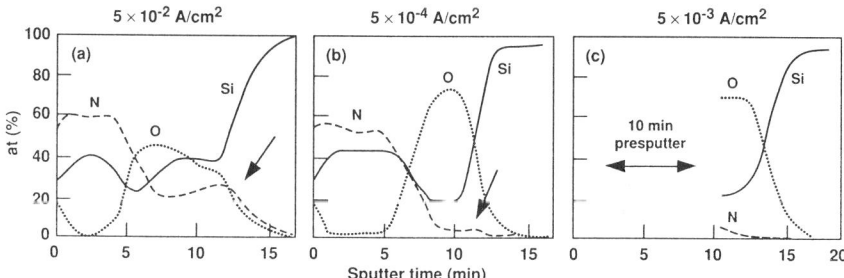

Figure 22 Auger depth profiles of a silicon nitride/silicon dioxide film composite on a silicon substrate indicating electron-beam-induced nitrogen pileup. The atomic % of nitrogen remains present into the SiO$_2$ regions and indicates a peak at the SiO$_2$/Si interface for high-beam currents (a) and is fully eliminated only with a removal of the Si$_3$N$_4$ film prior to electron beam exposure (b)(c).

C. X-Ray Photoelectron Spectroscopy (XPS)

The incident electron beams utilized for EDX, Auger, and SEM analyses rely on electron energies in the range of 10 keV to induce x-ray or secondary electron generation. Another mode of material excitation can be achieved with "soft" x-rays whose energies are only about 1.2–1.5 keV. Such irradiation induces less damage to the material, causes less charging, yet initiates the emission of photoelectrons. X-ray photoelectron spectroscopy (XPS, also termed ESCA: electron spectroscopy for chemical analysis) can provide extensive information regarding the nature of chemical bonding of a material surface [12]. The technique is very much a surface analysis method, monitoring at best only the top 12 nm. The most useful XPS data tend to be realized, however, from only the first 5 nm. In comparison to Auger, the XPS technique is superior for analyzing insulator films, but is generally inferior for depth profiling due to its larger spot size requirements.

1. Photoelectron Energies

The energies of the photoelectrons are defined by the incident x-ray energies and the specific binding energies of an electron in an atom's orbital shell. These are interrelated according to:

$$E = hV - E_B(i) - \phi, \tag{6}$$

where E is the emitted photoelectron energy, hV the incident x-ray energy (1253.6 eV for Mg Kα, 1486.6 eV for Al Kα), $E_B(i)$ the binding energy of the electron in the ith orbital of a material, and ϕ a work function term. Each element will emit specific photoelectron energies, which in essence allows "fingerprinting" of a material (with appropriate correction factors) against published values. Due to the energy-specific character of each orbital electron of an element, XPS can distinguish between different materials containing similar elements.

2. XPS Data

The binding energies of electrons for a particular element will vary with its molecular environment. Hence, chemical shifts from its elemental reference can be observed for different crystalline states, oxidation states, or compound compositions. A representative XPS survey scan spectrum is shown in Fig. 23 for a silicon surface with a (100) oriented substrate and covered by a 1.5-nm-thick native SiO_2 film. The observed carbon signal represents surface contamination from the ambient environment. (The spectrum also includes an oxygen Auger signal at about 950 eV. Auger signals are an inherent "by-product" of the XPS excitation process.) The different silicon signal peaks relate to the atomic function of the detected elements. The Si_{2p} peak and its adjacent secondary signals are expanded in Fig. 23b. The peak at about 103 eV is due to the SiO_2 film,

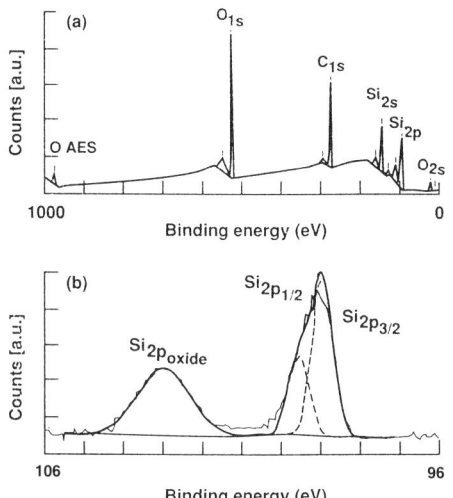

Figure 23 XPS spectra of a Si(100) surface. (a) Full line spectrum; (b) expanded details of Si_{2p} peak structure.

whereas the combined peak at about 99 eV consist of two separate peaks: silicon $Si_{2p1/2}$ and $Si_{2p3/2}$, which relate to the respective spin quantum numbers of the electrons.

XPS data become quite useful in monitoring the effects of surface-cleaning processes and potential film residue buildup, once the native silicon dioxide surface film has been etched off. In the case of a plasma etchant, such as $CF_4/CHF_3/O_2$, various fluorine-containing carbon residues can accumulate on the silicon surface if inadequate etching conditions are utilized. This is shown in Fig. 24. The C_{1s} peak in the upper curve is expanded in the lower spectrum to show the etchant residues to consist of CF_x (x = 1, 2, 3) and $-C-CF_x$ groups, in addition to C-H, C-C, and Si-C bonded fragments. The residue composition is predominantly a Teflon-like fluorocarbon polymer. Its thickness can be estimated from the XPS Si signal attenuation, and in this instance, it is about 4 nm. This fluorocarbon film formation is decisively influenced by the F/C ratio in the plasma. Surface etching occurs with a high F/C ratio, while depositions are observed with a low ratio. The addition of H_2 or CHF_3 enhances the formation of HF, thereby increasing the C concentration in the plasma. With O_2 additions, C is converted to CO and CO_2 to lower its plasma concentration to increase the F/C plasma ratio and the silicon etch rates.

The chemical shift of the spectra, due to its chemical composition, can be useful in the identification of organic films. Carbon, in various bonding configurations with hydrogen, oxygen, and fluorine, is shown in the polymer spectra

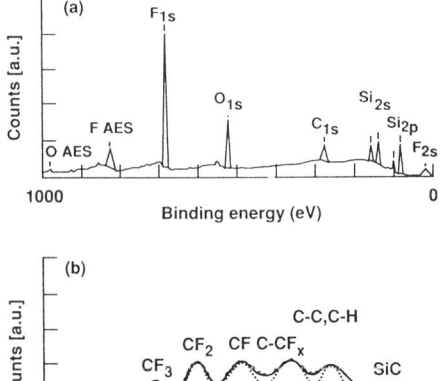

Figure 24 XPS spectra of a plasma-etched silicon surface with a $CF_4/CHF_3/O_2$ plasma. (a) Full line spectrum; (b) expanded details of the C_{1s} peak area.

of Fig. 25. With increased electronegativity, the binding energy shifts to higher levels. Thus energy of the aliphatic carbon-carbon bonds of polyethylene at 285 eV (trace 1) shifts to 286 eV with the addition of a single oxygen and hydrogen, and to 289 eV for the COO group in polyethylene teraphthalate (trace 2). Fluorinated polymers shift the spectra peaks to 291 eV and 292 eV for single or double CF_2 groups, respectively.

D. Secondary Ion Mass Spectroscopy (SIMS)

The various functional electrical characteristics of electronic devices are established through the intentional introduction of specific impurities into selected device regions. Not only must these impurities be present at specific concentration levels, but they also need to require predetermined impurity profiles, which may range from 10^{14} to 10^{21} atoms/cm^3 (or 10^{-15} to 10^{-19} g). Means to precisely verify such profiles over the full concentration range, especially within very shallow regions, are quite limited.

The most accurate method is based on the controlled removal of the material surface in conjunction with monitoring of the ionized atoms generated by the process. This secondary ion mass spectroscopy (SIMS) technique has an atomic sensitivity range for a number of elements reaching down as low as 10^{-14} atoms/cm^3. The mass spectrometer analysis provides elemental coverage from hy-

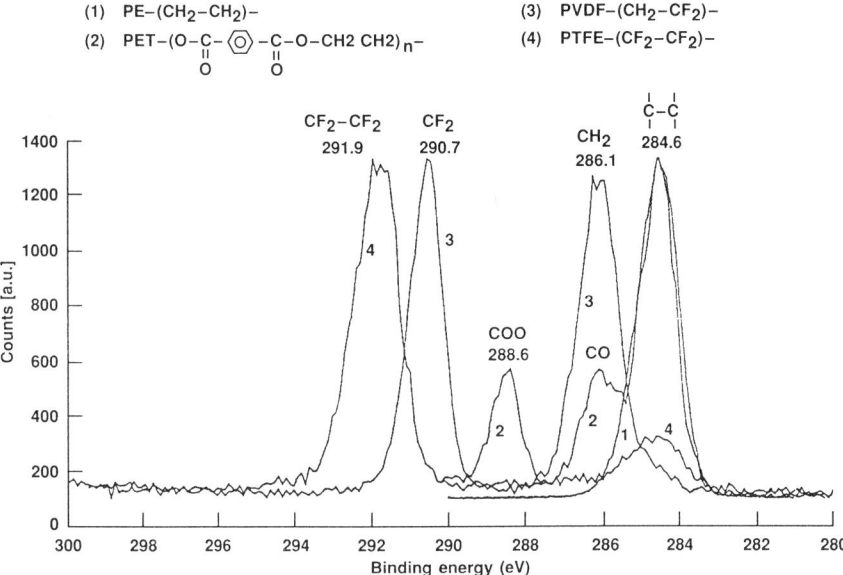

Figure 25 High-resolution XPS spectra of the C_{1s} region for different H-, O-, and F-containing polymers indicating carbon "chemical shifts" due to the nature of different carbon bondings.

drogen to uranium, although not to the same degree of sensitivity for each element.

1. SIMS Operational Principles

A focused energetic primary ion beam (1.0–20 keV) is directed onto a material and absorbed in the upper layers of that material. During the penetration, the ion transfers energy to the material atoms. The atoms at the surface will absorb sufficient energy to detach from the material matrix and be ejected. Sputtered-off atoms consist predominantly of neutral atoms or molecules. However, a small fraction of them are positively or negatively charged ions and can therefore be analyzed by a mass spectrometer. Several incident primary ion beams are utilized. Nonreactive ion beams (i.e., Ar^+) are employed for chemical information, such as for pure metals. The addition of oxygen greatly enhances the production of positive secondary ions, whereas cesium increases the yield of negative ions. These different primary beams are selectively used to optimize the analytical sensitivities.

The detection limits of SIMS depend on several parameters.

1. The volume of material and, hence, the number of host/matrix atoms removed for each data point
2. The concentration of the element or elements of interest
3. The probability that a given atom of interest will be ionized upon sputtering ("the secondary ion yield")
4. The percentage of the emitted secondary ions reaching the detector (the spectrometer extraction efficiency and transmission)

The ratio of secondary ion yields relative to the primary ion beam can be of the order of 10^{-4} to 10^{-1}, while the transmissions efficiency of modern ion probes ranges from 5 to 40%. Under these conditions, an analytical volume contained in a 150-µm-diameter and 100-Å-thick sample will permit the detection of 10^{14} atoms/cm^3 for elements such as B, In, Cr, Mn, As, and the alkali metals. The detection limit of contaminants such as H_2, C, O_2, and N_2 is influenced by the quality of the instrument's residual vacuum. Measurable levels of 10^{16} to 10^{17} atoms/cm^3 are attainable with ultra-high-vacuum conditions [13]. Depth analysis (profiling) is provided when the analytical volume progresses with the continuous erosion of material due to the sputtering process. The technique can obtain a complete mass spectrum for every 50–2500 Å, with the specific sensitivity being dictated in part by the operating conditions of the tool. Alternatively, selective monitoring of one or more specific elements can yield the corresponding concentration profiles. Typical sampling spot sizes range from 25 µm to 50 µm.

2. SIMS Analyses

SIMS impurity profiles are shown in Fig. 26 for the case of two different arsenic implants into a silicon substrate. The precalculated impurity peaks of these implants were intended to be near 5×10^{19} atoms/cm^3 and 1×10^{17} atoms/cm^3, respectively. The actual As detection is done by monitoring the $^{75}As^{28}Si^-$ ions rather than the simple $^{75}As^-$ ions, since due to a lower electron affinity (EA) of negatively ionized arsenic in comparison to that of silicon (viz., $EA_{As} - 0.8$ eV, $EA_{Si} = 1.4$ eV), the secondary ion yield for $^{75}As^{28}Si^-$ ions is significantly higher. Hence, the $^{75}As^{28}Si^-$ signal sensitivity is increased by as much as a factor of 3. The upper concentration profile in Fig. 26 maintains a Gaussian distribution down to about 5×10^{17} atoms/cm^3 and levels off at 5×10^{16} atoms/cm^3. In the second (lower) concentration profile, the leveling off occurs at the arsenic background level within the sample of 3×10^{15} atoms/cm^3. The relatively deep tailing off of these profiles into the silicon substrate is related to ion implantation channeling effects into the crystalline silicon matrix.

The SIMS technique can also be used to monitor changes of an impurity profile within a sample and assist in the understanding of anomalous electrical effects in device structures. A case in point is the migration of fluorine atoms

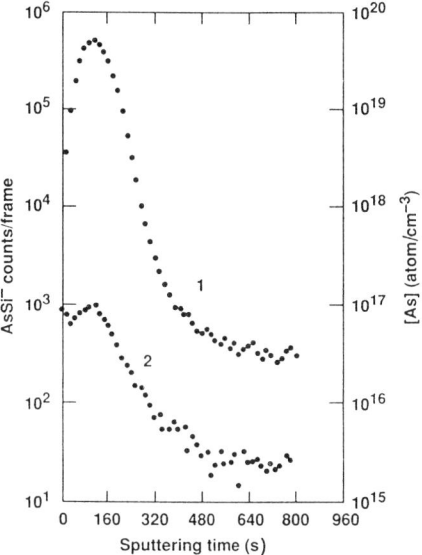

Figure 26 SIMS depth profile analyses of arsenic implants into silicon at (1) 5×10^{14} atoms/cm^2 and (2) 1×10^{12} atoms/cm^2 fluences.

in boron-implanted silicon structures during annealing conditions. The implant process results in the formation of an amorphous region in the silicon substrate. An anneal cycle at 550°C electrically activates the implanted boron atoms via an epitaxial recrystallization. However, a damaged region (due to the implantation process) near the crystalline-amorphous interface remains and appears to act as a gettering sink for fluorine, which is a constituent of the BF$_2$+ implant species precursor. The progression of the fluorine peak formation under different anneal conditions is shown in Fig. 27. In practical terms, the peak location has been used as a sensitive marker to define the extent of the lattice recrystallization under various implantation and annealing conditions.

V. DEVICE FAILURE ANALYSIS METHODOLOGY

The chemistry and physics of microelectronic device reliability were discussed in Chapter 2, Section VI, including some of the material-oriented causes for possible device failures. Chapter 8 touched on the electrochemical forces that can result in material anomalies and undesirable structural or chemical effects. Despite continuous scrutiny and frequent monitoring of all manufacturing processes and the maintenance of extreme process discipline, both particle and chem-

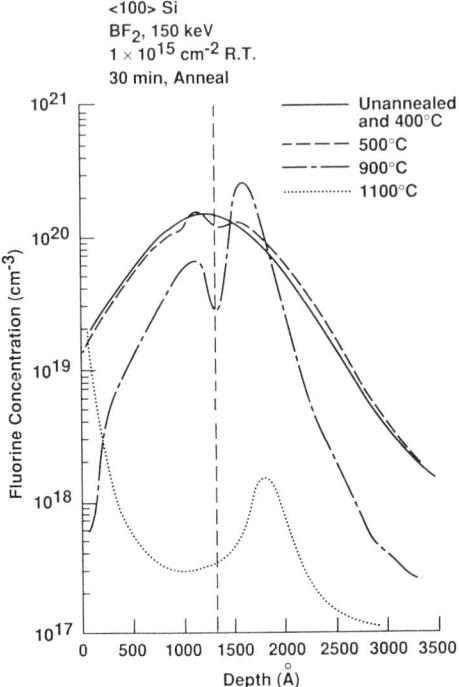

Figure 27 SIMS depth profiles of fluorine for a boron implant into a silicon substrate covered with a 1300-Å-thick SiO_2 film. Heat anneal cycling creates a fluorine peak due to the gettering effect at the amorphous/crystalline silicon interface region.

ical contaminations do at times creep into a device chip structure and instigate a device failure. The challenge of a postmortem analysis of such a chip is then focused on finding that possible one random and isolated defect or contaminant, so as to hopefully identify the source and nature of the defects and to initiate the necessary action to prevent its recurrence in subsequently manufactured products. These efforts most often fall on the shoulders of the material microanalysts, who will rely on the full arsenal of analytical tools for that investigation. To be efficient in this endeavor, such device failure analyses are organized around a disciplined sequence, commencing with optical observations and progressing to structural and microanalytical techniques. More often than not, several tools are applied to complement one another and strengthen specific analysis results. A general outline of a failure analysis flow is shown in Fig. 28.

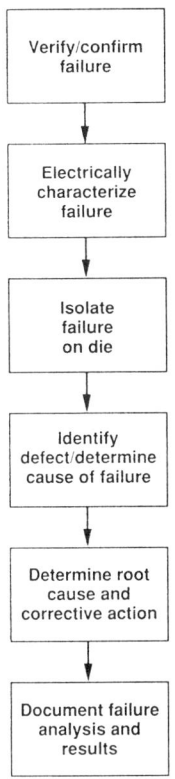

Figure 28 Key analysis steps generally utilized for device failure analysis techniques.

A. General Considerations

The first step in the investigation of a failed device is to electrically confirm the failure with the use of production tests and test equipment. For all products, these include tests for shorts, opens, leakage, basic function, and parametrics. The generated datalog provides information on where in the test program the part failed (e.g., which vector and how far into the vector the failure is detected, and which pins failed, and under what test conditions).

If the part fails for opens/shorts or leakage, the pins are analyzed on a curve tracer. For all other types of failure, the device is further analyzed on a general-purpose functional tester or a special engineering test system, to more specifically locate the failure on the die. This is called fault isolation or fault location. For memory products, this is usually straightforward, requiring relatively simple equipment and software algorithms that raster the memory array. For logic products, such as the microprocessor, fault isolation can pose a significant challenge.

The electrical fault isolation is typically followed by visual examinations of the chip. However, the chip is normally hermetically sealed in some form to isolate it from the ambient environment of a package. This package must be opened without damaging the functionality of the circuit or changing the failure mode of the device, and this is done with a variety of techniques. For plastic packages, jet or spray etchers are primarily used to remove the plastic materials. In the case of ceramic-based packages, the two pieces of ceramic material sealed together with glass need to be separated, either by cracking the glass seal between specifically shaped vise blades or by sawing through the seal with a diamond saw. Sidebrazed or laminated ceramic packages can be opened by removing the hermetically sealed gold-plated lid and by heating the package to about 400°C to melt the seal, or by using a sharp blade to pry the lid apart from the package.

Once the package is delidded, the chip can be further analyzed, either electrically or physically. In general, gross leakages measured on the external package pins may be due to an electrostatic discharge, electrical overstress, or various types of contamination. These leakages can be visualized with a night vision technology-like emission microscope, i.e., one that senses heat. It can therefore localize the electrical abnormality and indicate the area of the device that should be further investigated and physically unlayered.

1. Physical Unlayering

To effectively apply many of the various failure analysis methods and to identify the specific root cause failure mechanism, it is frequently necessary to first expose the region of interest by the selective removal of overlying layers, or by precisely sectioning the device. Conventionally, this preparation is achieved through a combination of metallographic sample preparation and chemical stripback techniques. At times, it is even desirable to modify circuits on a microscale as an aid to failure analysis, or to verify that a particular design modification may fix a given failure mode. Chemically selective reactive ion etching (RIE) is the preferred technique for passivation layer removal, for example using a low-pressure gas mixture of carbon tetrafluoride and oxygen. The RIE technique is also useful to "decorate" various metal defects (i.e., hillocks, voids, precipitates), and for uncovering particles to allow EDX analyses. An alternate material-stripping technique involves the use of FIB, which, as discussed above, permits very precise material removal in submicron regions or around particles, or high-fidelity sectioning.

The sample of interest is now ready to be further analyzed with the various optical, physical, and chemical analysis tools as outlined in the previous sections, in addition to possibly others. (For example, material stress issues in device structures can be assessed with Raman spectroscopy.) The sequence of an evaluation is dictated more or less by the particular problem to be investigated; however, the nondestructive methods tend to be applied first, followed by destructive techniques, if necessary. Furthermore, complementary or corroborating data with various techniques are normally obtained to arrive at unambiguous results. To that end, certain tools are equipped with several different detectors, allowing the analysis of a very specific area by different methods. SEMs typically are equipped with EDX detectors for chemical microanalysis by x-ray spectroscopy. These may then be followed by Auger and XPS profiling and TEM microstructure and morphology analyses. Many more extensive details of failure analysis techniques can be found in the literature [14].

VI. SUMMARY

The ability to analyze partial atomic layers, submicron material areas, and complex multicomponent structures by means of a host of highly sophisticated optical, chemical, and physical tools is the basis for much of the dramatic progress realized by the electronic industry since the 1950s. Most of the sophisticated tools applied to material problems of the 1990s did not exist two or three decades earlier. And the ongoing rate of progress will continue to place a high demand on even more improved capabilities and resolutions. Inevitably, additional new methods and new tools will be required. These will be essential in the support of new development efforts, while simultaneously, the very same tool set will be as essential for efficient and effective failure analysis efforts in support to pinpoint product deficiencies and, in turn, process deficiencies in the manufacturing environment. Failure analysis constitutes a critical link leading to improved product reliability of sophisticated electronic products.

EXERCISES

1. A new semiconductor company has designed an advanced process flow utilizing tungsten lines as interconnects. A SEM image taken after final processing (with passivation/topside) at 5 keV shows only a smooth undulating surface. The same sample imaged at 30 keV exhibits bright stripes that look a lot like the pattern of the final metal layer. Explain why this happens.
2. Both SEM and TEM can be used to measure grain size. Explain the difference between grain size and grain size distribution. Suggest a method for measuring grain size distribution from a TEM image.
3. A process engineer requests a BPSG thickness measurement by TEM. The BPSG is deposited on thermal oxide. Although this might be possible, the

contrast between the layers would not be very strong. Why? Hint—what crystal structure is BPSG?
4. A polysilicon deposition engineer requests a measurement of the depth of amorphization of the polycrystalline region, which has been implanted with arsenic. Explain why TEM is a good method to make this measurement.
5. BPO_4 crystals are notorious for forming in BPSG films. Explain why TEM is a good method to conclusively image and identify these crystals.
6. A customer requests an EDX analysis of an SiO_2 film. An experienced microanalyst performs the analysis of the sample in which the spectrum shows sodium contamination. Another EDX spectrum was run under high-beam current. The customer insists that sodium cannot be in the sample, because the manufacturing line would have been shut down. What happened?
7. Detecting boron in BPSG films is tricky by EDX when it is present in low atomic concentrations (<4 wt%). Since the boron's x-ray also has a very low energy (183 eV), it can only travel about 1000 Å in oxide films before being completely absorbed. Explain why EDX is not a good method for determining total boron concentration.
8. An investigator is examining a thin film of aluminum in an Auger instrument, which is also provided with an x-ray spectrometer. The incident 10-keV electron beam produces an Auger spectrum containing a strong signal from Al and from oxygen. The x-ray detector detects only Al, even though it is capable of detecting the oxygen signal as well. Explain this observation.
9. The analysis of TiN is difficult by EDX, because the titanium Lα line overlaps with the nitrogen Kα line. What benefit could be derived from the use of the XPS technique?

REFERENCES

1. M. Pluta, *Advanced Light Microscopy*, Elsevier, Amsterdam, 1988.
2. D. Newbury, D. C. Joy, P. Echlin, C. E. Fiori, and J. I. Goldstein, *Advanced Scanning Electron Microscopy and X-Ray Microscopy*, Plenum Press, New York, 1986.
3. R. M. Anderson, ed., *Specimen Preparation for Transmission Electron Microscopy II*, Vol. 199 in MRS symposium proceeding series, 1990.
4. L. M. Gignac and A. L. Edel, 17th International Symposium for Testing and Failure Analysis, Los Angeles, CA, November 1991, p. 41.
5. L. Reimer, *Transmission Electron Microscopy: Physics of Image Formation and Microanalysis*, Springer-Verlag, Berlin, 1983.
6. K. Nikawa, K. Nasu, M. Murase, S. Inoue, T. Kaito, and T. Adachi, New application of focused ion beam technique to failure analysis and process monitoring of VLSI, Proceedings of the 1989 International Reliability Physics Symposium, 1989, pp. 43–52.

7. H. Iwasaki, Y. Ikku, T. Kaito, and T. Adachi, Combined analysis of Cross-sections using an integrated FIB-EF-EDS instrument, Proceedings of the 40th Applied Physics Soc. Jpn. Spring Meeting, 1993, pp. 569.
8. R. Woldseth, *X-Ray Energy spectrometry*, Kevex Corporation, San Carlos, CA 1973.
9. M. P. Seah, *Methods of Surface Analysis* (J. M. Walls, ed.), Cambridge University Press, Cambridge, 1989.
10. D. Briggs and M. P. Seah, *Practical Surface Analysis*, Vol. 1, Wiley, New York, 1990.
11. Y. E. Strausser, D. Franklin, and P. Courtney, *Thin Solids Films*, 1981, pp. 84, 145.
12. D. T. Clar, *Handbook of X-Ray and Ultraviolet Photoelectron Spectroscopy* (E. D. Briggs, ed.), Hayden, London, 1977.
13. W. Katz and J. G. Newman. Review the fundamental of SIMS, *MRS Bull.* 12 (6): 40 (1987).
14. A. H. Landzberg, ed., *Microelectronics Manufacturing Diagnostics Handbook*, Van Nostrand Reinhold, New York, 1991.

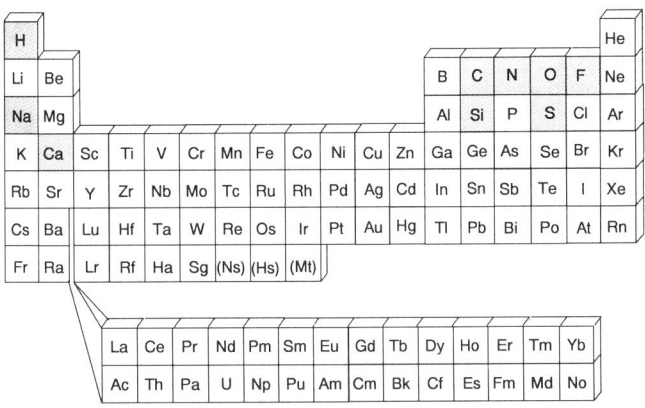

10
Polyimides as Dielectrics and Optical Interconnects

Tohru Matsuura *Nippon Telegraph and Telephone Corporation, Musashino-shi, Tokyo, Japan*

I. INTRODUCTION

A large family of organic polymers has potential applications as dielectric materials in microelectronics and as optical interconnects in optoelectronic devices, or even as conductive polymers. Their attractiveness stems primarily from their favorable electrical properties (i.e., high resistivity or good conductivity, low dielectric constant), their mechanical properties (i.e., lightweight, flexibility), and because of their ease of processing. Unfortunately, however, many of these polymer candidates are marginal with respect to one additional property fundamental for most of the foregoing device applications—thermal stability above 300°C.

This thermal requirement is driven by the need for the polymer material to survive wire bonding and metal deposition processes, which are integral to the fabrication of electronic devices. As indicated in previous chapters, many efforts are made to perfect lower-temperature processes and to achieve advanced material and process integrations desirable for continuing advancement in electronic device fabrications. Much progress has been made; however, the efforts have not generally resulted in process temperatures at which many organic polymers could become useful. A more specialized class of organic polymers—polyimides—have many of the above mentioned desirable physical and electrical properties, while also exhibiting thermal stabilities to above 450°C.

Microelectronic device fabrication commenced with the fabrication of discrete devices, followed by integrated devices, and later by large-scale integration (LSI) of devices. This progress has resulted in higher device packing density, higher device and circuit speed, and lower fabrication costs. Devices have also been fabricated for optical applications (i.e., light emitting diodes, lasers). Further advances in device structures are on the horizon, and it is anticipated that future optical and electrical devices will be merged to create optoelectronic integrated circuits and optoelectronic multichip modules, in conjunction with very-large-scale integration (VLSI) devices. Figure 1 represents a concept of an optoelectronic interconnect device. A large amount of information may be transferred onto the chip by light signals passing through optical fibers and waveguides. The light signals are converted into electrical signals by a GaAs optic/electronic converter. High-speed signal processing is performed by an electronic VLSI chip.

Much progress has also been made in the area of organic insulators. The first aromatic polyimides (Kapton) were commercialized by duPont in 1960 [1] and were developed for various aerospace applications. The Kapton chemical structure is

Its properties are summarized in Table 1 [2]. Kapton is physically stable over a wide range of temperatures (-270–$400°C$). Besides its aerospace applications, its excellent heat resistance, chemical resistance, and toughness make it also suitable for use as insulation films, in flexible printed circuit boards, and for tape automated bonding applications. However, the trend for more-advanced microelectronics and optoelectronics devices brings along the need for even better performing materials and less expensive manufacturing processes.

Polyimides can satisfy many of these needs, and thus new polyimide coatings have been developed for various microelectronic and optoelectronic applications. A typical polyimide coating is the commercially available Pylalin (produced by duPont), which is a poly(amic acid) solution of N-methyl-2-pyrrolidone (NMP).

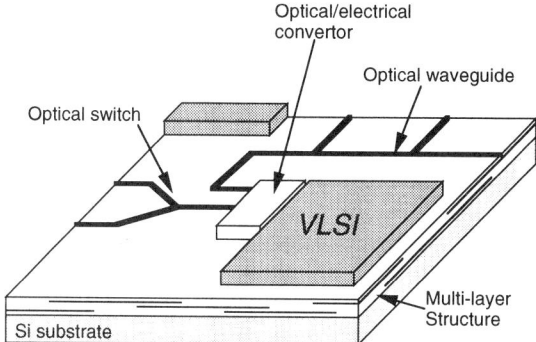

Figure 1 A schematic concept of an optoelectronic interconnect.

Table 1 Properties of Kapton Type H[a]

Tensile strength	25 ksi
Ultimate elongation	70%
Tensile modulus	430 ksi
Density	1.42
Melting point	None
Zero strength temp. (20 psi load/15 s)	815°C
Glass transition temp.	385°C
Cut-through temp.	435°C
Thermal coefficient of expansion	2.0×10^{-5} °C^{-1}
Shrinkage	0.3%
Oxygen index	37%
Dielectric strength (1 mil)	7000 V
Dielectric constant (1 kHz)	3.5
Dissipation factor (1 kHz)	0.003
Volume resistivity (at 50% Ω cm RH)	10^{18}
Corona start voltage (at 50% RH)	465 V
Surface resistivity	10^{16} Ω
Moisture absorption (at 50% RH, 23.5°C)	1.3%

[a]Typical values for a 25.4 μm film at 25°C.
Source: Ref. 2.

The molecular structure of this poly(amic acid) is

[Chemical structure diagram of poly(amic acid)]

It is converted by heating into the same polyimide structure as Kapton via cyclohydration. Another commercial polyimide, PIQ (polyimide-isoindoloquinazolinedione, produced by Hitachi Chemical), is also used as a dielectric coating. This material has a ladder structure incorporated into the polyimide to give it additional thermal stability:

[Chemical structure diagram of PIQ with R_1 and R_2 groups]

PIQ can withstand processing temperatures typically associated with microelectronic processes, such as Al annealing (360° for 30 min), Au–Si alloying (430°C for a few minutes), and die bonding (500°C for a few seconds) [3]. This chapter will concentrate on the process and material issues of a variety of polyimides, using the polyimide PMDA/ODA prepared from pyromellitic dianhydride and 4,4'-oxidianiline as a primary example.

II. POLYIMIDES AS DIELECTRICS AND OPTICAL INTERCONNECTS

A. Application Areas

As indicated, polyimides have several favorable properties, making them attractive for several different, diverse applications: as interlevel metal dielectrics, protective coatings, alpha-ray shielding, and waveguides.

Interlayer Dielectric. A multilevel metal structure is key to a VLSI device structure (Fig. 9, Chapter 1). Such structures require an interlevel dielectric to serve as electrical insulation between the upper and lower wiring levels. The dielectric may be a silicon dioxide or a polyimide film. The dielectric material must be easily processable so as to create a flat surface for subsequent upper metal level wiring layers and to allow through-holes fabrication (vias; see Fig. 9, Chapter 1), which serve the role of interconnection points for the metal levels. Polyimide insulation layers with good planarity, and prepared from polyimide precursor solutions, are often more suitable for forming these relatively thick (2–3 μm) interlevel dielectrics in comparison with conventional chemical vapor deposition SiO_2 techniques.

Protective Overcoat. Many high-performance devices must be effectively sealed off from the ambient environment to assure the prevention of moisture or contaminant exposure. Si_3N_4 is typically used as that protective layer; yet it may contain pinhole defects. The addition of an overcoating polyimide film is an effective way to seal off these pinhole defects, shield against exposure to moisture, and contribute to the overall mechanical protection of the device chip or module.

Alpha-Ray Shielding. Despite all efforts in synthesizing and using chemicals and materials of the highest purity (Chapter 2), some low-level impurities often remain. Alpha radiation, generated by radioisotopes of thorium and uranium present in some of the materials (i.e., inorganic fillers in epoxy resins) and used to package electronic devices or modules, have been shown to initiate soft errors in memory devices. Soft errors are transient defects and thereby cause a non-repetitive memory cell failure of the affected memory device. A thick high-purity polyimide layer protects these devices from exposure to alpha radiation by absorbing their short-range penetration.

Optical Waveguides. Optical telecommunication, which allows the transmission of a very large amount of information by light signals, is rapidly replacing conventional electrical telecommunications. Silica is a typical optical material for optical fibers and waveguides. It has excellent optical transparency over a wide range of wavelengths from ultraviolet to near-infrared. Ongoing advances in the optical properties of certain polymers, such as polymethylmethacrylate, polystyrene, and polycarbonate have been targeted for use as plastic optical fibers or waveguides. However, this set of polymers, and many others (Table 2), do not possess the necessary thermal stability (>400°C), as required for the more challenging optical interconnect applications on electronic device chips or modules. With the rapid pace of progress in the electronic technologies, the design concept of Fig. 1 will inevitably become a manufacturable physical reality in the not too distant future. This need dictates continued efforts to develop materials with higher thermal stability, while retaining full process compatibility with traditional integrated circuit fabrication methodologies to assure their device reliability [4]. Optical waveguides have been fabricated from optical quality polyimides [5].

Solar Cells and Thermal Control Systems. Although the optical transparency of most polyimides is not very high, optically transparent polyimides are desirable in the aerospace field as coating materials on specific space components, such as solar cells and thermal control systems (Fig. 2).

Table 2 Thermal Degradation of Polymers

Polymer	Structure	$T_{d,0}$ (°C)	$T_{d,1/2}$ (°C)
Poly(ethylene)	—CH$_2$-CH$_2$—	380	404
Poly(propylene)	—CH$_2$-CH(CH$_3$)—	320	387
Poly(styrene)	—CH$_2$-CH(C$_6$H$_5$)—	327	364
Poly(vinyl chloride)	—CH$_2$-CH(Cl)—	170	270
Poly(tetrafluoro ethylene)	—CF$_2$CF$_2$—	-	509
Poly(vinyl alcohol)	—CH$_2$-CH(OH)—	220	274
Poly(acrylo-nitril)	—CHCH$_2$— with C≡N	290	450
Poly(methyl methacrylate)	—CH$_2$-C(CH$_3$)(COOCH$_3$)—	280	337
Poly(butadiene)	—CH$_2$CH:CH CH$_2$—	280	407
Poly(ethylene oxide)	—CH$_2$-CH$_2$-O—	-	345
Poly(ethylene telephthalate)	—CH$_2$-CH$_2$-O-CO-C$_6$H$_4$-CO—	380	450
Poly(ε-caproamide) [Nylon 6]	—NHCH$_2$(CH$_2$)$_3$CH$_2$-CO—	350	430
Poly(p-phenylene telephthalamide)	—CO-C$_6$H$_4$-CO-NH-C$_6$H$_4$-NH—	~447	~527
Polyimide [PMDA/ODA]	(PMDA-ODA structure)	450	~567

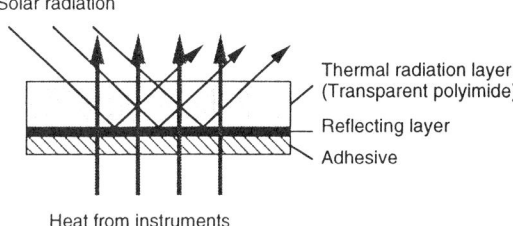

Figure 2 Cross-sectional diagram of a thermal control system.

B. Chemical and Physical Properties

The properties possessed by various polyimides are far-ranging, and many have become useful for specific applications. Those with the appropriate and desirable properties do exhibit several advantages over their corresponding inorganic material counterparts, such as SiO_2 and Si_3N_4. However, some of the other material properties may be somewhat inferior, although not detrimental.

Ease of Processability. Polyimides are relatively easy and cheap to fabricate by room-temperature spin-coating of its polymer precursor solution onto a wafer (or substrate) and by heat treatment (300–400°C) techniques. This contrasts with the need for sophisticated vapor deposition or sputtering tools operating at high temperatures (400–900°C) when forming SiO_2 or Si_3N_4 layers. Beside thick-film formation (5–20 μm), the polyimides may also be used for microprocessing to create polyimide layers over a wide range of thicknesses (0.1–100 μm) by means of chemical content variations and control over the spin-coating tool process parameters. The spin-coating technique also allows film thickness modifications by adjustments of the solution's fluidity. Through-hole (via) formation between wiring levels is readily achieved by means of either wet etching polyimides with KOH or by dry etching (reactive ion etching) using oxygen.

Material Property Tailoring. Both SiO_2 and Si_3N_4 have generally fixed physical and chemical property characteristics. In the case of polyimides, these can be significantly modified through the synthesis of different molecular structures and chemical content. Very specific characteristics can be realized to enhance their use for unique applications.

Chemical Stability. Most electrical or optical devices and packages are frequently exposed during their fabrication to immersion in organic solvents (such as acetone or alcohol for the purpose of cleaning and rinsing), in acids for wet etching and neutralization, and in specific polar solvents contained in photoresists. With the exception of soluble polyimides, the general class of applicable polyimides is unaffected by these chemicals.

Chemical Purity. Ionized impurities in insulation materials, such as Na⁺, can cause serious device operation degradation, if not complete device failures. This kind of contamination can normally be avoided with proper material and process control of high-purity polyimides.

Thermal Stability. A select number of polyimides are thermally stable with respect to the typical heat treatment (~300°C) encountered during metal deposition and processing operations, and during wire bonding. Their thermal stability is, however, inferior to that of SiO_2 and Si_3N_4 layers.

Low Dielectric Constant. The continuous progress in device miniaturization greatly contributes to higher levels of integration and high electrical signal propagation. The use of lower dielectric materials can also contribute to these device features. The dielectric constant of polyimides ranges between 2.5 and 3.8, while it is 3.8 for SiO_2 and 7.0 for Si_3N_4. The relationship between propagation delay (τ) and the dielectric constant (ε) of an insulating material is given by

$$\tau \propto \varepsilon^{1/2} \tag{1}$$

τ is also proportional to the total length of circuit wiring. Hence, shorter circuit wiring reduces the delay and increases the speed of a device. To more effectively minimize the overall wiring length, advanced and highly integrated devices require more and more wiring, which must be placed closer and closer together. However, nearby neighboring wires will begin to cause electrical interference with each other (crosstalk) as their spacing decreases. The crosstalk can be dramatically reduced, or even inhibited, with significantly lower dielectric constant materials.

Mechanical Stability. Polyimides are sufficiently flexible to absorb internal stresses that may be produced during the device or module fabrication process. The polyimides can also be tailored to match the thermal expansion characteristics of the underlying materials. Furthermore, polyimides are sufficiently resilient to serve as protective coatings against external physical impacts.

Optical Transparency. The effectiveness of polyimides as optical interconnect rest heavily on its transparency to specific light wavelengths. Molecular composition modifications allow the optimization of specific materials to create transparent "windows" at desirable wavelengths. This tailoring is relatively easy for these organic materials, while at the same time retaining many of the other required material properties.

III. POLYIMIDE CHEMISTRY

Aromatic polyimides are generally synthesized from benzenetetracarboxylic dianhydride and aromatic diamine. The polyimide is an alternating arrangement of dianhydride and diamine units, the basic synthesis and structure of which is:

PMDA + ODA → PMDA/ODA Poly(amic acid) (Polymerization, Solv.: DMAc, Room Temp)

PMDA/ODA Poly(amic acid) → PMDA/ODA Polyimide (Imidization, -H$_2$O) (2)

In this particular synthesis, the pyromellitic dianhydride (PMDA) is the benzenetetracarboxylic dianhydride, and 4,4'-oxidianiline (ODA) the aromatic diamine. The naming of the resulting polyimide is normally derived in the form of an acronym from their monomers precursors. Hence, the polyimide synthesized from the PMDA and ODA monomers is PMDA/ODA. Figure 3 summarizes a number of monomers with which one can form a variety of aromatic polymers via the combination of any of the R_1 tetracarboxylic dianhydrides with any of the R_2 diamines. As should be expected, the synthesis of polyimides from different monomers will result in different thermal, electrical, chemical, mechanical, and optical properties. Due to the great variety of possible monomer combinations, the opportunity to seek out unique properties for a polyimide are quite favorable. Extending this synthesis option by combining several monomers of tetracarboxylic dianhydride and diamine allows for the formation of a variety of copolymers with additional unique or expanded properties.

Referring back to Eq. (2), equal molar quantities of the benzenetetracarboxylic dianhydride and the aromatic diamine are polymerizable at room temperature in a polar solvent, such as N,N-dimethylacetamide (DMAc), N-methyl-2-pyrrolidone (NMP), and dimethylsulfoxide (DMSO). Poly(amic acid)s are also

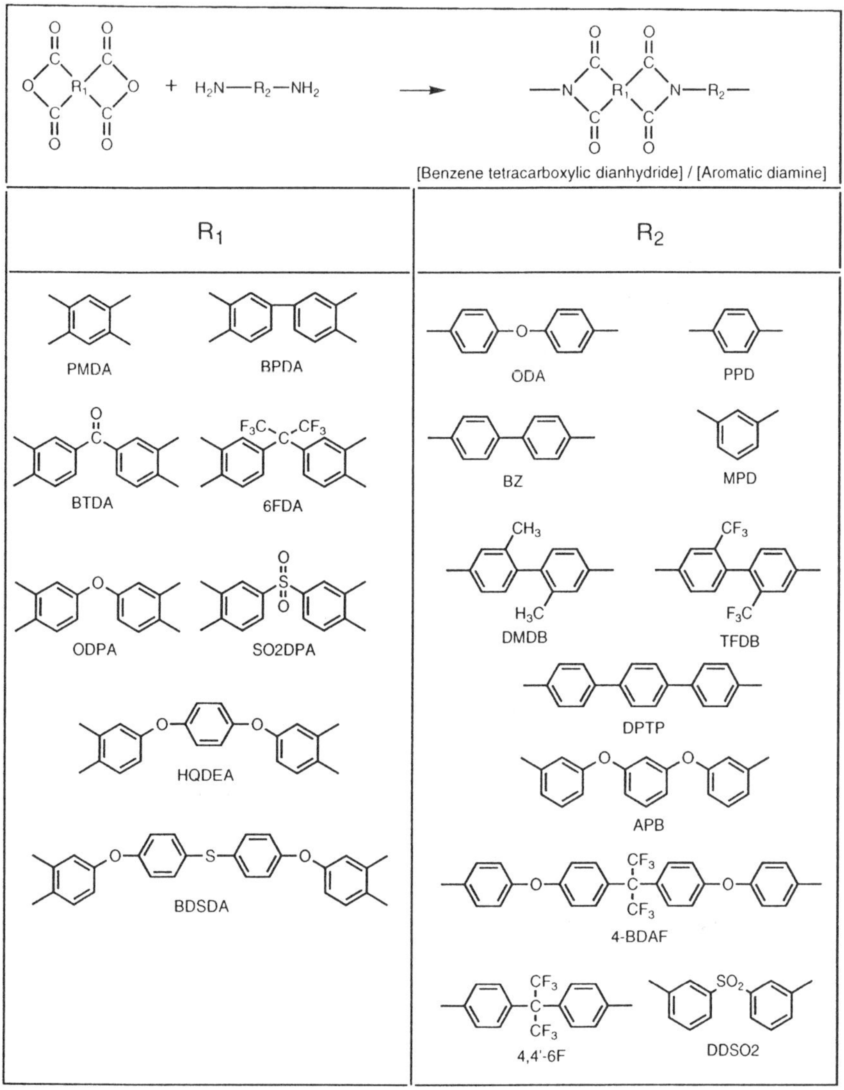

Figure 3 Structures of various aromatic polyimides.

soluble in these organic polar solvents, such as DMAc and DMSO. These polar solvents serve to dissolve the intermediate products of the poly(amic acid)s and subsequently aid in the polymerization reaction. Polar solvent concentrations of 10–15 wt % are sufficient for realizing most polymerizations. The process temperature will influence the rate and the extent of the polymerization, with reaction temperatures of 15–75°C being typical. Above 100°C, polyimide precipitation can occur through cyclization. Solution mixing is also critical to assure polymerization and molecular weight consistency. As the reaction progresses, the solution's viscosity increases due to its molecular weight increase. The polymerization is reversible, since the reactants (monomers) and the product (amic acid) are in equilibrium.

Subsequent to the polymerization, the poly(amic acid) is converted into an insoluble polyimide via a cyclodehydration reaction by means of heating or treatment with a dehydrating agent, such as acetic dianhydride. This transformation is called "imidization." As this reaction proceeds, the color of the poly(amic acid) tends to darken in the case of the PMDA/ODA system from a pale yellow to a deep yellow or orange. Other systems exhibit similar changes. Imidization temperatures generally range from 100 to 250°C and require 15 to 60 min.

The progress of the imidization reaction, as well as kinetic investigations, is typically monitored by infrared (IR) spectroscopy. Figure 4 shows the IR spectra for the PMDA/ODA poly(amic acid) and the resultant polyimide. The N–H absorption band at 3.08 μm of the poly(amic acid) is replaced with imide bands at 5.63 and 13.85 μm.

The more specific reaction mechanism of the polyimide formation is:

Intermediate compound (1) is created by the interaction of the amino group with the dianhydride ring. A subsequent transfer of a hydrogen atom and the opening of the dianhydride ring produce the poly(amic acid) structure. An additional hydrogen atom transfer from the N atom to the COOH group, and the N attachment to the second carbon atom forms the cyclic intermediate (2). Upon dehydration, the polyimide is produced.

Although most common commercially available polyimides are prepared by the two-stage synthesis, polymerization and imidization, others have been based on the direct reaction between benzenetetracarboxylic dianhydrides and diisocyanate [6] via the removal of CO_2, as shown by

$$\text{Benzenetetracarboxylic dianhydride} + \text{Aromatic diisocyanate} \xrightarrow[-CO_2]{\text{Polycondensation}} \text{Polyimide} \quad (4)$$

The solvents used for this reaction are DMAc or DMSO with the addition of a small amount of H_2O and thermal activation. This direct route is suitable for soluble polyimides, whereas insoluble polyimides would have a tendency to precipitate during the polymerization reaction. The inhomogeneity created by such precipitation can interfere with the progress of the polymerization process, especially in the case where high-molecular-weight polyimides are desired.

IV. PROCESS CONSIDERATIONS

The general process sequence and process requirements for forming polyimide films are quite similar to those used for the resist process in lithography technologies (Chapter 6, Sec. II.D). However, the film thicknesses are normally quite different. In the case of lithographic films, the thicknesses tend to range from 1.0 to 3.0 μm, whereas for polyimide films, the range extends from 5 to more

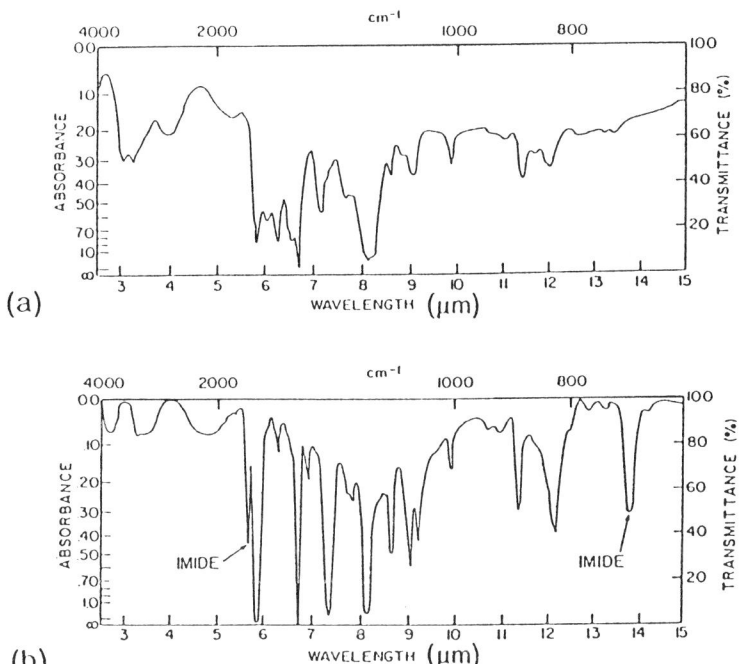

Figure 4 Infrared spectra of (a) PMDA/ODA poly(amic acid) and (b) polyimide films 2.5 μm thick. The poly(amic acid) film was dried for 2 h at 80°C. The polyimide film was heat ramped to 300°C over a 45 min period, then held at 300°C for 1 h. (From Ref. 1.)

than 20 μm. A specific thickness is mainly dictated by the particular application. Also, for those applications which require polyimide patterning (i.e., as for metal interlevel layers), the overall process would include the additional basic lithography resist process steps. Thus, summarizing, such a processing sequence, applicable for film formations on semiconductor wafers, consists of the following:

1. Wafer surface priming with an adhesion promoter
2. Layer application with spin-coating of poly(amic acid) solution
3. Prebaking at about 100°C
4. Photoresist application
5. Mask aligning, exposing, and developing
6. Etching of exposed poly(amic acid) film pattern
7. Photoresist stripping
8. Postbaking (imidization at about 300°C)

This sequence is schematically outlined in Fig. 5a. In the case of nonpatterning applications (i.e., films used for encapsulation or surface protection), process steps 4–7 are not required, and the thickness uniformity tolerances are less critical. This is, however, not true for patterned polyimide films; in fact, absolute thickness control, as well as thickness uniformity across the full wafer, is of utmost importance to achieve the necessary device performance consistency for all device chips across that wafer.

A. Adhesion

Excellent adhesion of the polyimide films to the underlying surface (most often Si, SiO_2, Si_3N_4, or a metal) is of primary importance for long-term functional microelectronic or optoelectronic devices. These devices often operate in extreme environmental surroundings or experience frequent temperature cycling of up to 100°C during operations. Such exposures place major demands on the physiomechanical properties of the interfaces between the polyimide and adjoining surfaces. Poor adhesion will result in delamination, spalding, or void formation, and some of the implications of these types of physical defects were discussed in Chapter 8. To mitigate the occurrence of such defect generation, the surfaces on which the polyimides are to be fabricated are first treated (primed) with an adhesion promoter (step 1 in process flow). In general, directly applied polyimides do not adhere well to the traditional inorganic surfaces. To overcome this deficiency, aminosilanes are used to serve as effective adhesion promoters [7]. They are applied by spin-coating, and heat treated with adhesion promotion techniques similar to those utilized in resist film applications (Chapter 6, Sec. D2). These adhesion promotion films are very thin (10–100 Å), since they merely act as a chemical interface between two different materials. The adhesion mechanism for the case of PMDA/ODA being applied to a silicon surface is, in outline:

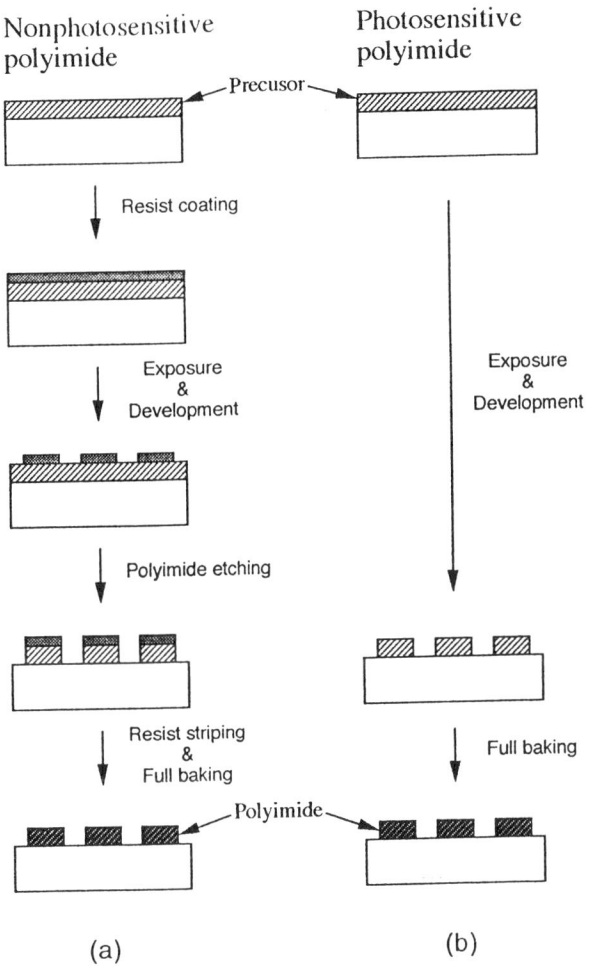

Figure 5 Comparison of (a) the photopatterning method for conventional polyimide and (b) direct patterning using a photosensitive polyimide.

The 3-aminopropyltrimethoxysilane attaches to available hydroxy groups (−OH) on the oxidized silicon surface and is oligomerized. The aliphatic amine portion of the attached aminosilane interacts with the poly(amic acid), which is subsequently converted to the polyimide.

Alternatively, adhesion of a polyimide can also be improved with the introduction of siloxane segments directly into the polyimide backbone [8]. Its preparation scheme is described by:

Polyimides

$$\text{(6)}$$

This polyimide siloxane is a block copolymer prepared from benzophenonetetracarboxylic dianhydride (BTDA) reacting with methylenedianiline (MDA) and bis-gamma aminopropyltetramethyldisiloxane. Its adhesion is even superior to that of PMDA/ODA with an adhesion promoter, when applied to Si, SiO_2, or Si_3N_4 surfaces, and it retains those qualities even in boiling water.

B. Viscosity

The range of thicknesses and the uniformity across the wafer are primarily managed by the adjustment of the poly(amic acid) solution viscosity and the control of the spin-coating process parameters. The specific thickness, as dictated by the need of the particular application, rests on the ability to optimize a complex interdependency among viscosity, solution concentration and density, solvent volatility, spin speed, spin time, substrate (wafer) radius, and temperature.

The polyimide thickness increases with increasing solution viscosity. Sufficiently high viscosity, however, deteriorates the coating uniformity due to the hindrance of the solution from efficiently flowing off the wafer's edge. Instead, it tends to hang on to those edges like cotton candy. This results in an edge buildup, and thus a poor wafer surface uniformity from center to edge. The dependence of the intrinsic viscosity $[\eta]$ is proportional to the average molecular weight [Mw] of the poly(amic acid) in solution, and it is shown in Fig. 6 for the case of a PMDA/ODA solution. This relationship fits the Mark–Houwink equation:

$$[\eta] = 1.85 \times 10^{-4} \, Mw^{0.80} \tag{7}$$

Increasing the solution concentration of the precursor monomers can increase the poly(amic acid) solution viscosity. However, obtaining higher concentration poly(amic acid) solutions requires more efficient cooling to compensate for the exothermic heat of polymerization reaction. As indicated by the data in Table 3, the low molecular weight [as observed by low inherent viscosity (n_{inh})

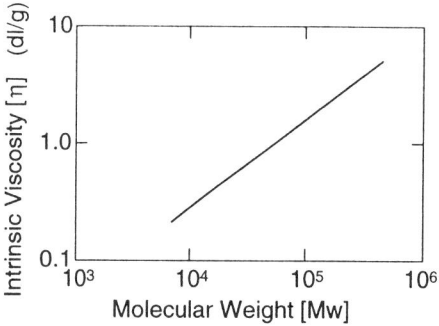

Figure 6 Intrinsic viscosity versus weight-average molecular weight for PMDA/ODA poly(amic acid). (From Ref. 2.)

Table 3 Effect of Polymerization on Inherent Viscosity of PMDA/ODA Poly (Amic Acid)

Polymerization condition		
Temp. (°C)	Time (min)	Inherent viscosity[a]
25	120	4.05
65	30	3.47
85–88	30	2.44
115–119	15	1.16
125–128	15	1.00
135–137	15	0.59

[a]Determined at 0.5% concentration at 30°C.
Source: Ref. 1.

poly(amic acid)] is produced above 75°C. Above 100°C [1], cyclization causes polyimide precipitations.

The limitations of uniform film formations due to high inherent viscosity solutions can be partially overcome by process parameter adjustments. Increasing the solution temperature decreases the viscosity. The length of spin time may also be beneficial in some instances to allow more "flow time". Adjustments in spin speed is another alternative. In general, the higher the spin speed, the thinner the wafer due to the increased efficiency of the centrifugal force pushing the solution off the wafer. However, this effect gets again moderated with increased solution viscosity. Additive multiple thin layers may be necessary in some instances to realize acceptable thick polyimide films.

C. Planarity

Another aspect in polyimide processing is that of film planarity, already touched on during the discussion of viscosity. In that case, the reference was with respect to global wafer uniformity from center to edge. However, local uniformity over surface topography is just as critical, especially for multilevel metal structures. High-density device integration depends on the presence of planar surfaces as the buildup of metal and insulator structures progresses. Figure 7 illustrates the locally influencing planarization relative to thickness parameter of the metal height (t_1) and the polyimide variation (t_2). The degree of that planarization is then defined by

$$\text{Degree of planarization} = 1 - \frac{t_2}{t_1} \tag{8}$$

Planarization of 1 (i.e., $t_2 = 0$) is ideal. Subsequent to the spin-coating, the solvent in the poly(amic acid) solution is removed during the postbake heating stage (process step 8) via evaporation, causing the film to shrink. The resultant thickness of the polyimide tends to be proportional to its concentration in the precursor solution, and the shrinkage can introduce new surface nonplanarities. Low-concentration solutions create poor planarity: t_2 becomes large, and the

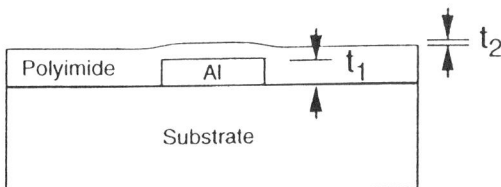

Figure 7 Thickness parameter relationship defining the degree of planarization.

region between metal lines will exhibit concaved surfaces, with its center region possibly being thinner than t_1. Furthermore, the wider the spacing between metal lines, the thinner t_2 becomes. This pattern density sensitivity complicates process control, and higher concentration solutions are required. However, its value cannot be increased too much, since then the "flow smoothening" effect is retarded. Again, much trial and error is needed to achieve optimized planarity. The successful process conditions—that is, overall evening out of irregular surface heights—need to take into account the desired film thickness, pattern geometry, and pattern density effects.

D. Moisture Sensitivity

Polyimides, once adequately cured, are in general physically and chemically stable. On the other hand, the polyimide precursors (poly(amic acid)) are quite sensitive to hydrolysis. Their viscosity gradually decreases with time, due to the hydrolysis-reaction-based depolymerization. Higher water content in solution will accelerate that decrease. Such time-dependent viscosity instability greatly complicates day-to-day process consistency in a manufacturing environment, where process stability is extremely critical. Hence, the storage of the polyimide precursor solution without exposure to water contamination (moisture) becomes a critical material management issue in order to prolong the shelf life of the solution.

However, despite rigorous process control, the actual shelf life of a poly(amic acid) solution is relatively short—a few months, even when placed in refrigerated storage. Efforts to overcome this material weakness have been pursued, but the successes are limited. The most promising route has been in the development of a series of soluble polyimides. Of these, those polyimides which contain fluorine or bent $-C(CF_3)_2-$ structures in their chemical and molecular makeup remain soluble in organic solvents. To form the soluble polyimide requires only the solvent removal, as opposed to the production of water in the case of the polyimide imidization reactions from poly(amic acid) solutions. Although the moisture sensitivity is reduced for the fluorine-containing polyimides, they are limited in their applications for most microelectronic processes, since they are easy to be soluble in organic solvents and these are ubiquitous in cleaning and rinsing processes.

A second approach for moisture-resistant polyimide preparations relies on poly(amic akyl ester):

Polyimides

[Diagram: Diester diacyl chloride + ODA (Diamine), R: Alkyl group, Polymerization in Solv.: DMAc, Room Temp → PMDA/ODA Poly(amic alkyl ester) → Imidization (−ROH) → PMDA/ODA Polyimide]

(9)

Although the poly(amic alkyl ester) is more stable against water than poly(amic acid), the film formation during the imidization is accompanied by both a large reduction in thickness and a concurrent large amount of stress, generated due to the removal of the bulky alcohol group.

E. Photosensitivity

It is often beneficial, and in special instances preferable, to pattern the polyimide film directly as opposed to creating the pattern with an overlying photoresist film (Fig. 5a). Direct patterning simplifies the overall process greatly, as indicated by the outline in Fig. 5b. Such photosensitive polyimides have been formulated, and they have become an important addition to the polyimide process capabilities, particularly when the polyimide application is to serve as an interlevel dielectric layer.

Photosensitivity is achieved with the appropriate introduction of photosensitive groups into the poly(amic acid) structure [9], the chemical principle of which is, in outline:

(10)

The photoreactive methacrylate ester group is introduced into the polyimide precursor solution, which is then applied as a coating to the wafer. Ultraviolet light exposure with a wavelength of less than 500 nm forms the insoluble crosslinked poly(amic acid) intermediate. Postexposure development removes the unexposed (noncrosslinked) portion of the coating. The remaining crosslinked intermediate is imidized with heat curing at 400°C for about 1 h to form a highly heat-resistant polyimide relief structure, ready for further device processing.

V. MATERIAL PROPERTY CONSIDERATIONS

As outlined in Sec. II, the effective use and specific applications of polyimides are greatly dependent on one or several of the polyimides' properties. The ability

A. Thermal Stability

The primary criterion for choosing a particular aromatic polyimide over other organic polymers is its excellent thermal stability at temperatures typically above 300°C. This stability tends to arise from the polyimide's rigid phenyl and imide ring structure. Polyimide thermal stability values are derived from thermogravimetric analysis (TGA) studies. Some representative values are summarized in Table 2. The thermal decomposition behaviors (TGA curves) of three PMDA-based polyimides are shown in Fig. 8. All remain stable up to about 500°C (less than 1.5% weight loss), after which they decompose at different rates. They remain thermally stable for months at elevated temperatures in air—PMDA/ODA at 275°C for more than a year, at 300°C for more than a month.

B. Thermal Expansion

If at all possible, it is desirable to match the thermal expansion characteristics of the polyimide with the material with which it will be in contact. This is, however, an almost impossible task to achieve with one particular polyimide, since the complex electronic device structures consist of a several different inorganic materials, each with a different thermal coefficient of expansion. The

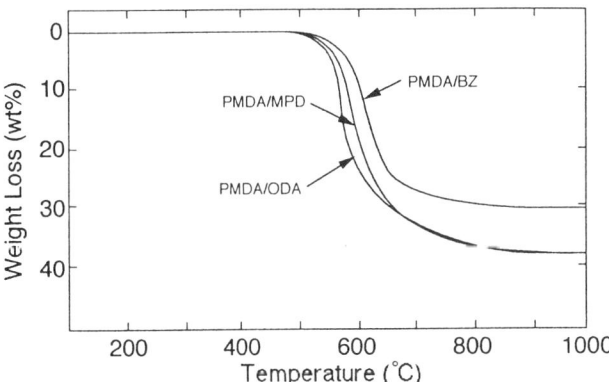

Figure 8 Thermogravimetric analysis (TGA) curves of various polyimides prepared from PMDA at a heating rate of 3°C/min in a dry helium atmosphere (MPD: *m*-phenylenediamine, BZ: benzidine). (From Ref. 1.)

thermal coefficient of expansion (TCE) is defined as a material elongation per degree, when heated. Figure 9 is a thermomechanical analysis (TMA) of a PMDA/ODA film. The slope of this curve represents the TCE of this film, in this case $2 \times 10^{-5}/°C$ over the range of 50–300°C. The range scale of polyimide TCEs are compared with inorganic materials and metals in Fig. 10. The greater the difference in thermal expansion between adjoining materials, the greater the possibility of material peeling or cracking, or structure bending during any heat cycling during the device fabrication process or subsequent device operation.

Since the molecular chains in most polymers, including conventional polyimides, are bent and tangled, these materials usually exhibit higher TCEs ($\sim 10\times$) than most inorganic materials, but they can be quite compatible with the TCEs of metals (Fig. 10). Efforts to develop low-thermal-expansion polyimides (TCE below $10^{-5}/°C$) to match with other inorganic materials have resulted in such materials. These consist of linear polymer backbones and with rigid-group molecular construction [10]. The latter contain phenyl and imide rings—a fortunate situation, since such structures also enhance these materials

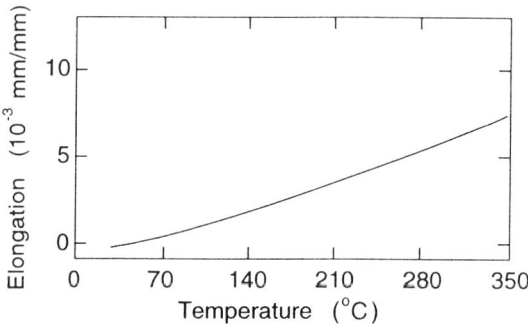

Figure 9 Thermomechanical analysis (TMA) curve of PMDA/ODA polyimide.

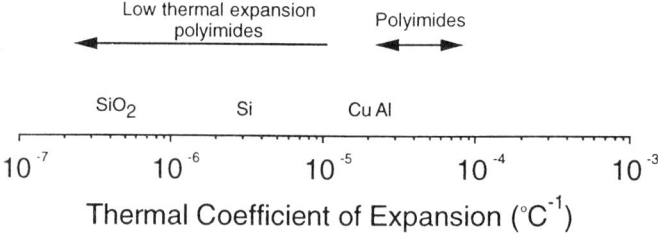

Figure 10 Thermal coefficients of expansion of various microelectronic materials. (From Ref. 10.)

Table 4 Thermal Coefficients of Expansion of Various Polyimides

Polymer[a]	TCE ($10^{-5} \times °C^{-1}$)
PMDA/MPD	3.20
PMDA/BZ	0.59
PMDA/DPTP	0.56
PMDA/ODA	2.16
PMDA/4-BDAF	4.57
BPDA/PPD	0.26
BPDA/MPD	4.00
BPDA/BZ	0.54
BPDA/DPTP	0.59
BPDA/ODA	4.56
BPDA/4-BDAF	5.61
BTDA/PPD	2.10
BTDA/MPD	2.94
BTDA/BZ	2.17
BTDA/DPTP	1.83
BTDA/ODA	4.28
BTDA/4-BDAF	5.47

[a]See text and glossary for explanation of acronyms.
Source: Ref. 10.

with better thermochemical stability. Table 4 list the TCEs of various aromatic polyimides. Of these, the polyimide combinations of PMDA or BPDA (biphenyl-tetracarboxylic dianhydride) with the aromatic diamines PPD (p-phenylene diamine), BZ (benzidine), and DPTP (diamino-p-terephenyl) exhibit TCEs of less than $2 \times 10^{-5}/°C$. The actual elongation measurements based on a thermomechanical analysis (TMA) for the PMDA/ODA are shown in Fig. 10. The measurements relate to the in-plane elongation, the slope of which is the in-plane TCE. The vertical component may be slightly different.

C. Dielectric Constant

As indicated in the introduction, reducing the dielectric constant of insulation materials enhances the ability to fabricate higher device integration and speeds up the signal propagation. Desirable values should be below 3.0.

Conventional polyimides have dielectric constants of above 3.0. The dielectric constant decreases with decreasing molecular polarization, and fluorination has been found to be the most effective means to influence this material parameter. Table 5 lists the dielectric constant for a number of polyimides with different molecular structures measured at 10 GHz. The nonfluorinated polyimides, such as PMDA/ODA has a dielectric constant of 3.2, while those containing $-C(CF_3)_2-$ groups exhibit lower values. The highest fluorine-containing material 6FDA/4,4'-6F (prepared from 2,2-bis(dicarboxyphenyl)hexafluoropropane dianhydride, 6FDA, and 2,2-bis(4-aminophenyl)hexafluoropropane, 4,4'-6F), exhibits the lowest dielectric constant of 2.39 [11]. Introducing perfluoroalkyl side chains into the polyimide backbone is a useful means of achieving such high-fluorine-content materials. However, their thermal stability decreases with such long side chains, as shown in Table 6. These measurements were made at 1 KHz, and the listed polymer decomposition temperatures ($T_{d,0.1}$) represent a measured weight loss of 10% in a dry nitrogen atmosphere. The decrease of the dielectric constant by 0.5 is associated with a thermal stability loss of more than 100°C. On the other hand, fluorinated polyimides are much more resistant to moisture uptake due to the fluorine hydrophobicity (Table 7). Hence, a lower-dielectric-constant polyimide containing fluorine structures will not be as significantly affected by exposure to high moisture environments as normal

Table 5 Dielectric Constants of Polyimides

Polymer[a]	Dielectric constant at 10 GHz[b]
PMDA/ODA	3.22
BTDA/ODA	3.15
ODPA/ODA	3.07
HQDEA/ODA	3.02
BDSDA/ODA	2.97
6FDA/ODA	2.79
6FDA/DDSO2	2.86
6FDA/APB	2.67
6FDA/4-BDAF	2.50
6FDA/4,4'-6f	2.39

[a]Films were desiccated prior to testing.
[b]Measurements were done at room temperature and approximately 25% RH.
Source: Ref. 12.

Table 6 Characteristics of Polyimides with Fluorinated Alkoxy Side Chains

R_f	Fluorine content (%)	$T_{d,0.1}$ (°C)	Dielectric constant at 1 kHz
H	0.0	589	3.5
$OCH_2(CF_2)_3F$	23.6	495	3.3
$OCH_2(CF_2)_6F$	34.6	471	3.0
$OCH_2(CF_2)_7F$	37.3	459	3.0
$OCH_2(CF_2)_{10}H$	42.4	457	3.0

Table 7 Dielectric Constants of Polyimide Films Measured at Various Relative Humidities

	Dielectric constant at 1 MHz		
Polymer	30–35% RH	47–50% RH	100% RH
PMDA/ODA	3.38	3.44	3.85
6FDA/4-BDAF	2.77	2.81	2.80
ODPA/4-BDAF	3.08	3.12	3.17
PMDA/4-BDAF	2.93	2.97	3.01
BTDA/4-BDAF	3.03	3.09	3.15
BDSDA/4-BDAF	3.08	3.14	3.16

Source: Ref. 12.

polyimides. For example, under high relative humidity (RH) conditions the fluorinated polyimide 6FDA 4-BDAF's dielectric constant increases only by 1% in comparison with its nonfluorinated counterpart PMDA/ODA, which experiences a 14% increase (Table 7). Poly(tetrafluoroethylene) (PTFE) is another material with a high fluorine content. It also has a large waterproofing effect.

D. Optical Transparency

Optical transparency of polyimides is dictated by their light absorption characteristics. For polyimides to be useful in optical applications, they must exhibit high transparencies in the visible and/or infrared regions. In other words, preferably they should be colorless. A number of such materials have been developed for several applications. Some are used in space components as protective layers in solar cells and thermal control systems. Others, transparent in the visible region, allow ready lithography mask alignments through polyimides (serving as dielectric films) to underlying structural elements. Still others, those with transparency in the visible and near-infrared regions, can serve in optical interconnect configurations. For example, the near-infrared light at 1.3 and 1.55 μm is practical for long-distance optical telecommunication systems. A typical method to measure the optical transmission loss at visible wavelengths is shown in the schematic of Fig. 11. A laser light is coupled into a 10 μm polyimide film through a prism, and passes the polyimide film. That portion of the light which is scattered out of the film is monitored, and the optical loss is calculated from a plot of the scattered light intensity versus the propagation length. Table 8 summarizes the optical loss of a number of different polyimides. The optical loss relationship to polyimide structure is similar to that of optical transmission.

Visible wavelength absorption is caused by electronic charge transfer within the polyimide structure. The transparency increases with decreasing molecular interactions and with shorter intramolecular π-conjugations. Comparative transparencies of 5.1–6.4 μm thick polyimide films are shown in Fig. 12. The conventional PMDA/ODA has low optical transparency due to its dark yellow coloration. It is semitransparent. As the molecular structures simplify from the

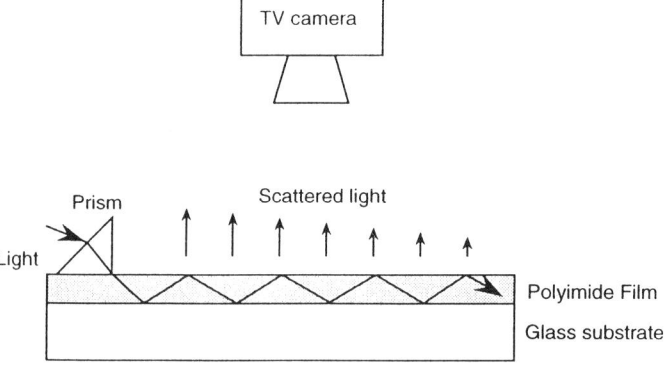

Figure 11 Experimental setup for optical loss measurement.

Table 8 Optical Loss of Various Polyimide Films

Polyimide	Optical loss (dB/cm)[a]	Cure temp. (°C)
PMDA/DMDB	36	350
6FDA/ODA[b]	10	300
6FDA/DMDB	5.5	350
PMDA/TFDB	4.3	350
6FDA/4,4'-6F[b]	3	300
6FDA/TFDB	0.7	350

[a]Wavelength, 0.633 μm.
[b]From Ref. 4.

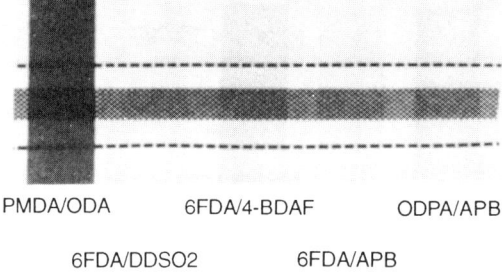

PMDA/ODA   6FDA/4-BDAF   ODPA/APB

6FDA/DDSO2   6FDA/APB

Figure 12 Optical transparency of various polyimide films. (From Ref. 12.)

$-C(CF_3)_2-$ and $-SO_2-$ groups, and shifts to ether $(-O-)$ linkages, optical transparency increases [12].

With respect to the actual light absorption mechanisms, the visible region differs from that of the near-infrared. The light absorption of most organic polymers is based on the vibration of chemical bonds. The third harmonics of both the C–H and O–H bond vibrations are located in the near-infrared wavelength region. These C–H absorptions can be shifted to longer wavelengths by replacing the hydrogen with deuterium or fluorine. Figure 13 compares the fundamental stretching bands and their harmonics for these three bonds, shifting the C–H third harmonic at 1.2 μm to 1.4 μm for C–D and to 2.8 μm for C–F bonds, and thereby causing the C–H based polyimide to become transparent at 1.3 μm [13]. Low-loss optical fibers at 1.3 μm have been fabricated with either deuterated or fluorodeuterated polystyrenes [14]. Further transparency into the 1.7 μm range can only be achieved with fluorinated structures.

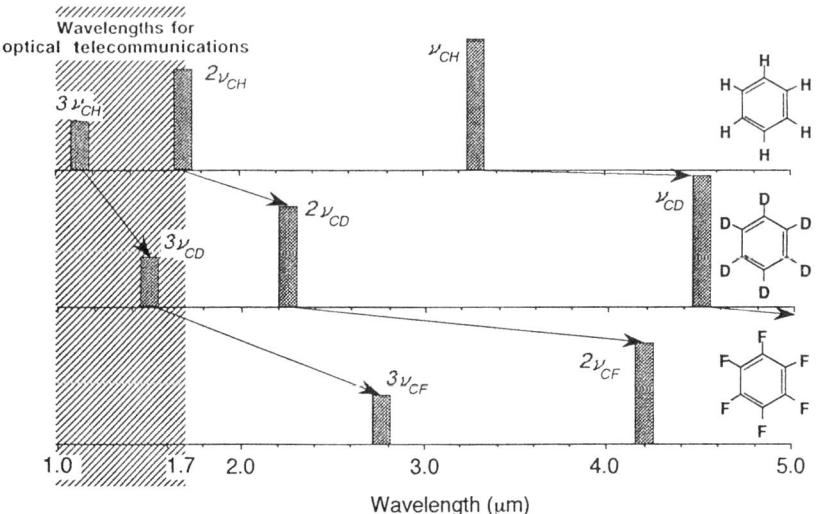

Figure 13 Schematic representation of the C–H, C–D, and C–F fundamental stretching bonds and their harmonic absorption wavelengths. (From Ref. 17.)

E. Refractive Index

Another important property for most optical materials is the refractive index. Optical interconnections between materials with different refractive indices can cause a light signal loss due to light scattering and reflections. Tight control over the refractive index of a material is of critical importance to assure optimum optical operations, especially when they are to operate as single-mode optical waveguides used for optical communications. Such waveguide devices consist of a high-refractive-index core material embedded in a slightly lower refractive-index cladding to assure single-mode operation and low optical loss. The schematic in Fig. 14a outlines the fabrication sequence for generating an optical waveguide device, and Fig. 14b is a cross-sectional micrograph of such structure. The rectangular 8×8 μm core is completely embedded in the cladding layer, the differences of refractive index between these two materials being less than 0.4%. Precise control of the refractive index difference enhances the single-mode device operations.

The average refractive index of a family of polyimides decreases with increased fluorine content, due to changes in molecular polarization (P). The relationship between the refractive index (n) and P follows the Lorentz–Lorenz formula:

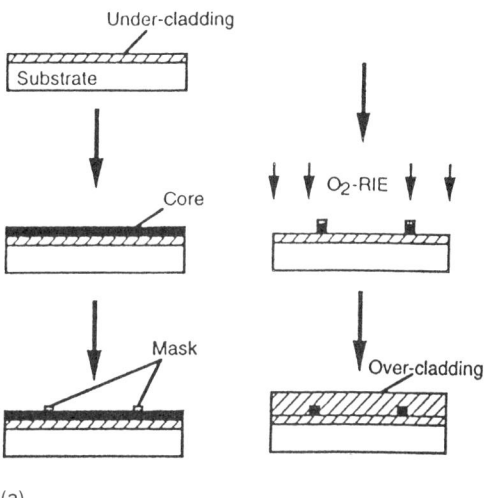

(a)

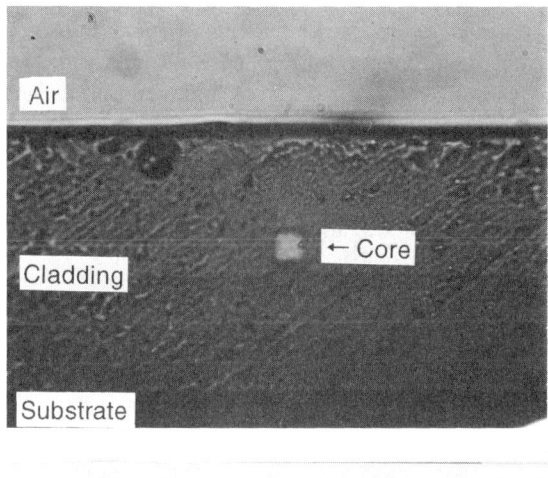

(b)

Figure 14 Single-mode waveguide: (a) outline of process flow; (b) cross-sectional micrograph of 8 × 8 μm imbedded core.

$$R = \frac{4\pi}{3} NP = \frac{n^2 - 1}{n^2 + 2} \frac{M}{\rho}$$

$$n = \left(\frac{1 + P \cdot \frac{\rho}{M}}{1 - P \cdot \frac{\rho}{M}} \right)^{1/2} \tag{11}$$

where R is the molecular refraction, N the Avogadro number, M the molecular weight, and ρ the density. The value for P (ρ/M) decreases with increasing fluorine content, and hence n decreases.

Adjustment of the refractive index is achieved by copolymerizing two fluorinated polyimides, each having different refractive indices. This effect is shown in Fig. 15, in which the molecular content of the copolyimides covers the whole composition range and offers an opportunity to achieve refractive index values ($n_{\parallel}$) between 1.520 and 1.573.

Deposited polyimide films exhibit large optical anisotropy; that is, the in-plane refractive index parallel to the substrate surface ($n_{\parallel}$) is different from that of the out-of-plane value ($n_{\perp}$). The difference between these two values is known as the birefringence and is caused by the molecular orientation of the polyimide. The polymeric structure in the vicinity of the polyimide/substrate interface is influenced by chemical and adhesion forces, in which the polymer chains are forced into a more ordered orientation. Material curing and annealing shifts the refractive index upward (Fig. 16) due to film densification and thickness reduc-

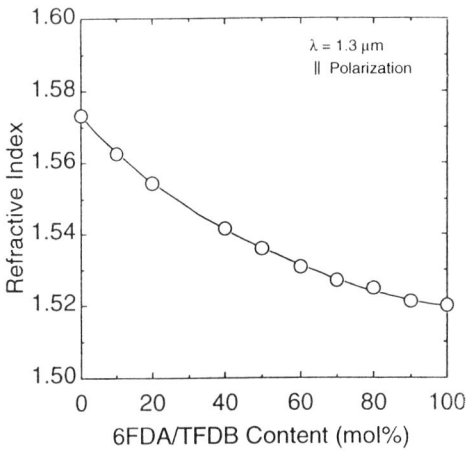

Figure 15 Refractive index of fluorinated copolyimides.

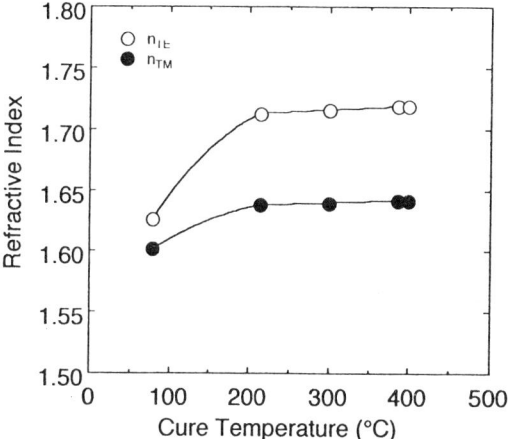

Figure 16 Refractive indices as a function of curing and annealing conditions for converting PMDA/ODA poly(amic acid) to polyimide.

tion during the imidization process. The birefringence increases by a factor of 2.5 as a result of more molecular orientation parallel to the surface.

VI. POLYIMIDE PROPERTY MODIFICATIONS

Even though the general class of polyimides has generally attractive thermal, physical, and chemical properties, most of them require some degree of modification to tailor those properties for specific applications. In most instances, any one particular property is easily modified via changes in the polyimide chemical content. However, the change of one property normally results also in shifts of most of the other physical properties. Process optimization to accommodate these synergistic changes has been the foundation of most of the synthesis development activities of polyimides in the context of their applications for high-performance microelectronic and optoelectronic devices, an example of which is reviewed below for a particular fluorinated polyimide family.

Property modifications rely on three primary principles:

1. Realize lower-dielectric-constant material, higher water absorption resistance, better optical transparency, and lower refractive index through increased fluorine content.
2. Create lower thermal expansion and higher thermal stability mainly via rigid rodlike structure formation.
3. Adjust thermal expansion and refractive index through copolymerization.

Again, it must be pointed out, that each of these parameter modifiers is not the only means to affect the property. They are, however, good starting points. The level of fluorine content, as opposed to most other chemical constituents, tends to have the largest impact on the material property shifts, while chemical group structures and copolymerization compositions allow additional fine-tuning.

A. Copolymerization

The synthesis scheme for copolymerization is given by

$$\text{(12)}$$

This synthesis follows the same two-step process outlined in Eq. (2). Depending, however, on the constituent compositions, a variety of possible polyimides are realized. The individual homo polyimides PMDA/TFDB and 6FDA/TFDB are of course formed in the absence of the other respective tetracarboxylic dianhydride.

The basis for most copolymerizations is an attempt to blend or combine the properties of the individual polymers. In the case of the two diamines, 6FDA and PMDA, the high fluorine content of 6FDA fosters structure flexibility, while fluorine-free PMDA enhances rigid-rod structure characteristics. The third monomer (TFDB) in Eq. (12) exhibits a combination of rigid-rod structures and additional fluorine [15]. Combining these three monomers to various degrees of molecular concentrations results in a variety of copolyimides displaying appropriate shifts in various properties over a large range. Table 9 summarizes the key properties of the respective homo polyimides for this family of copolyimides. Being able to achieve a full range of copolyimides with this particular set of monomers is somewhat unique, in that the TFDB maintains a high polymerization activity despite the presence of fluoro substitutions in its structure. In poly(amic acid) synthesis containing other fluoro-containing diamines, the amino group exhibits greater electron affinity toward the fluoro group, thereby reducing the polymerization reactivity. This in turn restricts the ability to realize the full range of potential copolyimide properties.

B. Modifications of Physical Properties

The fluorine is responsible for a much lower water absorption, as expected. The low moisture absorption of 0.2% for 6FDA/TFDB is about six times lower than observed for commercial Kapton, which minimizes shifts in the dielectric constant under high-humidity conditions.

The change in thermal expansion over the full copolyimide range for the foregoing two homo polyimides is shown in Fig. 17, along with the values of several inorganic materials. By blending high thermally expanding 6FDA/TFDB (TCE of $8.2 \times 10^{-5}/°C$) with PDMA/TFDB (TCE of $-5 \times 10^{-6}/°C$), any value

Table 9 Characteristics of Two Fluorinated Polyimides

	6FDA/TFDB	PMDA/TFDB
Fluorine content (%)	31.3	23.0
Intrinsic viscosity (dL/g)[a]	1.00	1.79
Polymer decomposition temp. (°C)[b]	569	610
Glass transition temp.[c]	335	> 400
Dielectric constant[d]		
Dry condition	2.8	3.2
Wet condition (50% RH atmosphere)	3.0	3.6
Water absorption (%)[e]	0.2	0.7

[a]Poly(amic acid) measured in DMAc at 30°C.
[b]10% weight loss in N_2 atmosphere.
[c]Measured by differential scanning calorimetry.
[d]At 1 MHz.
[e]After 3 days.
Source: Ref. 15.

within this range can be realized in order to best match the copolyimide to the appropriate microelectronic inorganic material. The relatively large difference between the homo polyimides is contributed to the more flexible 6FDA/TFDB structure due to the looser molecular packing of the $-C(CF_3)_2-$ groups.

Another aspect of the thermal expansion properties is shown in Fig. 18. The temperature-driven elongation curves indicate that the copolyimide containing 10 mole % 6FDA/TFDB (imidized at 350°C on a silicon wafer, cooled, and subsequently peeled off the wafer) exhibits little change during its two temperature excursions. Since the TCE of the 10% composition matches that of silicon,

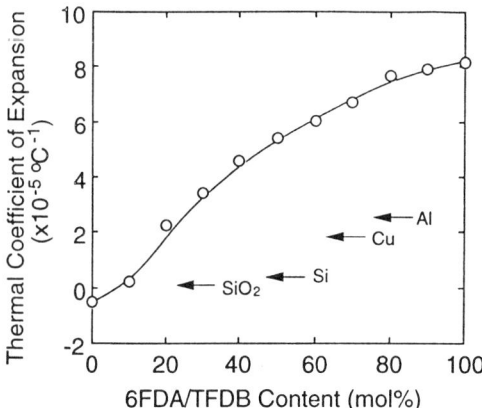

Figure 17 Thermal coefficient of expansion for fluorinated copolyimides, measured as the mean between 50 and 300°C. (From Ref. 16.)

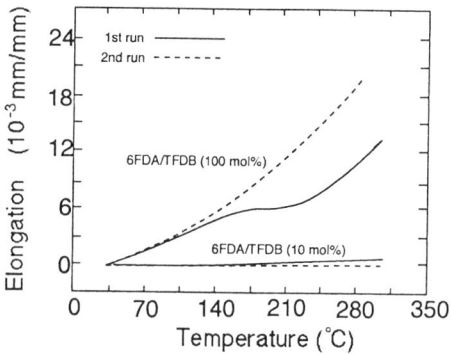

Figure 18 Thermomechanical analysis (TMA) curves of fluorinated copolyimides. (From Ref. 16.)

no stress was created during the postimidization cool-down, and hence no change in the TCE. On the other hand, the 100% 6FDA/TFDB polyimide sample contracted significant stress due to its large TCE difference with silicon. Subsequent heat cycling released this stress, as exhibited by the observed increase in elongation.

C. Modification of Visible Transparency

These same fluorinated polyimides and the full range of their copolyimides have high optical transparencies at the visible wavelength. A 10 µm thick 6FDA/TFDB film is colorless after a 350°C imidization process. Its UV-visible transmission spectra is compared in Fig. 19 with three other variants of this polyimide family. The DMDB-based polyimides have structures in which the $-CF_3$ group in the TFDB is replaced with a $-CH_3$ group. This substitution shifts the absorption cut-off to higher wavelengths by about 50 nm. This is accompanied by a loss of transmittance in the visible region to below 90% (the residual 10% absorption of the 6FDA/TFDB is due to fresnel reflections at the polyimide/air interface). The favorable transparency of the highly fluorinated polymer relies on the combined effects of the fluorine's general influence on inhibiting molecular interactions, the $-C(CF_3)_2-$ groups in the 6FTA reducing the π-conjugation in the polymer chain, and the inside $-CF_3$ groups in the TFDB preventing the biphenyls from lying in the same plane. High transmission corresponds to low optical loss and this inverse relationship is observed for this family of polyimides (compare Table 8 and Fig. 19).

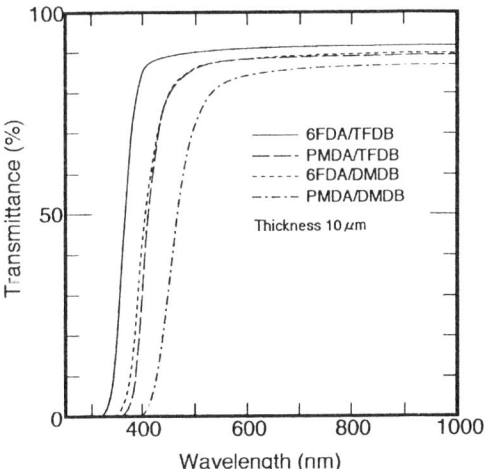

Figure 19 The UV-visible spectra of various polyimide films.

D. Modification of Near-Infrared Transparency

In contrast to the visible light absorption being caused by a charge transfer mechanism, the absorption of near-infrared is initiated by the harmonics of molecular vibration. The near-infrared absorption spectrum for 6FDA/TFDB in a 10% acetone-d_6 is shown in Fig. 20. Three peaks are observed: the third harmonic of the C–H bond stretching vibration ($3\nu_{CH}$, 1.1 µm), a combination of the C–H bond's second harmonic stretching vibration and its deformation vibration ($2\nu_{CH} + \delta_{CH}$, 1.4 µm), and the C–H bond's second harmonic stretching vibration ($2\nu_{CH}$, 1.65 µm). These absorptions reduce the transparencies at the desirable telecommunication wavelengths of 1.3 and 1.55 µm, although the optical loss at 1.3 µm is estimated to be less than 0.1 dB/cm. This is significantly less than what is observed for conventional polymers (i.e., polymethylmethacrylate and polystyrene, due to their larger C–H bond content). Also, as indicated in Fig. 20, the presence of water in polyimides enhances the light absorption at 1.4 µm due to the O–H bond's second harmonic stretching vibration.

To minimize absorption in this wavelength region, polyimides are modified through additional fluorine content. Perfluorinated polyimides exhibit very high transparency over a large range of the near-infrared region, and they remain adequately stable at higher temperatures. They are synthesized from tetrafluoro-*m*-phenylenediamine (4FMPD) and 1,4-bis(3,4-dicarboxytrifluorophenoxy) te-

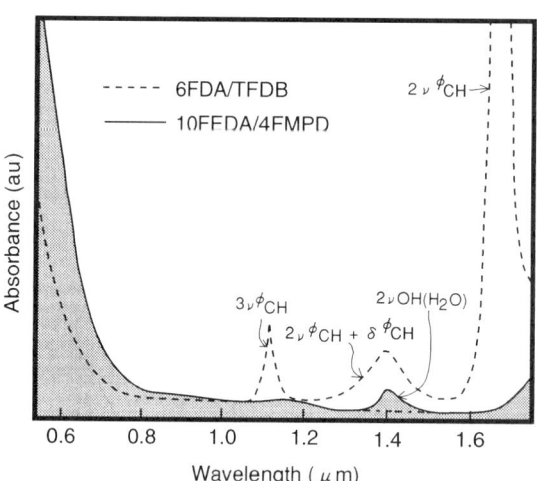

Figure 20 Light absorption spectra of fluorinated polyimide (6FDA/TFDB) and perfluorinated polyimide (10FEDA/4FMPD) solutions. (From Ref. 17.)

trafluorobenzene dianhydride (10FEDA) [17]:

$$\text{10FEDA} + \text{4FMPD} \xrightarrow[\text{Solvent: DMAc}]{\text{Room temp}} \text{10FEDA/4FMPD Poly(amic acid)} \xrightarrow[-H_2O]{\Delta} \text{10FEDA/4FMPD Polyimide} \quad (13)$$

The synthesis of polyimide is challenging with most perfluorodiamines in which the fluorine atoms are attached at the ortho (o) or para (p) positions, due to their low reactivity with amino groups. This is less complicated with the meta (m) positioned fluorine in 4FMPD. Combining it with the long monomer 10FEDA (which was specifically designed for perfluorinated polyimide synthesis) results in flexible, high-molecular-weight polyimides. The flexibility is mainly due to the ether linkages. The improved light absorption spectra of 10FEDA/4FMPD is also shown in Fig. 20. Except for a small absorption peak at 1.4 μm (due to moisture either on the film surface or absorbed by the remaining solvent), this perfluorinated polyimide has no substantial absorption over the entire near-infrared range. It is thermally stable, with its $T_{d,0.1}$ being at 501°C, and its T_g at 301°C.

This improved transparency reflects itself in optical loss measurements of optical waveguides using this material. Figure 21 shows the optical loss of polyimide waveguide using the fluorinated copolyimides of 6FDA/TFDB and PMDA/TFDB. Note the transparent "windows" at 1.3 and 1.55 μm with measured losses of less than 0.3 dB/cm.

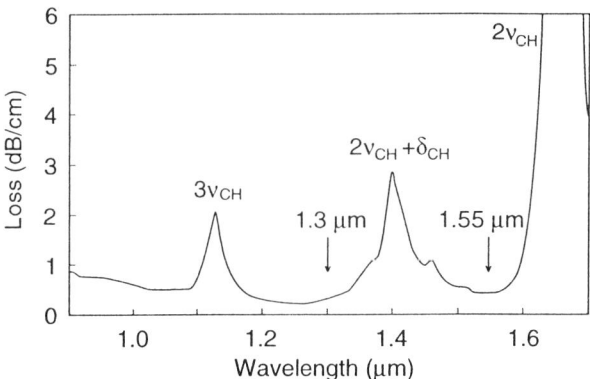

Figure 21 Loss spectrum of a single-mode waveguide. (From Ref. 5.)

VII. SUMMARY

Although polyimides are easily processable, and synthesis efforts can create a large family of polyimides with modified mechanical, physical, and optical properties, the future demands for higher-performance microelectronic and optical devices will pose continued challenges to the synthetic polymer chemist. Improved heat resistance is becoming increasingly important. Optical interconnects have unique advantages over electrical interconnects, and thus the integration of optoelectronic with microelectronic devices will require significant new development efforts to create successful new families of polyimides and other polymers.

EXERCISES

1. Aromatic polyimides are synthesized from aromatic diamines and benzenetetracarboxylic dianhydrides. Give the synthesis scheme for the polyimide starting from benzophenonetetracarboxylic dianhydride (BTDA) and *m*-phenylene diamine (MPD).

2. It is crucial for silicon devices to prepare the polyimide layer without stress. Give the polyimide structure of a non-copolyimide suitable for this use.

3. The polyimide 6FDA/TFDB is synthesized from 6FDA (Mw = 444) and TFDB (Mw = 320). It has a low dielectric constant and low water absorption, because of its high fluorine content. Calculate the fluorine content of this polyimide using the given molecular weights of the monomers.

4. Give relative polymerization reactivity for the following three aromatic diamines in poly(amic acid) synthesis.

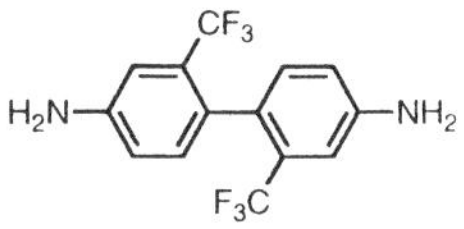

(a) TFDB

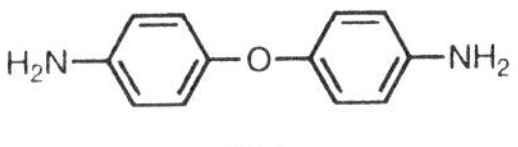

(b) ODA

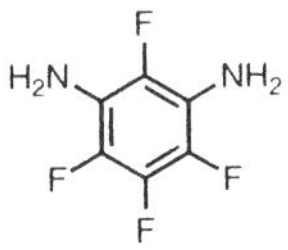

(c) 4FMPD

5. Polymerization from the poly(amic acid) of an aromatic diamine and benzenetetracarboxylic dianhydride in organic polar solvents is carried out at ambient temperature. Why? What happens at higher temperatures?

6. Discuss the molecular structures of new polyimides from the viewpoints of (a) application to microelectronics (low dielectric constant, thermal expansion control, low water absorption, high thermal stability, etc.); (b) application to optical interconnects (optical transparency, refractive index control, high thermal stability, etc.).

GLOSSARY

Structure (Benzenetetracarboxylic dianhydride)	Acronym	Composition name
	PMDA	pyromellitic dianhydride
	BPDA	3,3'4,4'-biphenyltetracarboxylic dianhydride
	BTDA	3,3'4,4'-benzophenonetetracarboxylic dianhydride
	6FDA	2,2-bis(3,4-dicarboxyphenyl) hexafluoropropane dianhydride
	ODPA	4,4'-oxydiphthallic anhydride

Structure (Benzenetetracarboxylic dianhydride)	Acronym	Composition name
	SO2DPA	3,3'4,4'-sulfonyldiphthalic anhydride
	HQDEA	1,4-bis(3,4-dicarboxyphenoxy)benzene dianhydride
	BDSDA	4,4'-bis(3,4-dicarboxyphenoxy) diphenylsulfide dianhydride
	10FEDA	1,4-bis(3,4-dicarboxytrifluorophenoxy) tetrafluorobenzene dianhydride

Structure (Aromatic diamine)	Acronym	Composition name
H₂N–⟨⟩–O–⟨⟩–NH₂	ODA	4,4'-oxydianiline
H₂N–⟨⟩–NH₂	PPD	*p*-phenylenediamine
H₂N–⟨⟩–NH₂ (meta)	MPD	*m*-phenylenediamine
H₂N–⟨F₄⟩–NH₂	4FMPD	tetrafluoro-*m*-phenylenediamine
H₂N–⟨CH₃⟩–⟨CH₃⟩–NH₂	DMDB	2,2'-dimethyl-4,4'-diaminobiphenyl
H₂N–⟨CF₃⟩–⟨CF₃⟩–NH₂	TFDB	2,2'-dis(trifluoromethyl)-4,4'-diaminobiphenyl

Structure (Aromatic diamine)	Acronym	Composition name
H₂N–⟨⟩–CH₂–⟨⟩–NH₂	MDA	4,4'-methylenedianiline
H₂N–⟨⟩–⟨⟩–NH₂	BZ	benzidine
H₂N–⟨⟩–⟨⟩–⟨⟩–NH₂	DPTP	4,4''-diamino-*p*-telephenyl
H₂N–⟨⟩–O–⟨⟩–O–⟨⟩–NH₂	APB	1,3-bis(aminophenoxy)benzene
H₂N–⟨⟩–O–⟨⟩–C(CF₃)₂–⟨⟩–O–⟨⟩–NH₂	4-BDAF	2,2-bis[4-(4-aminophenoxy)phenyl] hexafluoropropane
H₂N–⟨⟩–C(CF₃)₂–⟨⟩–NH₂	4,4'-6F	2,2-bis(4-aminophenyl) hexafluoropropane
H₂N–⟨⟩–SO₂–⟨⟩–NH₂	DDSO2	4,4'-diaminodiphenylsulfone

REFERENCES

1. C. E. Sroog, A. L. Endrey, S. V. Abramo, C. E. Berr, W. M. Edwards, and K. L. Oliver, *J. Polym. Sci., Part A 3*:1373 (1965).
2. C. E. Sroog, *J. Polym. Sci., Macromol. Rev. 11*:161 (1976).
3. H. Satou, H. Suzuki, and D. Makino, in *Polyimides* (D. Wilson, H. D. Stenzenberger, and P. M. Hergenrother, eds.), Blackie, Glasgow and London, 1990, pp. 227–251.
4. R. Ruiter, H. Franke, and C. Feger, *Appl. Opt. 27*:4565 (1988).
5. T. Matsuura, S. Ando, S. Matsui, S. Sasaki, and F. Yamamoto, *Electron. Lett. 29*: 2107 (1993); C. T. Sullivan, *SPIE 994*:92 (1988).
6. W. J. Farrissey, L. M. Alberino, and A. A. R. Sayigh, *J. Elast. Plast. 7*:285 (1975).
7. J. Greenblatt, C. J. Araps, and H. R. Anderson, Jr., in *Polyimides* (K. L. Mittal, ed.), Plenum Press, New York and London, *1*, 1984, pp. 573–588.
8. G. C. Davis, B. A. Heath, and G. Gildenblat, in *Polyimides* (K. L. Mittal, ed.), Plenum Press, New York and London, *2*, 1984, pp. 847–869.
9. H. Ahne, H. Krüger, E. Pammer, and R. Rubner, in *Polyimides* (K. L. Mittal, ed.), Plenum Press, New York and London, *2*, 1984, pp. 905–918.
10. S. Numata, K. Fujisaki, D. Makino, and N. Kinjo, *Proceedings of the Second International Conference on Polyimides, Recent Advances in Polyimide Science and Technology*, (W. Weber, M. R. Gupta, eds.) Society of Plastics Engineers, New York, 1987, pp. 164–173.
11. A. K. St. Clair, T. L. St. Clair, and W. P. Winfree, *Polym. Mater. Sci. Eng. Proc. 59*:28 (1988).
12. A. K. St. Clair, T. L. St. Clair, and K. I. Shevket, *Polym. Mater. Sci. Eng. Proc. 51*:62 (1984).
13. S. Ando, T. Matsuura, and S. Sasaki, *ACS Symposium Series 537, Polymers for Microelectronics*:304 (1994).
14. T. Kaino, *Appl. Phys. Lett. 48*:757 (1986).
15. T. Matsuura, Y. Hasuda, S. Nishi, and N. Yamada, *Macromolecules 24*:5001 (1991).
16. T. Matsuura, N. Yamada, S. Nishi, and Y. Hasuda, *Macromolecules 26*:419 (1993).
17. S. Ando, T. Matsuura, and S. Sasaki, *Macromolecules 25*:5858 (1992).

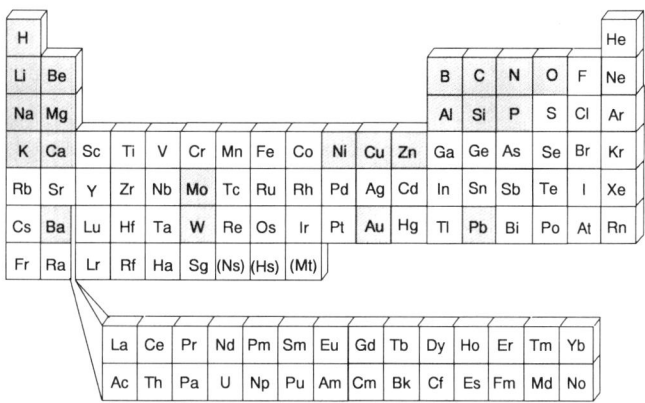

11
Ceramic Materials

Charles H. Perry   *IBM Corporation, Hopewell Junction, New York*

I. INTRODUCTION

Ceramic materials along with wood and metal have been used from the dawn of civilization. It was observed that the properties of fired clay were ideal for containing liquids. Vessels made from clay were sturdy, did not deteriorate, and did not absorb moisture. Similarly, ceramic materials were excellent for construction. Today, these properties of ceramics are referred to as mechanically rigid, chemically stable, and impermeable.

With the discovery of electricity, additional properties of ceramics were observed. Early investigators noticed that glass was an excellent dielectric. This led to the use of glass and ceramic materials in scientific and engineering applications such as the Leyden jar and as insulators in telegraph and electricity transmission. The use of ceramic materials in electrical and electronic applications has since grown to where ceramics are ubiquitous in everyday life. Their unique properties have found applications from wristwatches to appliances and automobiles, as well as in consumer electronics, communications, and computers.

Ceramic and glass materials are nonmetallic and inorganic. Although they are typically oxides of either metals or nonmetals such as Al_3O_3 and SiO_2, others can include oxides of alkali, alkali earth, and transition metals. In fact, depending on the application, almost any oxide of the elements in the periodic table may be found as constituents in some ceramic or glass composition. Examples include lead oxide in low-temperature (<400°C) sealing glasses useful for hermetic seals as discussed in Chapter 2, and cerium oxide in oxygen sensors for automobile engine emission control systems. Since the early 1980s, another family of ceramic materials has found wide application. These materials are metal and nonmetal nitrides and carbides. Of these, AlN in particular has been extensively developed for those electronic packaging applications requiring high thermal conductivity.

While ceramics and glasses may possess the same chemical composition, some of their physical properties may differ due to differences in the physical arrangements of their atomic components. In ceramics, the atoms are organized in a long-range periodic crystal structure (on the order of 100 Å or greater depending on the material) like repeating tiles on a floor. However, since glass is a supercooled liquid and is in a metastable state, its structure does not have long-range crystal order, but a long-range random order. The differences in the bond lengths and atomic spatial interrelationships for quartz (crystalline SiO_2) and amorphous or glassy SiO_2 are illustrated in Fig. 1, and these characteristics will influence, limit or define certain material processes.

The most desirable properties of ceramics for electronic applications are summarized in Fig. 2. Compared with organic materials such as plastics, ceramics offer long-term superior resistance to corrosion and high temperature. Ceramics are also impermeable to liquids and gases. In fact, a fully densified ceramic body is impermeable to any molecule, including helium. This characteristic becomes important for those applications in which mechanical stability and insensitivity to environmental conditions are required. The impermeability prevents the possibility of warping or swelling of the insulating material.

The overall wiring density achievable with ceramic packaging can be greater than that for plastic packaging. This is a direct result of the ceramic packaging process fabrication techniques. While 95% of the nonsophisticated commercial integrated circuit (IC) products do use plastic packaging, unique high-performance, high-power IC devices require much greater wiring density, better thermal and mechanical stability, and better thermal management capability—all offered by ceramic packaging technology. Such high-performance ICs and associated ceramic packaging are essential for mainframe and mid-sized computers, massively parallel processors, high-performance workstations, and in some personal

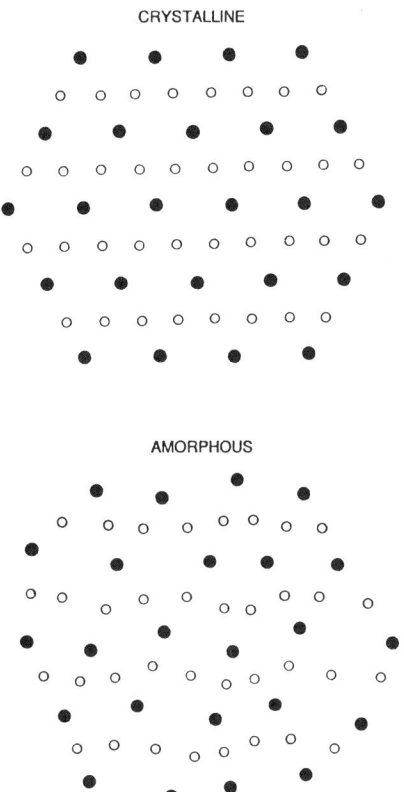

Figure 1 Two-dimensional schematic representation of crystal and amorphous SiO_2. ●: silicon atoms; ○: oxygen atoms.

computers. Ceramic packaging products requiring the same degree of material stabilities, but not necessarily the same high wiring density, are found on automobile engines and in high-performance aircraft. Table 1 lists a comparison of typical characteristics of plastic and ceramic packagings.

Traditional ceramic materials have been employed for many years to fabricate large quantities of discrete passive electronic elements such as resistors, capacitors, and inductors. These continue to be made today. However, it has been realized since the early 1970s that active devices (i.e., transistors and integrated circuits) could also greatly benefit from specific ceramic carriers (or packages) when these can be incorporated as functional interfaces for electrical connections, power transmission, and device cooling. This led to special ceramic and glass package technologies. They have progressively become more sophisticated

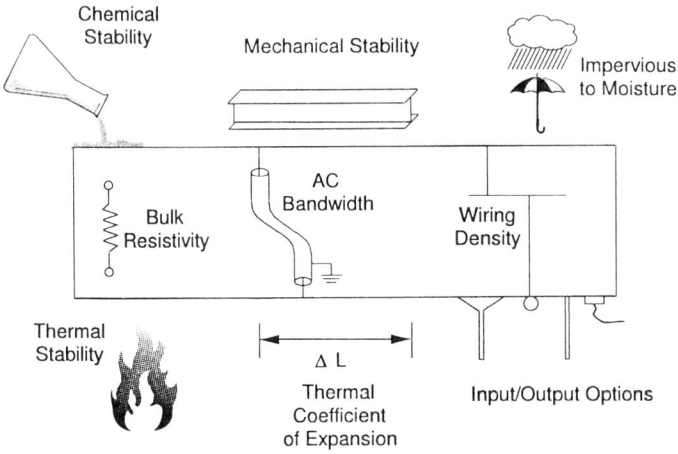

Figure 2 Desired properties of ceramics for electronic applications.

Table 1 Properties of Ceramic and Organic Cards

Property	Low/high temperature ceramic	Epoxy/polyimide organic
Via diameter (μm)	75	100
Linewidth (μm)	75	60
Grid (μm)	225	1200
Dielectric constant	5 to 9	4 to 5.5
Dissipation factor	0.001 to 0.002	0.01 to 0.03
Resistivity (Ω cm)	10^{16}	10^{14}
Thermal conductivity (W/mK)—no metal	5 to 20	0.2 to 0.3
Water adsorption (%)	None	0.15 to 0.3
Maximum use temperature (°C)	400	150 to 200
Thermal coefficient of expansion (ppm)	3 to 6	16 to 20

to match the growing needs of the increasingly complex integrated electronic device technologies. These new ceramic carriers not only contained the required basic interface features, they also exhibited high reliability and high performance along with the ability to absorb mechanical shock and to protect against a number of environmental factors.

The development of ceramic packages since the 1970s can be viewed in terms of three phases. From the beginning, it was recognized that circuit chips could be denser and faster if multiple transistors were placed on one chip carrier to provide the electrical interconnections, and if resistor elements could be incorporated into that same carrier structure. Such "thick-film" resistors offered the flexibility to tailor the resistance for optimum electrical performance with laser trimming, subsequent to attaching the transistor onto the carrier package. The thick-film resistor technology was quickly followed by the development of a thick wiring process technology. Their joint incorporation onto an insulating ceramic body was the beginning of the successful development of high-performance computers. Examples of early ceramic packages are shown in Fig. 3.

The device chip technology rapidly progressed from individual discrete transistors to integrations of tens, then hundreds, and even thousands of transistors on a single silicon chip. The requirements for the number of input/output (I/O) connections to address the various transistors increased accordingly. However, this presented a physical problem: the full set of connecting wires or solder balls, with diameters of about 75–100 µm, required more space than was physically available on the phase one ceramic package. Hence, new carrier design concepts were required for this next phase, some of which were touched on in Chapter 2. The key feature of these new packages was centered on a horizontal spreading out of the array of electrical connections into a larger dimensional matrix, which then offered the ability for pin or flex cable connections. This

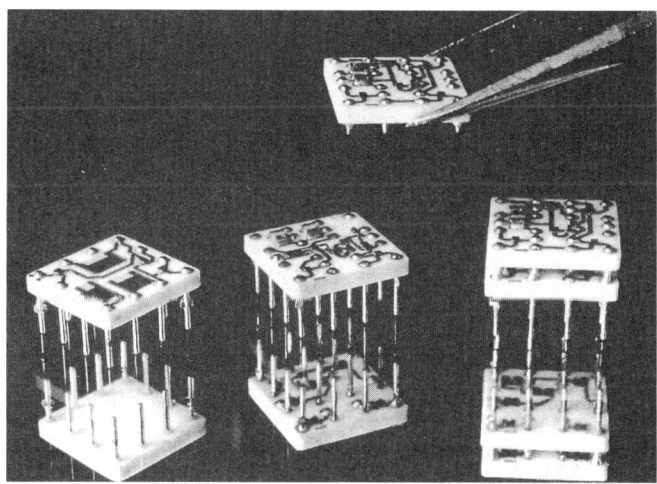

Figure 3 Early ceramic packaging using thick-film fired metallurgy and integrated resistors.

function of the package is known as "fan-out" or space transformation. An example of a package fan-out to a pin array is presented in Fig. 4.

The third phase of the ceramic package development relates to the requirements of wiring multiple integrated logic circuits and memory device chips together at very close proximity (to create maximum performance), as indicated in Fig. 5. To accommodate such very-high-density electrical interconnections, the modern supporting packages evolved into three-dimensional multilevel wiring ceramic or glass packages. These multichip supporting modules (MCMs) can typically contain any number of wiring layers. In high-speed device applications, the three-dimensional wiring package is a collection of transmission lines with controlled resistance, inductance, and capacitance. All of these functions are designed into a ceramic package material whose dielectric is a suitable media for the wiring network. A cross section of a high-performance MCM ceramic package is shown in Fig. 6. The wiring network is composed of "Z" connections (called vias) as well as "X" and "Y" connections. This type of package can wire up to 100 integrated circuits and memory chips with hundreds of I/O connections per device. On the bottom surface of the package is an array of pins for connecting the MCM to the next level of wiring.

Note that the dramatic progress made in ceramic technology, as it relates to ceramic electronic packaging, is closely allied with a similar rate of progress in new understanding of many basic and applied chemistry elements and chemical

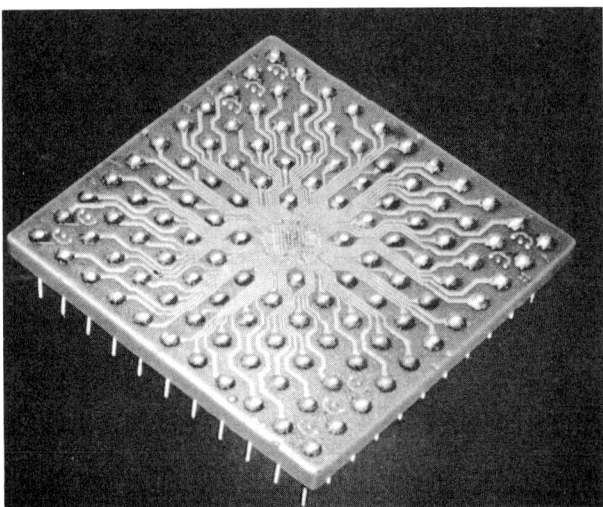

Figure 4 Ceramic package illustrating grid expansion from the chip I/O pitch to the pin I/O pitch.

Ceramic Materials

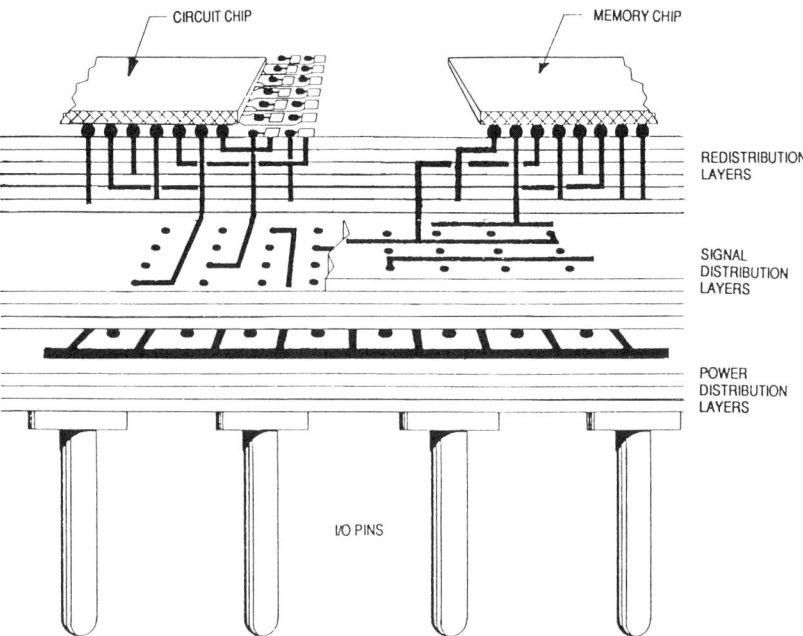

Figure 5 Multichip module (MCM) cross section, showing internal wiring layers and chip interconnections.

engineering concepts—from raw materials to reliable and manufacturable processes of finished products. Al_2O_3 is the most commonly used material in ceramic packaging. It may be pure alumina, or it could be a mix that includes a wide variety of additives to suit a particular process or product requirement. The interconnecting metals can range from refractory elements, like W and Mo, to noble elements, like Au, Ag, and Cu. The specific fabrication process will depend greatly on the product requirements, a list of which is given in Table 2. As can be seen, the considerations for a given product can be extensive. An apparently trivial requirement, such as substrate color, can result in significant development efforts to identify an appropriate color additive that is compatible with the ceramic and metallurgy in microstructure, metal/ceramic interface properties, and substrate strength. Some of these considerations will become more apparent in the following sections describing the fabrication processes and their associated chemistry.

Given the wide spectrum of the various ceramic packaging, and the even wider set of possible oxides and metals involved in fabricating such packages, it would be counterproductive to exhaustively detail the various chemistries that

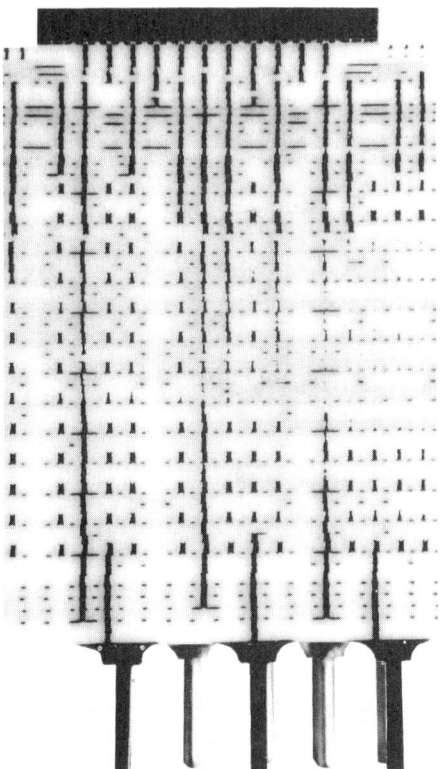

Figure 6 Cross section of 66-layer MCM package showing vertical via columns and portions of X–Y wiring interconnecting the chips and pins. (Courtesy of IBM.)

are applicable to the specific processes and products. Many of these details can be found in Ref. 1. However, regardless of a specific product, a specific raw material, or a specific process, there are a number of common considerations applicable to all systems. This chapter will discuss, with the aid of the multilayer ceramic (MLC) process as an example, various aspects of organic–organic, organic–inorganic, and inorganic–inorganic reactions. A basic understanding of these interactions will facilitate their applications to many other ceramic materials systems and fabrication processes.

Table 2 Ceramic Packaging Product Requirements

Cost
Thermal coefficient of expansion
Strength
Dielectric constant
Color
Electrical conductivity
Thermal conductivity
Wiring density
I/O system
Surface finish
Dimensional control
Integrated resistors/capacitors
Operating environment (temperature, atmosphere)

II. ELEMENTS OF MODERN ELECTRONIC PACKAGE FABRICATION

The fabrication of ceramic packaging can be divided into two categories based on the method used to fire or sinter the thick-film metals: post-fire or co-fire. Post-fire involves forming and firing a ceramic body in the absence of a metal. The metal is subsequently applied to the fired ceramic in a paste or ink form, and the combined system is then fired for a second time. Post-fire technology was developed first and is used in lower-cost, low-wiring-density applications. The co-firing process involves the concurrent firing of both the ceramic and metal, bringing along more process complexity. Co-firing of ceramic packaging is usually required in multilayer wiring application requirements.

The ensuing discussion will revolve around the co-firing fabrication methods, since the applicable chemical considerations generally also encompass those of the post-firing process. Similarly, the overall chemistry is more complex for the co-firing process—from the initial raw materials to the actual firing requirements. And the interactions between the various material components in the ceramic green body (unfired material), the metal paste, and the firing profile tend to be more critical and more challenging to control. Figure 7 is a summary of the process fabrication steps required for a multilayer electronic ceramic package.

Ceramics and glass are refractory, brittle materials that cannot be easily heated and formed into shapes. A_2O_3, for example, has a melting temperature of 2015°C. Even with additives like B_2O_3, melting temperatures much below 1000°C are difficult to achieve for any high-percentage alumina mixtures. For

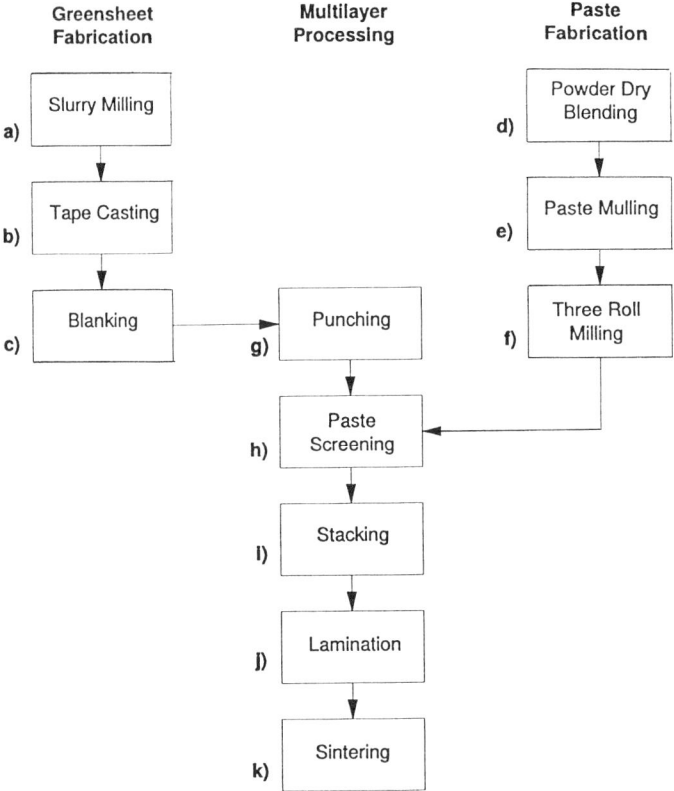

Figure 7 Sequence of the major processing elements to fabricate a multilayer electronic package.

this reason, ceramics are first transformed and processed as fine powders in an organic or aqueous solvent carrier to form a ceramic slurry.

To achieve the desirable ceramic particle size range, the ceramic material first undergoes a milling process [2], of which there are several techniques. All of them utilize some physical agitation system in which the ceramic is mixed with an organic liquid carrier or solvent, and a milling media (usually alumina). The milling media reduces the average ceramic or glass particle size by agitation in that it induces fractures on the particle surface due to mechanical impact. The average particle size reduces with milling time and material removal. Typically, the desired ceramic powder particle sizes are on the order of a few microns, but they may need to be submicron for specific applications. The finely ground ceramic powder–containing slurry is mechanically separated from the grinding media and dried. This dried powder (or cake) is then further processed, the details of this processing dictated by the ultimate end product. For most ceramic packaging applications, the cake is mixed for a second time (often in the same type of mill) with a liquid carrier system to form a new slurry (Fig. 7a). This second slurry is usually more complex than its predecessor. Unlike the first milling, whose primary function is to create uniform powder particles, the second milling sequence is aimed to assure a desired degree of particle dispersion and agglomeration within the slurry, and it is critical for subsequent casting processes. Thus, the second slurry mixture typically contains, in addition to solvents, other materials such as coagulents, or deflocculants, plasticizers, lubricants, and binders.

The degree of dispersion within the slurry is influenced by steric hinderances, electric charging, and van der Waals forces. Their combined effects can create particle–particle attractions to form larger aggregates (called granules), or the particles can be fully dispersed into a colloidal suspension. With the proper additions of coagulants or deflocculants, such effects can be offset. For processes like dry pressing and extrusion, appropriate agglomeration is assisted by the use of coagulants. In the case of tape casting, complete dispersion is desired, and this is achieved with deflocculents.

Subsequent to the second milling process, the freshly formed ceramic/organic slurry is pumped out of the mill and onto a caster (Fig. 7b). With a continuous caster, the ceramic mixture is formed into thin long sheets, or greensheets. The slurry is actually deposited onto a moving polymer carrier foil, as illustrated in Fig. 8. Thickness control of the greensheet is determined by the viscosity of the slurry, the distance between the doctor blade and the carrier, the speed of the carrier, and the temperature/atmosphere profile of the casting oven. During this

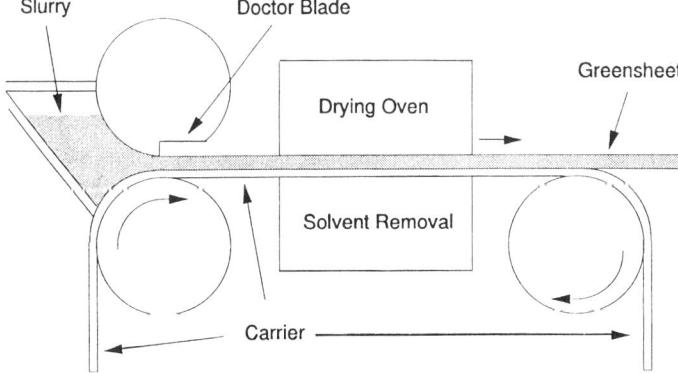

Figure 8 Schematic of a caster for fabricating greensheets from a slurry.

casting the solvent is removed from the slurry, leaving the ceramic or glass particles suspended in an organic binder matrix. The greensheet is then stripped from the carrier and picked up on a large spool. Greensheet width can vary from 100 mm up to 1 m, and their thickness can range from 0.025 to 1.0 mm. Length is arbitrary and depends on the slurry volume and type of caster, but it is typically in the range of 50 to 100 m. The specific dimensional properties of the greensheet are dictated by the desired product, as well as the tooling used in the subsequent fabrication processes. The formed greensheet is first unwound from the spool and inspected for any imperfections. This is followed by cutting the sheets into squares. The inspection data is factored into the cutting process to avoid inclusions of defect areas. Greensheet cutting utilizes a punch and die method called blanking. The operation includes the simultaneous formation of small holes on the edge of the greensheets to serve as references for indicating sheet thickness and casting direction (Fig. 7c).

A somewhat similar process sequence, as used to fabricate ceramic powders, is also exercised to suspend metal powders in a suitable carrier or vehicle to form a conductive paste (Fig. 7d–f), although with a different tool set. The dispersion of the metal into an organic vehicle facilitates the production of a screenable paste or "ink," which is used to form the electrical conduction lines within ceramic modules. The starting metal powders are in most instances produced via precipitation methods, although plasma-spraying techniques are used as well. As with ceramic particles, the size of the metal particles tends to be in the micron range. To tailor the rheological and flow properties of the metallic paste, most of the same types of additives involved in the preparation of ceramic slurries are utilized. However, it is often necessary to introduce several other additives as well to affect the metal sintering characteristics, adhesion, or resistivity. The additives are usually also powders, and the premixing is carried out in a dry blender. The powder composition is then mixed with the organics in a mulling machine quite similar to a mixer used to make bread dough. The next operation is the milling of the paste on a three-roll mill, as shown in Fig. 9. This operation not only mixes the paste further, but—most importantly—breaks up any agglomerated metal particles. Since many metal particles are malleable, their agglomerated particles cannot be broken up as is the case in a ceramic milling operation. It is also undesirable to have flattened metal particles or platelets within the paste. Hence, the powerful shearing action of the three-roll mill is used to deagglomerate the metal particles.

To achieve the three-dimensional wiring network as shown in Fig. 5, it is necessary to create an array of holes in the greensheet (Fig. 7g). This can be accomplished with a punch and die, with a laser, or with e-beam. The diameter of the individual holes is typically one-half of the greensheet thickness. The next process step is paste screening (Fig. 7h), which involves the creation of specific metal paste patterns on the surface of each greensheet. Simultaneously, the metal

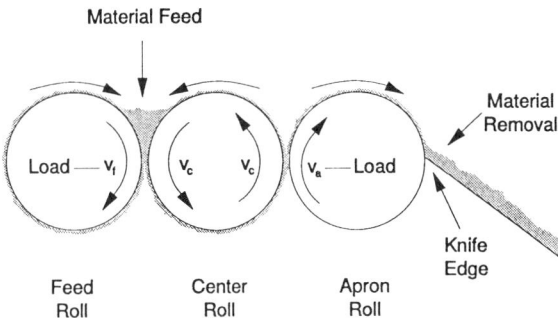

Figure 9 Schematic of a three-roll mill used for mixing a metal paste and breaking up agglomerated metal particles, in which the rotational velocities satisfy $V_f < V_c < V_a$ (V_f = feed; V_c = center; V_a = apron).

paste is forced into the low diameter/greensheet thickness aspect ratio holes. To assure efficient hole filling, the screening process is exercised under pressure. Alternatively, the paste would have to have a low viscosity, which is not always practical. This screening and hole filling process with metal paste is called personalization. It can be accomplished in one or two steps, depending on the tooling.

Personalized greensheets are then stacked (Fig. 7i) in the correct order and with the correct orientation to assure the formation of a three-dimensional conductive network during the subsequent sintering. Stacking relies on the use of alignment holes. These are located at the sheets' edges. With the proper vertical feature integrity, the key requirement of multilayer ceramic technology is met: layer to layer alignment of the metal-filled holes or vias. Alignment is a sensitive function of the mechanical expansion or contraction of the greensheet during the hole formation (punching) and screening operations. These in turn depend predominantly on the organics used both in casting the sheet and in the paste, on the operation temperature and humidity, and on various other processing variables. Even with all sheet-related variables under control, the organics within the paste can induce greensheet swelling or contraction subsequent to screening. These materials systems must therefore be engineered for compatibility to minimize sheet dimensional movement. The necessary degree of control required depends on the via diameter and overall size of the greensheet. For example, a greensheet with dimensions of 150 mm on each edge, has a 100 mm long diagonal from the center of the sheet to a corner via. For acceptable via to via alignment, the movement of the greensheet from center to edge must be less than one-third of a via diameter, or 0.04 mm for a 0.12 mm diameter via in a 0.24 mm thick greensheet, (i.e., ±0.4%). Achieving this degree of control requires careful selection of both the casting and the paste's organic systems, as well as exceptional process control in casting and screening.

After stacking, the assembly of greensheets is laminated at an elevated temperature and pressure. The temperature is generally slightly above the glass transition temperature of the binder in the greensheet, and the pressure can vary from 30 to 300 kg/cm^2.

The final process involves the firing, or sintering, of the laminated stack of "soft" greensheets (Fig. 7k). During sintering, the organic materials are removed, and the ceramic and metal particles are fused within a dense insulating matrix to transform into a tilelike hard ceramic with a fully embedded three-dimensional conducting network. This is accomplished by the combined control over temperature and atmosphere. It may also require in some instances special tooling fixtures to maintain the physical integrity of the stack. The success of the sintering process depends on two key aspects: presintering process and material control, and excellent sintering process parameter management. The former represents a culmination of maintaining absolute control over the ceramic and metal melting temperatures, particle size distributions, additives, organic systems for greensheets and paste, casting conditions, paste properties, screening variables, and lamination conditions and avoidance of any trace impurity introduction into any of the inorganic or organic materials. Having achieved control over all these variables, it becomes critical to maintain the same degree of discipline during the last phase of the process—that of sintering. For this reason, the combination of time, temperature, and atmosphere (or the so-called sintering profile) must be carefully designed and controlled to guarantee high product process yields and product reliability.

III. CHEMICAL CONSIDERATIONS IN RAW MATERIALS PROCESSING

All of the major process elements associated with modern ceramic packaging technologies, as outlined in Sec. II, rely heavily on various material–material interactions. These range from the surface chemistry of powders and the chemical material compatabilities of greensheet constituents, to the chemical interaction requirements between the greensheets and metal paste. A good understanding of these interaction characteristics and their process implications are the basis for any effective development efforts to achieve and further advance the ceramic packaging technologies.

A. Particle Surface Chemistry

Unlike most reactions that proceed between gas and liquids, the interactions of a solid material with another solid material, a liquid, or a gas are primarily dictated by the surface characteristics of the solid material and by its available surface area. Ceramic packaging technologies are primarily based on solid material surface interactions. To facilitate those solid surface reactions, the materials of interest are used in their powder form to dramatically modify their net volume/surface ratio. For example, 1 cm^3 of material has a surface area of 6 cm^2. If that same material is diced up into micron-sized cubes to realize 10^{12} cubes, a total surface area of 6×10^4 cm^2 would be generated, while its net total volume remains the same. Hence, the volume/surface area ratio of 1:6 has been changed to 1:10^4 along with the unique behavior of the produced powder in comparison with its solid form counterpart.

There are a number of theories and models dealing with the surface chemistry of materials. The most general one is based on acid–base interactions. Acids and bases are traditionally defined in terms of the types of ions donated in an aqueous solution. Acids donate hydrogen ions, while bases accept them. A more encompassing definition of acids and bases is that of accepting or donating electrons; that is, a Lewis acid is an electron acceptor, and a Lewis base is an electron donor. This definition can apply to the simplest acid–base interaction, such as a positive hydrogen ion (electron acceptor) reacting with a negative hydroxide ion (electron donor). The usefulness of this acid–base definition becomes apparent when considering interactions between materials not positively or negatively charged. Consider the reaction of trimethylboron and trimethylamine:

$$\begin{array}{c} CH_3 \\ \cdot\cdot \\ CH_3 : B \\ \cdot\cdot \\ CH_3 \end{array} \quad + \quad \begin{array}{c} CH_3 \\ ^{xx} \\ {}^{x}_{x} N {}^{x}_{x} CH_3 \\ _{xx} \\ CH_3 \end{array} \quad \rightarrow \quad \begin{array}{c} CH_3 \; CH_3 \\ \cdot\cdot \; ^{xx} \\ CH_3 : B \, {}^{x}_{x} N {}^{x}_{x} CH_3 \\ \cdot\cdot \; _{xx} \\ CH_3 \; CH_3 \end{array} \quad (1)$$

Both are neutral molecules. However, the boron in the trimethylboron molecule shares three pairs of electrons with the three methyl groups, and it would "prefer" to accept an additional two for completing an octet. This represents an electron-accepting Lewis acid site on the molecule. The nitrogen atom in the amine structure has an unshared pair of available electrons, making that portion of the molecule, and therefore the molecule itself, an electron-donating Lewis base.

A key consideration for any ceramic or metal particle surface is its Lewis acid–base character. It can be defined by the so-called isoelectronic point (IEP). The IEP is determined by suspending a ceramic powder into an aqueous solution under a potential gradient. When the nature of the powder surface is electron donating (basic), hydronium ions from the aqueous solution will be attracted to the powder particle, permit the powder to acquire a positive charge, and are then able to migrate in the electric field. By increasing the pH with the addition of hydroxide ions, a competition for the hydronium ions is set up between the particle's basic surface and the hydroxide ions in solution. With further increases in pH, a condition is reached where the ceramic particles cease to migrate. This represents the IEP for the powder. The IEPs for some ceramic powders are shown in Table 3.

The actual surface chemistry of ceramic or metal particles is considerably more complex than the simplified surface acid–base character. The particle surface can become charged through either the loss of surface ions to the medium or by preferred adsorption of species from the medium. Furthermore, the surfaces of "dried" ceramic particles very often exhibit a monolayer of water and/or residual solvents (retained from the milling process), thereby creating opportunities for particle charging.

The presence of such charged particles tends to foster particle agglomeration—an undesirable situation. It can be overcome, or at least controlled, with the selection of an appropriate chemical media to enhance particle–particle repulsion via a combination of electrostatic and steric mechanisms [3]. These chemical modifiers can be various ionic or polyelectrostatic solvents that change the character of the particle surface. There also exists another class of nonionic materials, which contain both acidic and basic functional groups on the same molecule. These Lewis acid or base sites can interact with the particle surface and serve as molecular attachment sites. Amino alcohols are an example of this type of dispersing additive. As shown in Fig. 10, the basic amine group of 2-amino-2-methyl-1-propanol attaches to the acidic SiO_2 surface particles. The alcohol group of the molecule, on the other hand, seeks out water molecules while also providing greater steric hindrance.

B. Greensheet Chemistry Compatibilities

Good ceramic particle dispersion is important, but it is not the only factor in selecting a chemical medium. Just as critical is the selection of binders, plasticizers, and solvents to formulate a suitable casting medium for the ceramic greensheets. Table 4 lists a typical slurry composition prior to casting. After casting most of the solvents are removed, with the resultant greensheet composition consisting of typically 90% (by weight) dispersed ceramic particles in a 10% (by weight) binder and plasticizer matrix. Thus, the material compatibility must be maintained for these subsequent "dried" composition constituents, although the primary compatibility issues are already dictated by the slurry needs.

Slurries often contain two solvents. These may be high- and low-boiling solvents to assist in achieving the desired greensheet microstructures and, in turn, good mechanical properties. Another combination may be that of acidic and basic solvents to enhance the net compatibility of all the slurry components with each other. The degree of material compatibility can be estimated by the

Table 3 Isoelectric Points of Oxides

Material	Nominal composition	IEP
Quartz	SiO_2	2
Soda-lime glass	$1.0\ Na_2O \cdot 0.58\ CaO \cdot 3.7\ SiO_2$	2 – 3
Cordierite	$5\ MgO \cdot 2\ Al_2O_3 \cdot 2\ SiO_2$	2 – 3
Titania	TiO_2	4 – 6
Mullite	$3\ A_2O_3 \cdot 2\ SiO_2$	6 – 8
Alumina	Al_2O_3	8 – 9
Magnesia	MgO	12

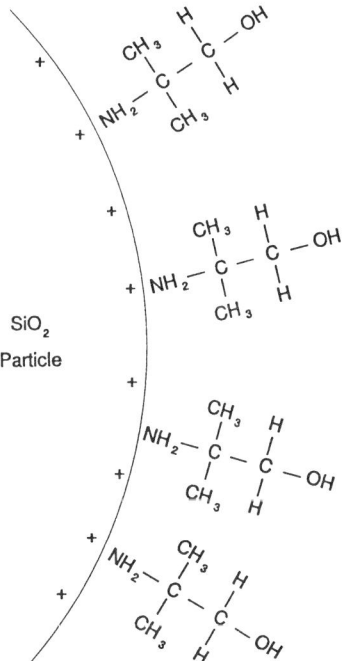

Figure 10 Schematic of the amino alcohol's basic amino group attachment to the acidic SiO_2 surface.

acid–base character of the slurry components. Mixing acids and bases results in the generation of heat, the quantity (kcal/mol) of which is a measure of their interaction strength. It is calculated from the exothermic heat of mixing ($-\Delta H^{ab}$)

$$-\Delta H^{ab} = C_A C_B + E_A E_B \tag{2}$$

where C and E are constants related to the covalent and electrostatic character of the interaction, and A and B refer to the acids and bases in the slurry [4]. Table 5 lists the values of C and E for several solvents, polymers, and silica.

Table 4 Composition of a Typical Ceramic-Based Slurry

Material	Function	Weight %
Ceramic powder	Substrate material	60.0
Fish oil	Deflocculant	1.0
Methyl isobutyl ketone	Solvent	24.0
Methyl alcohol	Solvent	8.0
Polyvinyl butyral	Binder	2.4
Polyethylene glycol	Plasticizer	2.6
Octyl phthalate	Plasticizer	2.0

Table 5 Electrostatic (E) and Covalent (C) Constants (kcal/mol)$^{1/2}$ for Various Materials

Acids	E_A	C_A
Phenol	4.33	0.44
Chloroform	3.31	0.16
Tert-butyl alcohol	2.04	0.30
Water	2.45	0.33
Silica	4.39	1.14
M-florophenol	4.42	0.51
Bases	E_B	C_B
Triethylamine	0.99	11.09
Pyridine	1.17	6.40
Ethyacetate	0.97	1.74
PMMA	0.68	0.96
Benzene	0.49	0.71
Dioxane	1.09	2.38

Strong interactions of the ceramic particles with both the binder and plasticizers are necessary for establishing good greensheet mechanical properties. An example is the interaction between the basic polymer binder polymethylmethacrylate (PMMA) and acidic SiO_2 powder. Using Eq. (2), the exothermic heat of reaction for PMMA and silica is calculated to be 4 kcal/mol, a value well within the typical range of 2–8 kcal/mol of other acid–base systems. In order to minimize strong competitive reactions of a solvent for PMMA and silica, a preferred solvent should be either neutral or acidic for which the heat of mixing remains less than about 2 kcal/mol. Otherwise, a strong basic solvent would compete with basic PMMA for interaction with the acidic silica, while a strongly acidic solvent would interact with the PMMA and thereby reduce the interaction of the PMMA with the silica. Similar considerations must be given to all other slurry component interactions so as to maximize the desirable chemical reactions, while simultaneously satisfying the various physical property needs.

Another indicator of binder compatibility with the ceramic powder is based on quantitative binder adsorption on the ceramic powder surface, since surface adsorption is the precursor to adhesion. It is measured in grams of polymer absorbed per surface area of powder, or g/m^2. For example, the adsorption of PMMA onto the surface of SiO_2 powder can be monitored as a function of the acid–base character of the solvent system [5]. In carbon tetrachloride (a neutral solvent), 12×10^{-4} g/m^2 of PMMA is adsorbed onto the silica powder. In acidic solvents of methylene chloride and chloroform, the PMMA adsorption is 5×10^{-4} and 1×10^{-4} g/m^2, respectively, the least adsorption being with the more acidic chloroform. Basic solvents have similar effects: in a benzene environment, the adsorption is 6×10^{-4} g/m^2, whereas in the more basic dioxane, it is 1×10^{-4} g/m^2. The adsorption characteristics are explained on the basis of available adsorption sites. The basic carbonyl oxygen of PMMA's repeating ester groups form acid–base bonds with acidic solvents, thereby reducing the number of basic polymer sites available for acidic silica powder interactions. On the other hand, basic solvents will interact with the acidic silica powder, reducing the availability of their surface sites for polymer adsorption. Combining the appropriate pair of solvents can offset these effects.

Subsequent handling and processing of greensheets require that they not be brittle and exhibit sufficient toughness. These mechanical properties are greatly influenced by the degree of interactions between the remaining binder and plasticizer with the ceramic powder, after the solvent removal. For example, after the removal of methyl isobutylketone (MIBK) and methanol from a slurry containing an acidic polyvinylbutyral powder and 88 wt % of either powdered alumina (basic) or glass (weakly acidic), 250 µm thick films were cast [5]. The specimens were tested in an Instron, and stress–strain curves for each film were determined. As shown in Fig. 11, the toughness of the alumina-containing film

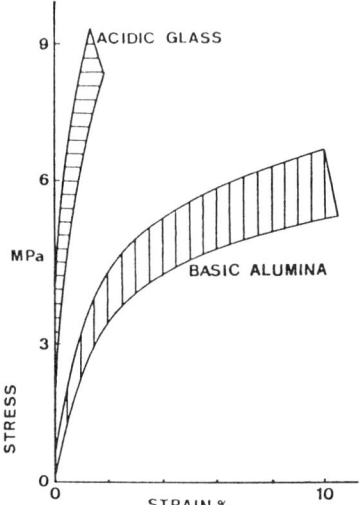

Figure 11 Instron stress–strain curves for 250-μm greensheets using polyvinylbutyral (Butvar) binder and alumina or glass filler. Acidic Butvar binder interacts more with basic alumina filler than acidic glass filler, yielding greensheets that are much tougher due to enhanced acid–base interactions. (From Ref. 5, reprinted with permission by *Journal of Adhesion Science and Technology*.)

is six times greater than that of the glass-containing film. This difference is due to enhanced acid–base interaction between the alumina filler and polymer.

C. Metal Paste–Greensheet Interactions

The same chemical considerations given to particle dispersion and to the organic and inorganic component compatibility in the slurry (so important for quality greensheet formation) apply equally well to the formulation of the metal paste. While the discussion of greensheet formation was centered on Lewis acid–base interactions, a review of the interaction issues between solvents, binder, and other components in the paste benefits from a different analysis viewpoint. In this case, it is important to understand the type and degree of the paste component interactions with the greensheets, which can greatly affect the latter's physical properties.

Generally, the interaction properties and chemistry of any organic material are dominated by its associated functional structure groups. Two commonly used polymers in greensheet processing are polymethyl methacrylate (PMMA, known commercially as Plexiglas or Lucite) and polyvinylbutyral (Butvar, a tough gum material used in laminating safety glass). The structures of these materials are shown in Fig. 12, and consistent with polymeric materials, they consist of repeating monomer structural units to give each material different chemical characteristics. The chemistry of PMMA is controlled by the monomer-containing ester groups, whereas Butvar's chemistry is more complex with its C–O–C acetal and four carbon butyral chain combination.

While polymers are characterized by repeating identical monomeric units, it is also common to synthesize polymers with different monomeric units. Copolymers have two or more different monomeric units. Depending on the way the monomers are arranged on the polymeric chain, the copolymers can be in alternating, random, or block positions. In the last case, groups of identical units of each monomer alternate on the chain. The polymer structure can also be a combination of chain and branching units. The chemistry of the monomers not only determines how the polymer will react with other materials, but also how the polymer will interact with itself—known as crosslinking. The crosslinking

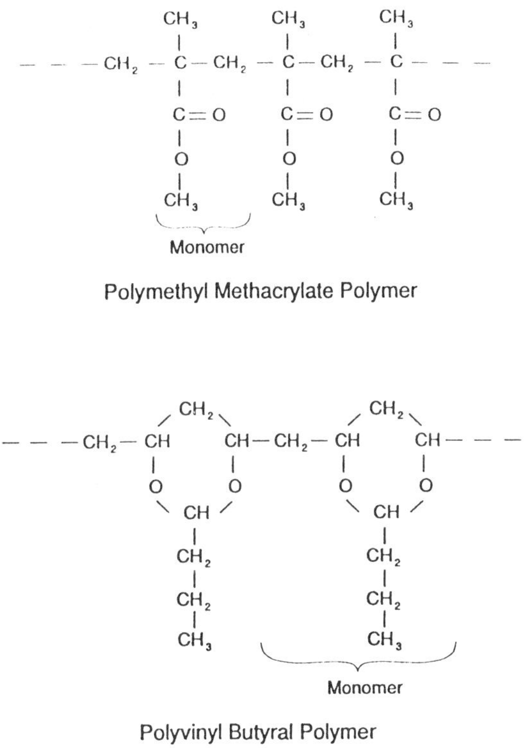

Figure 12 Molecular structures of polymethyl methacrylate and polyvinylbutyral polymers showing repeating monomers.

is important in both the greensheet and the metal paste fabrication, in that the viscosity can be strongly modified by the degree of this polymer behavior.

Hence, the control over, or the absence of, crosslinking reactions is a practical process issue. This can be further influenced with the introduction of different additives. For example, a paste vehicle system with good viscosity characteristics when combined with metal powder "A," may crosslink when utilized with metal powder "B" and result in an unsatisfactory paste viscosity and other associated process parameters. A full discussion of polymer synthesis, interactions, and crosslinking is beyond the scope of this chapter. However, the foregoing comments are intended to illustrate the possible issues encountered in developing a compatible greensheet and paste technology for MLC processing, and to recall that all interactions ultimately depend on the functional group chemistry of the materials involved.

D. Estimating Paste Solvent/Sheet Binder Interactions

The paste solvent interaction with the greensheet's binder and plasticizer must be controlled to assure good greensheet dimensional stability. A typical paste is 80–90 wt % solids and 10–20 wt % organics. Paste composition comprises a solvent, binder, surfactant, inorganic solids, and an organic additive with which to adjust the shear rate (or thixotropic properties) of the paste. It is desirable to have adequate flow of the paste for screening X–Y features and vias, yet minimize the spreading of the paste after screening. Thus, the thixotropic additive functions as a slight gel or colloidal network in the paste to inhibit gravity flow and stabilize the screened dimensions of the paste. In multilayer package technology, the paste stabilization is further enhanced by absorption of the paste solvent into the greensheet structure. Although this absorption assists in good screening control, the paste solvent must not induce swelling of the greensheet polymers. Such swelling can cause greensheet stability problems.

The considerations related to the interactions between the paste solvent and the binder and plasticizer of the greensheet centers on the polymer's solubility parameters. The general concept of "likes dissolve likes" applies, and the first order of solvent and polymer selection is based on their polarity and nonpolarity character. The solubility parameters are, however, also influenced by dispersion and hydrogen-bonding forces. The total material solubility factor (δ^2) is defined by

$$\delta^2 = \delta_d^2 + \delta_p^2 + \delta_h^2 \tag{3}$$

where δ_d, δ_p, and δ_h are the respective solubility contributions due to dispersion, polarity, and hydrogen bonding. In general, in order that polymer P be soluble in solvent S, their corresponding total solubility parameters should be nearly the same, so that

$$\delta_P - \delta_S \leq 0 \tag{4}$$

The specific solubility components are related to the cohesive energy of a material, and in many instances these often become difficult to determine experimentally. However, a number of methods have been described that allow the generation of these values theoretically with varying degrees of success [6]. Such data do yield sufficient information for predicting the solubility of a polymer–solvent system, and they indicate possible solvent–polymer effects like polymer swelling.

Each of the solubility parameters for a given material is calculated from its total molar attraction constant (F) and its molar volume (V) according to

$$\delta = \frac{F}{V} \tag{5}$$

F and V are derived from the summation of the various atomic or molecular group constituent contributions, making up a particular material structure. Thus, the dispersion components are generated as follows:

$$\delta_d = \frac{\Sigma F(d_i)}{V} \tag{6}$$

$$\delta_p = \frac{\sqrt{\Sigma F^2(P_i)}}{V} \tag{7}$$

$$\delta_h = \sqrt{\frac{\Sigma/E(h_i)}{V}} \tag{8}$$

The various constants for several of these material constituents, in addition to the respective molar volumes and Lewis acid/base character, are summarized in Table 6. The latter values are listed to indicate the acid/base character of typical organic functional groups.

Applying the information in Table 6 to butyric acid,

```
      H   H   H   O
      |   |   |   ||
  H - C - C - C - C -OH
      |   |   |
      H   H   H
```

the following information is derived:

Structure group	V	$F(d_i)$	$F^2(P_i)$	$E(h_i)$
CH_3	33.5	420	—	—
CH_2	16.1	270	—	—
CH_2	16.1	270	—	—
$\begin{matrix} O \\ \parallel \\ -C-OH \end{matrix}$	28.5	530	176,400	10,000
	94.2	1490	176,400	10,000

Summing the molar volume and the respective component parameters, the total solubility factor can be calculated:

$$\delta_d = \frac{1490}{94.2} = 15.8 \ \frac{J^{1/2}}{cm^{3/2}} \tag{9}$$

$$\delta_p = \frac{\sqrt{176,400}}{94.2} = 4.5 \ \frac{J^{1/2}}{cm^{3/2}} \tag{10}$$

$$\delta_h = \sqrt{\frac{10,000}{94.2}} = 10.3 \ \frac{J^{1/2}}{cm^{3/2}} \tag{11}$$

$$\delta = \sqrt{\delta_d^2 + \delta_p^2 + \delta_h^2} \tag{12}$$

$$\delta = 19.4 \ \frac{J^{1/2}}{cm^{3/2}} \tag{13}$$

The reported experimental values for the solubility parameter of butyric acid range from 18.8 to 23.1 $J^{1/2}/cm^{3/2}$. Thus, the calculation method will yield fairly good agreement with experimental values. The information in Table 6 is a limited set of the type of information available for predicting polymer properties [7]. It becomes quite convenient and useful for assessing solubility parameters of new materials.

IV. SINTERING CONSIDERATIONS

The primary design focus of the full MLC material component system centers around the various sintering process requirements. Besides defining the sintering temperature and atmosphere conditions, as dictated by the particular MLC metal network constituency, other material parameter requirements must also be considered for optimized material compatibility. Some of the parameters include thermal coefficient of expansion (TCE), modulus of rupture (MOR), thermal conductivity, and dielectric constant. These drive much of the presintering op-

Table 6 Attraction Constants (F), Hydrogen-Bonding Energy (E), Molar Volume, and Lewis Acid/Base Character for Organic Structure Groups

Structural group	$F(d_t)$ ($J^{0.5}$ cm$^{1.5}$/mol)	$F(p_t)$ ($J^{0.5}$ cm$^{1.5}$/mol)	$E(h_t)$ (J/mol)	Molar volume (cm^3/mol)	Lewis acid/base
$-\overset{\mid}{\underset{\mid}{C}}-$	−70	0	0	−19.2	Neutral
$-\overset{\mid}{\underset{H}{C}}-$	80	0	0	−1.0	Neutral
$-\overset{H}{\underset{H}{C}}-$	270	0	0	16.1	Neutral
$H-\overset{H}{\underset{H}{C}}-$	420	0	0	33.5	Neutral
$-NH_2$	280	—	8,400	19.2	Basic
$-\overset{O}{\underset{}{\overset{\parallel}{C}}}-H$	470	800	4,500	22.3	Acidic
$-\overset{O}{\underset{}{\overset{\parallel}{C}}}-OH$	530	420	10,000	28.5	Acidic
$-\overset{O}{\underset{}{\overset{\parallel}{C}}}-O-$	390	490	7,000	18.0	Basic
$-\overset{O}{\underset{}{\overset{\parallel}{C}}}-$	290	770	2,000	10.8	Basic
$-OH$	210	500	20,000	10.0	Acidic

eration development so that acceptable metal and ceramic structures will be formed during the sintering process. Figure 13 outlines the various chemical interactions that occur among all the materials and processes involved, all of which need to come together during sintering.

A. Sintering Profile

Figure 14 outlines the four segments of the sintering process (the sintering profile). It includes the key variables involved in each segment, as well as the change in average porosity of the ceramic/metal system during sintering. The

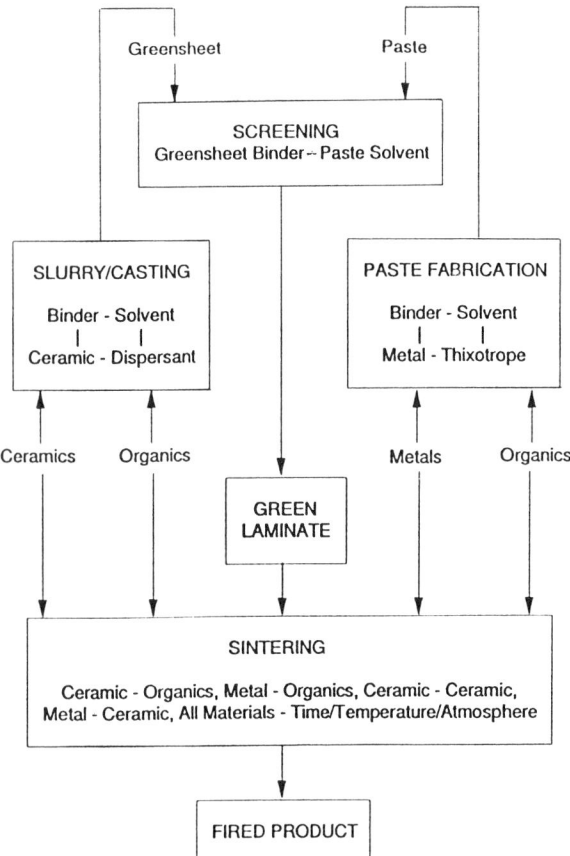

Figure 13 Schematic of multilayer ceramic processing elements involved in materials interactions.

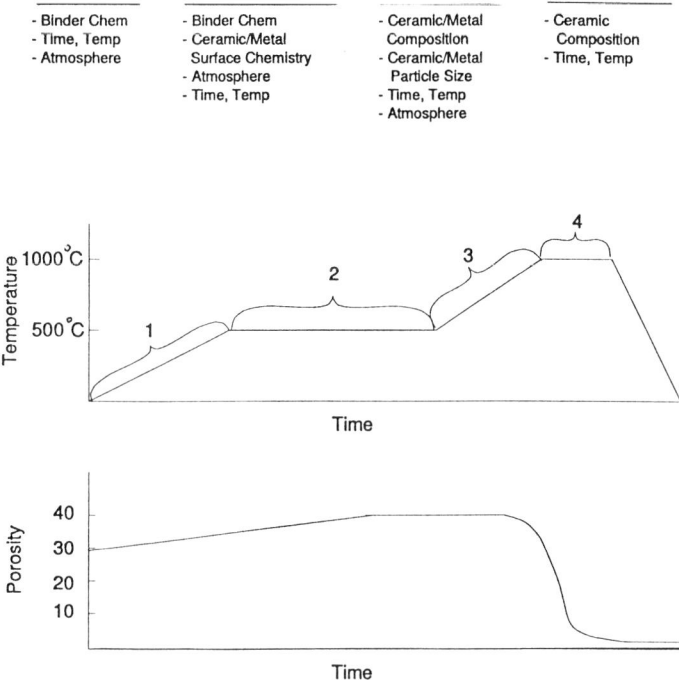

Figure 14 Schematic sintering profile and average porosity of a glass–ceramic package relative to key variables during the four main sintering segments.

variation of porosity with sintering time and temperature is one means of monitoring the shrinkage of the composite system. It is the aim to have the metal and ceramic not only sinter at the same time, but also at the same rate.

The pyrolysis segment in Figure 14 leads to a residual carbon "char." This process is influenced by the nature of the ceramic–organic and metal–organic chemistry during the thermal degradation of the polymers. The residues are then removed by carbon oxidation, the efficiency of which can be influenced by the ceramic and metal surface chemistry. After complete carbon removal, the temperature is further increased so that the combined effects of the particle surface energy and the thermal energy induce diffusion activities and produce a dense ceramic body with a conducting metal network. Maintaining the ceramic system at the high temperature (segment 4) causes material crystallization. The crystallization process involves some kinetics and fosters the development of the desired microstructure. At the various phases of the sintering process, the surface chemistry of the ceramic and metal particles can affect the sintering temperatures and the rate of sintering, as well as the degree of bonding and interdiffusion. These elements will be given further considerations with examples of different material systems.

B. Metal Network

For effective ceramic process technology development, the most fundamental of all the different product requirements listed in Table 2 is that of electrical conductivity. This parameter is determined by the specific metal network in the MLC electronic package. Two network groups exist on the basis of the particular metal in question: Packages that contain tungsten or molybdenum are considered high-temperature systems, while those with copper or other noble metals are referred to as low-temperature systems. Traditionally, W and Mo have been the preferred metals due to their relatively low thermal coefficient of expansion (TCE), making them compatible for economically viable ceramic systems. Both metals are refractory, and they are therefore co-sintered with the ceramic in the range of 1600 to 1700°C. Since M and Mo easily oxidize at these high temperatures, the development of reducing sintering atmospheres that permitted carbon oxidation and ceramic/metal sintering without oxidizing of the metal was possible.

In the case of gold, the entire sintering profile can be carried out in air, which greatly simplifies the process. However, since gold has a lower melting temperature, the compatible ceramics must also be processable at correspondingly lower sintering temperatures. Such ceramics exist, and some packages have been produced. But for obvious reasons, the use of precious metals as the conducting network in an MLC package has not been widely practiced.

Copper is a better conductor choice in an MLC package. It is economical, has excellent conductivity, and sinters at a relatively low temperature. It has, however, a high TCE, and it oxidizes readily. The successful introduction of copper into ceramic packages presented major development challenges in identifying the appropriate process methods to accommodate those two properties.

In co-fired copper situations, ceramics must sinter to full density at less than 1000°C, yet not sinter significantly at temperatures below 800°C. The latter condition is dictated by the need to keep the ceramic network porous at the lower temperatures, to allow the mass transport of the sintering atmosphere into the body of the package. In turn, this permits the controlled burning out of any residual carbon constituents left behind from the organic components of the greensheet and paste fabrication processes. Thus, the process must maintain a sufficiently low oxygen partial pressure for the carbon oxidation to proceed, but keep the copper within the network in a reduced form. Although the thermodynamics under these conditions are favorable for the carbon oxidation, the kinetics will be very slow. To enhance the kinetic process, a high-humidity sintering atmosphere was developed, along with new sintering furnaces and advanced metals as liners and fixtures. The desirability of copper in high-performance MLC electronic packages has been instrumental in the production of many new copper-compatible, low-temperature sintering co-firable ceramics since 1985 by several commercial companies.

C. Insulator Body

The selection of a copper-compatible insulator body is made on the basis of the sintering temperature and atmosphere, its dielectric constant, thermal conductivity, MOR, TCE, and color. Some of these requirements seem to be conflicting, as in the case of the dielectric constant, TCE, and sintering temperature of SiO_2. SiO_2 has the lowest dielectric constant (3.8) and TCE (5.5×10^{-7}) of any simple oxide. Both are desirable material characteristics. However, SiO_2 is a refractory material and requires sintering temperatures above 1500°C. Such high temperatures are incompatible with a copper-based metal network. Other material candidates are necessary to reduce the sintering temperatures to below 1000°C, even though any such materials will invariably contribute to increasing both the dielectric constant and the TCE. The challenge was therefore to find means of adequately lowering the sintering temperature without increasing the other parameters above acceptable limits. Two different approaches have evolved. One is based on a liquid-phase-assisted system, in which a low-melting glass is mixed with a crystalline refractory oxide to form a mixed-phase glass and ceramic system. Most commercial Al_2O_3 systems use this technique. The second approach relies on the crystallization of a single-phase glass or glass–ceramic system [8]. In such systems, the glass components are crystallized after sintering. The resultant glass crystal properties structure determine various bulk characteristics of the glass–ceramic material.

The chemistry for either the glass–ceramic system or the liquid–phase glass system may be considered from two different viewpoints: (1) how do the glass component interactions influence the bulk properties, or (2) how do the thermodynamics affect the glass components? In a metal network–ceramic package system, both aspects must be considered.

1. Liquid-Phase Glass Systems

The processability of a particular ceramic material can actually be modified quite easily through appropriate glass composition formulations. In turn, the resultant glass mixtures exhibit correspondingly modified bulk properties. These process and property shifts are normally rather straightforward, as for example in the mixing of SiO_2 and Al_2O_3. A 50–50 molar composition of these two materials in a powdered form will melt at a considerably lower temperature compared with powdered Al_2O_3 by itself (~1700°C). The bulk properties of the sintered material are also shifted. Thus, the dielectric constant of 6.8 is halfway in between the individual values of the two oxides (i.e., 3.8 for SiO_2; 9.8 for Al_2O_3), as is its TCE of 30×10^{-7} (compared with 5.5×10^{-7} for SiO_2; 60×10^{-7} for Al_2O_3). Good handbooks exist covering a wide variety of glass properties as functions of glass composition [9], but it is usually not possible to formulate a sufficiently suitable glass system based only on such data. It remains a process that is usually carried out empirically. The handbooks can provide a starting point; however, it is not unusual to test hundreds of slight variants of glass compositions before a good composition candidate is identified, one that exhibits most of the proper sintering and material properties. Hence, once a particular composition is formulated offering the desired physical properties, the glass mixture is milled into a powder and then followed with an evaluation of its sintering characteristics.

The sintering conditions of a glass material are greatly influenced by the glass's viscosity. This in turn is a function of composition and process temperatures. The reduction of the glass viscosity at a given temperature is achieved with the addition of various oxides, such as alkali or alkali earth oxides, or oxides like B_2O_3. Effective liquid-phase-assisted sintering relies on such oxide additions in order to lower the viscosity to about 1000 poise—a fluid state corresponding to the viscosity of honey.

At room temperature, the viscosity of window glass is in excess of 10^{16} poise and is considered as being rigid. Nevertheless, large window pane glass plates can manifest flow over time (years), and the thickness at the bottom of the glass plate will be measurably thicker than at the top. Glass is therefore a metastable supercooled liquid and continues to flow even at room temperature. As temperature increases, the viscosity of a glass will decrease until the material begins to measurably flow under the force of gravity. This occurs at around 10^8 poise and 750°C for soda-lime, or window, glass, as shown in Fig. 15.

In general, powdered glass of any type will start to sinter when its viscosity is less than 10^{10} poise, although the specific sintering temperatures will be further dictated by its specific particle size. Submicron particles start sintering at considerably lower temperatures due to the very high surface energy of such particles. For example, refractory materials like Al_2O_3 with a particle size of several microns require sintering temperatures in excess of 1700°C, while sintering occurs near 1500°C for powders with a particle size of 0.2 μm.

The onset and completion of sintering is further sensitized by the temperature profile ramp rate, as well as by the sintering hold time. A fast ramp rate increases the apparent sintering temperature; a slow ramp rate has the opposite effect. The viscosity reduction due to higher sintering temperatures corresponds to shorter densification completion times. The presence of water also lowers the glass viscosity at a given temperature, and thereby the required sintering temperature. Considered from another viewpoint, water increases the rate of sintering at a

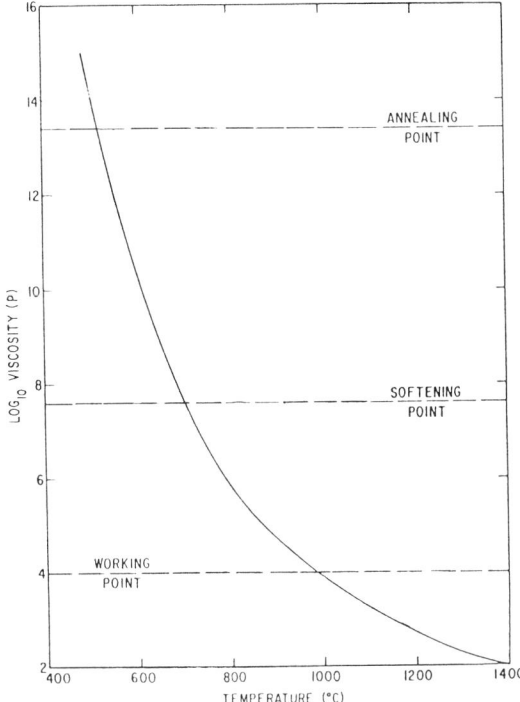

Figure 15 Viscosity of soda-lime, or window, glass as a function of temperature.

specific temperature. Water tends to adsorb into the molecular structure of a glass matrix during various process phases and can subsequently react with other glass components or contaminants (i.e., carbon). To assure repeatable and stable sintering conditions, it becomes critical to either prevent the adsorption of water, or to carefully have it diffuse out of the glass structures. This can occur at elevated temperatures, which, however, must remain well below the glass sintering temperatures.

2. Glass–Ceramic Systems

Glass–ceramic formulations must satisfy the same requirements as other glasses with regard to viscosity and sintering temperature. However, since the glass–ceramic is a single-phase system and the refractory crystal phase is not present, the necessary viscosity for sintering is around 10^{10} poise. The processing commences with the powdering of the glass followed by mixing with organics to form a slurry, and casting into greensheets in the usual manner. The temperature profile is set to first allow the sintering of the composition as a glass at lower temperatures by viscous flow, after which the temperature is increased to initiate glass crystallization. Crystallization (also known as devitrification) is normally not desired in general glass production, and additives are used to stabilize the glass and to maintain it in the vitreous state. In the glass–ceramic technology, however, devitrification is desired. Hence, in this instance, certain additives are mixed with the glass composition to purposely induce crystal formation of a particular type. This results in a material with a completely crystalline nature and essentially no glassy phase. As an example, beta-spodumene glass ceramics are made using a glass in which the intended beta-spodumene stoichiometry composition is Li_2O–Al_2O_3–$3SiO_2$, and it is formulated from a 1:1:3 molar ratio of powdered Li_2O, Al_2O_3, and SiO_2. Heating this stoichiometric composition in a platinum crucible to 1500°C results in a liquid glass. A slow cooling rate from the melt temperature will crystallize the glass into the crystalline beta-spodumene material. Alternatively, rapid cooling methods (quenching) are also employed. The process effectively freezes in the high-temperature liquid structure to produce normal glassy materials at room temperature. Rapid quench rates up to millions of degrees per second have been used to prepare glassy metals and even amorphous or glassy solid water. Less dramatic quenching methods involve the pouring of the molten glass into water or through water-cooled stainless steel rollers.

The desired sintering and postcrystallization bulk properties of the basic 1:1:3 glass composition are modified with the addition of other glass composition materials (i.e., Na_2O, P_2O_3, MgO, CaO, B_2O_3). The balancing of all the property requirements as a function of oxide additives is similar to those of glass compositions, as discussed earlier. Again, handbooks and literature surveys give guidelines, but ultimately a new composition is the result of empirical studies. It is not uncommon for an additive like P_2O_5 to be added to aid sintering, only to find it causes premature glass crystallization in the sintering cycle, and thus prevents the glass from reaching full density. When a glass commences crystallization before densification, the viscosity rapidly increases, stops the viscous flow, and negatively affects the sintering process.

Only in the simplest systems is it possible to predict the resultant properties of the fired glass or glass–ceramic material, or the modifying effect that a given additive will have on the glass viscosity with temperature. Some oxides are known as network formers. They tend to generate stable glass networks that tend not to crystallize even with slow cooling from the melt temperature. Examples are SiO_2, B_2O_3, and P_2O_5. Although P_2O_5 by itself can induce premature crystallization, when it is mixed with other oxides, it can function as a glass former and actually inhibit crystallization. Such reaction behavior is not unusual in glass and ceramic chemistry. Na_2O and CaO, when mixed with SiO_2, allow a much lower sintering temperature at the expense of lower chemical stability and increased TCE. In this case the additional oxides act as flux or network modifiers. Al_2O_3 and transition metal oxide systems crystallize from the melt temperature and can only be made glassy by extreme quench rates. Mixing such oxides with a low-melting glass former (like P_2O_5) results in higher viscosity and a more refractory glass. P_2O_5, mixed with TiO_2 in equal molar amounts, results in a glass that is stable and has a melting temperature of more than 1200°C (compared with the melting of P_2O_5 at less than 600°C). The viscosity of the glass former can be either increased or decreased depending on the oxide additive. In practice, glass mixtures of more than one oxide are formulated to achieve the desired properties.

The foregoing discussions of either the liquid mixed-phase glass–ceramics system or the single-phase glass–ceramics suggest that their respective chemistry is very complex. In glass-based systems, random rearrangements of component oxides can occur (sometimes referred to as scrambling reactions) and make the prediction of the product difficult, since there are so many possible combinations. Effects of contaminants can also be significant. Similarly, glass–ceramic materials crystallization is greatly influenced by the base component oxides and secondarily by other minor components. In many three-component, or ternary, systems (like cordierite or beta-spodumene), various possible crystal structures can occur. The specific structures are difficult to predict without extensive investigations, as are the relationships between composition, crystal structure, and properties.

V. MELTING AND SINTERING THERMODYNAMICS AND KINETICS

At high temperatures, materials tend to react with each other, the degree driven by temperature, atmosphere, and the nature of the materials. General predictions of high-temperature reactions, and the ability to control them, can be made based on knowledge of the thermodynamic properties of the materials involved. In glass chemistry, the compatibility of the component oxides with respect to their oxidations and reductions at the glass-melting temperature must be well understood. They must then also be well controlled. This applies equally to the sintering process. Here, the residual carbon, for example, must be oxidized, while preventing the simultaneous oxidation of the metal within the package module. Finally, thermodynamic concepts are also helpful in understanding various fundamental aspects of glass to metal adhesion at elevated temperatures.

Based on the classical Gibbs relation of

$$\Delta G = \Delta H - T \Delta S \tag{14}$$

in which ΔG is the Gibbs free energy, ΔH the enthalpy, T the absolute temperature, and ΔS the entropy, a reaction is thermodynamically spontaneous if ΔG

is negative, it is in equilibrium if ΔG is equal to zero, and it will not proceed at all if ΔG is positive. The relationship in Eq. (14) implies that the entropy term (i.e., the irreversible loss of available energy) will tend to dominate the sign of the free energy in high-temperature processes.

Tables and graphs have been developed giving the standard free energy of formation of oxides [10]. Note, however, that predicting whether a reaction will take place based on thermodynamic data does not mean the expected reaction will be observed. Such a reaction may, for example, be too slow for observation in a practical time period; this might be due to poor chemical kinetics (i.e., reactant activity much less than unity) or mass transport constraints in gaseous reactants and products.

A. High-Temperature Glass Chemistry

Figure 16 is a compilation of the standard free energy of formation of oxides as a function of temperature. All reactions are normalized to one mole of gaseous oxygen, and all reactions assume an activity of unity for all reactants. The figure includes a vertical line on the left side of the graph marked with an O (which represents O_2) and H (representing H_2). The two points, in conjunction with the scales for the H_2/H_2O ratio (top, right, and bottom) and the partial pressure scale for O_2 (right and bottom), allow the determination of the oxygen partial pressure requirements for specific oxidation and reduction reactions. For example, a straight line drawn from H to the H_2/H_2O scale at 10^{-4} (top of graph) would intersect just below the line for the copper/copper oxide reaction at 800°C. This indicates that an atmosphere of 100 ppm of H_2 in H_2O would prevent copper from oxidizing at 800°C.

Similarly, a line drawn from O through the same point just below the copper/copper oxide line at 800°C intersects the oxygen partial pressure scale close to 10^{-10}. This implies that if an oxygen-containing gas passed through a fine copper mesh at 800°C, the oxygen content in the gas would be reduced to below 10^{-10}.

Figure 16 also includes data allowing the determination of whether a given material will be oxidized, reduced, or in equilibrium at a given temperature and atmosphere. For example, the oxidation of Ca to CaO is the lowest curve on the graph. This means that Ca has the most reducing power for any other metal or nonmetal in Fig. 16, making CaO the most stable oxide. The oxidation of Cu to Cu_2O (near the top of the graph) indicates that Cu is the least reducing material, and Cu_2O is an unstable oxide easily reduced by any material below it on the graph. The free energy of formation of P_2O_5 at 1000°C is shown to be about -85 kcal. This suggests that it would not be productive to use titanium metal mixed with P_2O_5 to make a titanium–phosphate glass. The formation of TiO_2 at 1000°C is about -160 kcal. The difference of -75 kcal indicates Ti

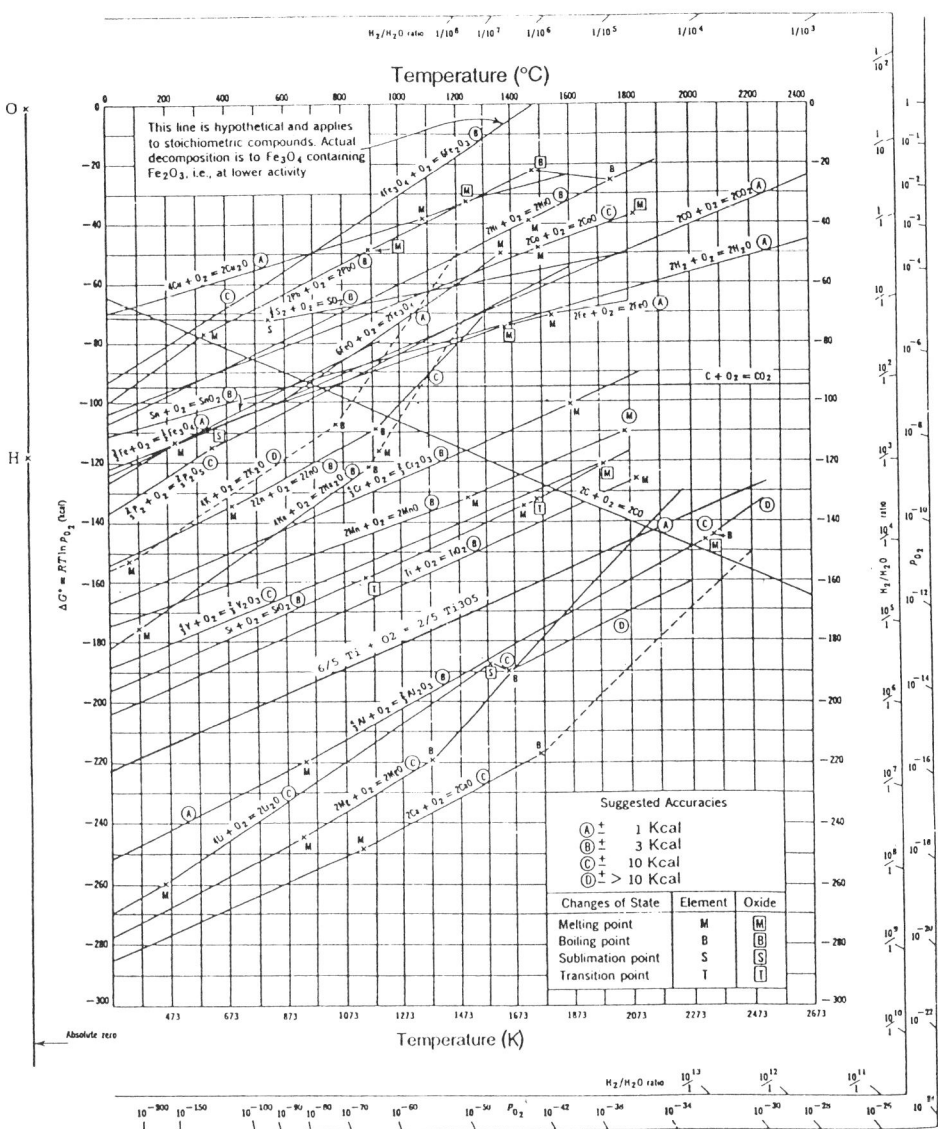

Figure 16 Standard free energy of formation for various oxides as a function of temperature.

metal to be strongly reducing relative to P_2O_5 at 1000°C. It would therefore react with P_2O_5 to form lower oxides of phosphorus as it is itself oxidized.

As mentioned earlier, a reaction may be favored thermodynamically, yet not proceed. Or it may proceed very slowly, due to insufficient reactivity of the reactants. For example, in the formation of titanium–phosphate glass, which is to contain a high Ti^{3+} ion concentration for magnetic or electrical reasons, powdered Ti metal or Ti_2O_3 would, chemically, seem to be good starting candidates. However, based on their thermodynamic properties, both would be unacceptable. Thus, an indirect process route must be pursued to formulate the desired glass composition.

Titanium (like vanadium) forms a series of nonstoichiometric oxides, trititanium pentoxide (Ti_3O_5) being one of them. It can be readily prepared by heating the correct amount of Ti metal and TiO_2 in vacuum at 1200°C. Unlike the rapid reaction of Ti metal or Ti_2O_3 in the presence of P_2O_5, a mixture of Ti_3O_5 and P_2O_5 heated to 1000°C results in only a very slow reduction reaction of P_2O_5 by Ti_3O_5. In fact, even maintaining the mixture at 1500°C for up to 10 min still does not result in any significant reduction of the phosphorus pentoxide. The inactivity is due to the more stable Ti^{3+} ions in the Ti_3O_5 crystal structure, and this allows, therefore, for the preparation of a high concentration of Ti^{+3} in a phosphate glass. The standard free energy of formation of Ti_3O_5 is below that of TiO_2. Thus, thermodynamically, a mixture of these two oxides could reduce the TiO_2 to lower oxides of titanium. However, the reduced Ti^{3+} activity in Ti_3O_5 again profoundly minimizes the kinetics of the reaction.

B. Carbon Oxidation

Another application of the data in Fig. 16 relates to the oxidation of residual carbon in the sintering profile as shown in Fig. 14. The pyrolysis process phase in the sintering process reduces the total organic content within the greensheets and paste from about 10% by weight to below 0.5%. This remaining lower percentage of organics is in the form of a carbon char and must be removed or reduced to below 0.1% via carbon oxidation to form CO or CO_2. Otherwise, the dielectric constant of the ceramic will increase significantly. The carbon oxidation is accomplished by adjusting the partial pressure of oxygen during the binder burn-out (BBO) segment of the sintering cycle. The oxygen partial pressure required to achieve this is dictated by the package's metal system. As already discussed, the data from Fig. 16 regarding the copper/copper oxide oxidation reaction as a function of the oxygen partial pressure suggests that at 800°C, the oxygen partial pressure would be below 10^{-10}. This is significantly above the C/CO_2 and C/CO reaction lines. Therefore, an oxygen partial pressure of 10^{-10} will keep copper reduced but is thermodynamically favored to oxidize carbon. The necessary time at temperature to reduce the residual carbon from

0.5% to below 0.1%, however, is dependent on the carbon activity, the precursor chemistry for the oxygen generation (i.e., CO/CO_2, H_2/H_2O, oxygen in an inert gas, or some other method), and to a lesser degree the gas diffusion characteristics within the furnace configuration. Since the carbon oxidation kinetics are generally much slower than the diffusion of the reactants into (and their products out of) the porous substrate, gas diffusions are usually not the rate-limiting step.

C. Densification Chemistry

After the carbon removal process, the ceramic and metal in the package have not yet been significantly densified. Indeed, one of the biggest challenges in the fabrication of multilayered ceramics is to avoid as much as possible any premature ceramic and metal sintering (densification), so that the residual carbon can first be efficiently removed while the ceramic–metal body is still in an open porous state. The process condition can be further complicated by the desire to accelerate the carbon oxidation as much as practically possible by operating at the highest possible temperatures, without initiating any sintering activities. As discussed in Sec. IV, both the liquid-phase-assisted and the glass–ceramic sintering processes are sensitized by the glass viscosity as a function of temperature. Adsorbed water generally reduces the viscosity of glass and would induce earlier sintering if left in the glass. Also, it is preferable to control the sintering onset and its rate by means of the glass composition rather than as a function of water content, since in some systems adsorbed water can react with residual carbon in the glass and cause bubbles in the glass structure. Water could also simply vaporize at higher temperatures, creating the same effect.

The overall sintering driving force is due to the reduction of the surface energy in both the ceramic and metal materials. The onset of sintering must be carefully controlled so that the ceramic and metal start to sinter at approximately the same time. The rate of sintering for both materials must also be similar. Otherwise, the resultant product flatness will be nonuniform, or exhibit high X–Y distortion. Normally, it is the metal sintering characteristics that are matched to those of the ceramic. For example, in the high-temperature Mo and W metallurgy systems, the sintering onset and its rate are matched to that of the ceramic by adjusting the metal particle sizes, although it can also be achieved with a mix of additives (like powdered glass). These same methods also apply for copper densification.

D. Metal–Ceramic Adhesion

During the final stages of densification, the metal–ceramic interface is developed. This intermaterial adhesion can be fostered in two ways. The first relies on the existence of a glass in the paste to react cohesively with the ceramic and to create a mechanical interlock between the metal conductor and surrounding ceramic in the package body. This method is generally used to initiate metal paste adherence to a fired ceramic, as in post-fire processing. However, in this case, the relatively fluid glass within the paste is intended to wet an already fired ceramic at a temperature that is significantly lower than the ceramic's sintering temperature. On the other hand, in a co-fire technology, the metal–ceramic bond must be formed at much higher temperatures, in which the viscosity of the bonding glass in the paste and the viscosity of the ceramic body are similar.

The second method of influencing metal–ceramic adhesion is based on the generation of an intermediate phase between the metal and an oxide component in the ceramic. The formation of that phase is a function of the glass or crystallized ceramic composition and the oxygen partial pressure, with the latter defining the formation of metals and nonmetal oxides. It has been suggested that copper adhesion to alumina, or an alumina-based ceramic, relies on the formation of a copper aluminate intermediate phase [11]. While there may be some debate as to the exact nature of such an intermediate phase, the copper–glass adhesion is clearly a function of the oxygen partial pressure. However, since the oxygen partial pressure is not always easy to control in the sintering profile, glass–ceramic adhesions are more commonly produced with glass-assisted bonding rather than the intermediate-phase methodology.

VI. SUMMARY

This chapter outlined some of the key challenges of MLC processing. Not all of them have been addressed. Contamination is an unavoidable problem in all aspects of processing—from raw materials through sintering. Contaminants in the parts per million (ppm) level can have significant (negative or positive!) impact on the metal or ceramic microstructure, the crosslinking in the slurry or paste systems, and the metal/ceramic sintering characteristics. It is not unusual for a raw material from a given supplier to work well and give the desired results, while the same raw material from other suppliers cannot be made to work. This points to the need for meaningful specifications for the raw materials, as well as for all aspects of the process. In a manufacturing environment, where thousands of ceramic substrates are produced daily, good specifications and absolute process controls to ensure those specifications are critical. Such specification development can take years as a technology matures and as each aspect of the process is fine-tuned for maximum yield. This requires a full understanding of not only the major components of each raw material and how they interact, but also how minor components at the ppm level can enhance or harm the process. The necessary understanding comes with experience and with the utilization of the most advanced analytical chemistry tools to diagnose the contaminant's effects.

Another chemical challenge involves the postsintering substrate process. If the manufactured substrate is going to be subject to wet processing, it must be hermetically sealed so that processing fluids can not enter and cause reliability problems. Also, some thought should be given to the allowable limits for wet processing such as pH, water content, or the use of surfactants. Water is very corrosive to glass and ceramics and can enhance fracture due to stress corrosion, especially near a bonded metal–ceramic interface. Some applications require the machining of the ceramic substrate by means of lapping and polishing, grinding of edges, or sizing and drilling holes. The effects of such wet processing on the ceramic and metal under these conditions must be understood and controlled.

Subsequent to sintering, the ceramic package must be plated with a suitable metal (or series of metals) to allow wire bond, chip attach, pin attach, soldering, etc. This is usually achieved with a combination of Ni and Au. The specific plating process is generally a function of the particular surface to be plated, as well as the chemistry of the plating bath. The surfaces are usually not simple pristine surfaces of Mo, W, or Cu. Porosity in the sintered metallurgy can cause problems, as can residual glass and other contaminants. Again, it becomes necessary to anticipate the postsintering process needs when developing the package materials system.

Future directions in MLC packaging will be driven by the needs for higher density and lower cost. The highest density ceramic MLC package in manufacturing (as of 1993) contains 75 μm lines and vias on a 225 μm pitch (distance center-to-center of adjacent vias). By the mid- to late 1990s, the wiring density will require 40 μm lines and vias on a 125 μm pitch. Since the internal conductors in a high-performance package are transmission lines with controlled impedance, the dielectric thickness must proportionately become thinner with the reduction in line spacing. This means that future greensheet thicknesses must be significantly reduced. The combination of smaller vias, higher metal paste wiring density, and thinner greensheets requires advances in greensheet stability and better design of organic processing aids in both paste and greensheet fabrication. Alternatively, it may drive entirely new fabrication processes to accommodate the need for thin, high-density personalized layers and new methods for aligning, stacking, and laminating the layers.

In 1992, the average price of a high-performance multilayer ceramic substrate was around $5.00 per square inch. To remain competitive with other packaging alternatives (i.e., polymers), the price will have to approach 50 cents a square inch by 1998. A 10× reduction in price over six years and a nearly 2× improvement in density provide major challenges for future multilayer ceramic materials and processes. This will drive a continued need for a greater understanding of the chemistry involved, and further optimization of the relevant chemical and chemical engineering principles to develop the most efficient manufacturing processes.

EXERCISES

1. What is the maximum X–Y dimensions for active area in millimeters for an MLC substrate fabricated from 0.18 mm thick greensheets that have a stability of plus or minus 0.05%? 0.1%?
2. Which solvents in Table 5 would inhibit the adsorption of PMMA onto SiO_2 more than chloroform and dioxane? Why?
3. A 2-pentanone, 4-hydroxy has the structure

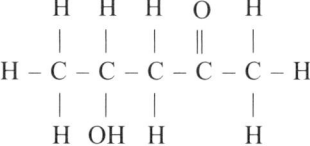

   Using solubility parameters and Table 6, estimate the solubility of 2-pentanone, 4-hydroxy in ethyl alcohol and acetone.
4. Lead oxide is mixed with silica to produce a low-viscosity glass for liquid-phase-assisted sintering of alumina. Determine if this glass system is compatible with copper metal at 800°C during carbon oxidation in the sintering cycle.
5. What is the oxygen partial pressure and hydrogen/water ratio to keep iron reduced at 1600°C but be strongly oxidizing to carbon?

REFERENCES

1. R. R. Tummala and E. Rymaszewski, eds., *Microelectronic Packaging Handbook*, Van Nostrand Reinhold, New York, 1989.
2. T. C. Patton, *Paint Flow and Pigment Dispersion*, John Wiley and Sons, New York, 1979, pp. 410–467.
3. J. S. Reed, *Introduction to the Principles of Ceramic Processing*, John Wiley and Sons, New York, 1987, pp. 123–151.
4. R. S. Drago, G. C. Vogael, and T. E. Needham, *J. Am. Chem. Soc. 93*:6014 (1971).
5. F. M. Fowkes, *J. Adhesion Sci. Tech. 1*:7 (1987).
6. D. W. Van Kievelen, *Properties of Polymers*, Elsevier Scientific Publishing Company, Amsterdam, 1980, pp. 129–172.
7. J. Bicerano, *Prediction of Polymer Properties*, Marcel Dekker, New York, 1993.
8. G. Partridge, *Glass Technol. 35*:116 (1994).
9. N. P. Bansal and R. H. Doremus, *Handbook of Glass Properties*, Academic Press, New York, 1986.
10. T. B. Reed, *Free Energy of Formation of Binary Compounds: An Atlas of Charts for High-Temperature Chemical Calculations*, MIT Press, Cambridge, Massachusetts, 1971.
11. Y. Yoshino and T. Shibata. *J. Am. Ceram. Soc. 75*:2756 (1992).

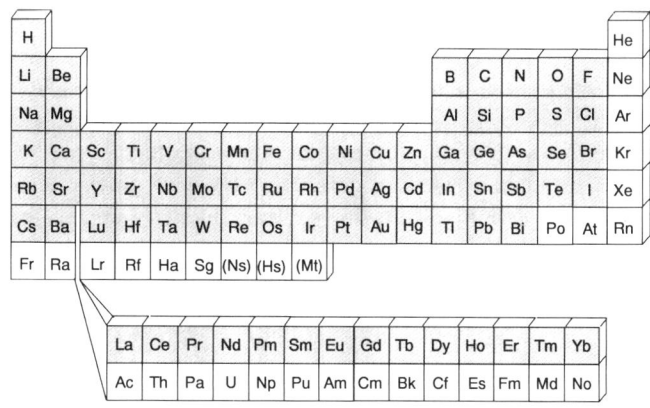

12
Chemistry of the New Superconductors

Frank J. Adrian[*]   *The Johns Hopkins University, Laurel, Maryland; and Technical University of Darmstadt, Darmstadt, Germany*

I. INTRODUCTION

Superconductivity is a novel state of matter entered by many conductors when cooled below their superconducting transition temperatures (T_c) [1]. The zero electrical resistance, for which superconductivity is best known, is utilized in high-field superconducting magnets and some electronic devices. Equally important, however, as shown by the celebrated Bardeen, Cooper, Schrieffer (BCS) theory of superconductivity, it is a novel state of matter in which quantum mechanics rules on a macroscopic scale of the order of many atomic distances rather than the atomic and subatomic dimensions that are its usual domain [2]. The large-scale quantum coherence of this state not only renders it immune to

---

[*]Retired: Olney, Maryland

the collisions of current-carrying electrons with lattice vibrations and impurities that create resistance in normal conductors, it gives this state fundamentally and practically important magnetic properties as well. Best known of the magnetic properties is the repulsion or exclusion of an external magnetic field (Meissner effect), especially as it manifests itself in levitating a magnet above a superconductor (or vice versa) at the point where the force from the energy required to expel the field equals the downward force of gravity. The Meissner effect is utilized in definitive tests for superconductivity and in magnetically levitated frictionless bearings. Equally important are strong magnetic field effects on the currents in circuits containing superconductors, which are used to measure extremely small magnetic field changes.

Chemistry has long played an important role in superconductivity, beginning with the binary and ternary intermetallic compounds and alloys. As shown in Fig. 1, depicting the history of superconductivity in terms of T_c vs. time, these materials were an important advance with substantially higher T_c than the elemental superconductors. Equally important, they had sufficient magnetic field

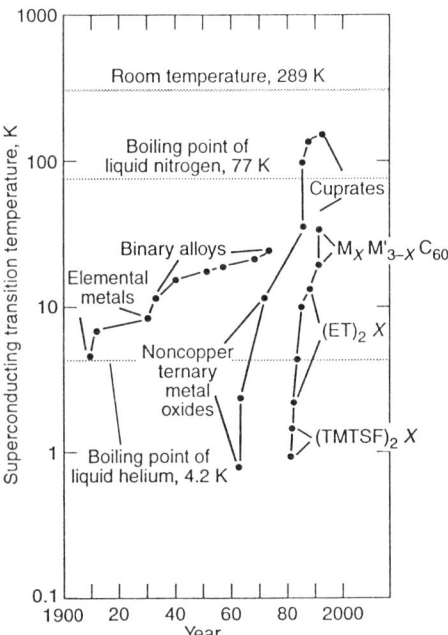

Figure 1 History of superconductivity as a plot of superconducting transition temperatures versus time in various classes of superconducting materials. (Adapted with permission from Ref. 3. Copyright 1992, The American Chemical Society.)

tolerance for application in high-field magnets. (As discussed in detail later, the energy required to expel a magnetic field limits the ability of superconductors to carry field-generating currents.) Chemistry was even more vital in the development of the so-called "synthetic metal" superconductors, such as the metal oxides [4] and organic materials [5] in Fig. 1. Nonetheless, its importance was fully established only in 1987 when discovery of superconductivity above liquid nitrogen temperature (77 K) in a class of copper-containing metal oxides ignited a revolution in this fascinating but aging field, where progress, at least as measured by increases in T_c, had been on hold for more than a decade [3,6]. A second front in this revolution opened in 1991 when salts of certain alkali metals and the exciting carbon allotrope buckministerfullerene (C_{60}) were found to superconduct at unprecedently high temperatures for organic materials [3,7].

These developments present both immense opportunities and challenges. The obvious utility of superconductors that do not require cooling by liquid helium, with its attendant cost and maintainence problems, is accompanied by the need to fabricate wires and other superconducting elements out of brittle ionic solids and also, perhaps, deal with complex organic compounds that are hard to prepare on a large scale and often are unstable in the presence of air and moisture. Understanding the electronic structures and mechanism(s) of conductivity and superconductivity in these materials, which could facilitate finding and utilizing new, even better, superconductors, is beset by an old nemesis of theoretical chemistry and physics: correlation of electron motions in many-electron molecules and solids by their coulomb repulsions. The nature of these problems, and the general experience that as one is resolved others take its place, should keep superconductivity in novel conductors at the frontier of science for the foreseeable future. For certain, progress in this field, which straddles the boundaries between chemistry and physics, will require a highly interdisciplinary effort, in which chemistry and chemists will play an important role.

II. SUPERCONDUCTING MATERIALS

A. The Conventional Superconductors

These materials are the elemental metals and intermetallic compounds and alloys, which, as shown in Fig. 1, dominated superconductivity for the first 50 years following its discovery in mercury in 1911. Many of these compounds have the general formula A_3B, where A is a transition element from Periods 4–6 of the periodic table, while B can be either a nontransition element from Groups III–V or a transition element from Periods 4–6. Some alloys of metals from neighboring groups in the periodic table are superconducting at somewhat higher temperatures than either component, e.g., $Nb_{0.75}Zr_{0.25}$ (T_c = 11 K) vs. 9.26 K for Nb. Other superconductors of this period include ternary metal compounds, compounds of metals and nonmetallic elements such as nitrides, carbides, etc., and graphite intercalation compounds.

With the exception of the graphite intercalation compounds, all the foregoing materials can be regarded as conventional metals, whose characteristic features are listed and compared with those of the synthetic metals in Table 1. They are relatively simple materials whose conductivity is due to weakly bound, readily delocalized valence electrons of metallic elements.

B. The New Superconductors—Synthetic Metals

As also shown in Fig. 1, superconductivity research changed abruptly around 1960, at which time a limit appeared to have been reached in the T_c of the intermetallic compounds and alloys. Interest then focused on superconductivity in inorganic conductors, and in organic conductors as well. These materials are generally referred to as synthetic metals, a term indicative both of their non-occurence in nature and the considerable synthetic effort usually required to prepare them. Their complexity is reflected in the sharp contrast between their properties, also listed in Table 1, and those of conventional metals. Conductivity in these nonstoichiometric, mixed-valence materials is due to the mobility of electrons among fractionally oxidized and/or reduced moieties, which usually are nonmetallic species. Of special importance here, many of these materials are high-T_c superconductors, defined rather arbitrarily as having a transition tem-

Table 1 Comparison of the Properties of Conventional and Synthetic Metals

Property	Conventional metals	Synthetic metals
Primary location of the charge carriers	On metallic atoms	On nonmetallic atoms and molecules, e.g., oxygen atoms and the π-orbitals of conjugated molecules
Valency	Elements or stoichiometric compounds with classical valences	Nonstoichiometric, mixed-valence compounds
Average number of charge carriers per site	Integral (≥1)	Fractional (≤1)[a]
Crystal structure	Simple: Cubic or hexagonal	Complex: Tetragonal, rhombic or monoclinic[a]
Isotropic conductors?	Yes	No; usually conduct along only one or two crystal axes[a]

[a]The cubic, three-dimensionally conducting fullerides are exceptions to these rules.

perature above that of the highest-T_c intermetallic compound, Nb_3Ge with $T_c \simeq$ 23.2 K.

1. Inorganics

The synthetic inorganic conductors and superconductors are strongly ionic compounds [4]. Metal oxides are the commonest and the most prominent because of the high-T_c cuprates; however, some borides, sulfides, and selenides are also superconductors. In these compounds the mixed valences and corresponding fractional oxidation/reduction charges result from doping, that is, partially replacing one metallic component with another of higher or lower valence, or by varying the oxygen content. Among the non-high-T_c oxides are (1) the hexagonal tungsten bronzes Rb_xWO_3 ($x \simeq 0.16$–0.33) with a maximum $T_c = 7$ K for $x = 0.25$; (2) $Li_xTi_{3-x}O_4$, which has the spinel crystal structure of $MgAl_2O_4$, and a maximum T_c of 14 K for $x = 1$; and (3) $BaPb_{1-x}Bi_xO_3$, which crystallizes in a somewhat distorted version of the cubic perovskite structure, and has a maximum T_c of 13 K for $x = 0.3$.

Table 2 lists a number of high-T_c oxide superconductors. With one exception, $Ba_{1-x}K_xBiO_3$, they are all cuprates. Like its low-T_c predecessor $BaPb_{1-x}Bi_xO_3$,

Table 2 Inorganic Oxide Superconductors, Mostly Cuprates

Compound	Designations	Charge carrier	CuO_2 layers w/o intervening oxygens	T_c (K)
$(Ba_{1-x}K_x)BiO_3$		hole		30, $x = 0.4$
$(La_{2-x}Sr_x)CuO_4$	214, (T)	hole	1	38, $x = 0.15$
$(Nd_{2-x}Ce_x)CuO_4$	(T')	electron	1	24, $x = 0.15$
$YBa_2Cu_3O_{6+\delta}$	Y123, YBCO	hole	2	93, $\delta = 1$; 58, $\delta = 0.5$
$YBa_2Cu_4O_8$	Y124	hole	2	80
$Y_2Ba_4Cu_7O_{15}$	Y247	hole	2	40
$Bi_2Sr_2CuO_6$	Bi2201	hole	1	0–20
$Bi_2Sr_2CaCu_2O_8$	Bi2212	hole	2	85
$Bi_2Sr_2Ca_2Cu_3O_{10}$	Bi2223	hole	3	110
$Tl_2Ba_2CuO_6$	Tl2201	hole	1	80
$Tl_2Ba_2CaCu_2O_8$	Tl2212	hole	2	108
$Tl_2Ba_2Ca_2Cu_3O_{10}$	Tl2223	hole	3	127
$TlBa_2CuO_5$	Tl1201	hole	1	0–20?
$TlBa_2CaCu_2O_7$	Tl1212	hole	2	80
$TlBa_2Ca_2Cu_3O_9$	Tl1223	hole	3	122
$HgBa_2CuO_4$	Hg1201	hole	1	94
$HgBa_2CaCu_2O_6$	Hg1212	hole	2	?
$HgBa_2Ca_2Cu_3O_8$	Hg1223	hole	3	133

$Ba_{1-x}K_xBiO_3$ has the distorted perovskite crystal structure. It is a three-dimensional conductor, whereas the cuprates conduct well only in two dimensions. Most cuprates are ternary or quarternary metal oxides, in which one metal is a trivalent species from Periods 5 or 6 of the periodic table or a lanthanide-series rare earth (mercury is an exception), one or two others are divalent alkaline earths, and in all cases, one metal is the copper that forms the essential CuO_2 layers wherein resides their conductivity and superconductivity. The other components form the intervening metal and metal-oxide layers, which not only separate and support the CuO_2 layers, they, most importantly, serve as charge reservoirs that introduce mobile charges by partially oxidizing or reducing the CuO_2 layers. For example, replacement of trivalent lanthanum by divalent strontium or barium in $La_{2-x}M_xCuO_4$, whose crystal structure is shown in Fig. 2a, removes electrons from the stoichiometric CuO_2^{2-} layers of undoped, insulating La_2CuO_4, thereby creating mobile charge-carrying holes (missing electrons) in them. On the other hand, replacing trivalent neodymium, samarium, or praeseodymium with quadrivalent cerium in $Ln_{2-x}Ce_xCuO_4$, whose crystal structure is shown in Fig. 2b, adds mobile electrons to the CuO_2 layers. In $YBa_2Cu_3O_{6+\delta}$, the charge-carrying holes are introduced into the CuO_2 layers by increasing the oxygen content from six to seven. These oxygens are added to CuO chains, which are devoid of oxygen in the insulating compound $YBa_2Cu_3O_6$, and fully oxygenated in $YBa_2Cu_3O_7$, whose crystal structure is shown in Fig. 3. In some of the more complex cuprates, the charge carriers are partly or wholly due to

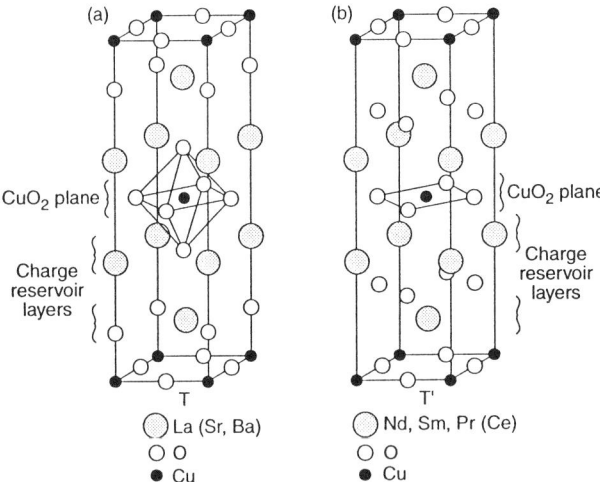

Figure 2 Crystal structures of (a) the hole superconductor $La_{2-x}M_xCuO_4$, where M ≡ Sr, Ba, and (b) the electron superconductor $Ln_{2-x}Ce_xCuO_4$, where Ln ≡ Nd, Sm, Pr.

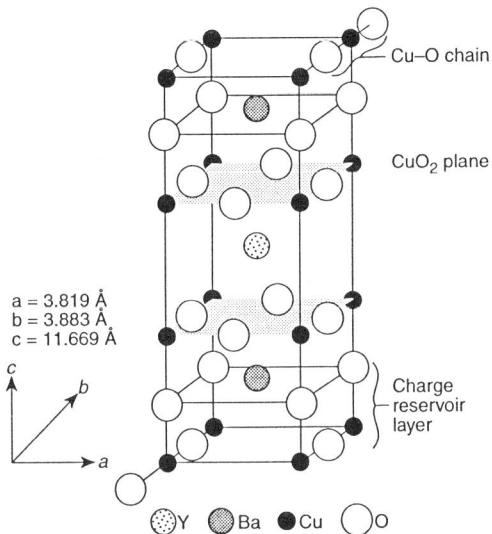

Figure 3 Crystal structure of $YBa_2Cu_3O_7$. The $YBa_2Cu_3O_{6+\delta}$ compounds have the same structure, except the oxygen content of the CuO chains decreases as δ decreases until they are devoid of oxygen in $YBa_2Cu_3O_6$.

various defects, such as vacancies, and substitutional and interstitial ions, in the crystal structures corresponding to their nominal (i.e., ideal) compositional formulas. Among these compounds are two families of thallium-containing cuprates with the nominal formulas $Tl_2Ba_2Ca_{n-1}Cu_nO_{2n+4}$ and $TlBa_2Ca_{n-1}Cu_nO_{2n+3}$, whose crystal structures are shown in Fig. 4. Also, there is an analogous series of bismuth-cuprates with the nominal formulas $Bi_2Sr_2Ca_{n-1}Cu_nO_{2n+4}$, and a mercury-based family with the formulas $HgBa_2Ca_{n-1}Cu_nO_{2n+2}$. In these compounds, T_c increases with n, which is the number of CuO_2 layers that are separated only by layers of alkaline earth ions. However, preparation of the $n \geq 4$ members is extremely difficult and may not yield significant increases in T_c.

The dopant-induced fractional oxidations and reductions are not necessarily confined to the CuO_2 planes, but the wide variety of charge-reservoir structures indicates that, except for a possible role in the low conductivity along the axis perpendicular to the CuO_2 planes, charge carriers outside the CuO_2 planes are not required for conductivity and superconductivity in the cuprates.

A common nomenclature for the cuprate superconductors is to specify the trivalent element and the proportions of the metallic components. For example, the $YBa_2Cu_3O_{6+\delta}$ family is denoted Y123 or sometimes Y123O(6 + x), and

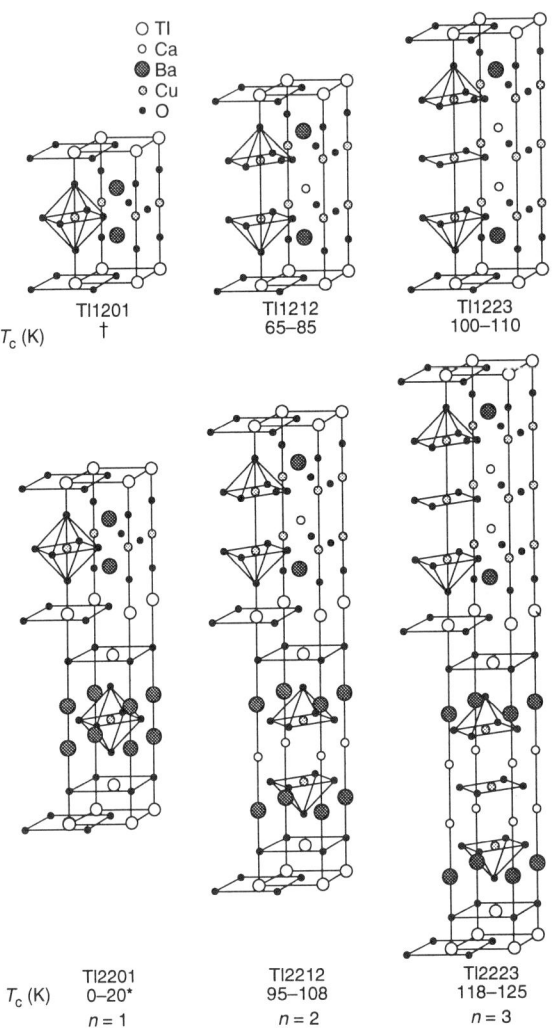

Figure 4 Nominal crystal structures of the families of thallium-containing cuprates. Top: The $TlBa_2Ca_{n-1}Cu_nO_{2n+3}$ family where $n = 0, 1, 2$ going from left to right. Bottom: The $Tl_2Ba_2Ca_{n-1}Cu_nO_{2n+4}$ family where $n = 0, 1, 2$ going from left to right. The $Tl_2Ba_2Ca_{n-1}Cu_nO_{2n+4}$ structures consist of two $TlBa_2Ca_{n-1}Cu_nO_{2n+3}$ slabs that are displaced relative to each other along a (110) axis by a distance $\sqrt{a}/2$, where a is the lattice constant in the basal plane. (Reproduced from Ref. 8. Copyright 1988, The American Institute of Physics.)

the $Tl_2Ba_2Ca_{n-1}Cu_nO_{2n+4}$ and $TlBa_2Ca_{n-1}Cu_nO_{2n+3}$ families are denoted Tl22($n-1$)n and Tl12($n-1$)n, respectively. The specification for $La_{2-x}M_xCuO_4$, which, along with $Ln_{2-x}Ce_xCuO_4$, is an exception in that the undoped parent compound is a binary oxide, includes the oxygen content and is 214.

2. Organics

The organic conductors and superconductors are mostly charge-transfer compounds of fixed composition in which both the donor and acceptor can be organic molecules, or one component can be an inorganic ion [5]. Their fractional oxidation/reduction states and corresponding mobile charges are due to the limited abilities of the donor to contribute electrons and/or the acceptor to receive them. For example, until the advent of the alkali metal fullerides, nearly all organic superconductors were charge transfer salts of a tetrathiafulvalene or a tetraselenafulvalene donor (D) and a complex inorganic anion (A) with the charge formula $(D_2)^+A^-$. Table 3 lists some of the most important organic superconductors of this type. The first family to be discovered, salts of tetramethyltetraselenafulvalene (TMTSF) with complex inorganic anions, have the crystal structure illustrated for $(TMTSF)_2ClO_4$ in Fig. 5. These compounds are predominantly one-dimensional conductors in the direction of the donor stacking axis. Only $(TMTSF)_2ClO_4$ is superconductive at ambient pressure ($T_c = 1.2$ K), but several are superconductive under pressure.

A second family, with a wider range of crystal structures and many more ambient pressure superconductors including some with T_c slightly above 10 K, is salts of *bis*-ethylenedithio-tetrathiafulvalene (ET or, alternatively, BEDT-TTF). Some, like $(ET)_2ReO_4$, are one-dimensional conductors with crystal struc-

Table 3 Selected $(TMTSF)_2X$ and $(ET)_2X$ Organic Superconductors

Compound	T (K)	Pressure (kbar)
$(TMTSF)_2ClO_4$	1.2	0
$(TMTSF)_2PF_6$	1.0	7
$(TMTSF)_2ReO_4$	1.4	12
$(TMTSF)_2FSO_3$	2.5	>6
$(ET)_2ReO_4$	1.5	5
β-$(ET)_2I_3$	1.6	0
β-$(ET)_2IBrI$	2.8	0
β-$(ET)_2IAuI$	4.9	0
β^*-$(ET)_2I_3$	8.1	0.5, temporarily to produce the metastable β^* phase
κ-$(ET)_2Cu(NCS)_2$	10.5	0
κ-$(ET)_2Cu[N(CN)_2]Br$	11.6	0
κ-$(ET)_2Cu[N(CN)_2]Cl$	12.8	0.3

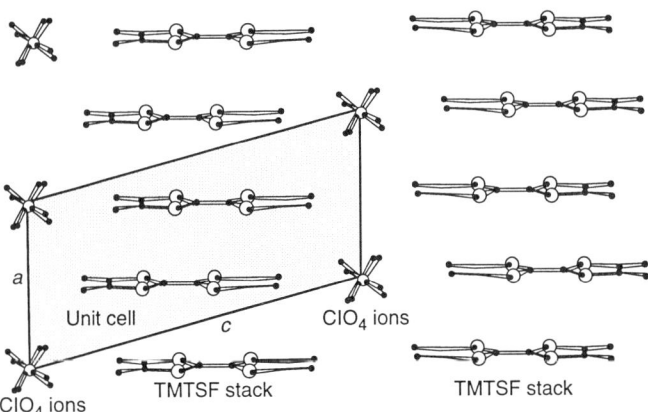

Figure 5 Crystal structure of $(TMTSF)_2ClO_4$. (Adapted with permission from Ref. 3. Copyright 1992, The American Chemical Society.)

tures similar to those of the $(TMTSF)_2X$ compounds. The β-crystal phase, which features linear triatomic anions such as I_3^-, and the κ-crystal phase, with polymer-forming anions such as $Cu(SCN)_2^-$, are two-dimensional conductors in which layers of ET molecules, whose molecular planes are approximately perpendicular to the layer plane, alternate with anion layers. The κ-phase, shown in Fig. 6 for κ-$(ET)_2Cu(SCN)_2$, and the β-phase differ in the arrangement of the ET molecules. T_c tends to be highest in compounds with polymeric anions, as for example, $(ET)_2Cu(SCN)_2$ and $(ET)_2Cu[N(CN)_2]Br$, with T_c = 10.5 and 12 K, respectively.

In addition to having the highest T_c save for the inorganic cuprates, the alkali metal fullerides are unique both for integer ion charges, corresponding to the charge formula $(M^+)_3(C_{60})^{3-}$, and for being three-dimensional conductors, with the cubic crystal structure shown in Fig. 7. This structure is formed by intercalating the alkali ions into the tetrahedral and octahedral sites of the face-centered cubic lattice of C_{60}. However, Li_3C_{60} and Na_3C_{60} do not exist, because these alkali ions are too small to be contained within all interstitial sites of the C_{60} lattice. Various alkali metal fullerides, which include a number of mixed compounds $M_{3-x}M'_xC_{60}$, are listed as a plot of their T_c vs. lattice constant in Fig. 8. The simple, approximately linear dependence of T_c on lattice constant in these compounds will be discussed in more detail in Sec. VII.

Organic polymers, such as polyacetylene, usually owe their conductivity to dopants that add electrons or holes to a conjugated polymer chain. Superconducting polymers would be of great fundamental and practical interest, but among the many conductive polymers known as of 1994, only the inorganic

New Superconductors 511

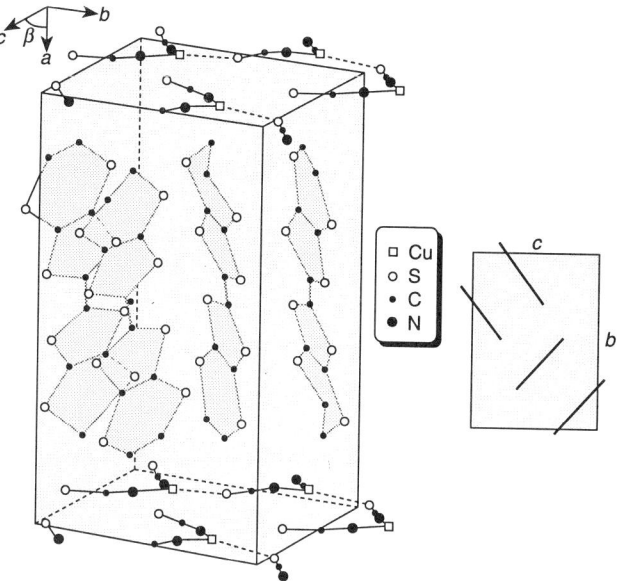

Figure 6 Crystal structure of $\kappa\text{-}(ET)_2Cu(SCN)_2$. (Reproduced from Ref. 9. Copyright 1992 by the AAAS.)

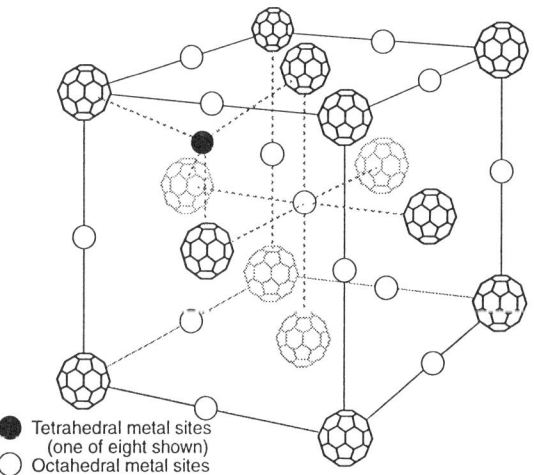

Figure 7 Crystal structure of the alkali metal fullerides $M_{3-x}M'_xC_{60}$. (Adapted with permission from Ref. 3. Copyright 1992, The American Chemical Society.)

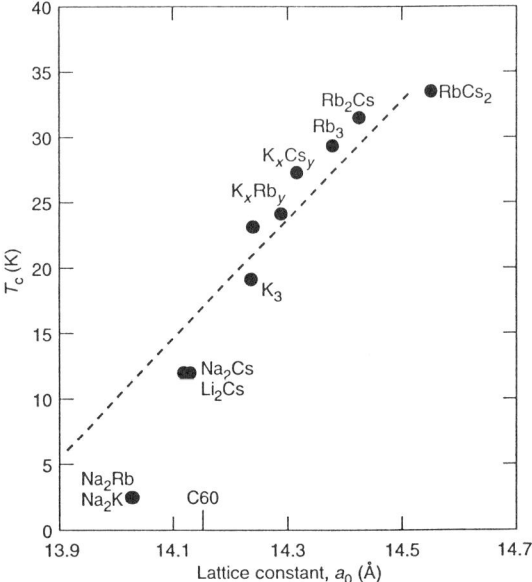

Figure 8 Relationship between T_c and the lattice constant in the $M_{3-x}M'_xC_{60}$ superconductors. The dashed line is a linear fit to the data. (Reprinted with permission from Ref. 10. Copyright 1992, Macmillan Magazines Limited.)

polysulfur nitride $(SN)_x$ is superconductive, at the impractically low T_c of 0.3 K.

III. STRUCTURE/CONDUCTIVITY RELATIONS IN SYNTHETIC METALS

The BCS theory shows that the superconducting state is a fluidlike state, produced by the simultaneous formation of pairs of conduction electrons, known as Cooper pairs, and condensation of these pairs into a quantum liquid in which all the pairs effectively occupy a single energy level that is stabilized by their mutual interactions. (Note that as pairs the electrons are freed from the Pauli exclusion principle prohibiting more than one electron from occupying a single energy state.) The driving force behind the pairing, in conventional superconductors at least, is an indirect attraction between the electrons that is transmitted by the polarizations these electrons induce in the crystal lattice.

A. Experimental Methods for Structure Determination

Almost every possible experimental method has been involved to some extent in investigating the synthetic metals and superconductors, but some have been used far more than others. Even these can only be described in cursory fashion here. (Reference 11 describes many of these methods.)

Electrical resistance measurements on synthetic metals, although conceptually simple, face both technical and fundamental problems. The former are attaching electrical leads to small fragile samples, and measurements on electrically discontinuous, inhomogeneous samples that may contain only a few percent of a superconducting phase. The fundamental problem is that observation of the Meissner effect by magnetic susceptibility measurement is required to distinguish between superconducting and other resistance-changing transitions. Alternating-current measurements of resistance, particularly in the microwave region, have long been used to overcome the former problems. A refinement of the microwave technique, called magnetically modulated microwave absorption (MAMMA), combines the resistance and Meissner measurements [3]. It can be carried out using an electron spin resonance spectrometer—a readily available laboratory instrument, used in other phases of superconductivity research—whose primary function is to investigate paramagnetic molecules and solids by observing their microwave absorption in magnetic fields. With minor modifications this instrument can measure, as a function of temperature, both the sample resistance and a magnetically modulated resistance that varies synchronously with a small oscillating magnetic field. The former will change with any resistance-changing transition, but the latter will change only if it is a superconducting transition, which is affected by a changing magnetic field through the Meissner effect.

Although indispensable to superconductivity research, X-ray crystallography and neutron diffraction techniques for determining the structures of crystalline solids are thoroughly discussed elsewhere. Furthermore, modern computer-based instrumentation makes this task fairly routine once a suitable sample is available. Single crystals of the inorganic oxides large enough for detailed crystal structure determination, and also measurements of anisotropic quantities such as conductivity, can be grown from metal oxide melts in many cases, but usually only after their composition has been determined and a preliminary structure of the material has been obtained from the powder or small-crystal samples. This task, and the exciting days after the liquid nitrogen barrier was broken, are summarized by an account of the isolation of the superconducting component of the original multiphasic preparation of $YBa_2Cu_3O_7$ and the determination of its crystal structure [12]. Electrochemical preparation of the organic charge-transfer conductors usually produces adequate single crystals.

Nuclear magnetic resonance (NMR) is an especially important method for investigating both the electronic structure of superconducting materials and various properties of their superconducting state. This technique, in which radio-frequency energy stimulates and observes transitions between different quantized orientations of nuclear spin magnetic moments in a strong magnetic field, can yield a detailed picture of the electronic environment in the immediate vicinity of that nucleus. Equally useful are measurements of the relaxation rate, that is, the rate at which the nuclear spins return to thermal equilibrium after being energized by a radio-frequency pulse. They can provide detailed information about the conducting and superconducting states because the relaxation is due to magnetic interactions between the nuclear spins and the conduction electrons.

Electron spectroscopic methods are of special importance in investigating conductive materials, whose opacity greatly limits conventional optical spectroscopies. One of these is photoemission spectroscopy (PES), better known to chemists as electron spectroscopy for chemical analysis (ESCA). In this method, the energy distribution of electrons excited from a sample by soft X-ray photons provides information about its composition and electronic structure. Another is electron energy loss spectroscopy (EELS). This method observes electronic transitions indirectly from the resulting loss in energy of a monoenergetic beam of electrons impinging on the sample.

B. Conduction Bands

Conductors characteristically have valence (conduction) electrons that only partially fill the available energy levels of one or more groups of band states [3,4]. These band states and their energies, which are the crystal lattice analog of molecular orbitals (MOs) and MO energy levels in molecules, result from interactions between the atomic/molecular orbitals of the component atoms/molecules of the crystalline material, just as MOs are formed by interaction of atomic orbitals within a molecule. The number of band states equals the number of interacting orbital states, just as with MO states. However, because there are a great many more interacting states in a crystal than in a molecule, bands are a virtual continuum of very closely spaced states lying between the lowest and uppermost states, rather than a discrete set of molecular energy levels. Of course, the conduction-band ground state is formed by placing pairs of electrons of opposite spin in levels of increasing energy until all electrons are accommodated, just as with MOs in molecules. The energy of the highest occupied states of the band is known as the Fermi energy (E_F) or Fermi level. The interactions responsible for superconductivity occur within a narrow energy region near E_F known as "the foam on the Fermi sea," but the lower energy conduction states play an important if passive role in supporting the "foam" region through the Pauli exclusion principle, which keeps the interacting electrons in the "foam" from sinking into these lower energy states.

An example of special relevance for the cuprates with their CuO_2 conduction planes is the conduction bands of a two-dimensional array of atoms/molecules, each of which interacts with four nearest neighbors [3]. As shown in Fig. 9, the lower band is formed from the highest occupied atomic/molecular orbitals (HOMO) of the interacting atoms/molecules and the upper band is similarly formed from their lowest unoccupied atomic/molecular orbitals (LUMO). Because of the dense packing of the band states, the bands are most conveniently described by a plot of the band state energy versus the density of states or the number of states in a unit energy interval. This continuity, or near energy degeneracy, of the band states enables electrons in a partially filled band to accommodate the kinetic energy acquired during conductive motion in an electric field by moving from their original band states into slightly higher energy states.

Conversely, conduction cannot occur if, as shown in Fig. 9a, the lower band is completely filled and the upper band is empty, which is the usual situation in nonmetallic inorganic and organic solids, formed from closed-shell atoms, molecules, or ions. The fractional reduction or oxidation of such an insulator, discussed in Sec. II.B, creates a conductor by respectively adding electrons to the upper band, as shown in Fig. 9b, or removing them from the lower band,

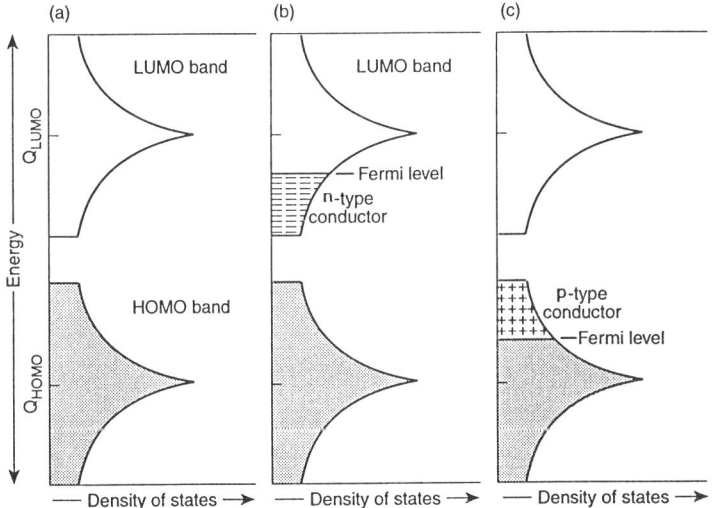

Figure 9 Lower (HOMO) and upper (LUMO) conduction bands of a two-dimensional array of interacting atoms or molecules. (a) Insulator with a filled lower band and an empty upper band; (b) n-type conductor produced by adding electrons to the upper band; (c) p-type conductor produced by removing electrons from the lower band. (Adapted with permission from Ref. 3. Copyright 1992, The American Chemical Society.)

as shown in Fig. 9c. In the former case, the charge carriers are electrons and the material is a n(egative)-type conductor; in the latter case, the charge carriers are positively charged holes (i.e., missing electrons in the formerly filled band) and the material is a p(ositive)-type conductor. The electronic structures of the high-T_c synthetic metals, discussed next, illustrate these principles well, although not without some important complications from electron correlation effects that are either minimal or absent in conventional metals.

C. Electronic Structures of the High-T_c Superconductors

1. The Cuprates

A vital and novel feature of the cuprates' electronic structure is the square-planar-coordinated open-shell Cu^{2+} $(3d)^9$ ions in the CuO_2 planes, into which the charge carriers are introduced by doping. At first thought the missing electron or hole in the filled Cu $(3d)^{10}$ shell should yield a half-filled conduction band, but as is typical of synthetic metals, it actually produces a type of insulator called a Mott–Hubbard insulator [3,4]. This occurs because the valence interactions between orbitals on adjacent sites, which enable the charge carriers to move between sites, usually are so much weaker in these materials than they are in conventional metals that they cannot overcome the large electron repulsion energies in states with two charge carriers on a single atomic or molecular site. Such states *must occur during conductive motion when there is an average of one charge carrier per site.*

Experiment, especially NMR, and theory have shown that the missing electron or hole is in the Cu $3d_{x^2-y^2}\sigma$ orbital, whose lobes point at the nearest neighbor O^{2-} ions in the CuO_2 plane. This configuration maximizes both the coulomb interactions of the positively charged Cu^{2+} $3d$ hole with the four nearest O^{2-} ions in the CuO_2 plane (the apical O^{2-} ions found in many of the cuprates are considerably more distant) and its valence interactions with their $2p\sigma$ orbitals. These valence interactions are weak nonetheless, and they yield only a slight delocalization of the Cu^{2+} holes onto the neighboring O^{2-} ions. However, this delocalization does produce weak indirect covalent interactions, of the order of 2–3 kcal/mol, between neighboring Cu^{2+} ions. These interactions result in an antiferromagnetic state in which the electrons spins of neighboring Cu^{2+} ions are antiparallel. This antiferromagnetic spin alignment frequently occurs in situations where each atom or ion in a solid has a large number of potential bonding partners because the antiferromagnetic "bond," although weaker than a true covalent bond, has the advantage that it can be formed with all bonding partners simultaneously.

Experiment and theory also agree that the doping-induced conduction holes reside primarily on the O^{2-} ions, which is consistent with the avoidance of the same high-energy Cu^{3+} states required for conductive hole motion in the half-filled Cu band. The more difficult question of whether these charge-carrying holes reside in oxygen $2p\sigma$ orbitals pointing at their nearest Cu^{2+} neighbors, which would maximize valence interactions, or $2p\pi$ orbitals oriented perpendicular to the Cu–O–Cu axis, which would minimize coulomb repulsions between these positively charged holes and the neighboring Cu^{2+} ions, was resolved in favor of the $p\sigma$ orbital by ^{17}O nuclear magnetic resonance, and by the polarizations of the O $1s \rightarrow 2p$ transitions observed by electron energy loss spectroscopy.

The electronic structure of the excess-electron cuprate superconductors, $Ln_{2-x}Ce_xCuO_4$, is somewhat simpler, as the added electrons must partially fill the Cu^{2+} $3d_{x^2-y^2}$ holes of the undoped Ln_2CuO_4 compound.

In both cases, however, one has the very complex electronic structural problem of mobile holes moving in a network of fixed Cu^{2+} holes, with which they have spin-dependent valence interactions.

Indicative of this complexity is the variation of the properties of $La_{2-x}Sr_xCuO_4$ with increasing dopant and charge-carrier concentration, shown in Fig. 10. At low carrier concentration ($0 \geq x \geq 0.05$), the long-range antiferromag-

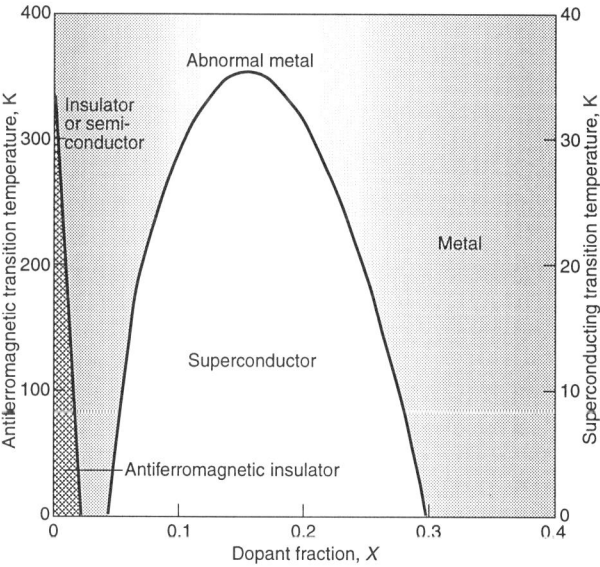

Figure 10 Temperature versus composition phase diagram showing the variation of the properties of $La_{2-x}Sr_xCuO_4$ with Sr content. (Reprinted with permission from Ref. 7. Copyright 1992, The American Chemical Society.)

netism of pure La_2CuO_4 disappears, although some short-range antiferromagnetic order remains, and the material is a semiconductor. For $0.05 \geq x \geq 0.3$, it becomes a superconductor with an unusual metallic state (as discussed in Sec. VII.B.1) and a superconducting transition temperature that increases with x to a maximum at $x \simeq 0.15$ and then falls back to zero. Finally, for $x \geq 0.3$, the material is a rather normal, nonsuperconducting metal. This competition between superconductivity and other states (in this case, antiferromagnetism and a normal metal), which can be shifted in favor of one or the other by charge-carrier concentration and other structural factors, is characteristic of superconductivity, especially in the high-T_c materials. Understanding the reasons for this is a major problem in superconductivity theory, whose solution may resolve the problem of the superconducting mechanism itself.

2. The Organics

The $(TMTSF)_2X$ and $(ET)_2X$ Compounds. These compounds typically have quarter-filled conduction bands, with an average of one conduction hole per two donor sites. Roughly, they fall into two structural classes according to whether or not the donors form one-dimensional (1-D) stacks, as in all the TMTSF compounds, or two-dimensional (2-D) planes.

The major complication in these compounds, especially for those with 1-D stacks or quasi-1-D stacks with only weak interstack interactions, is the tendency for the even spacing of the molecules along the stacking axis to distort to an unevenly spaced configuration as the temperature is lowered. This transition, called a Peierls transition, localizes the electrons or holes in the resulting dimers, trimers, etc., thus producing an insulator [3,4]. Often the spins of these localized electrons then form an antiferromagnetic state. This tendency toward Peierls transitions is a major reason that there are relatively few superconductors among the organic charge-transfer compounds, particularly pure organics, such as the first organic conductor tetrathiafulvalenium-tetracyanoquinodimethanide (TTF-TCNQ), whose organic acceptor stacks are especially prone to Peierls transitions.

In the quasi-one-dimensional conducting $(TMTSF)_2X$ compounds, the competition between a superconducting transition and transition to an insulating antiferromagnetic state is often decided by such structural factors as the anion or pressure. Indicative of this is the temperature–pressure phase diagram for $(TMTSF)_2PF_6$, shown in Fig. 11. Below 7.5 kbar pressure, the transition is to an insulating, antiferromagnetic state at a temperature that decreases with increasing pressure. Above 7.5 kbar, the transition is to a superconducting state, whose transition temperature also decreases with pressure. In $(TMTSF)_2ClO_4$, which is the only zero-pressure superconductor in this family, interactions between the donors and the ClO_4^- ions clearly play a role in preventing the transition to the insulating state. If $(TMTSF)_2ClO_4$ is cooled rapidly enough to

New Superconductors 519

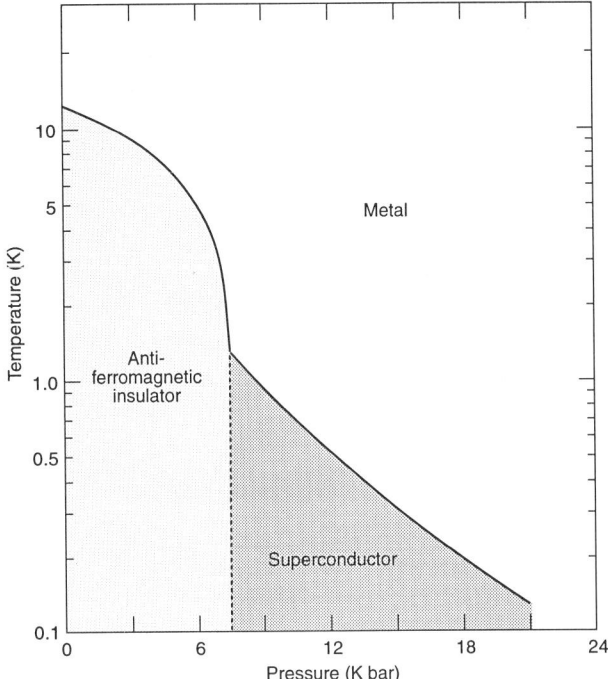

Figure 11 Temperature versus pressure phase diagram of $(TMTSF)_2PF_6$. (Drawn from data of Ref. 13.)

prevent the transition of the ClO_4^- ions from an orientationally disordered to an ordered state at 24 K, then the compound becomes an insulator rather than a superconductor upon further cooling.

The two-dimensional conductors, such as κ-$(ET)_2Cu(SCN)_2$, are less prone to Peierls transitions, because it is harder for lattice distortions to localize electrons that are free to move in two dimensions. Accordingly, there are more, and higher-T_c superconductors in this class. Nonetheless, they exhibit a wide range of structure- and pressure-dependent behavior.

The Fullerides. Nuclear magnetic resonance has shown that the alkalis are present as cations, indicating complete charge transfer to the charge formula: $(M^+)_3C_{60}^{3-}$. The nearly spherical C_{60} molecule is conveniently regarded as a large atom when describing its ionic compounds, whereas a description better suited to its covalent, bond-breaking, or addition reactions is a rolled-up graphite sheet, with five-membered rings incorporated into the six-membered-ring graphite structure to achieve the closed surface. The large-atom description reflects the analogies between the symmetry classifications and energy degeneracies of the π molecular-orbital states of C_{60} and those of atomic orbitals. For example, the lowest unoccupied molecular orbital (LUMO) of C_{60}, which receives the electrons donated by the alkali metals, is a triply degenerate, atomic p-like orbital [7]. (Other C_{60} orbitals are analogous to atomic s, d, and f atomic orbitals.) Like the ground state of nitrogen and other atoms with a $(p)^3$ electron configuration, the ground state of the C_{60}^{3-} trianion is "^4S," in which the three added electrons minimize their coulomb repulsions by occupying different sublevels of the triply degenerate LUMO, and the spins of these electrons combine additively for a total spin of 3/2 in accordance with Hund's rule of maximum spin multiplicity. M_3C_{60} is an isotropic conductor because these isotropic "^4S" C_{60}^{3-} ions interact equally well in all directions. Although the LUMO conduction band of C_{60} is half filled in the $M_3^+C_{60}^{3-}$ compounds, they are not Mott–Hubbard insulators, because the large size of C_{60} minimizes the coulomb repulsions in the C_{60}^{4-} states that necessarily arise during charge transport.

The six electrons donated to each C_{60} by the alkalis in the M_6C_{60} compounds completely fill the LUMO, and so M_6C_{60} is an insulator.

In the alkaline earth fullerides, Ca_5C_{60} (T_c = 8.4 K) and Ba_6C_{60} (T_c = 7 K), the divalent alkaline earth atoms contribute enough electrons to completely fill the LUMO of C_{60}, and so their conductivity and superconductivity must be due to partial filling of the next lowest unoccupied C_{60} orbital. This also is a triply degenerate orbital.

IV. THE SUPERCONDUCTING STATE

A. Phonon-Mediated Electron Attraction and Cooper Pairing

Figure 12 is a schematic illustration of how the interaction between the conduction electrons and the vibrations of the crystal lattice leads to an indirect Cooper pairing attraction between these electrons. The process, often called electron–phonon- or just phonon-mediated pairing (from the term *phonon*, which refers to a quantized vibrational excitation of the lattice), starts when one conduction electron polarizes the lattice by attracting the nearby positive ions toward it. This electron then moves on, but the heavy, slowly moving ions retain their distorted positions long enough to attract a second electron to this site, which is energetically favorable for an electron. This retarded nature of the phonon-mediated electron attraction is important because it enables this weak attractive force to overcome the much stronger coulomb repulsion between the electrons.

Even so, these attractive interactions are very weak. They owe their effectiveness to the fact, discovered by Cooper, that the individual interactions be-

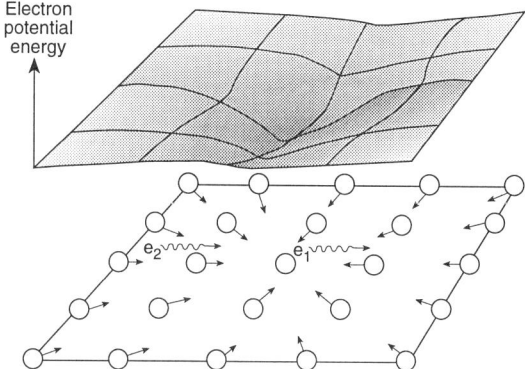

Figure 12 Indirect attraction between conduction electrons mediated by the electron–phonon interaction. (Reprinted with permission from Ref. 3. Copyright 1992, The American Chemical Society.)

tween a great many electronic states near the top of the conduction band can combine additively to form a bound electron-pair state, known as the Cooper pair.

B. The BCS Model

Bardeen, Cooper, and Schrieffer then showed that, upon cooling to a temperature where thermal fluctuations can no longer disrupt Cooper pairing, many pairs simultaneously form and condense to produce the superconducting state, whose quantum mechanical coherence further increases the average binding energy of the individual Cooper pairs. The Cooper pair binding energy per electron, also known as the superconducting energy gap, is temperature-dependent, varying from zero at T_c to a maximum value of $\Delta(0)$ at 0 K. In the simplest version of BCS theory, the reduced energy gap, $2\Delta(T)/k_B T_c$ where k_B is Boltzmann's constant, is a universal function of the reduced temperature, T/T_c, shown in Fig. 13. This theory also gives the following expressions for $\Delta(0)$ and T_c [4]:

$$\Delta(0) = 2E_V \exp \frac{-1}{D(E_F)V} \qquad T_c = 1.14 \frac{E_V}{k_B} \exp \frac{-1}{D(E_F)V} \qquad (1)$$

Here, E_V is the average excitation energy of a vibration involved in the Cooper pairing, $-V$ is the energy due to the attractive interaction between any two electron states, and $D(E_F)$ is the density of states at Fermi level. The strong dependence of $\Delta(0)$ and T_c on $D(E_F)$ reflects the fact that attractive interactions between many electrons contribute to the stabilization of the Cooper pairs.

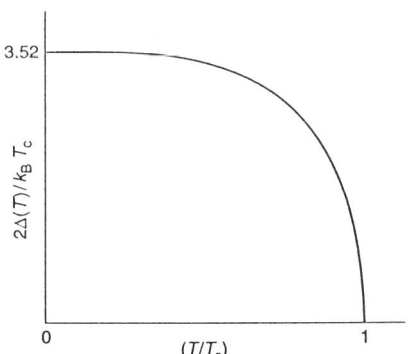

Figure 13 Universal relationship between the reduced superconducting energy gap T_c and the reduced temperature in a weakly coupled superconductor.

C. Fundamental Superconductor Properties and Parameters

Apart from their zero resistance, superconductors have a wide variety of fundamentally and practically important properties. Foremost among these are their unique magnetic properties, which are the basis of many applications and, also, limitations that must be dealt with in their use. Coherent quantum mechanical tunneling between superconductors gives superconducting junctions novel and useful properties, especially when combined with magnetic field effects.

1. Magnetic Properties of Superconductors

The novel magnetic behavior of superconductors is due to the requirement that the angular momenta of all Cooper pairs in the superconducting condensate have the same quantized value, combined with the fact that a magnetic field acting on a moving charge generates an angular momentum (as occurs in electric motors) [2,3]. This quantization requirement is not unique to superconductors—it is the basis of the Bohr model of hydrogen and other atoms—but the magnetic contribution to angular momentum is small except in extended quantum systems such as a superconductor. The resulting quantization condition on the magnetic flux, that is, the product of the area within any closed path inside a superconductor and the average field inside the path and normal to the area, can be shown to be

$$\phi = \frac{chn}{2e} = n\phi_0 \qquad (2)$$

where n is a nonnegative integer, c is the velocity of light, h is Planck's constant,

$2e$ is the Cooper pair charge, and ϕ_0 ($= hc/2e = 2.068 \times 10^{-7}$ gauss cm²) is the quantum of flux.

Because superconductors are macroscopic quantum states, this flux quantization condition applies to and can be observed in tangible superconducting rings. If the flux does not satisfy Eq. (2), a supercurrent will flow in the ring, producing an additional field that adjusts the flux up or down to the *nearest quantized value*. This effect was observed in 1961.

The Meissner effect is a consequence of flux quantization. Inside any superconductor in a magnetic field there will always be some closed path small enough that the quantized flux value closest to the actual flux determined by the path is zero. Thus, the superconductor must adjust the flux in this region to zero. The argument will then apply to a slightly larger path around the original path, and so on, until the field has been completely expelled from the superconductor except for a very thin surface layer. A supercurrent flowing in this layer produces the magnetic field that cancels the applied field within the superconductor. The thickness of this layer is an important superconductivity parameter, known as the field penetration depth (λ), or just the penetration depth. Typical values of λ at 0 K range from a few hundred to a few thousand Angstroms.

The energy required to expel a magnetic field limits the magnetic field a superconductor can tolerate. In simple, so-called Type I superconductors, T_c decreases with increasing field, as shown in Fig. 14a, until it reaches 0 K at the critical field (H_c) where the field expulsion energy ($VH_c^2/8\pi$, where V is the volume of the superconductor) exceeds the stabilizing energy of the superconductor. In Type I superconductors, which includes most elemental metals, H_c and the currents that generate it are so small (typically a few hundred gauss, as found in small permanent magnets) that these superconductors have few applications.

Fortunately, there is another class of superconductors, called Type II, that remain superconducting in the presence of large magnetic fields and currents. Like Type I materials, these superconductors completely expel a magnetic field up to some relatively low value, known as the lower critical field (H_{c1}). However, as shown in Fig. 14b, they remain superconducting far above H_{c1} by entering a mixed state in which the magnetic field penetrates the superconductor in more or less equally spaced cylindrical regions called flux tubes or vortices, while remaining superconducting everywhere except in the cores of the flux tubes.

The radius of the nonsuperconducting core at the center of each flux tube is determined by another fundamental parameter, the coherence length (ξ). This is the distance over which quantum mechanical coherence is maintained in the superconducting state, or equivalently, the distance over which any change in the superconductor, including loss of superconductivity in a magnetic field, must

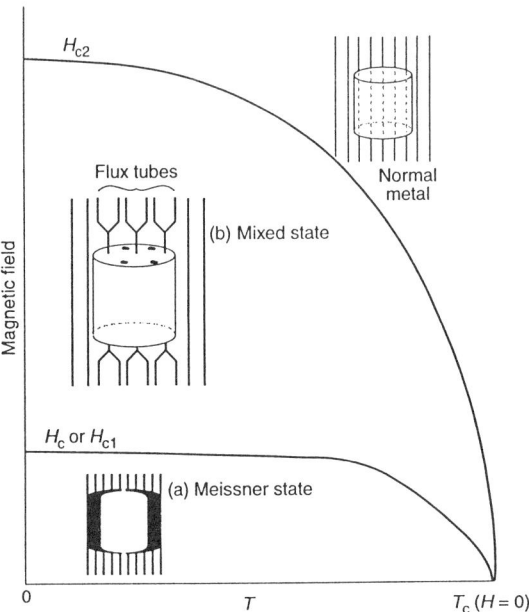

Figure 14 Dependence of the critical magnetic fields on temperature: (a) the Meissner state; (b) the mixed-state of Type II superconductors. The figure is not drawn to scale; H_{c2} is usually very much larger than H_{c1}. (Reprinted with permission from Ref. 14. Copyright 1993 Macmillan Magazines Limited.)

extend. It also is the size or spatial extent of the Cooper pairs. The mixed state is energetically favorable when the magnetic field penetration depth, which determines the radii of the flux tubes, is larger than this coherence length, because then it saves the energy required to expel the field from the entire superconductor at the cost only of the binding energy of the relatively few Cooper pairs within a coherence length of the flux tube axes.

The number of flux tubes increases with increasing field until, at the upper critical field H_{c2}, their nonsuperconducting regions overlap and superconductivity ends. A rough estimate of H_{c2} is that it yields half a flux quantum through an area determined by the coherence lengths, at which point it is energetically more favorable for the superconductor to revert to the normal state rather than to increase this flux to a full quantum as required by Eq. (2). However, coherence lengths are so short in Type II superconductors, particularly in the high-T_c cuprates (3–30 Å), that the upper critical fields are extremely large, up to megagauss.

On the other hand, the coherence lengths of Type I superconductors are much longer than the magnetic field penetration depth. These materials cannot preserve superconductivity above H_c by forming a mixed state because, for each flux tube formed, superconductivity is lost in a cylindrical region that extends far outside each flux tube, to encompass the entire superconductor.

Superconductors with moderate to high T_c will usually be Type II. The reason is that the positional uncertainty of the Cooper pairs (Δx), which determines the coherence length, varies inversely with spread or uncertainty in their momentum (Δp), in accordance with the Heisenberg uncertainty principle ($\Delta p \, \Delta x \simeq h/2\pi$). The uncertainty in the Cooper pair momenta increases with T_c, and in high-T_c materials it is large enough to produce the required short coherence lengths. Short coherence lengths are a mixed blessing, however; as will be seen later, they complicate the fabrication of superconducting devices.

Although Type II superconductors tolerate large magnetic fields by forming flux tubes, the flux tubes themselves pose another obstacle for applications. The problem is magnetic interactions between the flux tubes and supercurrents, which lead to energy-dissipating motion of the flux tubes. As discussed in more detail in Sec. VI, this limits high-current applications to temperatures considerably below T_c. Adventitiously or deliberately added impurities or defects can be helpful if they impede the motion of the flux tubes more than they damage the intrinsic properties of the superconductor.

2. Quantum Coherence and Superconducting Junctions

Junctions between a superconductor and a normal metal or between two superconductors are important both for research on superconductors and in devices. As illustrated in Fig. 15a, these junctions usually are tunneling junctions in which the two components are separated by an insulating barrier that is thin enough for the charge carriers on opposite sides of the junction to be coupled by overlap of their electronic wave functions in the barrier region. This allows the charge carriers to "tunnel" through the barrier that they lack the energy to surmount.

Measurement of current versus voltage in a superconductor–metal junction determines the superconducting energy gap. This is because passage of a current from a superconductor, where it is carried by Cooper pairs, into a normal metal, where it is carried by single electrons, requires a voltage large enough that the potential energy change on passing from the superconductor to the metal compensates for the loss of the Cooper pair binding energy.

A Josephson junction is formed by two superconductors separated by a thin insulating barrier [3]. It has very unusual properties because the carrier states coupled through the barrier are not single-particle states but macroscopic quantum states containing many Cooper pairs. Thus, the junction current is determined not only by the Cooper pair densities in these states but also by their

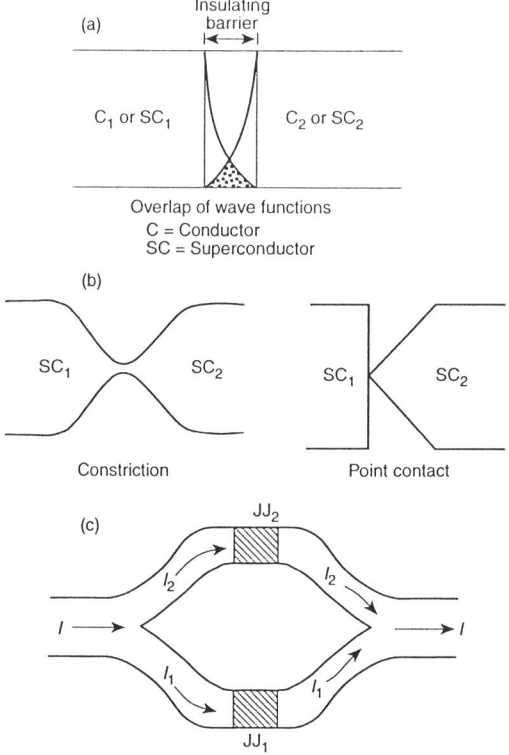

Figure 15 Superconducting junctions: (a) tunneling junction; (b) weak links or Josephson junctions at constrictions in a bulk superconductor or point contacts between grains in a granular superconductor; (c) superconducting quantum interference device (SQUID) formed by two Josephson junctions in parallel. (Reprinted with permission from Ref. 3. Copyright 1992, The American Chemical Society.)

quantum mechanical phases, giving it such unique properties as the alternating current (ac) Josephson effect, in which a constant (dc) voltage across the junction produces an alternating current through the junction. The frequency of this current (ν) varies linearly with the voltage (V), according to

$$\nu = 2\frac{eV}{h} = 483 \times 10^{12} V \text{ (Hz/V)} \tag{3}$$

where e is the electronic charge and h is Planck's constant. This effect has been used to show that the superconducting charge carriers are electron pairs, and in voltage-controlled ultrahigh-frequency generators and switches.

As shown in Fig. 15b, Josephson junctions can also be formed by a constriction in a superconductor or at a point contact between two superconducting grains if their crystal axes are misaligned, thereby creating phase differences. Such adventitious junctions, called weak links in the superconductor, can severely limit the supercurrent in granular superconductors.

Two Josephson junctions in parallel, illustrated in Fig. 15c, constitute a superconducting quantum interference device (SQUID). Because the supercurrents I_1 and I_2 through the two arms of the device are macroscopic quantum mechanical entities with both amplitude and phase, there can be interference between them when they rejoin to determine the total junction current I. This interference is similar to the interference between two light beams from a single source that travel different distances to reach the detector. In the SQUID the role of source-to-detector distance is played by a magnetic field that affects the transmission of the supercurrents through the two Josephson junctions. This device can measure very small magnetic field changes because the current is sensitive to very small flux changes of the order of a quantum of flux (2.068×10^{-7} gauss cm^2) through the area defined by the SQUID circuit.

D. Good Conductors Are Poor Superconductors

Although the BCS theory explains many properties of superconductors, it cannot predict T_c or even whether a given material will be a superconductor, because it is very difficult to determine the parameters, especially the interaction energy V, in Eq. (1). It does, however, support the empirical rule that good superconductors are most likely to be found among poor conductors, provided they do not undergo a phase change to an insulator when cooled. The previous section contains several examples of this competition between superconductivity and instability with respect to transitions to insulating and/or magnetic states. The reason for this is that the same strong electron–lattice interactions that promote Cooper pairing create electrical resistance by scattering the conduction electrons of the normal state, and they usually are the driving force behind the change in crystal structure involved in a transition to an insulating phase. Conductive dimensionality is also important—the higher the better—because, as noted earlier, much experimental evidence indicates that transitions to insulating states are more likely if the conduction electrons are already partially localized. Finally, a large density of states at the Fermi surface strongly favors superconductivity, but it also increases normal-state resistance because there are more states at the top of the conduction band for the conduction electrons to be scattered into.

The bright spot in this somewhat cloudy theoretical picture, at least for those more experimentally than theoretically inclined, is that creating synthetic metal structures that are likely to be good superconductors is largely a process of imaginative application of rather simple principles, combined with skill in preparative chemistry and perhaps a bit of serendipity.

V. PREPARATIVE CHEMISTRY

The tasks of preparative chemistry are the preparation of high-quality samples of known superconductors in forms suitable for experiment and applications, the development of process methods for industrial production, and the design and synthesis of new superconductors. The first two tasks are complicated both by the chemical and structural instabilities of these partly open-shell, free-radical-like materials and by their mechanical properties, which make them hard to shape in specific forms. These preparative challenges, combined with the aforementioned limited understanding of the interactions responsible for high-T_c superconductivity, make the final task even more difficult. In response to these problems, an increasingly wide variety of methods are being devised and applied.

A. Inorganic Superconductors

The metal-oxide superconductors are prepared by methods similar to those used for most multicomponent metal oxides, as discussed in Chapter 11. The usual procedure is heating finely ground and compacted mixtures of the appropriate metal oxides, or compounds that decompose to yield the oxides, to temperatures where reaction can occur by solid state diffusion. Often there are several cycles of heating with intervening regrinding and recompacting of the mixture, and finally annealing in oxygen. Better mixing, and a consequent reduction in reaction temperature, sometimes can be achieved by evaporation of a solution of the reactants, or even just a stirred suspension of the finely ground materials. These solution methods also can be useful for preparing the superconductors in particular shapes, such as thin films. Other methods for producing thin films of the oxide superconductors will be discussed later.

The reactions are carried out in standard, commercially available furnaces that can be programmed for controlled rates of both heating and cooling. Tube furnaces are more convenient for reactions carried out in flowing gases of variable composition, although it can be somewhat harder to regulate their temperature. Containers for the reactants typically are made of refractory metal oxides, most often high-density alumina (aluminum oxide). Metal containers are also used, especially gold crucibles, which minimize contamination of single crystals grown from mixtures of molten metal oxides.

Although some materials can be prepared by rudimentary application of these methods, they usually are multiphasic, as were the cuprates of the initial seminal discoveries. Sample impurity and inhomogeneity plagued early experiments and still can be a problem in more complex compounds, such as the bismuth and thallium cuprates. Two factors complicate the preparation of inorganic oxide superconductors, especially the cuprates.

First, as might be expected given the association of instability and superconductivity, these compounds often are only metastable, not thermodynamically stable, at room temperature. Metastability, i.e., kinetic stability, means conversion to thermodynamically stable compounds is negligibly slow under the conditions of use, although whether these conditions include the presence of air and moisture is questionable for some cuprates. This lack of thermodynamic stability rules out room-temperature preparation using solution precursors and, in some cases, requires preparation at temperatures higher than would be required for solid state diffusion among the metal oxide reactants.

Second, they tend to contain various crystallographic defects, especially substitutional impurities at metal ion sites, vacancies at oxygen sites, and interstitial oxygens [15]. One reason for this is that entropy stabilizes these defects at the high temperatures required to prepare these compounds. Another is that defect formation can compensate for mismatches in ion sizes, such as the sizes of the CuO_2 layers and the intervening metal and metal/oxygen layers in the cuprates. Thus, doping with metal ions or oxygen to achieve the desired charge carrier concentration may enhance defect formation. These defects can make the final compound quite sensitive to how the sample is cooled from its preparation temperature. If equilibrium is maintained by slow cooling, then the entropy-stabilized defects will be eliminated either by formation of separate phases or by ordering of the defects in a new phase that is different from the high-temperature phase. Rapid cooling, on the other hand, usually will accommodate the defects by producing a disordered structure with randomly distributed defects and/or multiple phases. This may be the best superconductor, however, because defects can play an important role in the superconductivity of these materials.

Preparation of compounds that are only metastable under the conditions of their use is quite common and follows one of three general routes [16]:

1. Equilibrium preparation under conditions of thermodynamically stability, followed by return to ambient conditions. This is usually the simplest method.
2. Preparation under equilibrium conditions of a precursor that can be doped or otherwise modified to produce the desired compound.
3. Preparation under nonequilibrium conditions. This is usually the most complicated method, as it involves controlling not only thermodynamic variables such as temperature and pressure, but also the kinetic variable time.

Often some or all steps of the multistep second and third routes can be carried out in a single heating and cooling cycle. However, finding the right temperatures and other variables, such as the rates of heating and cooling, etc., requires considerable trial and error. These preparative routes are illustrated for various cuprates in the following sections.

1. Route 1 Syntheses—BaPb$_{1-x}$Bi$_x$O$_3$ and La$_{2-x}$M$_x$CuO$_4$

The non-copper-containing superconductor BaPb$_{1-x}$Bi$_x$O$_3$ can be prepared by heating appropriate mixtures of the component metal oxides, or alternatively, carbonates or nitrates that decompose on heating to yield the oxides, in air to the temperature at which the single-phase solid solution is thermodynamically stable, followed by cooling. At this temperature the reaction is

$$\text{BaO} + (1-x)\text{PbO}_2 + x\text{BiO}_2 \xrightarrow[\text{O}_2]{\Delta} \text{BaPb}_{1-x}\text{Bi}_x\text{O}_3(s) \tag{4}$$

whereas at lower temperatures the reaction yields separate phases:

$$\text{BaO} + (1-x)\text{PbO}_2 + x\text{BiO}_2 \xrightarrow[\text{O}_2]{\Delta} (1-x)\text{BaPbO}_3(s) + x\text{BaBiO}_3(s) \tag{5}$$

As shown by a schematic phase diagram of this system in Fig. 16, the temperature separating the single-phase and two-phase regions is strongly dependent on the Bi concentration and is highest at the highest-T_c composition ($T_c \approx 13$ K at $x = 0.30$). Also, this composition is just short of the point ($x > 0.35$) at which the solid solution changes from metallic to insulating behavior. Both these

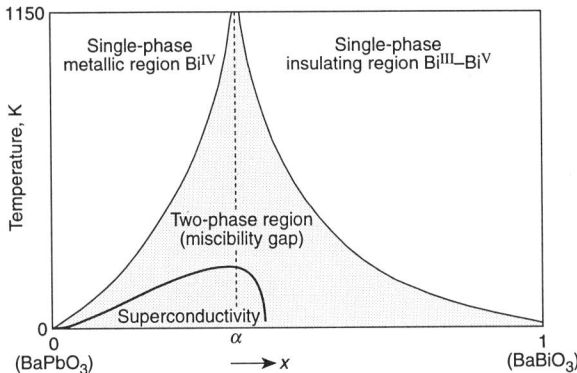

Figure 16 Temperature versus composition phase diagram for the BaPb$_{1-x}$Bi$_x$O$_3$ solid solution. The superconducting phase that exists as a metastable phase within the two-phase region is reached by cooling rapidly enough from the single-phase regions to avoid separation into the two phases. (Adapted from Ref. 16. Copyright 1991, The American Institute of Physics.)

factors illustrate the tendency of the highest-T_c compounds to be unstable; here, it is thermodynamic instability toward separation into $BaPbO_3$ and $BaBiO_3$ phases.

Attempts to modify the $BaBi_{1-x}Pb_xO_3$ superconductor were made prior to discovery of the $La_{2-x}M_xCuO_4$ compounds, but they were unsuccessful until 1988, when the first $Ba_{1-x}K_xBiO_3$ superconductor was prepared. The synthesis of this compound is difficult, one key element being to add excess alkali metal to the starting metal oxide mixture.

The doped lanthanum cuprates, $La_{2-x}M_xCuO_4$, can be prepared by heating stoichiometric quantities of La_2O_3, CuO, and the alkaline earth oxide or carbonate to around 1100°C in oxygen for 10 h or so and recooling to room temperature. It is also possible to dope La_2CuO_4 with excess oxygen to the formula $La_2CuO_{4+\delta}$ by heating it to around 600°C in oxygen at a few kbar pressure. The amount of excess oxygen introduced is estimated from the weight gain of the sample. The variation in properties with excess-oxygen-induced carrier concentration is similar to that obtained from alkaline earth doping, cf. Fig. 10, with transition temperatures around 30 K for $\delta \simeq 0.1\text{–}0.15$.

2. Route 2 Syntheses—The Excess Electron and Y123 Compounds

The excess electron superconductors $Ln_{2-x}Ce_xCuO_4$ (Ln denotes a rare earth such as Nd, Pr, or Sm) cannot be prepared by the method used for $La_{2-x}M_xCuO_4$, because there are no conditions under which the component oxides form a solid solution with the desired formula. However, the semimetallic, nonsuperconducting, oxygen-rich compound $Ln_{2-x}Ce_xCuO_{4+\delta}$ can be formed by this method. Reduction of this compound by annealing for 10 h at 1000°C in a stream of Ar/O_2 at very low oxygen pressure ($p_{O_2} \simeq 8 \times 10^{-5}$ atm), removes the excess oxygen, yielding $Ln_{2-x}Ce_xCuO_4$. The reactions in this case are

$$(1 - \tfrac{1}{2}x)Ln_2O_3 + xCeO_2 + CuO \xrightarrow[O_2]{\Delta} Ln_{2-x}Ce_xCuO_{4+\delta} \qquad (6a)$$

$$Ln_{2-x}Ce_xCuO_{4+\delta} \xrightarrow[Ar]{\Delta} Ln_{2-x}Ce_xCuO_4 + \tfrac{1}{2}\delta O_2 \qquad (6b)$$

The 93 K superconductors $RBa_2Cu_3O_7$, where R is yttrium or any of a number of lanthanide-series rare earths, can be prepared in what is effectively a one-step process by sintering stoichiometric mixtures of the appropriate oxides ($BaCO_3$, which decomposes on heating to give BaO, can also be used) at around 1000°C in flowing O_2 for 10 h or so and then cooling. It is likely, however, that this procedure actually involves the two reactions:

$$\frac{1}{2}R_2O_3 + 2BaO + 3CuO \xrightarrow[O_2]{\Delta} RBa_2Cu_3O_{6.5} \quad (7a)$$

$$RBa_2Cu_3O_{6.5} + \frac{1}{4}O_2 \xrightarrow[O_2]{cool} RBa_2Cu_3O_7 \quad (7b)$$

in which an equilibrium compound $RBa_2Cu_3O_{6+\delta}$ with $\delta \leq 0.5$ is formed at 1000°C and is converted to $RBa_2Cu_3O_7$ by uptake of additional oxygen during the process of cooling in flowing O_2. One indication of this is that high oxygen pressures ($p_{O_2} > 30$ at) do not favor formation of $YBa_2Cu_3O_7$, as would be expected if this compound were stable at 1000°C, but instead lead to formation of the superconducting compounds $YBa_2Cu_4O_8$ and $Y_2Ba_4Cu_7O_{15}$. $YBa_2Cu_4O_8$ has two adjacent CuO chain layers instead of the one present in $YBa_2Cu_3O_7$, while $Y_2Ba_4Cu_7O_{15}$ consists of alternating $YBa_2Cu_4O_8$ and $YBa_2Cu_3O_7$ slabs. The complexity of this system is indicated by the phase diagram in Fig. 17, which shows the different compounds formed at various temperatures and oxygen pressures.

Although the full range of oxygen-deficient compounds $RBa_2Cu_3O_{6+\delta}$ can be formed by adding limited amounts of oxygen to $RBa_2Cu_3O_6$, a better method is to remove oxygen from $RBa_2Cu_3O_7$ by heating at temperatures up to 900°C and quenching to room temperature. The oxygen content of these samples can be determined from the change in weight upon heating in a reducing atmosphere of 10–20% H_2, the reaction being

$$RBa_2Cu_3O_{6+\delta} + (2.5+\delta)H_2 \rightarrow 2BaO$$
$$+ 0.5R_2O_3 + 3Cu + (2.5+\delta)H_2O \quad (8)$$

Even so, the properties of $YBa_2Cu_3O_{6+\delta}$ for a given oxygen content can vary markedly, a situation that at first led to considerable confusion. The problem was traced to the distribution of the oxygen remaining in the CuO chains becoming increasingly random as the oxygen removal temperature is increased. If oxygen removal is effected at the relatively low temperature of 400°C by carrying out the reaction in a sealed tube with a zirconium "getter" to prevent back reaction of oxygen once it leaves the sample, then the $YBa_2Cu_3O_{6+\delta}$ sample is a mixture of only three phases: (1) the 93 K superconductor $YBa_2Cu_3O_7$, whose CuO chains are fully oxygenated; (2) the 53 K superconductor $YBa_2Cu_3O_{6.5}$, in which oxygens are removed from every other CuO chain, with a consequent doubling of the crystal unit cell along the *a* axis; and (3) the antiferromagnetic insulator $YBa_2Cu_3O_6$, whose CuO chains are devoid of oxygen. The influence of these three phases can be seen in the plateaus at 58 K

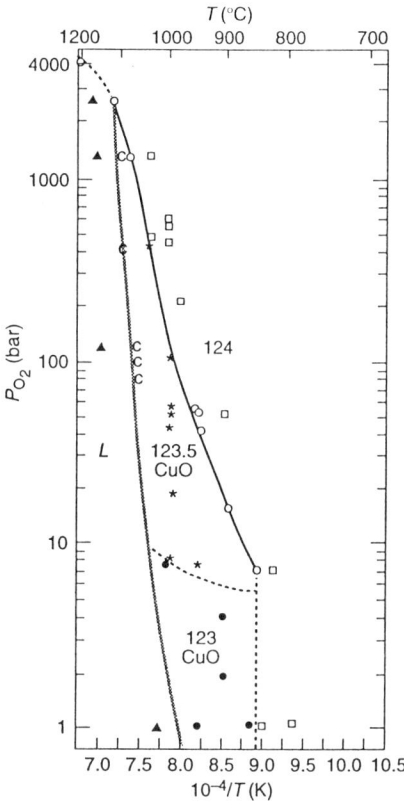

Figure 17 Pressure versus temperature phase diagram for the various yttrium–barium–copper-oxide superconductors. 123 is $YBa_2Cu_3O_7$, 123.5 is $YBa_2Cu_{3.5}O_{7.5} \equiv Y_2Ba_4Cu_7O_{15}$, and 124 is $YBa_2Cu_4O_8$. (Reprinted from Ref. 17. Copyright Physica C, 1989.)

and 93 K in the plot of T_c versus chain oxygen concentration in Fig. 18. However, the variation of T_c with δ in $YBa_2Cu_3O_{6+\delta}$ samples prepared by oxygen removal at higher temperatures depends idiosyncratically on temperature and other reaction conditions, and it does not exhibit the 58 K and 93 K plateaus. It has been proposed that the behavior shown in Fig. 18 is due to mixing of the three phases, whose relative amounts vary with oxygen content, on a near-atomic scale [15]. This is a complicated situation in which the T_c of $YBa_2Cu_3O_{6+\delta}$ is determined by microscopic local crystal domains, too small to be detected by X-ray or neutron diffraction crystallography, and whose structures are more complex and less symmetric than the overall crystal structure.

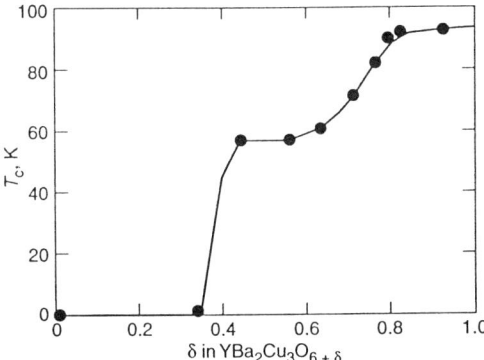

Figure 18 Variation of T_c with oxygen content in $YBa_2Cu_3O_{6+\delta}$ if the oxygen content is established by removing oxygen from $YBa_2Cu_3O_7$ at the lowest possible temperatures using oxygen "gettering" techniques. (Drawn from data of Ref. 18.)

3. Route 3 Syntheses—The Bi and Tl-Cuprates

These compounds are considerably harder to prepare than the simpler cuprates. Furthermore, their preparations tend to yield mixtures of phases, each of which deviates considerably from its nominal formula and crystal structure. This last feature likely plays an essential role in their superconductivity because the nominal formulas of the highest T_c materials, that is, $Bi_2Sr_2Ca_{n-1}Cu_nO_{2n+4}$ and $Tl_2Ba_2Ca_{n-1}Cu_nO_{2n+4}$, are stoichiometric compounds that should not have mobile charges in their CuO_2 layers. A major reason for the complex, ill-defined structures of these compounds is that the aforementioned size mismatches between the CuO_2 layers and the charge-reservoir layers are especially acute in them. For example, a typical Ti–O distance of 2.72 Å in $Tl_2Ba_2Ca_{n-1}Cu_nO_{2n+4}$ is much larger than the sum of the Tl^{3+} and O^{2-} ionic radii, which is 2.28 Å. Consequently, these compounds are so energetically unstable at all temperatures and pressures that vacancies, impurities, and/or substituent ions are required to provide stabilizing entropy at their elevated preparation temperatures. For example, addition of lead oxide to the Bi/Sr/Ca/Cu/O system makes it much easier to prepare the three-layer bismuth compound $Bi_2Sr_2Ca_2Cu_3O_{10}$.

Even so, many of these compounds can be prepared by a two-step process of heating and annealing in oxygen, but often with strong dependences on heating rates and durations that are indicative of nonequilibrium reactions. For example, the 127 K material $Tl_2Ba_2Ca_2Cu_3O_{10}$ can be prepared by heating a stoichiometric mixture of the oxides for 1 h at 890°C, but when the heating time is extended to 2 h the primary product is $Tl_2Ba_2CaCu_2O_8$, and $Tl_2Ba_2Ca_2Cu_3O_{10}$

is only a minor phase. This indicates that $Tl_2Ba_2Ca_2Cu_3O_{10}$ is not an equilibrium product of the reaction, only a kinetic intermediate.

Another nonequilibrium method of preparing these compounds is to build them up layer by layer, using the film preparation methods discussed next. This method is also of interest for preparing new compounds that have proven difficult to prepare by heating mixtures of finely ground and compressed metal oxides, as for example, preparing the $n > 4$ members of the $Bi_2Sr_2Ca_{n-1}Cu_nO_{2n+4}$ and $Tl_2Ba_2Ca_{n-1}Cu_nO_{2n+4}$ families with four or more CuO_2 layers separated only by alkaline earth ions.

4. Mercury-Containing Cuprates

A notable indication of the challenges involved in the design and preparation of cuprate superconductors is the long interval of five years between the discovery of the thallium-containing cuprate superconductors and an analogous family containing mercury, despite mercury being a tempting candidate as an immediate neighbor of thallium in the periodic table [19].

A novel feature in the preparation of these compounds is that instead of the usual procedure of reacting separate metal oxides, carbonates, etc., mercuric oxide is reacted with a multimetal oxide containing the other metallic components. For example, the first mercury-containing cuprate superconductor $HgBa_2CuO_{4+\delta}$ was prepared by the solid state reaction of stoichiometric mixtures of Ba_2CuO_3 with yellow mercuric oxide in an evacuated silica tube. All operations of grinding and mixing were carried out in a dry box under an inert atmosphere. The temperature of the tube, which was enclosed in a steel container, was raised to 800°C over a period of 5 h and then slowly cooled to room temperature over a period of 10 h. The Ba_2CuO_3 precursor used in the initial synthesis can be prepared by the reaction of stoichiometric quantities of BaO and CuO in oxygen at 930°C.

Somewhat surprisingly, given its preparation by reaction in an evacuated tube, there are strong experimental indications that the $HgBa_2CuO_{4+\delta}$ superconductor ($T_c \simeq 94$ K) contains excess oxygen, without which the stoichiometric $HgBa_2CuO_4$ compound should not have mobile charge carriers in its CuO_2 layers. One such indication is X-ray diffraction studies showing that the crystal structure of this compound has interstitial oxygens in the Hg plane of the nominal structure, which is derived from the $TlBa_2CuO_5$ (Tl1201) structure in Fig. 4 by removing the oxygens from the TlO layer and replacing the thalliums by mercury atoms. Also, annealing in oxygen at 500°C for 24 h increases T_c slightly to 95 K, and more significantly, a similar annealing in argon decreases T_c to around 59 K.

Analogous reactions between mercuric oxide and other precursors with the nominal compositions $Ba_2CaCu_2O_5$ and $Ba_2Ca_2Cu_3O_x$ (prepared by calcining mixtures of metal oxides or nitrates with the indicated metal compositions) yielded multiphasic mixtures whose superconductivity began at 133 K. The 133 K superconductor is most likely the $n = 3$ member of a series that, by analogy with the $TlBa_2Ca_{n-1}Cu_nO_{2n+3}$ family, has the general formula $HgBa_2Ca_{n-1}Cu_nO_{2n+2}$.

5. Superconducting Films

The preparation of thin films of the cuprate superconductors, which is vital to their application in electronics, typically involves synthesis under extreme nonequilibrium conditions. Usually, the film deposition substrate plays an important role by determining the initial atomic arrangements of the depositing film, which arrangement is preserved in successive layers. Unfortunately, this greatly restricts the choice of deposition substrates, usually to ones whose crystal structure and lattice spacings match those of the material being deposited.

Figure 19 shows the most widely used methods for producing thin films of electronic materials [20; see also 6, p. 64]. Extensively developed prior to the discovery of the high-T_c superconductors, they are (1) evaporation or sputtering,

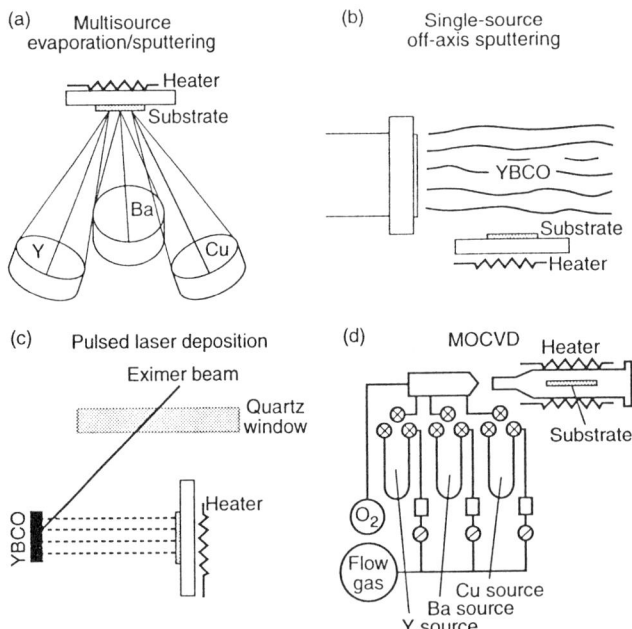

Figure 19 Various methods for producing thin films of electronic materials, including the cuprate superconductors. (Reprinted with permission from Ref. 20. Copyright 1989, *Solid State Technology*.)

using single or multiple sources, (2) pulsed laser ablation processing (LAP), and (3) chemical vapor deposition (CVD). Application of these methods to the cuprate superconductors has been quite successful, considering that these compounds are far more complex than most previous film materials such as elemental metals, semiconductors, and conventional superconductors. Films of $YBa_2Cu_3O_7$ (YBCO) are relatively easy to prepare because it has a well-defined, intrinsically nonstoichiometric crystal structure that needs neither metal dopants nor lattice imperfections to introduce charge carriers into its CuO_2 planes. The best YBCO films are single-crystal-like with the c axis perpendicular to the deposition substrate. They have critical temperatures up to 92 K, and critical current densities of 6×10^6 and 4×10^7 amp/cm^2 at 77 and 4 K, respectively. Superconducting films of $Bi_2Sr_2Ca_{n-1}Cu_nO_{2n+4}$ (BSCCO) and $Tl_2Ba_2Ca_{n-1}Cu_nO_{2n+4}$ (TBCCO) also have been prepared, although the greater complexity and multiphasic nature of these compounds makes it harder to obtain high-quality films containing only the highest-T_c phase.

Most of the preceding film growth methods can be used in two modes: (1) in situ growth in the presence of oxygen yields superconductive films directly; and (2) ex situ growth, in which the as-grown film requires postannealing in oxygen before it is superconductive.

In situ growth has the following advantages: (1) Lower growth temperatures can be used because only surface diffusion of oxygen is required. (2) Multilayer films of differing composition are readily grown. (3) The films have high critical magnetic fields and current densities, and low surface resistance, even if their T_c are sometimes a bit low. (4) Thick films, i.e., greater than the several thousand Angstroms limit for ex situ films, are readily grown because subsequent diffusion of oxygen into the sample is not required.

The disadvantages of in situ growth are as follows: (1) A more complicated apparatus is required. (2) It is difficult to heat the deposition substrate to the required temperature in the presence of oxygen. (3) It is difficult to grow large-area films of uniform composition and quality.

Evaporation Methods. Evaporation methods include both thermal evaporation, and sputtering by bombardment of the source material(s) with energetic ions from an electrical discharge in a gas such as argon.

Thermal evaporation methods generally use separate sources for each metallic component, as this allows compositions to be carefully controlled. For ex situ film growth BaF_2 is used because the resulting preanneal films are stable, whereas Ba tends to react with air to form barium carbonate, which is difficult to dissociate in the postanneal. Postannealing of films containing BaF_2 is a two-step process. The first step of heating at 850°C in moist oxygen hydrolyzes the BaF_2 and forms the incompletely oxygenated $YBa_2Cu_3O_{6+\delta}$. A second anneal at lower temperatures converts this to $YBa_2Cu_3O_7$.

The metallic sources used in thermal evaporation are incompatible with the presence of the 0.1–1.0 mtorr of O_2 required for in situ film growth, and so differential pumping is required to maintain this pressure at the deposition substrate, and a much lower O_2 pressure at the sources. The oxygen also complicates the required substrate heating (roughly 700°C for YBCO) by reacting with the hot surfaces. This, in turn, affects heat transfer between the heater and the substrate. Use of platinum heating elements, external heating lamps, and sealed heaters are ways around this problem, both here and for in situ depositions by other methods.

Use of alternative oxygen sources such as O_3 and NO_2, combined in some cases with photodissociation to yield highly active O atoms, allows the oxidant pressure to be reduced to a point where differential pumping is unnecessary. Another oxidation technique uses O atoms produced in a microwave discharge that is closer to the deposition substrate than the source(s), thus yielding negligible O concentrations near the source.

Sputtering methods work best if a single target containing all the component metals, or metal oxides, is used. Off-axis sputtering, which, as shown in Fig. 19b, places the deposition substrate outside the sputtering plasma, is best for multicomponent superconducting films. With the conventional in-line geometry, ion bombardment of the depositing film can physically damage it or alter its composition by selectively resputtering certain components. Off-axis sputtering can produce large uniform films, but deposition is very slow, of the order of a thousand Angstroms per hour.

Pulsed Laser Ablation Methods. Pulsed laser ablation processing (also known as laser deposition) is a popular alternative to single-source, off-axis sputtering for producing superconducting films. Here, an excimer laser, focused to an energy density of the order of 1 J/cm^2 or greater, ablates material from the source in the form of atoms, molecules, and ions. These species deposit on the heated deposition substrate in the presence of some 100 mtorr of oxygen, resulting in the preparation of an in situ film. Considerable experimentation is required to optimize these films, as their composition can vary with such parameters as the laser energy density, gas pressure, and substrate temperature. Film growth rate is fairly rapid, roughly a few tens of Angstroms per second, but the area of uniform deposition is small. However, larger uniform films can be achieved with a more complex apparatus that allows the deposition substrate to be translated and/or rotated during deposition.

Chemical Vapor Deposition. Chemical vapor deposition (CVD) or metal–organic chemical vapor deposition (MOCVD) (Chapter 5) is widely used for producing thin-film electronics elements because it yields high-quality individual layers and sharp transitions between layers of different materials. Also, large-area films are readily prepared. In this technique, precursor molecules containing the required metallic components are volatilized, if not already in gas form, and mixed into a flowing inert gas that transports them to the reaction chamber. There, they are thermally decomposed to yield metallic components that deposit on the substrate. Oxygen is readily introduced separately. A general problem with this method is film contamination by organic by-products of the thermal decomposition. Although the number of suitable high-T_c precursors containing the heavy metal atoms of the cuprate superconductors is limited, they do exist, often as chelated metal salts of diketones. One application used powdered Y-, Ba-, and Cu-tetramethyl-heptanedionates, which can be volatilized at 300°C.

Chemical deposition methods are also being investigated. One method forms YBCO films by spin casting a solution of the metal trifluoroacetates onto the substrate. This is heated at a relatively low temperature that decomposes the film into a mixture of metal oxyfluorides. The rest of the processing is similar to that used in the thermal evaporation method with BaF_2, that is, heating in moist O_2 to hydrolyze the fluorides and allow the YBCO film to grow. Annealing in pure O_2 completes the process.

B. Organic Superconductors

Organic conductors and superconductors are quite complicated materials with many classes of compounds and a wide range of compositional and structural modifications within a class. They have a correspondingly wide range of properties, including numerous alternatives to the superconducting state, and so considerably fewer superconductors are found among organic conductors than among inorganic conductors. Nonetheless they are important materials, not only as conductors and superconductors, but for a variety of other interesting optoelectronic properties as well. Insofar as superconductors are desired, however, an important part of the synthetic strategy is to maximize conductive dimensionality, either by forcing planar molecules hosting charge carriers to form sheets, as in κ-$(ET)_2Cu(SCN)_2$, or ideally, to locate the charge carriers on symmetric molecules, such as C_{60}.

1. Charge-Transfer Compounds with Tetrachalcagenafulvalene Donors

The importance of these donors, whose general structural formula is

$$\begin{array}{c} R_1 \\ \diagdown \\ C \end{array} \begin{array}{c} X \\ \diagdown \\ \end{array} \begin{array}{c} X \\ \diagup \\ C \end{array} \begin{array}{c} R_3 \\ \diagup \\ \end{array}$$

$$\|\quad C=C\quad \|\qquad X=S, Se, Te,$$

$$\begin{array}{c} \diagup \\ C \\ R_2 \end{array} \begin{array}{c} \diagdown \\ X \end{array} \begin{array}{c} \diagup \\ X \end{array} \begin{array}{c} \diagdown \\ C \\ R_4 \end{array}$$

stems from the relative ease of removing a chalcogen lone-pair electron, especially if the resulting positive charge can be delocalized among the π-orbitals of a pair of donors. This produces the fractional charge transfer required to avoid the Mott–Hubbard insulating state in the resulting charge-transfer salt.

Synthesis of the Donors. As befits their unique donor capabilities, the synthetic chemistry of the tetrachalcagenafulvalene donors is far too extensive for more than an overview of the general methodology here [5]. There are two general routes to the tetrachalcagenafulvalenes: (1) coupling reactions that join two 1,3-dichalcagenole-2-chalcagenones, and (2) direct synthesis.

The most versatile, and most frequently used, coupling reactions utilize trivalent phosphorus compounds, such as trialkyl or triaryl phosphines or phosphites, as the coupling reagent:

$$2 \quad \begin{array}{c} R-C-X \\ \| \quad \quad C=Y \\ R-C-X \end{array} \quad \xrightarrow{R'_3P \text{ or } (R'O)_3P} \quad \begin{array}{c} R-C-X \\ \| \quad \quad C=C \\ R-C-X \end{array} \begin{array}{c} X-C-R \\ \| \\ X-C-R \end{array} \quad (9)$$

$$X=S, Se; \quad Y=O, S, Se$$

This is a complicated reaction in which the phosphorus reagent first adds to the C=Y bond of one chalcagenone to form a phophonium intermediate, which initiates the coupling by reacting similarly with a second chalcagenone molecule. Elimination of $Y=PR'_3$ from this coupled intermediate followed by removal of the remaining bridging chalcogen by reaction with another molecule of the phosphorus reagent completes the reaction. Accordingly, the reaction conditions, phosphorus reagent, solvent, etc., required in specific cases is largely a matter of trial and error.

Most other coupling reactions, as well as noncoupling reactions, involve the formation and self-reaction of carbenes. A coupling reaction, often used for preparing tetrathiafulvalene (TTF) and some of its derivatives, is the oxidation of 1,3-dithiol-2-thione to a 1,3-dithiolium salt by a peroxyacid followed by deprotonation of the dithiolium salt by an amine or some other base to yield a carbene. This carbene reacts either with another carbene or dithiolium ion to yield TTF:

$$\begin{array}{c} C-S \\ \| \quad \quad C=S \\ C-S \end{array} \xrightarrow[\text{acetone}]{H_3CO_3H} \begin{array}{c} C-S \\ \| \quad \quad C^+ H(HSO_4)^- \\ C-S \end{array} \quad \xrightarrow{R_3N}$$

(A) (B)

$$\begin{array}{c} C-S \\ \| \quad \quad C: \\ C-S \end{array} \xrightarrow{+ \text{ C or B}} \begin{array}{c} C-S \\ \| \quad \quad C=C \\ C-S \end{array} \begin{array}{c} S-C \\ \| \\ S-C \end{array} \quad (10)$$

(C)

The reaction fails for selenium compounds because the 1,3-diseleno-2-selenones cannot be oxidized to the corresponding diselenolium salt.

A commonly used noncoupling route is reaction of 1,2-dichalcagenolates with ethylene dichloride, as, for example, the preparation of tetratellurafulvalene (TTeF):

$$2 \underset{H}{\overset{H}{>}}C{=}C\overset{TeLi}{\underset{TeLi}{<}} \xrightarrow{Cl_2C=CCl_2} \underset{H}{\overset{H}{>}}C{=}C\overset{Te}{\underset{Te}{<}}C{=}C\overset{Te}{\underset{Te}{>}}C{=}C\underset{H}{\overset{H}{<}} \quad (11)$$

The mechanism of this reaction is fairly straightforward: successive reactions of the four halogens of the ethylene dichloride molecule to form X−C bonds and LiCl. Preparation of the precursor is complicated, however, involving the following multistep reaction:

$$Me_3Sn{-}SnMe_3 + H{-}C{\equiv}C{-}H \xrightarrow{(Ph_3P)_4Pd} \underset{H}{\overset{H}{>}}C{=}C\overset{SnMe_3}{\underset{SnMe_3}{<}}$$

$$\underset{H}{\overset{H}{>}}C{=}C\overset{SnMe_3}{\underset{SnMe_3}{<}} \xrightarrow{n{-}BuLi} \underset{H}{\overset{H}{>}}C{=}C\overset{Li}{\underset{Li}{<}} \xrightarrow{Te} \underset{H}{\overset{H}{>}}C{=}C\overset{TeLi}{\underset{TeLi}{<}} \quad (12)$$

This procedure should also be applicable to the synthesis of tetraselenafulvalene (TSeF), but not TTF, because sulfur does not insert into the C−Li bond under these conditions. However, some TTF compounds can be prepared by analogous reactions, using disodium 1,2-dithiolate precursors prepared by the reaction of sodium with 1,2-dithiols.

Synthesis of the Donor Precursors. There are various synthetic routes to the 1,3-dichalcagenole-2-chalcagenones used in the coupling reactions. The unsubstituted compounds, 1,3-dithiole-2-thione and 1,3-diselenole-2-selenone, can be prepared by the reaction of sodium acetylide with carbon disulfide and sulfur or carbon diselenide and selenium, respectively, in an acidic medium:

$$NaC{\equiv}CH \xrightarrow[H^+]{CX_2,\ X} \underset{H}{\overset{H}{>}}C{=}C\overset{X}{\underset{X}{<}}C{=}X\ ;\ X=S,\ Se \quad (13)$$

Coupling of these compounds in a heated solution of triethylphosphite in benzene yields tetrathiafulvalene (TTF) and tetraselenafulvalene (TSeF), respectively. However, TTF is more conveniently prepared by the coupling reaction involving the 1,3-dithiolium salt, shown in Eq. (10).

Although it is often more convenient to prepare or obtain TTF or TSeF and then prepare the desired derivatives, some donors, including TMTSF and ET, are prepared by coupling the derivatized 1,3-dichalcagenole-2-chalcagenones. A compound that can be coupled by an excess of triethyl phosphite to yield TMTSF is prepared by the following sequence of reactions:

$$\text{\raisebox{-2pt}{\begin{tabular}{c}Se\\$\|$\end{tabular}}} \underset{}{\bigcirc\!\!-\!\!N\!-\!\!C\!-\!Se^-} + \underset{H_3C}{H\!-\!\!C\!-\!\!C\!\!\overset{O}{\underset{CH_3}{\diagdown}}} \xrightarrow{CH_2Cl_2} \underset{}{\bigcirc\!\!-\!\!N\!-\!\!\overset{Se}{\underset{\|}{C}}\!-\!Se\!-\!\overset{O\diagup\!\!\!\diagdown^{CH_3}}{\underset{H}{\overset{|}{C}\!\!-\!\!CH_3}}} + Br^-$$

$$\xrightarrow{H^+,\ \text{conc.}\ H_2SO_4}$$

(14)

$$S\!=\!C\!\!\underset{Se}{\overset{Se\diagdown\!\!\!\!\diagdown_{C\!-\!CH_3}}{\diagup\overset{\|}{\underset{CH_3}{C}}}} \xleftarrow[(pip H)^+ClO_4^-]{H_2S} \bigcirc\!\!-\!\!N^+\!\!=\!\!C\!\!\underset{Se}{\overset{Se\diagdown\!\!\!\!\diagdown_{C\!-\!CH_3}}{\diagup\overset{\|}{\underset{CH_3}{C}}}} + H_2O$$

where pip denotes piperidine. The diselenocarbamate starting compound can be prepared by reaction of carbon diselenide with piperidine in alcoholic potassium hydroxide solution.

The derivatized 1,3-dithio-2-thione precursors of the *bis*-(alkylenedithio)-tetrathiafulvalenes, of which ET is the most important, are prepared by chemical or electrochemical reduction of carbon disulfide to a dithiolate, followed by reaction with an alkyl dihalide:

$$4Na + 4CS_2 \xrightarrow{DMF} \underset{NaS}{\overset{NaS\diagdown\!\!\!\!\diagdown_{C\diagdown\!\!\!\!\diagdown S}}{\diagup\overset{\|}{\underset{S}{C}}}}\!\!C\!=\!S + Na_2CS_3 \xrightarrow{(CH_2)_nBr_2} (CH_2)_n\!\!\underset{S}{\overset{S\diagdown\!\!\!\!\diagdown_{C\diagdown\!\!\!\!\diagdown S}}{\diagup\overset{\|}{\underset{S}{C}}}}\!\!C\!=\!S \quad (15)$$

where DMF is dimethylformamide. Since the dithiolate intermediate is unstable, it usually is reacted with zinc chloride and then treated with tetraethyl-ammonium bromide to form a stable zinc chelate. This compound is then reacted with the alkyl dihalide to produce the 4,5-*bis*-(alkylenedithio)-1,3-dithiole-2-thione. Coupling of the $n = 2, 3$ members of this series of compounds by heating 30 min at 110°C in triethyl phosphite gave *bis*-(ethylenedithio)-tetrathiafulvalene (ET) and *bis*-(propylenedithio)-tetrathiafulvalene, respectively. However, this coupling reaction did not work with the methylene analog ($n = 1$) unless the thione sulfur was replaced by oxygen by oxidizing with mercuric acetate in a chloroform/glacial acetic acid solution.

Preparation of Derivatives of TTF and TSeF. Preparation of derivatives of tetrathiafulvalene and tetraselenafulvalene generally uses intermediates in which one or more of the vinylic hydrogens are replaced by lithium. Lithiation is readily accomplished, as these hydrogens are mildly acidic and readily undergo lithium–hydrogen exchange reactions with strong bases such as *n*-butyllithium or lithium diisopropylamide (LDA) (however, only LDA will lithiate tetraselenafulvalene) in tetrahydrofuran or ether at very low temperatures (ca. $-80°C$). All sites can be lithiated:

$$\text{(16)}$$

or, alternatively, the number of sites lithiated can be controlled to a considerable degree by varying the amount of the lithium reagent used. A number of reactions are available for preparing various derivatives from these lithiated intermediates. Among them are alkylation or arylation by reaction with alkyl or aryl halides, reactions with alkyl dichalcogenides, and insertion of Se or Te into the C−Li bond followed by the alkylation or halogenation reactions. These reactions are summarized below:

$$\text{(17)}$$

Y=Se, Te / X=S
Y=S, Se / X=Se

Although reaction conditions, yields, and feasibility vary considerably among the various tetrachalcogenofulvalenes (e.g., S will not insert into the C−Li bond of a tetrathiafulvalene, but it will for a tetraselenafulvalene), these reactions are the most general available for preparing tetrachalcogenofulvalene derivatives.

Preparation of the Charge-Transfer Salts. Some organic charge transfer salts can be prepared by direct reaction of the donor and acceptor dissolved in a suitable solvent with the temperature and concentration chosen so as to keep the reaction rate low enough that good single crystals are formed. Often the reaction rate is controlled by placing the solutions of donor and acceptor in separate arms of an H or U tube, with the tube joining these arms containing a region of pure solvent and/or a semipermeable frit through which reactants must diffuse in order to react. This method was used to prepare the first, nonsuperconducting, organic metals in which the donor was a tetrathiafulvalene, tetraselenafulvalene, or a simple derivative of one of these compounds, and the acceptor was tetracyanoquinodimethane or one of its derivatives.

In other cases, notably most salts of organic donors and inorganic anions, the respective oxidation and reduction potentials of the donor and acceptor are such that the direct reaction does not occur. One way to effect these reactions is to chemically or electrochemically oxidize the donor prior to reaction. However, the standard method is to carry out both the oxidation and reduction simultaneously in an electrochemical cell, typically the H cell shown in Fig. 20.

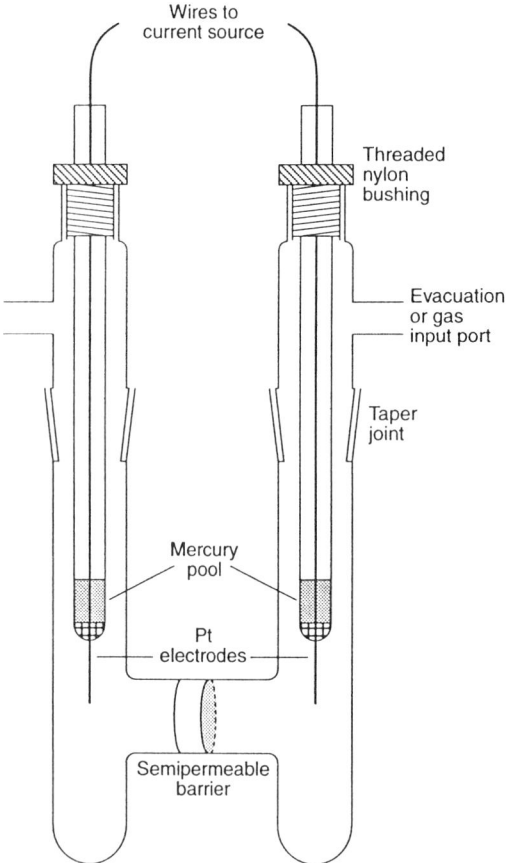

Figure 20 An electrochemical H cell for electrochemical growth of single crystals of conducting and superconducting organic charge-transfer compounds.

This procedure has several advantages, the greatest being that it yields good single crystals.

Using such a cell, salts of the $(TMTSF)_2X$ and $(ET)_2X$ families, where X is a complex inorganic anion such those listed in Table 3, were prepared by oxidizing the donor in the presence of an electrolyte consisting of a solution of $(n\text{-}Bu)_4NX$ in 1,1,2-trichloroethane or tetrahydrofuran. In cases where the anions are roughly the same size and shape, such as the linear triatomic ions I_3^-, I_2Br^-, $BrIBr^-$, and $IAuI^-$, it is possible to grow mixed crystals with two different anions. The reaction and crystal growth rate are controlled by the current density, which usually is kept so low (ca. 1 $\mu A/cm^2$) that the process takes days to weeks. A key to successful reactions and good crystals is use of the highest-purity reagents and solvents available, scrupulous cleaning of all parts of the electrochemical apparatus, and carrying out all procedures in an inert argon atmosphere and in the absence of light.

More complicated is synthesis of the $\kappa-(ET)_2X$ salts, where X^- is a polymer-forming anion composed of a metal, usually Cu, and halides or pseudo-halides such as CN^-, SCN^-, or $N(CN)_2^-$. These anions are usually produced by reaction of the metal halide with a halide anion in the electrocrystallization cell, e.g.,

$$CuSCN + SCN^- \rightarrow Cu(NCS)_2^- \tag{18}$$

However, these reactions often require considerable ingenuity, and trial and error. For example, the preparation of $\kappa-(ET)_2Cu(NCS)_2$ used a mixture of CuSCN, KSCN, and the cyclic polyether 18-crown-6, formula $(CH_2OCH_2)_6$, in 1,1,2-trichloroethylene containing a few to ten percent of anhydrous ethanol, with ET added to the anode chamber of the H cell. The unique feature of the process is the crown ether, which is believed to trap the K^+ ion, freeing its SCN^- to react with the very insoluble CuSCN to form $Cu(NCS)_2^-$, which is soluble. The ethanol may increase the solubility of the CuSCN somewhat. However, this procedure is highly specialized, as replacement of CuSCN by AgSCN yields the semiconducting compound $(ET)Ag_{1.6}(SCN)_2$, and $SeCN^-$ cannot be substituted for SCN^-.

The compounds $\kappa-(ET)_2Cu[N(CN)_2]X$ (X = Cl, Br, I) were prepared by electrochemical oxidation of ET in the presence of $(Ph_4P)N(CN)_2$ and the appropriate copper halide.

2. The Fullerides

The novel carbon allotrope C_{60}, with a nearly spherical icosahedral symmetry, was postulated in 1985 as the carrier of the peak at 720 in the mass spectra of the products from the laser vaporization of carbon. Verification of this structure and its preparation in chemically useful quantities came in 1990, when it was found that an electric arc between graphite electrodes in an atmosphere of 100 torr of helium produced a carbon soot that contained a few percent of C_{60} [21].

The solubility of C_{60} in benzene and other nonpolar solvents enabled it to be separated from the other insoluble constituents of the soot. Alternatively, the C_{60} could be sublimed away from the other constituents by heating the soot to around 400°C in a vacuum or an inert atmosphere. Final purification, especially separation from C_{70}, which is a closed cage structure, too, of lower symmetry than C_{60}, is done chromatographically on an alumina column with 5:95 toluene/hexane as the eluant.

Preparation of the superconducting alkali metal salts of C_{60} must be carried out under inert conditions, usually vacuum or a low pressure of helium, as these compounds react with air and moisture [7]. Polycrystalline samples are prepared by heating stoichiometric amounts of C_{60} and the alkali metal or metals to a temperature (200–400°C) sufficient to vaporize the alkali metal and activate its diffusion into the C_{60} lattice. In some cases, particularly when preparing mixed alkali metal fullerides $M_{3-x}M'_xC_{60}$, use of mercury amalgams of the alkali metals facilitates the reaction and helps maintain the desired alkali metal composition.

Films of C_{60}, and of C_{70}, can be prepared by sublimation under vacuum from an alumina crucible at 300–350°C. These films then can be reacted with alkali metal vapors to produce conductive compounds. Only C_{60} yields superconductors, however. Composition is controlled by monitoring the conductivity of the film during the reaction. The maximum conductivity corresponds to the M_3C_{60} compound. Continued alkali metal doping yields a mixture of M_3C_{60} and the insulating M_6C_{60}, with decreasing conductivity until the sample is entirely nonconductive M_6C_{60}.

As noted earlier, the mixed compounds Na_2MC_{60} (M = K, Rb, Cs), but not Na_3C_{60}, exist, because the larger alkali ion expands its interstitial sites enough to constrict the remaining sites to the point where they can contain the smaller Na^+ cation. Their preparation is similar to that of the other $M_{3-x}M'_xC_{60}$ compounds except for problems posed by the relatively low vapor pressures of sodium and its reactivity with pyrex. These problems have been overcome by using NaM alloys and carrying out the reaction either in quartz tubes or in stainless steel or copper reaction chambers. Another method is to obtain the sodium from thermal decomposition of such compounds as NaN_3, NaH, and $NaBH_4$.

The possibility of replacing some or all of the alkali cations by ammonium derivatives also has been explored. Ammonia enters Na_2MC_{60} (M = Rb, Cs) and K_3C_{60}, but not Rb_3C_{60}, upon exposure to 380 torr of NH_3 for several days, and then annealing at 100°C for a day [22]. The weight gain of the ammonia-treated samples corresponded to the formulas $(NH_3)_4Na_2MC_{60}$ and $(NH_3)K_3C_{60}$. These ammonia intercalations are readily reversed by gentle heating. X-ray structural data on the $(NH_3)_4Na_2CsC_{60}$ compound, combined with considerations of the relative sizes of the tetrahedral and octahedral interstitial sites of the C_{60}

lattice, suggest ammonia intercalation produces one $Na(NH_3)_4^+$ cation, which occupies the large octahedral interstitial sites of the C_{60} lattice, while the remaining Na ion and the Cs ion occupy the smaller tetrahedral interstitial sites. The lattice constant increases from 14.132 Å in Na_2CsC_{60} to 14.473 Å in $(NH_3)_4Na_2CsC_{60}$, with an increase in T_c from 10.5 to 29.6 K, consistent with the approximately linear increase in T_c with lattice constant shown in Fig. 8.

Similar procedures intercalate the alkaline earths Ca and Ba into C_{60}, except that the reactions are carried out in tantalum cells to avoid problems posed by the reactivity of Ca and Ba with pyrex and quartz. Also, the reactions require lengthy periods of heating—20 h at 550°C in the case of calcium, and for hours to weeks at above 600°C in the case of barium—combined with lengthy annealing because the alkaline earths diffuse only slowly into C_{60}.

VI. SUPERCONDUCTOR FABRICATION METHODS

Applications face the challenges of forming the brittle, polycrystalline ionic superconductors into wires, films, and other structures [23]. In some cases, the problems of reactivity with air and moisture also must be dealt with, especially with the organic superconductors.

Applications for high-T_c superconductors fall into the following three classes:

1. Low-power applications, at liquid nitrogen temperatures. These include passive electronic devices such as microwave cavities, waveguides, delay lines, etc., some sensors, and magnetically levitated frictionless bearings.
2. Low-power, but high-current-density, applications in active electronic devices. These range from simple applications like intercircuit connections, which require only a single superconducting layer, to complex devices involving two or more superconducting layers separated by intervening insulating, semiconducting, or metallic layers. These devices require superconducting films with high critical current densities at liquid nitrogen temperature.
3. Finally, there are true high-power applications such as superconducting magnets and electric transmission lines. The magnets require very high current densities (ca. 10^5 amperes/cm^2), ideally with liquid nitrogen cooling. Liquid helium cooling also would be commercially feasible, however, particularly if it enabled full utilization of the exceptionally large critical fields and current densities of the highest-T_c materials at 4.2 K. Power transmission requires considerably lower current densities (ca. 2×10^4 amperes/cm^2), but it must be achieved with liquid nitrogen cooling to be commercially feasible.

As noted in Sec. IV.C.1, the primary limitation on current densities in high-T_c materials is not their upper critical fields, but motion of the magnetic flux that penetrates the superconductor in the mixed state. This motion, known as the flux flow state, occurs when the thermally activated forces combined with the Lorentz force due to the interaction between the flux and the supercurrent exceed the forces holding the flux lines in fixed positions. The flux flow state is not superconductive because the flux motions dissipate energy. Below a certain temperature, known as the flux-lattice melting temperature (T_m), thermally activated motion of the flux lines decreases to a point where the magnetic repulsions between the flux lines can confine them to fixed, regularly spaced, positions. The superconductor then can carry a large current density that increases with decreasing temperature below T_m to a maximum at 0 K. Consequently, the important parameters for high-power applications are both T_c and T_m, which lies below T_c by amounts that vary considerably among the high-T_c superconductors, and with the purity and state of aggregation of the material. Fortunately, imperfections in a superconductor, such as defects, impurities, and grain boundaries between crystalline regions, can increase the effective T_m by trapping the flux lines at the imperfections. If such imperfections are not naturally present, they are commonly introduced deliberately. Notably, thin films, into which defects, grain boundaries, etc., are inevitably introduced during preparation, tend to have much higher critical current densities than defect-free materials such as single crystals.

A. Thin Films

The major problem in preparing thin films of the cuprate superconductors is that these films can be grown only on a limited number of substrates. Often these materials are not the final substrate onto which the superconductor is to be deposited, in which case the film-growth substrate must also be compatible with the final substrate.

In addition to having a crystal structure similar to that of the deposited superconductor, an ideal substrate has the following properties: (1) It is readily available as good, twin-free, single crystals. (2) It is free of phase transitions over the wide temperature range involved in processing and using the film. (3) Its thermal expansion coefficients match those of the high-T_c material. (4) It does not react with the constituents of the high-T_c material during deposition. (5) It is free of impurities, especially magnetic ones. Naturally, most substrates satisfying these requirements for cuprate films are themselves metal oxides, ideally with the perovskite crystal structure. Frequently used substrates are strontium titanate ($SrTiO_3$), lanthanum aluminate ($LaAlO_3$), magnesium oxide (MgO), and sapphire (Al_2O_3). Unfortunately, the best deposition substrates for the cuprates share their incompatibility with certain important final substrates such as silicon. This makes them unsuitable as a buffer layer between a cuprate film and the final substrate.

The best buffer layer materials have limited compatibility with both the cuprate and the final substrate. Two of these are yttrium-stabilized zirconia (YSZ) and cerium oxide, both with the cubic fluorite crystal structure. A useful feature of YSZ is its permeability to oxygen, so that the silicon at a YSZ/Si interface can be oxidized to SiO_2 if desired, after deposition of the YSZ layer.

The noble metals silver and gold do not react with the high-T_c superconductors, and so they can be used for making electrical contacts and for formation of superconductor/normal metal (SN) junctions. However, they cannot be used in multilayer devices such as superconductor/normal metal/superconductor (SNS) junctions, because the superconductor films cannot be grown on them.

Formation of the superconductor/insulator/superconductor (SIS) structures required for Josephson junctions and SQUIDs is especially difficult for the high-T_c cuprates because the thickness of the insulating layer must be of the order of the superconducting coherence length, which is of near atomic dimensions, that is, $\simeq 10$–30 Å in the CuO_2 plane and $\simeq 3$ Å perpendicular to this plane. Using the semiconducting compound $PrBa_2Cu_3O_7$, which has the same crystal structure as $YBa_2Cu_3O_7$, as the insulating layer, it is possible to form $YBa_2Cu_3O_7/PrBa_2Cu_3O_7$ superlattices and junctions in which the $PrBa_2Cu_3O_7$ layer is only one lattice spacing thick along the c axis. These have various applications, but even this spacing is too large for Josephson junctions, given the very short c axis coherence length.

An important method for creating Josephson junctions in thin films utilizes the fact that such a junction is formed naturally by two superconducting grains with misaligned crystal axes. The required misalignment can be achieved by growing the film on a substrate with a discontinuity, which can be achieved in several ways: (1) fusing together two differently oriented substrate crystals; (2) etching a step or similar discontinuity into the substrate crystal; (3) production of a grain boundary in the deposition substrate by depositing it on an underlying substrate with a discontinuity. For example, deposition of a $SrTiO_3$ film, which is an excellent substrate for deposition of $YBa_2Cu_3O_7$, onto a sapphire substrate partially covered with a thin layer of MgO, led to a $SrTiO_3$ film whose crystal axes above the MgO were oriented differently from those above the bare sapphire. The corresponding change in the crystal axis orientations of the deposited $YBa_2Cu_3O_7$ film produced the grain boundary required for the Josephson junction.

B. Superconducting Wires

The somewhat conflicting objectives here are (1) to form flexible wires from the superconducting ceramics, in which the individual superconductor grains are aligned well enough to minimize current-limiting weak links; (2) to introduce the defects needed to immobilize the magnetic flux lines within the superconductor. There are several ways of approaching this problem.

One is to fill a silver tube with a metal–oxide mixture that will yield the superconductor when heated. This tube is then deformed, by some combination of extrusion, rolling, and pressing, into a wire or a flat tape. Heating then forms the superconductor within a protective metal sheath. The resulting presence of a metallic conductor surrounding the superconductor is not a detriment; in fact, it is a necessity. All superconductors carrying high currents must be backed up by some sort of protective metal shunt to avoid a catastrophic buildup of heat should the superconductor somehow revert to the normal state. Although single-strand wires produced by this method will be inflexible and brittle if large enough to carry the required currents, bundling together a great many thin wires filled with the superconductor precursor and heating the composite wire yields a strong, flexible, multifilament wire, capable of carrying very high currents.

The bismuth–strontium–calcium–copper oxides are especially well suited to this method because they tend to form flat crystalline grains that align themselves well within the metal enclosure, particularly if it is a flat tape. The disadvantage of the bismuth-based cuprates is that they are especially susceptible to flux motion, and they must be cooled to around 25 K for applications involving strong magnetic fields.

The thallium–barium–calcium–copper oxides are less susceptible to flux motion and have even higher superconducting transition temperatures than the bismuth compounds, but their more three-dimensional grains neither fit nor align well within the constricted silver tube. An alternative approach to these materials is to deposit a film of them on an extended wire-shaped structure using the various film production methods discussed previously. However, this raises the problem of forming an extended wire from the brittle ceramic materials that are the best substrates for depositing the superconducting films.

Other approaches can be easier to implement, but they are less likely to produce wires with well-aligned grains. One is to alloy the metallic constituents of the superconductor with silver; this alloy then can be drawn into a wire or tape. This is followed by oxidation to create superconductor grains imbedded in the metal. Another method is spin casting a mixture of a powdered superconductor and a polymer into the desired shape, followed by rapid annealing in oxygen to burn off the polymer and form the superconductor.

Introducing flux-pinning defects into superconducting wires is a common practice, used for conventional superconductors as well as the cuprates. However, it is more difficult with the cuprates because the best defects will have dimensions of the order of their superconducting coherence lengths (ca. 10–30 Å). Such microscopic defects can be introduced by bombardment with energetic particles such as electrons, neutrons, or ions, but this requires a nuclear reactor or particle accelerator.

Given the relative newness of these superconductors compared with other electronic materials such as semiconductors, progress has been substantial, and chemical technologies have played an important role in this progress.

VII. STRUCTURE/PROPERTY RELATIONS IN HIGH-T_c MATERIALS

As noted earlier, the task of producing better superconductors by structural modifications of existing ones and finding new classes of superconductors is largely one of trial and error guided by empirical and semiempirical correlations between structure, properties, and high-T_c superconductivity. This section will discuss research directed toward refining these relations and discovering new ones.

A. Possible Routes to High-T_c Superconductivity

Although limited, theory does provide some general principles that can be useful both for interpreting experimental results and for guiding efforts to produce improved superconductors. According to Eq. (1), T_c increases with the excitation energy of the vibrations that mediate the Cooper pairing, the strength of the coupling between the charge carriers and these vibrations, and the density of states at the Fermi level. As noted earlier, many of these properties lead to relatively low normal state conductivities and a tendency toward structural instabilities.

The possibility of high-T_c superconductors with Cooper pairing mediated by high-energy electronic transitions rather than vibrations has been extensively considered, especially in the organics, beginning with the suggestion of a conductive polymer chain with Cooper pairing mediated by interaction of its conduction electrons with polarizable sidegroups [24]. A problem with this idea is that high-energy modes will be ineffective in promoting Cooper pairing if, as is more likely than with low-energy modes, polarization induced in them by their interaction with one conduction electron decays before it can attract a second electron. Nonetheless, numerous variants of this idea have been proposed for the cuprate superconductors, including the so-called magnon mechanisms, in which the charge carriers attract each other by the distortions they induce in the antiferromagnetic Cu^{2+} spin alignment. The magnon mechanism is more attractive than the others because it does not require high-energy magnetic excitations, and if it did, spin selection rules could slow their decay.

Another possibility, originally advanced to account for yesteryear's "high-T_c" compounds, such as Nb_3Sn and V_3Si, is a peak in the density of states near the top of the conduction band. This is a contender for the cuprate superconductivity mechanism because, as shown in Fig. 9, such peaks, known as van

Hove peaks, tend to occur in two-dimensional conduction bands near half filling. Even though the energy scale (E_V) is now determined by the rather narrow width of the van Hove peak rather than the larger vibrational energy, this mechanism can lead to a high T_c because of the exponential dependence of T_c on $D(E_F)$.

The M_3C_{60} compounds have many of the structural features required for high T_c by the BCS model. These include (1) a large density of states at the Fermi level, owing to the weak interactions between the large, widely separated C_{60} molecules and the consequent narrow conduction band; (2) the rich vibrational spectrum of C_{60} with intramolecular vibration frequencies ranging from 200 to 1500 cm^{-1}, and lower-frequency torsional and intermolecular vibrations; (3) the possibility that the conduction electrons are strongly coupled to some of these vibrations by Jahn–Teller effects in the electronically degenerate C_{60}^{2-} and C_{60}^{4-} ions that exist transiently during current flow.

B. Electrical Resistance

Electrical resistance in the normal state can provide important clues to superconductivity because both phenomena involve interaction of the conduction electrons with the lattice vibrations, or possibly with vibrations of electronic charge or electron spins in nonconductive electronic states.

1. The Cuprates

The temperature dependence of the cuprates' electrical resistance in their normal states is very unusual, and its relation to their superconductivity is a very active research topic. As shown in Fig. 21, the electrical resistance in the CuO$_2$ plane (ρ_{ab}) decreases linearly with decreasing temperature from well above room temperature down to T_c, even if it is as low as the 10 K of Bi$_2$Sr$_2$CuO$_{6+x}$, and often extrapolates to zero at 0 K. Although superficially metal-like, this linear ρ vs. T dependence is highly unusual in its persistence to temperatures where there are no thermally excited lattice vibrations to scatter the conduction electrons. As illustrated for copper in Fig. 21, the resistance of conventional metals has fallen almost to zero in this region. Another notable point is that the slopes of the ρ vs. T curves are nearly the same in all cuprates, indicating the resistance mechanism is intrinsic to the CuO$_2$ plane.

This behavior suggests that an electronic scattering mechanism is responsible for both the electrical resistance and superconductive Cooper pairing of the cuprates. The idea is that, whereas no lattice vibrations can be thermally excited at 10 K, there could be excitation of very-low-energy modes of an electronic subsystem. An attractive candidate is fluctuations of the axis along which the Cu^{2+} spins are antiferromagnetically ordered, because the energy of an antiferromagnetic state depends only slightly on the orientation of this axis. However, fluctuations of this axis could strongly scatter the conduction electrons that are

New Superconductors

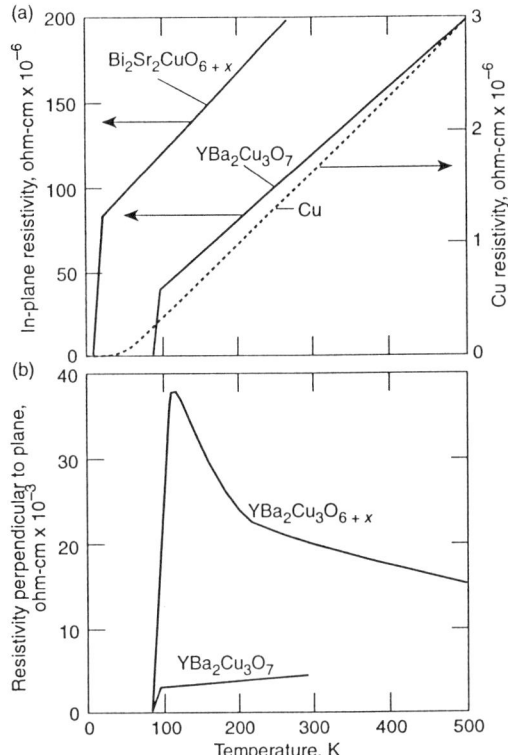

Figure 21 Electrical resistivity in the cuprate superconductors. (a) Resistivity in the CuO_2 plane. The resistivity of Cu is included for comparison. (b) Resistivity perpendicular to the CuO_2 plane in a granular sample of $YBa_2Cu_3O_7$ (upper line) and a single crystal of $YBa_2Cu_3O_7$ (lower line). (Adapted with permission from Ref. 3. Copyright 1992, The American Chemical Society.)

coupled to it by their strong, spin-dependent valence interactions with the Cu^{2+} spins.

As also shown in Fig. 21, the resistance perpendicular to the CuO_2 planes (ρ_c) is much higher than the in-plane resistance and, in many samples, exhibits the semiconductive behavior of increasing with decreasing temperature. However, ρ_c is considerably lower, and decreases somewhat with decreasing temperature, in the highest-quality, fully oxygenated, single crystals of $YBa_2Cu_3O_7$. This variability of ρ_c, often with little corresponding change in T_c, makes its role in superconductivity obscure, although several factors, such as the increase of T_c with n in $Tl_2Ba_2Ca_{n-1}Cu_nO_{2n+4}$ and similar compounds, suggests it has one.

The ratio of the out-of-plane to in-plane resistances varies greatly among the cuprates, ranging from about 10 in the best $YBa_2Cu_3O_7$ samples to greater than 10^4 in the bismuth-based compounds. This indicates that ρ_c increases with the separation of the CuO_2 planes.

2. The Organics

The anticorrelation between room-temperature conductivity and T_c predicted by BCS theory is present to a considerable extent in the organic superconductors with tetrachalcagenafulvalene donors. Compounds with T_c near 10 K tend to have room-temperature conductivities of 10 to 50 (ohm cm)$^{-1}$, while those with T_c near 1 K have considerably higher conductivities of 200 to 600 (ohm cm)$^{-1}$. However, there are exceptions to this trend, and the matter is further complicated by the complex temperature dependence of the resistance in many of these compounds. For example, in $\kappa\text{-}(ET)_2Cu(SCN)_2$ the resistance increases by a factor of 2.5 on going from 300 to 100 K, and then decreases rapidly to about one-tenth of its room-temperature value just prior to the superconducting transition.

The alkali metal fullerides exhibit a clearer anticorrelation between conductivity and T_c. The conductivities increase as decreasing lattice spacing increases the intermolecular interactions between the C_{60} conduction orbitals, thereby broadening the conduction band and decreasing the density of states at the Fermi level. Conversely, T_c decreases, as predicted by Eq. (1) and shown in Fig. 8.

C. Pressure Effects

The BCS theory predicts T_c should decrease with increasing pressure, which decreases intersite distances with the same effects just discussed.

This prediction is often contradicted by the cuprates and other oxide superconductors, very likely because the dominant effect of pressure in these materials is to change the charge-carrier concentration in a way that increases T_c. For example, the pressure-induced increase of T_c in $La_{2-x}Sr_xCuO_4$ led to attempts to duplicate its lattice-compressing effect by replacing lanthanum with the chemically similar but smaller yttrium. This, however, had the dramatic and unexpected consequence of producing the new $YBa_2Cu_3O_7$ structure. High pressure has been found to increase T_c from 133 to 155 K in the mercury compound $HgBa_2Ca_2Cu_3O_{8+\delta}$.

As shown in Fig. 11, pressure usually decreases T_c in the TMTSF- and ET-based superconductors, but the situation is complicated by pressure-dependent structural changes.

Very clear-cut BCS-like pressure effects are observed in the M_3C_{60} compounds, where the dependence of T_c on lattice constant (a_0), shown in Fig. 8, is the same irrespective of whether the changes in a_0 are due to pressure or smaller alkalis.

D. Compositional and Structural Modifications

Many examples of these investigations, ranging from the effects of compositional changes on T_c to creating new classes of superconducting materials, have already been discussed. Notable examples are chemical adjustment of the charge-carrier concentration in the cuprates, changing T_c by substitution of chemically similar but differently sized components in the fullerides, and changes in a nonsuperconducting moiety that changes the crystal structure, as in the creation of the κ-phase of the $(ET)_2X$ salts.

1. Isotopic Substitutions

These are aimed at identifying lattice vibrations as the source of Cooper pairing. The BCS theory predicts that T_c will increase with decreasing isotopic mass because this increases the vibrational energy (E_V) in Eq. (1). This effect actually was discovered in 1950 (prior to BCS theory) in the isotopes of lead, tin, and mercury, thereby confirming the idea that electron–phonon interactions played a key role in superconductivity. Unfortunately, it is now known that isotope effects can be considerably more complicated than indicated by simple BCS theory. Therefore, presence or absence of an isotope effect is an indication for or against involvement of lattice vibrations, but not definitive proof.

Consequently, the general absence of oxygen and copper isotope effects in the cuprate superconductors at those compositions which maximize T_c is not universally regarded as mandating a non-phonon-pairing mechanism. A further puzzle is that the isotopic mass effect often appears, sometimes even more strongly than predicted by BCS theory, when T_c is decreased from its maximum value by chemical modifications that change the charge-carrier concentration in the CuO_2 planes. Possible explanations for this complex behavior include the following: (1) Vibrational energy changes, especially in the O^{2-} ions, might affect the charge-carrier concentration in the CuO_2 planes. (2) If the high T_c is due to a van Hove peak in the density of states at the Fermi level, then the energy scale (E_V) is determined by the width of this peak, which is independent of isotopic mass, rather than a vibrational frequency.

On the other hand, the oxygen isotope effect in the noncuprate high-T_c compound $Ba_{1-x}K_xBiO_3$ is normal and supports oxygen vibrations as the source of superconducting electron pairing in this compound.

Isotopic substitutions in the ET-based superconductors have been limited to replacement of hydrogen by deuterium, with the anti- BCS result that T_c is somewhat higher in the deuterated compounds. This indicates that hydrogen vibrations play no role in mediating electron pairing in these materials. The effect may somehow result from deuterium substitution reducing the effective size of the ET molecules because the deuterium vibrations have lower amplitudes.

Several investigations of the ^{13}C isotope effect on T_c in the alkali metal fullerides have been carried out. The $^{13}C_{60}$ is prepared in the usual manner by electric arc vaporization of carbon rods that are wholly or part ^{13}C. The observed isotope effect supports a vibrational pairing mechanism in these compounds.

2. Chemical Modifications

These investigations are somewhat easier for the inorganic superconductors, where usually only elemental changes are required. Changing the inorganic component in organic superconductors, e.g., the inorganic anion in the $(ET)_2X$ compounds, or the alkali in the M_3C_{60} compounds, ranges from easy to quite difficult. Creation of new organic donors and/or acceptors that yield organic superconductors is usually an arduous process.

The effects of varying the charge carrier concentration in cuprates by partial replacement of the noncopper cations with others of different valence, or by varying the oxygen content, are similar to those shown in Fig. 10 for $La_{2-x}M_xCuO_4$, and also observed in the electron superconductors $Ln_{2-x}Ce_xCuO_4$, although it is not possible to vary the carrier concentration sufficiently in the other cuprates to span the entire range of behavior observed in the these compounds.

Other lines of investigation are (1) to completely replace one noncopper cation with another of the same valence, which presumably will not change the charge-carrier concentration in the CuO_2 planes; (2) to partially replace the Cu^{2+} ions with other metals; (3) to partially replace the oxygens with fluorine or sulfur. An extremely wide range of modifications of this type have been attempted without significantly increasing T_c, and in many cases T_c is substantially decreased or superconductivity is lost.

As noted in Sec. V it is possible to replace Y in $YBa_2Cu_3O_{6+\delta}$ with a number of trivalent rare earths without changing its superconducting properties appreciably. A notable exception is praeseodymium, which yields the insulator $PrBa_2Cu_3O_7$, presumably because the entry of Pr as tetravalent Pr^{4+} depletes the CuO_2 layer of charge-carrying holes. A puzzling feature is that many of these rare earth ions have open valence shells and, consequently, magnetic moments. Electron-spin-dependent interactions between magnetic impurities and superconducting electrons often destroy conventional metallic superconductivity by interfering with the spin alignment of the electrons in a Cooper pair, which is similar to that in a covalent chemical bond. One can argue that this does not occur in $YBa_2Cu_3O_{6+\delta}$ because the electronically active CuO_2 planes interact only weakly with the somewhat distant yttrium site, but this raises the question of how pairs of CuO_2 planes separated only by a metal layer interact. This question is pertinent because such paired planes tend to increase T_c, as is especially evident in the Tl-, Bi- and Hg-superconductors and is also possibly a factor in the T_c of $YBa_2Cu_3O_7$ being much higher than that of $La_{2-x}M_xCuO_4$.

Relative ion sizes are an important factor in most substitutions. If they differ too greatly, either the substitution cannot be performed or it will result in a different crystal structure, as was the fortunate case when it was attempted to replace lanthanum in $La_{2-x}Ba_xCuO_4$ with yttrium. Similarly, replacing La in $La_{2-x}M_xCuO_4$ by lanthanide-series rare earth ions, with smaller ionic radii, results either in a nonsuperconducting compound or a totally different superconductor, e.g., the electron superconductor $Ln_{2-x}Ce_xCuO_4$.

Also of great interest are the effects of replacing copper by various divalent metals, such as zinc and some $3d$ transition metals, and by trivalent species such as Al, Ga, and other $3d$ transition metals. These substitutions, which have been carried out mostly in $YBa_2Cu_3O_7$, show that the CuO_2 planes are vital to cuprate superconductivity. The key result was that trivalent metals, which are expected to substitute preferentially at the chain-Cu sites because crystal field calculations show that the negative potential from the surrounding ions is larger there than at the plane-Cu site, lowered T_c substantially less at a given concentration than did divalent impurities, which were expected to substitute at least as much for the plane as the chain coppers. The effect of zinc, which is the closest chemically to copper, is especially striking: in $YBa_2(Cu_{1-x}Zn_x)_3O_7$, T_c is 60 K and 0 K (nonsuperconductive) for $x \simeq 0.03$ and 0.1, respectively.

VIII. SUMMARY

It is especially difficult to predict the course of superconductivity research and development, given its history of "breakthroughs" interspersed between lengthy periods of slow, steady development. It is unquestionably an important field that both enriches our appreciation of the subtleties of nature and has applications vital to both science and our general welfare, for example, the superconducting magnets of nuclear magnetic resonance spectroscopy and medical imaging. Also important, its interdisciplinary character brings together scientists of diverse backgrounds, contrary to current trends toward increasing, often funding-driven, specialization. This approach is not only favorable for continued advances in superconductivity per se, but its discoveries are likely to have important consequences for related, and even seemingly unrelated, fields.

EXERCISES

Some of these problems have more than one reasonable answer, and there are disagreements as to which, if any, is right. The important thing, therefore, is to be able to reach such answers based on the principles in this chapter.

A table of ionic radii and ionization potentials of the elements, such as found in the *CRC Handbook of Chemistry and Physics* (R. W. Weast, ed.) will be helpful with some of these problems.

1. Using the $YBa_2Cu_3O_7$ lattice constants given in Fig. 3, and the values $a = b = 3.782$ Å and $c = 13.249$ Å for $La_{1.85}Sr_{0.15}CuO_4$, calculate the charge-carrier concentration in electrons/cm^2 in the CuO_2 planes if the dopant-induced holes are (a) distributed evenly over all oxygens, (b) distributed only over the CuO_2 oxygens. Suggest one experimental and one theoretical reason that charge-carrier concentration is not necessarily a good indicator of the potential of a conductor to be a superconductor.

2. Use simple structural diagrams of the CuO_2 plane to show that conductive charge motion in undoped La_2CuO_4 requires structures with either Cu^{3+} or O^- ions. Give a reason that the charge displacement $Cu^{2+} \ldots O^{2-} \rightarrow Cu^+ \ldots O^-$ is unlikely to produce conductive hole motion even if it requires less energy than the displacement $Cu^{2+} \ldots O^{2-} \ldots Cu^{2+} \rightarrow Cu^+ \ldots O^{2-} \ldots Cu^{3+}$. Why is this reason not a problem for oxygen holes introduced by doping?

3. From the phase diagram in Fig. 17, suggest conditions for growing single crystals of $YBa_2Cu_3O_7$ and $Y_2Ba_4Cu_7O_{15}$. Is an excess of CuO in the starting material likely to be helpful? Why is growing single crystals of $YBa_2Cu_4O_8$ likely to be very difficult?

4. Identify within the Tl2212 and Tl2223 crystal structures, shown in Fig. 4, the Tl2201 structure. What is the composition of the remainder of these compounds? From this, suggest a method of preparing the Tl2212 and Tl2223 compounds using one of the methods for growing thin films of the cuprates.

5. If $YBa_2Cu_3O_7$ is doped with Ca is it more likely to substitute for Y or Ba, and why? If it replaces Y, what is its likely effect in the double-doped compound $(Y_{1-x-y}Pr_xCa_y)Ba_2Cu_3O_7$, if the Pr dopant lowers T_c by entering the lattice as Pr^{4+}?

6. The one-dimensional organic conductor tetrathiafulvalenium-tetracyanoquinododimethanide (TTF-TCNQ) has the charge formula $TTF^{0.57+} - TCNQ^{0.57-}$. Suggest a reason that its conductivity decreases markedly at 60 K and decreases again to a completely insulating state around 40 K. Suggest a reason that the fluorinated analog of this compound, TTF-TCNQF, is an insulator at room temperature.

7. Suggest a synthetic procedure for preparing the selenium analog of ET, that is, *bis*-(ethylenediseleno)-tetraselenafulvalene, that does not use the reaction in Eq. (15). Can your method be used to prepare ET itself, and why or why not?

8. Select two metallic elements other than the alkali and alkaline earth elements that might produce conductive and superconductive M_xC_{60} compounds with the structure of the alkali metal fullerides. Give two properties of M used as a basis for your selections. Are any of your choices as good a candidate as the alkali metals themselves, and why or why not?

REFERENCES

With apologies to the many individuals whose excellent work is not directly cited, it is opted here wherever possible for a general reference, usually a book or review article, rather than an original research article. These are chosen to provide additional details and further references for individuals wishing to explore some topic in depth.

1. B. Schechter, *The Path of No Resistance*, Simon and Schuster, New York, 1990.
2. A. Leggett, in *The New Physics* (P. Davis, ed.), Cambridge, New York, 1989, pp. 268–288.
3. F. J. Adrian and D. O. Cowan, *Chemical and Engineering News*, December 21, 1992, p. 24.
4. A. K. Cheetham and P. Day, eds., *Solid State Chemistry Compounds*, Clarendon Press, Oxford, 1992.
5. J. M. Williams et al., *Organic Superconductors (Including Fullerenes) Synthesis, Structure, Properties and Theory*, Prentice-Hall, New York, 1991.
6. *Physics Today*, Special Issue: High Temperature Superconductivity, *44*(6) (1991).
7. R. C. Haddon et al., in *Fullerenes*, ACS Symposium Series *481*, 1992, p. 71.
8. S. S. P. Parkin et al., *Phys. Rev. Lett. 61*:750 (1988).
9. R. Fainchtein et al., *Science 256*:1013 (1992).
10. K. Tanagaki et al., *Nature 356*:419 (1992).
11. T. Kuwana, ed., *Physical Methods in Modern Chemical Analysis*, Vols. 1 and 2, Academic Press, New York, 1980.
12. R. B. Hazen, *The Breakthrough: The Race for the Superconductor*, Summit, New York, 1988.
13. D. Jérome and H. J. Schulz, *Adv. Phys. 31*:299 (1982).
14. D. J. Bishop, *Nature 365*:394 (1993).
15. J. D. Jorgensen, in Ref. 6, p. 34.
16. A . W. Sleight, in Ref. 6, p. 24.
17. J. Karpinski et al., *Physica C 160*:449 (1989).
18. R. J. Cava et al., *Physica C 165*:419 (1990).
19. L. Gao et al., *Physica C 213*:261 (1993), and references contained therein.
20. R. Simon, *Solid State Technol. 32*(9):141 (1989).
21. R. F. Curl and R. E. Smalley, *Scientific American*, October 1991, p. 54.
22. O. Zhou et al., *Nature 362*:433 (1993).
23. P. Yam, *Scientific American*, December, 1993, p. 118, reviews the problems and progress in this area. See also D. Larbalestier, in Ref. 6, p. 74.
24. W. A. Little, *Scientific American*, February 1963, p. 21.

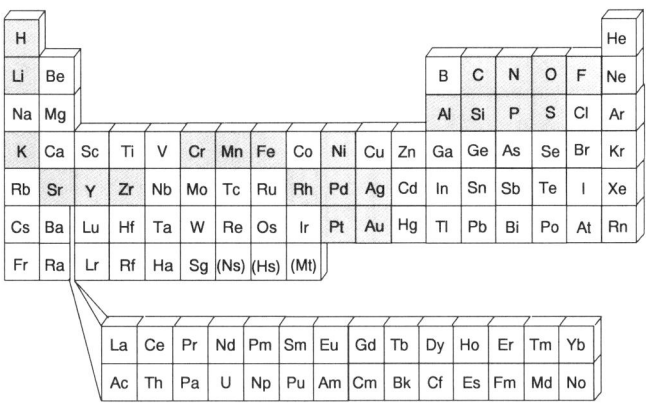

13
Fuel Cells

Nguyen Q. Minh *AlliedSignal, Inc., Torrance, California*

I. INTRODUCTION

A fuel cell is an energy conversion device that produces electricity (and heat) directly from a gaseous fuel by electrochemical combination of the fuel with an oxidant. The operating principles of a fuel cell are similar to those of a battery. However, unlike a battery, a fuel cell does not run down or require recharging; the fuel cell employs gases (from an external source) as reactants, and it operates as long as both fuel and oxidant are supplied to the fuel cell. A fuel cell consists of two electrodes (the anode and cathode) separated by an electrolyte (Fig. 1) [1]. In the operation of a fuel cell, fuel is fed to the anode, where it is oxidized, and electrons are released to the external circuit. Oxidant is fed to the cathode, where it is reduced, and electrons are accepted from the external circuit. The electron flow (from the anode to the cathode) through the external circuit produces direct current (dc) electricity. The electrolyte conducts ions between the

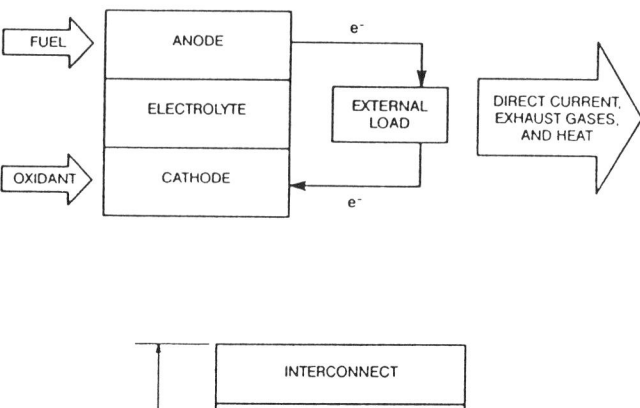

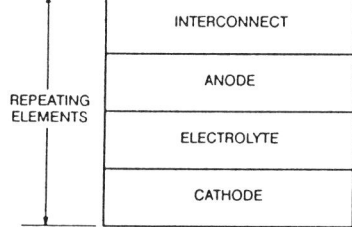

Figure 1 Fuel cell operation. (From Ref. 1.)

electrodes to balance the charge from the electron flow and complete the electrical circuit. Fuel cells are low-voltage devices (generally producing less than 1 V); therefore, for practical applications, fuel cells are not operated as single units. Rather, they are connected in electrical series to build voltage. A series of fuel cells is referred to as a stack. A component, called a bipolar separator or an interconnect, connects the anode of one cell to the cathode of the next in a stack (Fig. 1). Figure 2 shows a cross-sectional diagram of a typical fuel cell stack. The power produced by a stack is determined by the number (i.e., voltage) and the area (i.e., current) of the individual cells (single cells). Fuel cell stacks can be configured in series, parallel, both series and parallel, or as single units, depending on the particular application.

The principles of fuel cell operation were first reported by Sir William Grove in 1839. His fuel cell used dilute sulfuric acid as the electrolyte, and it operated at room temperature and ambient pressure. Since that time, fuel cell technology has expanded into nonaqueous and solid electrolytes and high-temperature and high-pressure operation. Early fuel cells were often small (laboratory-scale) in size, and their performance tended to degrade quickly with time. It was not until the 1950s when fuel cells of practical sizes were built and operated over prolonged periods. Since the 1960s, fuel cells have been routinely used as power sources in space flights. At the same time, fuel cells have been developed for

Fuel Cells 563

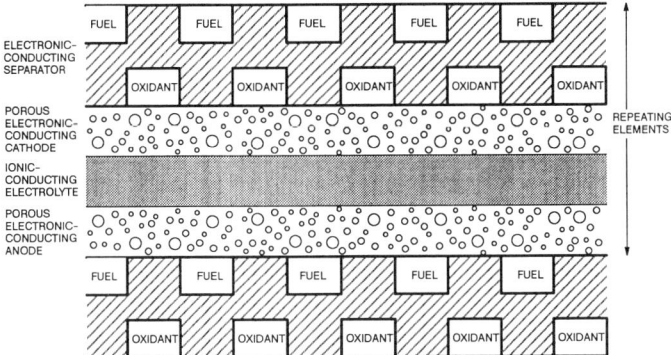

Figure 2 Cross-sectional diagram of a typical fuel cell stack.

commercial uses due to their clean and efficient generation of electric power from a variety of fuels. Fuel cell systems for co-generation applications have been commercially available since the early 1990s.

The key feature of a fuel cell is its high energy conversion efficiency. A fuel cell converts the chemical energy of fuel directly into electrical energy, and the usual losses involved in the conversion of fuel to heat, mechanical energy, and electrical energy are avoided. Consequently, a fuel cell can achieve a conversion efficiency significantly higher than those of conventional methods of power generation. In addition to the high conversion efficiency, fuel cells have the characteristics of environmental compatibility, modularity, siting flexibility, and multifuel capability:

1. *High conversion efficiency*: In a fuel cell, chemical energy of the fuel is converted directly to electrical energy; thus, the efficiency of fuel cell operation is not subject to the Carnot limitation. As a result, fuel cell efficiencies (45–60%) are considerably higher than those of other energy conversion methods (30–40%). In addition, the fuel cell produces high-quality by-product heat suitable for use in co-generation (simultaneous production of electrical power and heat) or bottoming cycles (use of by-product heat to generate additional electricity). Fuel cell efficiency is further increased when the by-product heat is fully utilized.
2. *Environmental compatibility*: Fuel cells are capable of using practical fuels as an energy source with insignificant environmental impact. Since fuel cell operation does not involve combustion, emissions of key pollutants are several orders of magnitude lower than those produced by conventional power generators (Fig. 3) [2]. Production of undesirable materials such NO_x, SO_x,

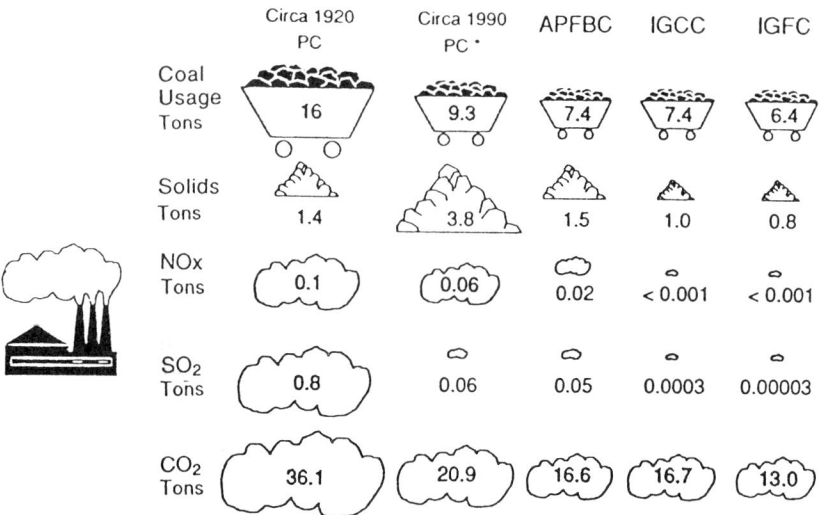

Figure 3 Comparison of emissions for fuel cell and conventional power systems (coal-based). Note: constant power output of 21.2 MW. APFBC, advanced pressurized fluidized bed combustion; IGCC, integrated gasification combined cycle; IGFC, integrated gasification fuel cell; PC, pulverized coal. *With flue gas desulfurization scrubbing. (From Ref. 2.)

and particulates is either negligible or undetectable for fuel cell systems in many applications.

3. *Modularity*: Fuel cells can be manufactured in modules. Thus, the size of a fuel cell system can be easily increased (or decreased) to match load requirements. Since the efficiency of a fuel stack is determined by the characteristics of the individual cell, efficiency of a fuel cell is largely independent of its size. As a result, fuel cells can maintain high efficiencies at part-load operation.

4. *Siting flexibility*: Because fuel cells can be made in a variety of sizes, they can be placed at different locations with minimum siting restrictions. Fuel cell operation is quiet because a fuel cell has no moving parts; the only noise derives from auxiliary equipment. Consequently, fuel cells can be easily located near points of use such as urban residential areas.

5. *Multifuel capability*: Fuel cell systems can be designed to use a variety of fuels. Suitable fuels include conventional fuels such as natural gas, alcohol, gasoline, and synthetic gas made from coal or plant matter. Certain types of fuel cells can process hydrocarbons internally and do not need expensive subsystems to process conventional fuels into simple forms.

Because of these desirable characteristics, fuel cells have been considered for a broad spectrum of electric power generation applications, including electric utility, co-generation, transportation, space, and marine uses. Fuel cell units can range from watt to multimegawatt levels (Table 1). Depending on the requirements of the application and the operating characteristics of the fuel cell, different types of fuel cells are selected for different applications. Fuel cells come in many different sizes. For example, a 25-W fuel cell (solid polymer fuel cell) is the size of a nickel–cadmium battery. Figure 4 is a photograph of a 5-kW power generator (using solid polymer fuel cells and including auxiliary equipment) [3]. The generator is 104 cm long, 80 cm wide, and 37 cm high, and weighs 150 kg. Figure 5 provides a pictorial view of a conceptual installed 200-kW co-generation system (using solid oxide fuel cells and including auxiliary

Table 1 Applications of Fuel Cells in Various Sizes

Fuel cell size	Examples of applications
10–500 W	Portable units, transmitter and communication devices
0.5–5 kW	Portable generators, navigational aids, wheelchairs
5–50 kW	Tractors, fork lift trucks, golf carts, electric vehicles, lighthouses, submerged vehicles, spacecraft
50–500 kW	Buses, heavy trucks, commercial co-generation systems, mining vehicles
0.5–5 MW	Locomotives, ships, submarines, on-site generators
5–50 MW	Dispersed power plants, industrial co-generation systems
50–500 MW	Central station power plants

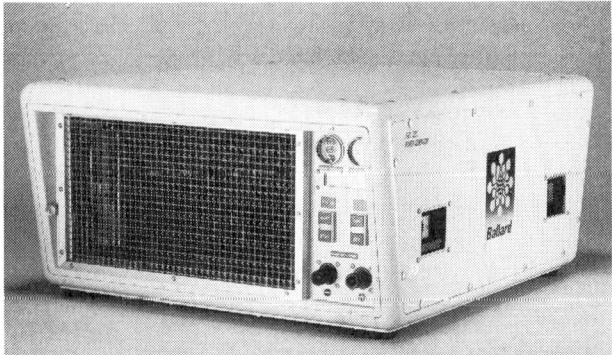

Figure 4 5-kW fuel cell power generator (using solid polymer fuel cells). (From Ref. 3.)

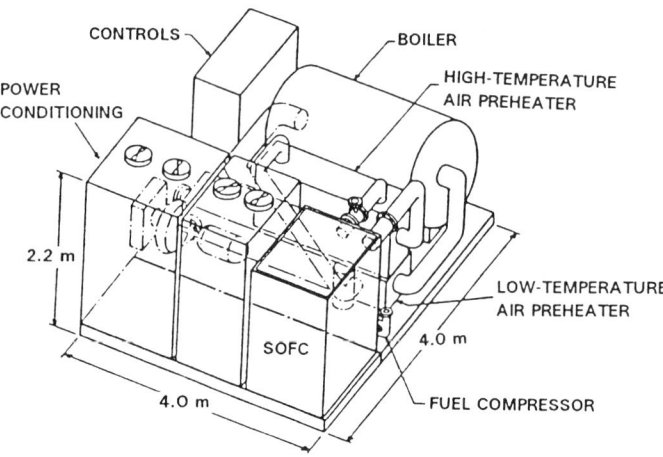

Figure 5 Pictorial view of a 200-kW fuel cell co-generation system (using solid oxide fuel cells). (From Ref. 4.)

equipment) [4]. The estimated weight of this system is 9300 kg, and the projected footprint is 16 m². Like any emerging technology, widespread commercial application of fuel cells strongly depends on their costs. Fuel cell costs are expected to fall as the technology advances and becomes more mature, and usage and production increases.

This chapter discusses in detail the principles of fuel cell operation and the materials and construction for various types of fuel cells. A fuel cell stack, in a broad sense, can be considered an electronic device because most of the components are made from electronic materials. As will be seen later, fuel cell technology, although different from the electronic technologies discussed in the previous chapters, has certain technical similarities. Fuel cell technology has adapted many processes and techniques that are common in the electronic industries. Some examples are thin-film fabrication methods, multilayer formation, and tape casting.

II. TYPES OF FUEL CELLS

Fuel cells use exclusively hydrogen as fuel, and oxygen as oxidant. In theory, any gas capable of being electrochemically oxidized or reduced can be used as fuel and oxidant in a fuel cell. However, hydrogen is exclusively used as fuel since it has high electrochemical reactivity and can be derived from common fuels such as hydrocarbons, alcohols, and coal. Oxygen is exclusively used as oxidant since it is readily and economically available from air. A fuel cell electrolyte must ionically conduct one of the elements present in the fuel and oxidant. Thus, for hydrogen fuel and oxygen oxidant, the electrolyte must conduct either hydrogen or hydrogen-containing ions (e.g., hydroxide ions) or oxide or oxygen-containing ions (e.g., carbonate ions, hydroxide ions). In this case, the reactions in the fuel cell consist of the oxidation of hydrogen at the anode and the reduction of oxygen at the cathode. The overall cell reaction yields water as the reaction product. Figure 6 shows, as an example, the reactions in a fuel cell having a proton-conducting electrolyte [1].

Fuel cells are commonly identified by the type of electrolyte used. For example, molten carbonate fuel cells use a mixture of molten carbonates as the electrolyte; alkaline fuel cells use an alkaline solution as the electrolyte. There are five main classes of fuel cells: phosphoric acid (PAFC), molten carbonate (MCFC), solid oxide (SOFC), solid polymer (SPFC), and alkaline (AFC) fuel cells. Among these fuel cells, the PAFC, MCFC, and SOFC have been developed for electric utility and co-generation sectors. The PAFC and SOFC are also being considered for transportation use. The SPFC has been developed mainly for space and transportation applications, and the AFC is an important power source for space flights. Table 2 summarizes the reactions in the various fuel cell types. Typical features and operational characteristics of the five types of fuel cells are listed in Table 3.

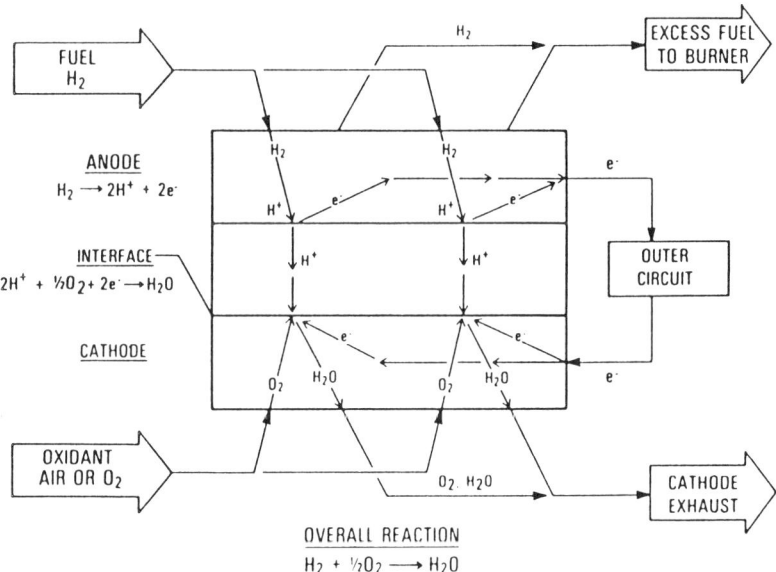

Figure 6 Schematic diagram of reactions in fuel cells based on proton-conducting electrolyte. (From Ref. 1.)

Table 2 Typical Reactions in Fuel Cells

Fuel cell type	Conducting ion	Anode reaction	Cathode reaction
PAFC	H^+	$H_2 = 2H^+ + 2e^-$	$\frac{1}{2}O_2 + 2H^+ + 2e^- = H_2O$
MCFC	CO_3^{2-}	$H_2 + CO_3^{2-} = H_2O + CO_2 + 2e^-$	$\frac{1}{2}O_2 + CO_2 + 2e^- = CO_3^{2-}$
SOFC	O^{2-}	$H_2 + O^{2-} = H_2O + 2e^-$	$\frac{1}{2}O_2 + 2e^- = O^{2-}$
SPFC	H^+	$H_2 = 2H^+ + 2e^-$	$\frac{1}{2}O_2 + 2H^+ + 2e^- = H_2O$
AFC	OH^-	$H_2 + 2OH^- = 2H_2O + 2e^-$	$\frac{1}{2}O_2 + H_2O + 2e^- = 2OH^-$

III. THERMODYNAMIC ASPECTS

The maximum electrical energy obtainable in a fuel cell at constant temperature and pressure is given by the Gibbs free energy change, ΔG, of the cell reaction:

$$\Delta G = \Delta H - T\Delta S = -zFE_r \tag{1}$$

where ΔH is the enthalpy change, T the temperature, ΔS the entropy change, z the number of electrons involved in the electrochemical reactions, F the Faraday constant, and E_r the thermodynamic or reversible voltage of the cell. At the standard state, the standard Gibbs free energy change of the cell reaction is given as

$$\Delta G^0 = \Delta H^0 - T\Delta S^0 = -zFE^0 \tag{2}$$

The reversible voltage of a fuel cell can be obtained from the Nernst equation. For example, for the following reactions in a fuel cell having an oxide-conducting electrolyte:

Anode $\qquad\qquad H_2 + O^{2-} = H_2O + 2e^- \qquad (3)$

Cathode $\qquad\qquad \frac{1}{2}O_2 + 2e^- = O^{2-} \qquad (4)$

Overall reaction $\qquad H_2 + \frac{1}{2}O_2 = H_2O \qquad (5)$

the Nernst equation is given as follows:

Table 3 Typical Features and Operational Characteristics of Various Types of Fuel Cells

Characteristic	Type of fuel cell				
	PAFC	MCFC	SOFC	SPFC	AFC[a]
Electrolyte	H_3PO_4	Molten Li_2CO_3-K_2CO_3	Solid Y_2O_3-stabilized ZrO_2 (YSZ)	Perfluoro-sulfonic acid membrane	KOH solution
Electrolyte support	SiC	$LiAlO_2$	None	None	Asbestos
Cathode	PTFE[b]-bonded Pt on C	Li-doped NiO	Sr-doped $LaMnO_3$	PTFE-bonded Pt on C	Pt-Au
Anode	PTFE-bonded Pt on C	Ni	Ni/YSZ	PTFE-bonded Pt on C	Pt-Pd
Interconnect/bipolar	Glassy carbon	SS[c] clad with Ni	Doped $LaCrO_3$	Graphite	Ni
Operating temperature	200°C	650°C	1000°C	80°C	100°C
Operating pressure	1–8 atm	1–3 atm	1 atm	1–5 atm	1–10 atm
Fuel	H_2	H_2, CO	H_2, CO	H_2	H_2
Oxidant	O_2	$O_2 + CO_2$	O_2	O_2	O_2
Contaminant tolerance	<1–2% CO <50 ppm sulfur	<1–5 ppm sulfur	<10–100 ppm sulfur	<50 ppm CO No sulfur	No CO_2, CO, No sulfur

[a] Compositions for AFC cell components are given as examples.
[b] PTFE = polytetrafluoroethylene.
[c] SS = stainless steel.

$$E_r = E^0 + \frac{RT}{2F} \ln \frac{P_{H_2} P_{O_2}^{1/2}}{P_{H_2O}} \qquad (6)$$

where R is the gas constant and P is the partial pressure of the subscript gaseous compound. The reversible voltage at the standard state can be expressed as

$$E^0 = -\frac{\Delta G^0}{2F} = \frac{RT}{2F} \ln K \qquad (7)$$

where K is the equilibrium constant of Eq. (5).

The reversible voltage is the maximum voltage that can be achieved by a fuel cell under specified conditions of gas composition, temperature, and pressure. The effect of gas composition on the reversible voltage can be derived from the Nernst equation. The effect of temperature and pressure can be analyzed based on changes in the Gibbs free energy with temperature and pressure. Thus, the following equations can be derived:

$$\left(\frac{\partial E_r}{\partial T}\right)_P = \frac{\Delta S}{zF} \tag{8}$$

$$\left(\frac{\partial E_r}{\partial P}\right)_T = -\frac{\Delta V}{zF} \tag{9}$$

where ΔV is the volume change. For the hydrogen/oxygen reaction, the entropy change is negative; therefore, the reversible voltage of the fuel cell decreases with increasing temperature. For the same reaction, the volume change is negative; therefore, the reversible voltage increases with increasing pressure.

Attempts to improve fuel cell performance using thermodynamic principles (e.g., increasing pressure) should consider other factors. An increase in the operating pressure improves fuel cell performance because of increased reactant partial pressure and increased mass transport rate. On the other hand, increase in pressure also imposes limits on material selection and causes other problems or concerns such as integrity of the gas seal and cell structure. An increase in the operating temperature is beneficial to fuel cell performance because of enhanced mass transfer, increased reaction rate, and usually, lower material resistance (and thus improved cell performance). Increases in operating temperature, on the other hand, may limit material choice and accelerate material-related problems such as interaction, degradation, and sintering.

In reversible operation, the heat absorbed by the fuel cell, Q, at constant temperature equals the change in entropy of the reactants:

$$Q = T \Delta S = \Delta H - \Delta G \tag{10}$$

If the change in entropy is negative, as is the case in most common fuel cell reactions, heat is generated by the cell. In practical fuel cell operation, this heat increases the operating temperature. Performance losses due to higher temperatures are generally avoided by cooling the cell to remove the heat generated.

IV. OPERATIONAL CHARACTERISTICS

Three important characteristics of a fuel cell during operation are polarization losses, power output, and cell efficiency.

A. Cell Polarization

The voltage E of an operating cell is always lower than the reversible voltage E_r. As the current is drawn from the fuel cell, the cell voltage falls due to polarization losses. Thus, the voltage of an operating cell is given as

$$E = E_r - \eta \tag{11}$$

where η is the cell polarization or overpotential.

The polarization of a cell is the sum of three main types of polarization: charge transfer or activation polarization η_A, resistance or ohmic polarization η_Ω, and diffusion or concentration polarization η_D. The activation polarization may be regarded as the extra potential (voltage) necessary to reduce the energy barrier of the rate-determining step of the cell reaction to a value such that the reaction proceeds at a desired rate. Activation polarization is related to current density, j, by

$$j = j_0 \exp\frac{\alpha \eta_A F}{RT} - j_0 \exp -\frac{(1-\alpha)\eta_A F}{RT} \tag{12}$$

where α is the transfer coefficient, and j_0 is the exchange current density. The transfer coefficient is considered as the fraction of the change in the polarization that leads to a change in the reaction rate constant. The exchange current density is the (equal) forward and reverse electrode reaction rate at the equilibrium potential. High exchange current density means high electrochemical reaction rate, and thus low activation polarization might be expected. When the electrode reaction is reversible, the second term on the right-hand side of Eq. (12) may be neglected, and Eq. (12) is reduced to the following equation (the Tafel equation):

$$\eta_A = a + b \log j \tag{13}$$

where a and b are constants (which are related to electrode material and type of electrode reaction).

The ohmic polarization is caused by resistance to conduction of ions (through the electrolyte) and electrons (through the electrodes and current collectors) and by contact resistance between cell components. The ohmic polarization is given as

$$\eta_\Omega = jR_i \tag{14}$$

where R_i is the cell internal resistance. (The ohmic polarization is commonly separated from other types of polarization and referred to as ohmic loss.)

The diffusion polarization appears when the electrode reaction is hindered by mass transport effects, i.e., when the feeding velocity of the reactant and/or the removing velocity of the reaction product from the electrode is slower than that corresponding to the discharge current density. When the electrode process is governed completely by diffusion, the limiting current density, j_L, is reached (characterized by a rapid drop in cell voltage). For an electrode process free of activation energy, the diffusion polarization is expressed as

$$\eta_D = \frac{RT}{zF} \ln \left(1 - \frac{j}{j_L}\right) \tag{15}$$

The diffusion polarization is dependent on the mass transport properties of the system; mass transport is a function of temperature, pressure, concentration, and the physical properties of the system.

Figure 7 shows a hypothetical voltage/current plot of an operating fuel cell. The figure illustrates the regions in which various types of polarization losses predominate. From Fig. 7, it can be seen that at low current densities, the major contribution to the cell voltage losses is from the activation polarization. As the current increases, the ohmic polarization dominates, as exhibited by the linearity in Fig. 7. When the current is high, the cell resistance is controlled by mass transport limitations, resulting in a rapid decrease in cell voltage.

In most typical fuel cells, because of the sophisticated construction of porous gas diffusion electrodes, the limiting current density is not reached, even at high currents. As a result, diffusion polarization is small. Also, cell components are usually made very thin to minimize ohmic polarization. Therefore, polarization loss of a typical fuel cell is mainly activation polarization. Material modifications and catalyst additions are commonly used to reduce activation polarization.

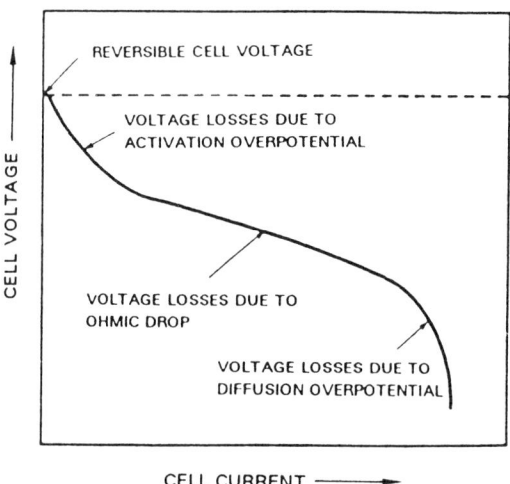

Figure 7 Voltage/current relation for an operating fuel cell.

B. Power Generation

The power output of a fuel cell, W, is the product of the voltage (E) and current (I):

$$W = IE \tag{16}$$

Qualitatively, one can predict the nature of the power/current relation by examining the behavior at low and high currents. At low currents, $I \to 0$, and at high currents, $E \to 0$. Hence, the power values approach zero at both ends. The current at which the power is at a maximum lies between these two extremes. Figure 8 is a qualitative presentation of the power/current relation in a hypothetical fuel cell. The hypothetical curve, corresponding to no activation and diffusion polarization losses (ohmic losses only) in the fuel cell, is also shown. It should be stressed that the power/current curve in Fig. 8 refers to ideal single-cell fuel cells in which planar electrodes are used. For cells constructed with sophisticated porous gas diffusion electrodes, the voltage/current is approximately linear. In such cases, the power/current relation tends to be parabolic.

It is difficult to obtain a general expression for maximum power when all forms of polarization are present. However, this may be done in the case where the cell voltage varies linearly with the current. In this case, the equation for the power is given by

$$W = I(E_r - IR_t) \tag{17}$$

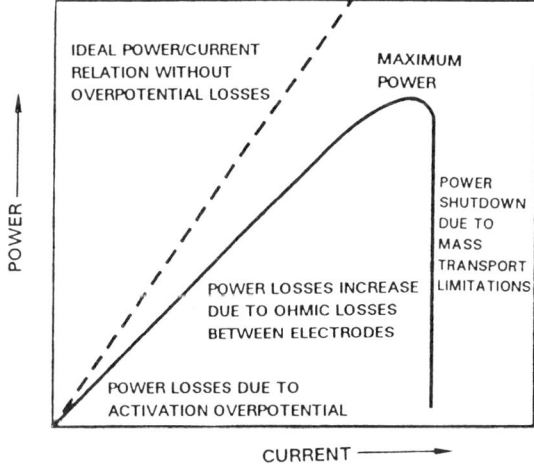

Figure 8 Power/current relation for an operating fuel cell.

where R_t is the cell total polarization resistance. Equation (17) shows that the power/current relation is parabolic. The conditions for the maximum power are

$$I = \frac{E_r}{2R_t} \tag{18}$$

$$E = \frac{E_r}{2} \tag{19}$$

The maximum power can be deduced as

$$W_{max} = \frac{E_r^2}{4R_t} \tag{20}$$

Thus, the electrolyte component of in this case, the maximum power is realized when the cell voltage is equal to one-half the reversible voltage.

Fuel cells are not always operated at the maximum power conditions. The conditions selected often depend on the particular application. For example, in electric utility applications, the power output of the fuel cell is preferred at the operating conditions where the cost of electricity produced is lowest.

C. Cell Efficiency

The efficiency ε of a fuel cell stack is the product of the thermodynamic efficiency ε_T, the voltage efficiency ε_V, the current efficiency ε_J (often referred to as the fuel utilization), and the heating value efficiency ε_H. The thermodynamic (or thermal) efficiency, the maximum or theoretical efficiency of a fuel cell, is equal to the ratio of the Gibbs free energy change to the enthalpy change of the cell reaction:

$$\varepsilon_T = \frac{\Delta G}{\Delta H} = 1 - \frac{T\Delta S}{\Delta H} \tag{21}$$

The voltage efficiency is defined as the ratio of the cell voltage under load to the equilibrium cell voltage:

$$\varepsilon_V = \frac{E}{E_r} \tag{22}$$

(Note that the equilibrium voltage, commonly referred to as the open-circuit voltage, may be different from the reversible voltage if there are side reactions, gas cross leakage, etc.) The current efficiency is a measure of the fraction of the electrochemically active fuel fed to the fuel cell that is actually consumed; it is given as

$$\varepsilon_J = \frac{j}{j_F} \tag{23}$$

where j_F is the amount of current produced for 100% conversion of the fuel. The heating value efficiency is the ratio of the enthalpy change of the cell reaction, ΔH, to the heating value of the fuel, $\Delta H'$:

$$\varepsilon_H = \frac{\Delta H}{\Delta H'} \tag{24}$$

This ratio is the direct measure of the fraction of heat energy that is available for conversion into electrical energy out of the total heat energy content of a given fuel gas. (For the hydrogen/oxygen reaction, the efficiency is referred to as low heating value, LHV, when the H_2O product is in the form of steam and as high heating value, HHV, when the H_2O is produced in the liquid state.)

For a fuel cell operated with a specific fuel and oxidant under defined conditions, the thermodynamic and heating value efficiencies are fixed. The fuel utilization (current efficiency) is relatively easy to control and is usually set to about 70–90%. Therefore, the cell efficiency is mainly determined by the voltage efficiency. Voltage efficiency of a fuel cell can be improved by reducing cell polarization losses.

V. CELL COMPONENTS

The main components of a fuel cell stack are the electrolyte, the anode, the cathode, and the separator or interconnect. Each component serves several functions in the fuel cell and must meet certain requirements. Each component must have the proper stability (chemical, phase, morphological, and dimensional) in oxidizing and/or reducing environments, compatibility with other components, and appropriate and adequate conductivity. The electrolyte and the separator must be impervious to both fuel and oxidant to prevent gas mixing, while the anode and cathode must be porous to allow gas access to the reaction sites. The anode and cathode must also have sufficient catalytic activity for the electrode reactions at the fuel cell operating conditions. The key requirements for the various cell components are summarized in Table 4. In addition to the requirements listed in Table 4, desirable properties for the components from practical viewpoints are high strength, ease of fabrication, and low cost. Also, depending on the type of fuel cell, other requirements may be needed. For example, the ceramic components of the solid oxide fuel cell must have similar coefficients of thermal expansion to avoid separation or cracking during fabrication and operation.

It should be noted that cell components may be composed of more than one material. For example, support structures are often required to contain liquid

Table 4 Requirements for Fuel Cell Components

Component	Requirements			
	Conductivity	Stability	Compatibility	Porosity
Electrolyte	High ionic conductivity; negligible electronic conductivity	Chemical, phase, morphological, and dimensional stability in fuel and oxidant environments	No damaging chemical interactions or interdiffusion with adjoining cell components	Fully dense
Cathode	High electronic conductivity	Chemical, phase, morphological, and dimensional stability in oxidant environment	No damaging chemical interactions or interdiffusion with adjoining cell components	Porous
Anode	High electronic conductivity	Chemical, phase, morphological, and dimensional stability in fuel environment	No damaging chemical interactions or interdiffusion with adjoining cell components	Porous
Interconnect/ separator	High electronic conductivity; negligible ionic conductivity	Chemical, phase, morphological, and dimensional stability in fuel and oxidant environments	No damaging chemical interactions or interdiffusion with adjoining cell components	Fully dense

electrolytes. Thus, the electrolyte component of a liquid electrolyte fuel cell usually consists of the liquid electrolyte and a support. Any material selected for use as an electrolyte support must meet many of the same requirements discussed earlier for a fuel cell electrolyte.

VI. PHOSPHORIC ACID FUEL CELL

The phosphoric acid fuel cell (PAFC) employs platinum catalyst supported on carbon with a PTFE (polytetrafluoroethylene) binder as the anode and cathode, and a silicon carbide (SiC) matrix filled with phosphoric acid as the electrolyte. Glassy carbon is commonly used as the separator to connect cells in electrical series in a stack. The PAFC operates at approximately 200°C and 1–8 atm (1–8×10^5 Pa). The operating pressure of the fuel cell depends on the application. On-site usage typically operates at atmospheric pressure, whereas electric utility or industrial usage operates at elevated pressure.

A. Electrolyte

Phosphoric acid, H_3PO_4, is the selected electrolyte for the PAFC, mainly because H_3PO_4 is the only acid suitable for use at an operating temperature of around 200°C. This operating temperature is preferred since it allows adequate CO tolerance, high oxygen reduction rates at the cathode, and efficient utilization of by-product heat. The acid is tolerant to CO_2 and has the required stability, conductivity, and low volatility under both high-pressure and high-temperature conditions.

The PAFC uses 100% H_3PO_4 as the electrolyte. Pure H_3PO_4 forms pyrophosphate polymer chains such as $-O-PO(OH)-O-PO(OH)-O-PO(OH)-$ and can self-ionize to give both anions and cations such as $-O-P(O)_2^--O$ and $-O-P(OH)_2^+-O$. This formation of polymer chains permits proton transfer via a Grotthus-type mechanism along the chains, thus giving high ionic conductivity even at 200°C. The presence of water in the acid decreases the conductivity. Pure H_3PO_4 solidifies at about 42°C with an accompanying volume increase. Frequent volume changes can damage the various cell components, leading to degradation in cell performance. Therefore, it is necessary to prevent electrolyte solidification by keeping the cell temperature above 42°C.

The phosphoric acid is contained in an inert electrolyte matrix (about 0.1–0.2 mm thick), which is highly porous, with small pores so as to be able to hold large quantities of electrolyte. The material for the matrix must be electronically nonconductive and chemically, mechanically, and morphologically stable in the fuel cell environment. SiC is the preferred matrix material for the PAFC. The matrix is typically made of about 90 wt % fine SiC powders (<10 μm particle size) bonded together with about 10 wt % PTFE.

Although phosphoric acid has low vapor pressure under fuel cell operating conditions, the fuel cell may have significant acid loss due to slow evaporation in long-term operation. The amount of acid loss depends on several factors and, in general, increases with higher gas flow rate and higher current density. Significant loss of the acid electrolyte may result in gas crossover through the matrix, causing cell performance to degrade. Several approaches have been developed to replenish acid loss in the matrix. For example, acid can be added at certain intervals under off-load conditions. The most common approach is to store excess electrolyte in electrode substrates (described below in Sec. B. Cathode). In this case, the pore size of the matrix is designed to be smaller than that of the substrate to ensure complete electrolyte filling in the matrix. This reduces the possibility of gas crossover under conditions of differential pressure.

B. Cathode

The PAFC cathode is composed of two layers: a catalyst layer to provide the active sites for the electrochemical reaction, and an electrode substrate to provide mechanical support to the thin catalyst layer. The catalyst layer of about 0.1 mm thickness is made of fine platinum crystallites dispersed on PTFE-bonded carbon support. The electrode substrate is made of carbon.

Platinum is the preferred catalyst material for the oxygen reduction in the PAFC because the metal has the required properties of electrocatalytic activity, stability, and electronic conductivity. Fine platinum particles are commonly used, often dispersed on high-surface-area carbon black. Carbon is chemically stable in the fuel cell environment and has sufficient electronic conductivity for use as support material. Figure 9 is a micrograph of a carbon-supported platinum catalyst [5]. The use of carbon to support the platinum catalyst particles dramatically reduces platinum loading without affecting electrode performance. For example, the Pt catalyst loading of PTFE-bonded carbon supported electrodes is about 0.25 mg/cm^2 compared with 9 mg/cm^2 for PTFE-bonded electrodes without carbon support.

The cathode pore structure is optimized to minimize liquid electrolyte movement. In the electrode, a three-phase (electrolyte, gas, electrode) interfacial region exists. To maintain the stable interfacial boundary, a hydrophobic phase such as PTFE is added to the catalyst/carbon mixture (the hydrophilic phase).

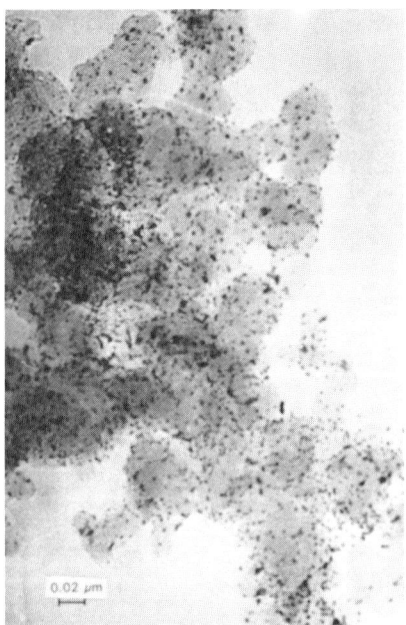

Figure 9 Micrograph of carbon-supported platinum catalyst. (From Ref. 5.)

The PTFE prevents the electrode from flooding (excessive filling of electrode pores by liquid electrolyte), while the hydrophilic areas are wetted with electrolyte and serve as the electrochemical active sites. The added PTFE also serves as a binder to maintain the integrity of the porous electrode structure. Between 10 and 60 wt % PTFE is often used.

The electrode substrate serves as a structural support for the catalyst layer and as the current collector, and it is porous to allow gas transport to the catalyst layer. The substrate material must be stable in the PAFC environment and electronically conductive, and carbon is the most common material. In one type of PAFC design, the substrate is ribbed to provide gas passages. In this case, the substrate is about 1.0–1.8 mm thick and 60–65% porous (20–40 µm pore size). The ribbed substrate is thick enough to be used to reserve excess acid electrolyte. In another PAFC design where the separator is ribbed, the substrate is made from thin carbon paper containing about 40 wt % PTFE. This carbon paper contains macropores with a mean pore size of about 12.5 µm and micropores with a median pore size of about 3.4 nm.

In a PAFC, the major source of polarization losses occurs at the cathode (Fig. 10) [6]. For example, at 180°C and 0.1 MPa, the polarization at PAFC cathodes (with Pt loading of 0.5 mg/cm^2) is about 560 mV at 300 mA/cm^2 in air. In comparison, the polarization at the anode under similar conditions is only about 12 mV, and the ohmic drop is about 36 mV. The high polarization at the cathode

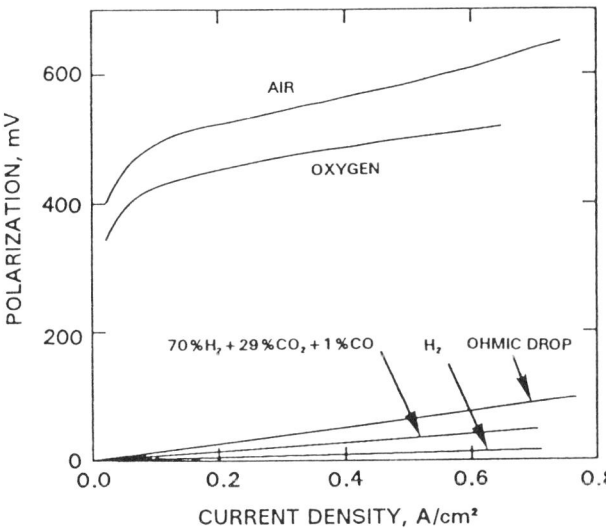

Figure 10 Contribution to polarization in PAFCs (0.5 mg/cm^2 Pt loading, 100% H_3PO_4, 180°C, 0.1 MPa). (From Ref. 6.)

is mainly due to the slow kinetics of oxygen reduction in phosphoric acid. It is well known that an increase in the cell operating pressure or temperature enhances the kinetics of the reduction, and thus the performance of the PAFC cathode. An increase in the oxygen partial pressure also improves the electrode performance. For instance, under the same conditions given in the foregoing example, the polarization of the PAFC cathode is about 480 mV in pure oxygen.

Under specified conditions of temperature and pressure, the kinetics of the oxygen reduction depends on a number of parameters such as electrolyte composition, catalyst characteristics, and electrode microstructure. The exchange current density (and thus the reduction rate) increases with a decrease in H_3PO_4 concentration (100 to 88%). The use of a more dilute acid electrolyte, however, tends to accelerate corrosion of the carbon support and makes water management more difficult. The oxygen reduction improves with increasing Pt loading (for low loadings or thin electrodes) (Fig. 11) [7]. Additional Pt loading above a certain level does not result in further improvements, since complete catalyst utilization is not realized as electrode thickness increases. The reduction of oxygen on small Pt particles in acid electrolytes is known to be a structure-sensitive or particle-size-dependent reaction. In general, the smallest possible Pt catalyst particles are used to provide the maximum activity per unit of mass, and the

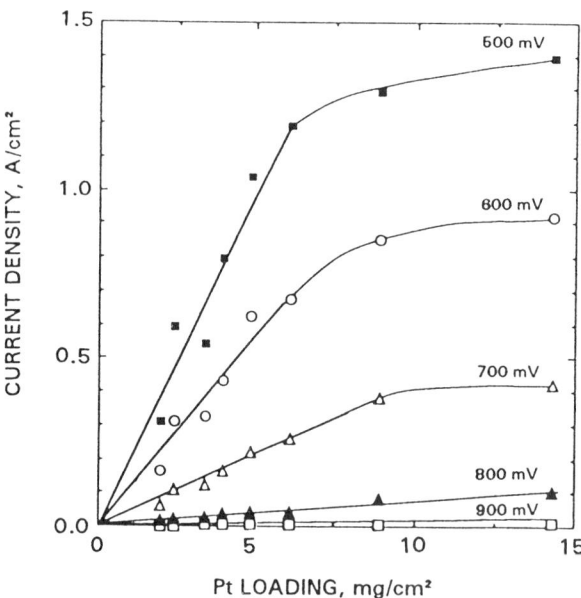

Figure 11 Effect of Pt loading on oxygen reduction at constant potential (air atmosphere, 85 wt % H_3PO_4, 120°C). (From Ref. 7.)

platinum in PAFCs is generally in the form of small clusters of about 1–5 nm. However, the effect of particle size on the electrocatalysis of oxygen reduction is still controversial. The maximum activity at a certain mass of platinum or surface area is not well defined. The microstructure of the cathode plays an important role in the gas transport and electrolyte wettability of the electrode, which in turn influence the oxygen reduction and thus the cathode performance. Electrodes with high PTFE content tend to be poorly wetted with electrolyte, and those with low PTFE content are prone to flooding. Experimental evidence indicates that a PTFE content of 30–40 wt % in the catalyst layer yields the optimal performance. The electrode substrate microstructure must thus be designed to prevent overpenetration of reactant gases, the best performance being obtained when 40–80% of the pore volume of the substrate is filled with electrolyte. Above 80% filling, diffusion of oxygen is limited, and the electrode performance decreases.

The performance of the PAFC cathode generally degrades with time, mainly because of sintering of Pt catalyst particles and corrosion of carbon support. During cell operation, high-surface-area Pt catalysts tend to sinter, i.e., the particles either coalesce to form large particles or transport (via a dissolution/precipitation mechanism) from smaller particles to larger ones, resulting in a decrease in electrochemical active sites. A loss in surface area of Pt is therefore detrimental to cathode performance. Various approaches have been considered to inhibit Pt sintering in PAFCs. For example, heat treatment of the catalyst appears to reduce the Pt sintering rate. Over long operating periods, the carbon support can corrode due to slow oxidation in hot H_3PO_4. The extent of the corrosion depends on the operating conditions (temperature, operating voltage, and acid concentration) and the nature of the carbon. The corrosion of carbon support will cause a loss of platinum crystallites. As carbon corrosion increases, the surface area of the support decreases, accompanied by a corresponding decrease in Pt active surface area.

C. Anode

The PAFC anode, similar to the cathode, consists of a catalyst layer and a carbon substrate. The catalyst layer is a PTFE-carbon with dispersed Pt catalysts. Under PAFC operating conditions, the oxidation of hydrogen is relatively rapid. The exchange current density for hydrogen oxidation (about 10^{-3} A/cm^2) is many orders of magnitude higher than that for oxygen reduction on Pt (10^{-9}–10^{-11} A/cm^2). Consequently, the anode polarization is only about 5 mV per 100 mA/cm^2.

Practical applications of PAFCs use gases other than pure hydrogen as the fuel (e.g., reformed natural gas), which may contain varying amounts of CO and other impurities and which can influence the performance of the PAFC anode. For example, at temperatures below 125°C, the anode is significantly poisoned by CO. The poisoning effect of CO is reversible and almost proportional to the CO content. It appears to be a simple competition with hydrogen for reaction sites, leading to a surface blockage due to the relatively strong chemisorptive bond of the CO molecule. For supported Pt catalyst, the nature of the support may play an important role in determining the poisoning severity. Thus, PAFC electrodes containing Pt catalysts supported on activated carbon have been found to be less prone to CO poisoning than those made from Pt supported on graphite.

Increasing the operating temperature above 125°C significantly reduces the CO effect. At 200°C, 1–2 vol % CO can be tolerated without adverse effects on the anode performance. The exact CO tolerance of the anode depends on the H_2/CO ratio at the gas exit and on the amount of sulfur present in the fuel. Typically, a PAFC will not be poisoned if the exit gas has a H_2/CO ratio of about 8 at a sulfur content (expressed as H_2S + COS) of up to 80 ppm. Higher sulfur content in the fuel will cause serious poisoning of the anode, since the sulfur compounds adsorb on Pt and block the active sites for hydrogen oxidation.

D. Separator

The PAFC separator is commonly made of high-density graphite or carbon, especially glassy carbon. Depending on the stack design, the separator may be ribbed with flow fields on both sides to provide gas channels.

E. PAFC Stack and Assembly

The PAFC stacks are commonly constructed from two types of cell configurations, ribbed separator type and ribbed substrate type. The latter configuration is generally preferred, especially in larger stacks, since this configuration promotes better uniform gas diffusion to the electrode, provides higher electrolyte reservoir capacity, and is more amenable to continuous manufacturing.

There are also two types of gas manifolding designs: internal manifolding and external manifolding. External manifolding designs have four gas manifolds attached to the sides of the stacks to distribute the fuel and oxidant and remove the exhaust gases. Internal manifolding designs have the gas-supplying manifold incorporated in the form of a duct inside the stack (see Fig. 16 of the molten carbonate fuel cell section for an example). The gas flow pattern is mainly crossflow in PAFCs.

In a PAFC stack, cooling plates are inserted (approximately one for every five cells) to remove heat generated in the fuel cell, the cooling method being either water or air (Fig. 12) [5]. Water cooling is more efficient than air cooling and, therefore, most suitable for large units. The air cooling system is simpler and often used for small units.

Fuel Cells

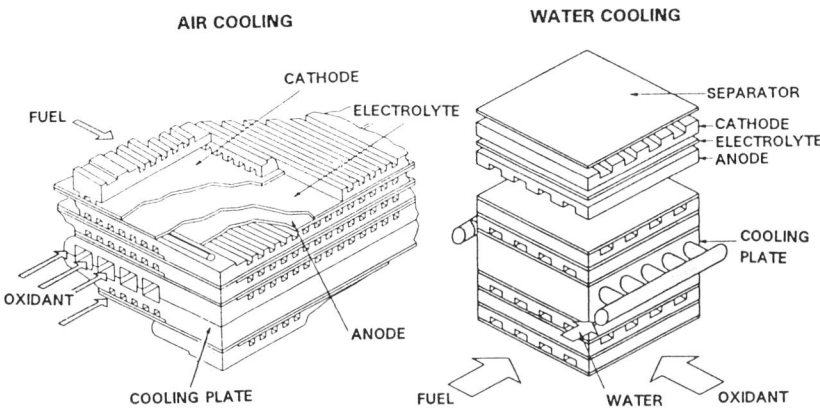

Figure 12 PAFC stack configurations with water cooling and air cooling systems. (From Ref. 5.)

The common fabrication process for the PAFC electrolyte matrix, cathode, and anode is by spraying the component ingredients onto the substrates followed by sintering. The H_3PO_4 electrolyte is added by impregnating the matrix and the electrodes with the acid. The PAFC assembly usually consists of stacking the cells, inserting cooling and current-collecting plates, adding clamping plates at the top and bottom of the assembly, and clamping the total assembly with clamping bolts. The clamping pressure is selected to minimize electrical resistances between cell components within the limit of material strength.

VII. MOLTEN CARBONATE FUEL CELL

The molten carbonate fuel cell (MCFC) consists of a Ni anode, a Li-doped NiO cathode, and a lithium aluminate ($LiAlO_2$) matrix (or electrolyte support) filled with a molten lithium and potassium carbonates electrolyte mixture (referred to as the electrolyte structure or the tile). Figure 13 shows examples of the cathode, anode, and electrolyte support microstructures [8]. In an MCFC stack, nickel-clad stainless steel separator plates connect cells in electrical series. The MCFC stacks are operated at a temperature of about 650°C and a pressure of 1–3 atm.

A. Electrolyte

In MCFCs, the molten carbonate electrolyte is contained in a $LiAlO_2$ matrix or electrolyte support, and it is retained in the ceramic matrix by capillary forces. This carbonate-filled ceramic matrix is commonly referred to as the electrolyte structure. The strength and stiffness of the electrolyte structure depend on the

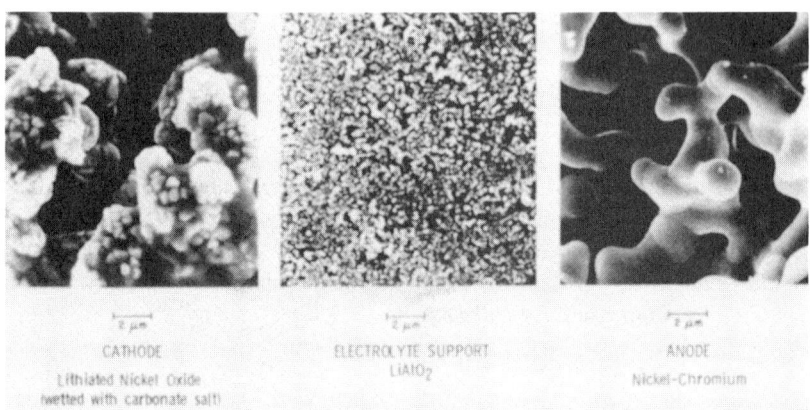

Figure 13 Micrographs of MCFC electrodes and electrolyte support. (From Ref. 8.)

relative amounts of molten carbonate and $LiAlO_2$. At low carbonate contents, the structure is rigid (essentially particle-to-particle contact); at high carbonate contents, the structure is fluid (a slurry); and in a narrow intermediate range, the structure is plastic. The composition of the plastic region is dependent on the particle size distribution and moves to higher carbonate contents for smaller particles. The electrolyte structure typically consists of about 45 wt % $LiAlO_2$ and 55 wt % carbonate. At the fuel cell operating temperature, the electrolyte structure is a thick paste, which provides gas seals (called the wet seal) at the edges of the cell. The MCFC electrolyte structure is fabricated by tape casting (Chapter 11, Sec. II).

1. Molten Carbonate Electrolyte

The molten carbonate electrolyte is an alkaline carbonate mixture and is liquid at the cell operating temperature. The basic properties that must be considered in the selection of a carbonate electrolyte are conductivity, gas solubility, surface tension, vapor pressure, and corrosivity. The preferred electrolyte, based on actual test results, is a binary lithium/potassium carbonate consisting of 62 mol % Li_2CO_3 and 38 mol % K_2CO_3; it is possibly not yet optimal.

Most of the carbonate is contained in the electrolyte support. However, the carbonate is also distributed in the anode and cathode, to allow the conducting of ions to and from the reaction site. The carbonate content in each component is primarily controlled by its relative pore size distribution, surface tension, wetting angles of the carbonate with the electrode material, and pressure difference between the anode and cathode compartments. These factors can be tailored and controlled to give appropriate carbonate distribution within the fuel cell.

Carbonate electrolyte loss is the most important cause of long-term performance degradation of the MCFC; electrolyte loss is a slow but continuous process, mainly due to corrosion and volatilization. The corrosion loss is largely limited to the initial stage of operation, and the rate is relatively steep. Volatilization loss occurs via the fuel and/or oxidant gases, and its rate is relatively slow but constant. Another important electrolyte loss is via leakage through the wet seal in externally manifolded stacks (caused by creepage, galvanic corrosion cells, and migration). However, this type of loss can be minimized by engineering design. The loss of carbonate electrolyte is commonly controlled by providing an excess initial charge of electrolyte. Intermittent additions of electrolyte can also be used to compensate for carbonate loss.

2. Electrolyte Support or Matrix

The MCFC electrolyte support is a matrix of fine ceramic particles (an aggregate of fine particles without interparticle bonding) that supports and contains the molten carbonate mixture. It must meet two conditions. The support material must be chemically inert against molten carbonates, adjacent cell components, and reactant gases, and it must be morphologically stable. And because the electrolyte structure is in contact with the porous anode and cathode, the matrix must have narrow pore size range (typically 0.1–0.5 μm) to control the distribution of the carbonate in the various cell components.

The size, shape, and distribution of the particles control carbonate retention, mechanical properties, and effective ionic conductivity of the electrolyte structure. The MCFC electrolyte supports have always been made from finely divided submicrometer $LiAlO_2$ particles. The most effective shape for the $LiAlO_2$ particles appears to be long rods or fibers of submicrometer diameter. Three allotropic forms of $LiAlO_2$ (α, β, and γ) have been reported; the γ form is the most stable in the MCFC environment. $LiAlO_2$ has a low solubility in the molten carbonates; therefore, particle growth and phase transformation occur very slowly if at all. Since changes in particle size do affect carbonate electrolyte distribution, any particle growth can affect the electrolyte structure. However, this has not been observed in MCFC operation.

B. Cathode

The material used for MCFC cathodes is NiO, which is a p-type semiconductor. When it is used as the MCFC cathode, the material is doped in situ with lithium (lithiation) from the Li_2CO_3 in the molten electrolyte. The lithiation significantly increases the conductivity of NiO. With an equilibrium lithium dopant content in NiO of about 2 at. % at 650°C, the conductivity of the material is about 5 Ω^{-1} cm^{-1}.

The NiO cathode is fabricated from a porous nickel plaque, which is oxidized and lithiated in situ. The porosity of the plaque before oxidation is about 70–

80%, with a mean pore size of 6–10 μm. After oxidation and lithiation, the NiO cathode appears to consist of agglomerates of NiO particles with the carbonate electrolyte filling the space between individual particles (micropores). The agglomerates are separated by gas-filled spaces (macropores), which correspond to the pores of the original metal plaque. Because lithiated NiO is completely wetted with the carbonate, the agglomerates are covered with a thin film of electrolyte. Since gases diffuse through the film to react at the electrode, it is important to avoid flooding the entire electrode. The MCFC NiO cathode is sensitive to both carbonate electrolyte flooding and to the degree of filling. It operates best within 15–30% filling of the total pore volume (Fig. 14) [9].

The polarization of the MCFC cathode is higher than that of the anode. Cathode polarization is affected by ohmic losses in both the liquid (electrolyte) and solid (electrode) phases as the cathode thickness increases. Diffusional losses in the gas phase also increase with cathode thickness, while the activation and liquid-phase diffusion losses decrease. The optimum cathode thickness is attained when the total losses are at a minimum, which for NiO cathodes is at a thickness of 0.8 mm. Temperature, pressure, and gas composition further affect cathode polarization. Of particular interest is the effect of CO_2 partial pressure, since a CO_2 partial pressure as low as possible is desired to decrease the NiO solubility and increase cell lifetime (as discussed later). In general, cathode polarization is linear with current density at high to moderate CO_2/O_2 ratios. At CO_2 partial pressure below 0.03 atm, a marked increase in polarization occurs when the current density exceeds 100 mA/cm^2.

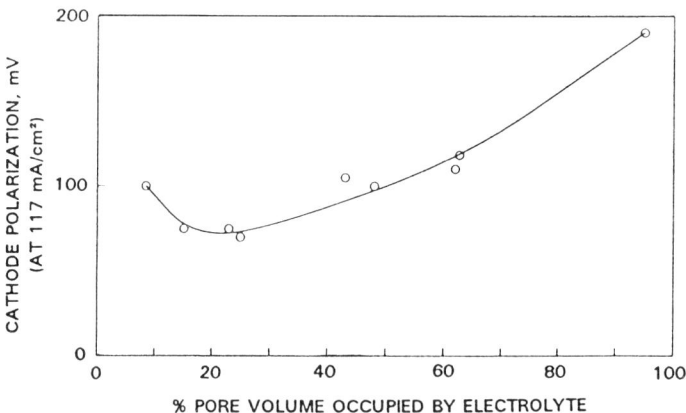

Figure 14 Effect of electrolyte filling on cathode polarization in O_2–CO_2 atmosphere (anode gas: 74.2% H_2–18.5% CO_2–7.3% H_2O, 650°C). (From Ref. 9.)

In the fuel cell cathode environment, NiO has a small degree of solubility in the carbonate electrolyte (about 10–15 ppm). However, the dissolved nickel ions diffuse, under a concentration gradient, from the cathode toward the anode. At some point between the electrodes and under the influence of reducing conditions caused by the anode gas, the dissolved nickel precipitates as nickel metal (Fig. 15) [10], creating a sink for the nickel ions, which facilitates further NiO dissolution. This can be a major life-limiting factor for the MCFC. The dissolution not only causes the loss of active material, it also leads to cell shorting.

The solubility of NiO is dependent on the basicity of the molten carbonate as defined by the following equilibrium:

$$CO_3^{2-} = O^{2-} + CO_2 \tag{25}$$

Depending on the carbonate basicity, the dissolution of NiO can be classified as acidic (low O^{2-}) and basic (high O^{2-}). In a relatively acidic electrolyte, NiO dissolves by dissociation:

$$NiO = Ni^{2+} + O^{2-} \tag{26}$$

In a basic electrolyte, NiO reacts with oxide ions to produce one of two forms of nickelate ions:

Figure 15 Micrograph of nickel deposit in MCFC electrolyte structure (cathode interface is at right). (From Ref. 10.)

$$NiO + O^{2-} = NiO_2^{2-} \tag{27}$$

or

$$2NiO + O^{2-} + \frac{1}{2}O_2 = 2NiO_2^- \tag{28}$$

Generally, the molten carbonate is relatively acidic, and the acidic dissolution of NiO is important. In the acidic dissolution region, making the carbonate electrolyte more basic decreases the solubility of NiO and thus is a means to reduce the NiO dissolution. Approaches to increase the electrolyte basicity, and thus lower NiO solubility, include use of low CO_2 partial pressure, high-Li_2CO_3 electrolyte, and electrolyte additives such as MgO or CaO.

C. Anode

The MCFC anode is 1 mm thick nickel with 60–70% porosity and a 5 µm median pore size. The anode is generally fabricated by sintering nickel powder. However, pure nickel also tends to sinter at fuel cell operating conditions, resulting in loss of surface area and pore growth. Additives have been used to control this sintering. For example, addition of a few percent chromium oxide to nickel has been shown to effectively prevent anode sintering by the formation of submicron $LiCrO_2$ on the nickel surface. The anode structure is also susceptible to deformation (creep) under compressive load during normal stack operation, i.e., shrinkage of anode thickness that occurs when cells are under load. The creep results in decreased porosity, increased contact resistance, and leaks. Oxide dispersion (e.g., Al_2O_3) can strengthen the nickel anode sufficiently to avoid creep.

The polarization of the nickel anode is small compared with that of the cathode, probably owing to the rapid kinetics of fuel oxidation. As a result, the nickel anode is relatively insensitive to the degree of filling by the electrolyte. This insensitivity is fortunate in that the anode can be used as an electrolyte reservoir, and thereby the long-term loss of electrolyte from corrosion and volatilization can be compensated for. Steam re-forming of methane and other light hydrocarbons is possible at the MCFC anode, because nickel is a catalyst for such re-forming reactions. The anode performance is very sensitive to sulfur and cannot tolerate more than 1–5 ppm of sulfur in the fuel gas without significant performance loss.

D. Separator

The separator provides cell-to-cell electrical contact in an MCFC stack. The material for the plate must be resistant to molten carbonate attack and compatible with other cell components. Stainless steel (SS) 310 is generally employed as a base metal for the separator plate due, in part, to its relatively low cost. The plates are coated with a thin film of molten carbonate and are in contact with the cathode and oxidant on one side and with the anode and fuel on the other side. The SS 310 has satisfactory corrosion resistance in the cathode environment, where a relatively stable and protective oxide scale forms in the presence of molten carbonate. However, this material is subject to hot corrosion in the reducing environment of the anode, this being due to oxide scale fluxing. The protective qualities of the scale are destroyed by the dissolution of the oxide in the carbonate and precipitation of the oxide elsewhere as a nonprotective porous mass. To minimize this effect, SS 310 plates are nickel-coated or -clad on the anode side for corrosion protection.

E. MCFC Stack and Assembly

The stack design is based exclusively on a planar geometrical configuration. It is an assembly of repeating unit cells in electrical series; each unit has a total thickness of about 5 mm. In addition to the unit cells and separators, MCFC stacks have bubble barriers and current collectors. The bubble barrier is a thin membrane positioned between the anode and the electrolyte to provide support and serve as a deterrent to gas crossover if cracks develop in the electrolyte structure. Nickel is the preferred material for the bubble barrier. It must contain pores that are filled with molten carbonate to provide continuous ionic transport. To be fully impregnated with carbonate, bubble barrier pores must be finer than the anode pores.

Current collectors are generally made of stainless steel (cathode) or nickel-plated steel (anode). The cathode current collector is a corrugated or perforated plate that also provides oxidant passages and cathode support. The anode current collector is a corrugated plate that is also used to provide fuel gas passages. In certain stack designs, the anode is ribbed to provide gas passages, and in this case, the separator plate serves as a current collector.

In an MCFC stack, cells can be supplied with reactant gases in crossflow, co-flow, or counterflow configurations. Gas manifolding can be either external or internal (Fig. 16) [11].

The MCFC electrolyte matrix and electrode sheets or tapes are made by tape casting. The method (a commonly used process in electronic and ceramic technology) involves forming a slip (or slurry) of powder dispersed in a solvent containing organic binders, plasticizers, and additives. The slip is spread on a flat surface to a controlled thickness using the knife edge of a doctor blade, and the solvent is allowed to dry to produce green (unfired) tapes. The organic binder in the tape is removed by heating to 400–500°C. (See Chapter 11, Sec. II.)

An MCFC stack is formed by stacking various components in proper order. The electrolyte is loaded into the electrolyte matrix and electrodes by adding

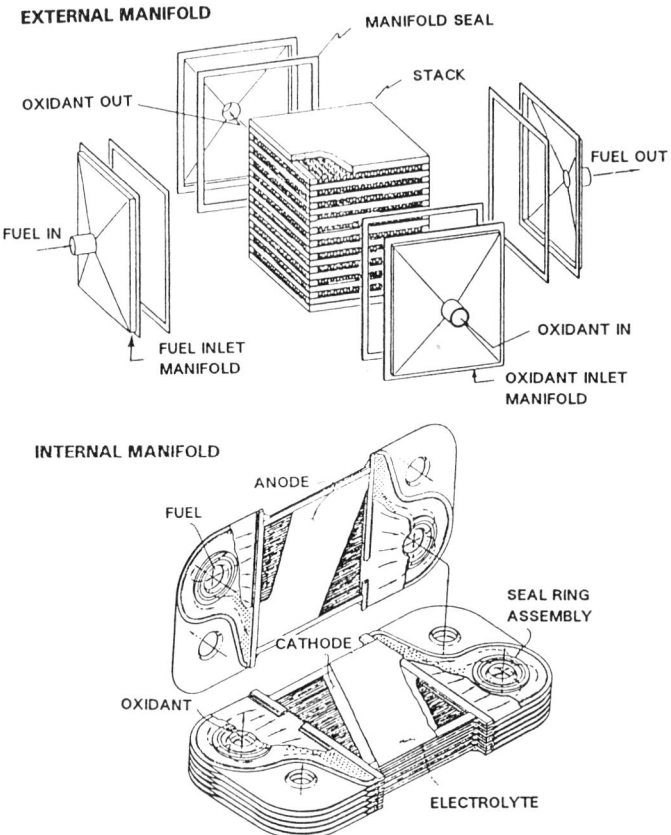

Figure 16 External manifold and internal manifold MCFC stacks. (From Ref. 11.)

carbonate in the gas passages, in the electrode tapes, or as one or more carbonate tapes. A constant compression is applied to the stack to minimize contact resistance between components.

VIII. SOLID OXIDE FUEL CELL

The solid oxide fuel cell (SOFC) is an all-ceramic device and uses exclusively oxide-ion conducting stabilized ZrO_2 as the electrolyte. The materials commonly used for the anode, cathode, and separator or interconnect are nickel/stabilized ZrO_2 cermet, doped $LaMnO_3$, and doped $LaCrO_3$, respectively. SOFCs are operated at about 1000°C and ambient pressure. Figure 17 shows examples of anode, electrolyte, cathode, and interconnect microstructures [12].

Fuel Cells

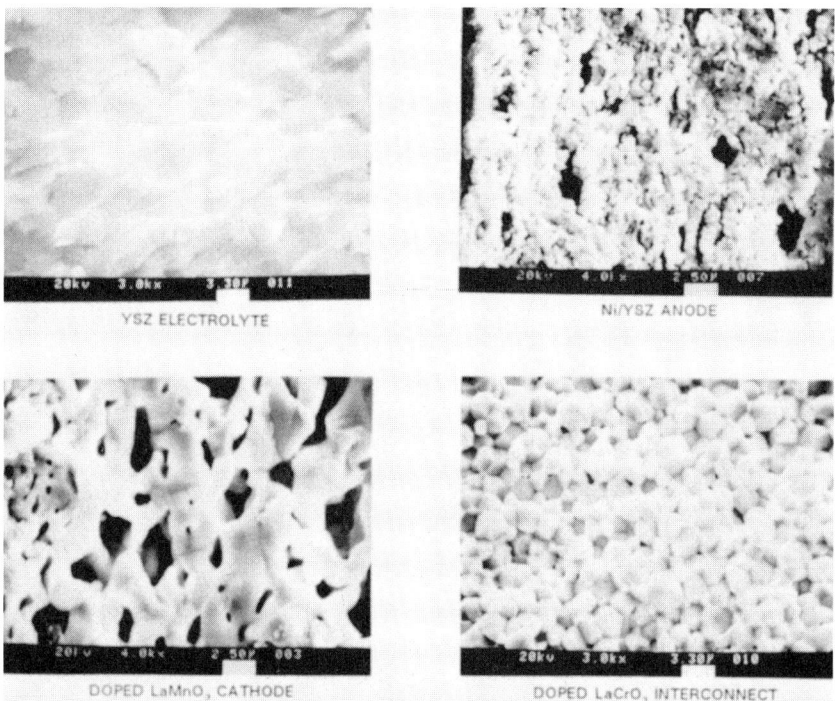

Figure 17 Microstructures of SOFC components. (From Ref. 12.)

A. Electrolyte

The electrolyte in a SOFC must have appropriate conductivity, proper stability (chemical, phase, morphological, and dimensional) in oxidizing and reducing environments, and chemical compatibility with other cell components at both operating and processing temperatures and atmospheres. The conductivity of the electrolyte has to be ionic over the range of oxygen pressures (1 to 10^{-18} atm) expected in operating fuel cells. The electrolyte must be dense to prevent mixing of the fuel and oxidant gases and, in addition, have thermal expansion that matches other components to avoid delamination or cracking of the ceramic components during fabrication and thermal cycling.

Stabilized ZrO_2, especially yttria-stabilized ZrO_2 (YSZ), possesses adequate oxide-ion conductivity in oxidizing and reducing atmospheres at 1000°C and has been used successfully as the electrolyte in SOFCs. ZrO_2, in its pure form, is not a good electrolyte, primarily because its ionic conductivity is too low.

Doping with lower-valence metal oxides such as Y_2O_3 stabilizes the high-temperature cubic fluorite structure of ZrO_2 and increases its oxygen vacancy concentration. This enhances ionic conductivity and leads to an extended oxygen partial pressure range of ionic conduction, making stabilized ZrO_2 suitable for use as an electrolyte in SOFCs. The ionic conductivity of Y_2O_3-stabilized ZrO_2 (8 mol % yttria) is about 0.1 Ω^{-1} cm^{-1} at 1000°C. The material has a thermal expansion coefficient of about 10.5×10^{-6} cm/cm K (from room temperature to 1000°C). Long-term behavior of stabilized ZrO_2 under fuel cell operating conditions has been extensively tested, and the required properties of the material have been shown not to be affected after long-term operation at 1000°C.

B. Cathode

Due to the high operating temperature of the SOFC, only noble metals and electronic conducting oxides can be used as cathode materials. However, noble metals such as platinum or palladium are unsuitable because of prohibitive cost, as are many oxides because of their thermal expansion mismatch with the zirconia electrolyte and their lack of conductivity. Among the suitable materials, strontium-doped lanthanum manganite (Sr-doped $LaMnO_3$) is most commonly used as SOFC cathode material.

Lanthanum manganite is a p-type perovskite oxide and shows reversible oxidation–reduction behavior. At high temperatures, the material can have oxygen excess, stoichiometry, or deficiency depending on oxygen partial pressure. For example, at 1200°C, the oxygen stoichiometry of $LaMnO_3$ ranges from 3.079 to 2.947 under oxygen partial pressures of 1 to $10^{-11.60}$ atm. $LaMnO_3$ begins to lose oxygen and experiences substantial decrease in conductivity at oxygen partial pressures $<10^{-12}$ atm (Fig. 18) [13]. In highly reducing conditions at 1000°C, $LaMnO_3$ decomposes into La_2O_3 and MnO. Thus, in the fabrication of the SOFC, it is desirable to process $LaMnO_3$ in oxidizing atmospheres.

The conductivity of $LaMnO_3$ can be enhanced by doping with a lower-valence ion such as strontium. The Sr-doped $LaMnO_3$ (10 mol % Sr) has a conductivity of about 130 Ω^{-1} cm^{-1} at 1000°C in air, sufficient for cathode material applications. The thermal expansion of this material is about 12.0×10^{-6} cm/cm K (from room temperature to 1000°C). Manganese in the $LaMnO_3$ is known to be a mobile species at high temperatures and can easily diffuse into the electrolyte and other cell components, changing the electrical characteristics or the structure of the cathode or the other components. However, at 1000°C, manganese migration is negligible, and there is no report on any significant manganese effects for cells operated up to tens of thousands of hours. At high temperatures, $LaMnO_3$ can react with YSZ to produce $La_2Zr_2O_7$. This reaction product is undesirable because its conductivity is about two and a half orders of magnitude lower than that of YSZ. Therefore, fabrication temperature is generally limited to below 1400°C to minimize this interaction.

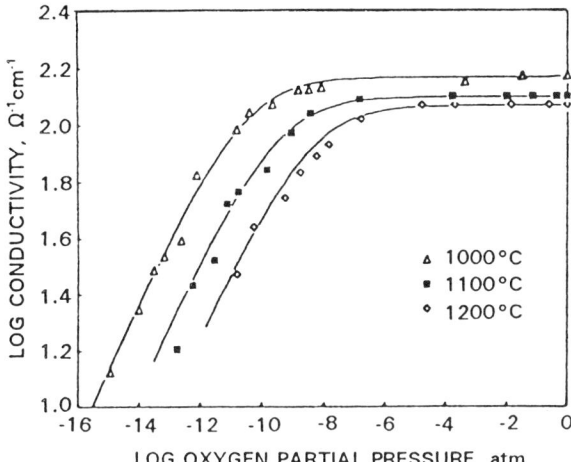

Figure 18 Conductivity as a function of oxygen partial pressure for $La_{0.8}Sr_{0.2}MnO_3$ at various temperatures. (From Ref. 13.)

In general, at the SOFC operating conditions, the oxygen reduction is slower than the hydrogen oxidation. However, the kinetics of the SOFC electrode reactions is sensitive to electrode morphology. Therefore, electrode polarization is very much dependent on how the electrode microstructure is fabricated.

C. Anode

Because of the reducing atmosphere of the fuel gas, metals can be used as anode electrode materials. Nickel and cobalt in particular have been found to be most suitable; nickel is commonly used because of its lower cost. The thermal expansion coefficient of nickel (about 14.6×10^{-6} cm/cm K) is considerably larger than that of the YSZ electrolyte (about 10.5×10^{-6} cm/cm K), and such significant thermal expansion mismatch between the anode and other components can cause delamination and cracking in the fuel cell. Also, nickel sinters at the cell operating temperature, leading to a decrease in active surface area and loss of porosity. These problems have been circumvented by incorporating YSZ particles in the anode to develop a skeleton of zirconia on which the metal is coated. The zirconia in the SOFC anode supports the fine nickel particles, inhibits coarsening of the metallic particles at the operating temperature, and provides an anode thermal expansion coefficient acceptably close to those of the other cell components. A compromise between conductivity and thermal expansion is required when the nickel loading in the anode is determined. Ni is needed above 30 vol % to maintain the required level of conductivity (300 Ω^{-1} cm^{-1} in a reducing atmosphere at 1000°C) while minimizing the degree of thermal expansion mismatch.

Nickel/YSZ cermet has been shown to possess sufficient catalytic activity to re-form hydrocarbons internally. The presence of sulfur-containing species in the fuel can have deleterious effects on performance of the anode, causing unacceptable loss of cell voltage. The exact sulfur tolerance of SOFC anodes at 1000°C is not well known (on the order of 10–100 ppm), although SOFCs are known to be more sulfur-tolerant than MCFCs.

D. Separator

The SOFC separator or interconnect must be stable in both oxidizing and reducing environments and must have as high an electronic conductivity as possible at 1000°C. The interconnect must also have low porosity to prevent gas cross-leaking between the electrodes. These stringent requirements eliminate all but a few metals and oxide systems from consideration. Doped $LaCrO_3$ has been commonly used as the interconnect, as it possesses adequate stability in the fuel cell environment, reasonable compatibility with other components, and good electrical conductivity. Various substitutions in the lanthanum chromite improve the thermal expansion match and conductivity under SOFC operating conditions.

$LaCrO_3$ is a p-type conductor due to holes in the $3d$ bond of the chromium ions. Under oxidizing conditions, substitution of a lower-valence ion on either the lanthanum or chromium sites of $LaCrO_3$ results in a charge-compensating transition of Cr^{3+} to Cr^{4+} ions, thereby enhancing the electronic conductivity of the material. For example, $LaCrO_3$ doped with 10 mol % Sr on the lanthanum site has a conductivity of about 25 Ω^{-1} cm^{-1}, and $LaCrO_3$ doped with 10 mol % Mg on the chromium site has a conductivity of about 10 Ω^{-1} cm^{-1} at 1000°C in air, compared with a conductivity of about 0.5 Ω^{-1} cm^{-1} for undoped $LaCrO_3$. The conductivity of doped $LaCrO_3$ is reduced about tenfold when the oxygen pressure is reduced from atmospheric pressure to the fuel condition. However, $LaCrO_3$ exposed to fuel on one side and to air on the other side has high enough conductivity to be used as the interconnect material in SOFCs. The thermal expansion coefficient of doped $LaCrO_3$ depends on the type and amount of dopant present and can be tailored to match those of zirconia. For example, the thermal expansion coefficient of strontium-doped material is about 10 to 11 × 10^{-6} cm/cm K from room temperature to 1000°C, matching that of the YSZ electrolyte.

$LaCrO_3$, like all chromium oxide compounds, does not densify at temperatures below 1700°C and oxygen activities above 10^{-9} atm (Fig. 19) [14]. The reason for the low sinterability is that $LaCrO_3$ appreciably volatilizes chromium oxides in oxidizing atmospheres, and because of the volatilization, the predominant mass transport during firing is an evaporation/condensation mechanism,

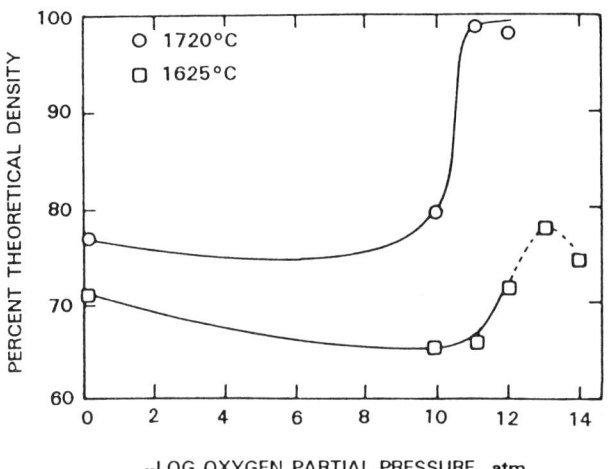

Figure 19 Sintered density of $La_{0.84}Sr_{0.16}CrO_3$ as function of oxygen partial pressure during sintering. (From Ref. 14.)

which leads to coarsening of the original particles without densification. Thus, for preparing dense $LaCrO_3$ interconnect, especially dense $LaCrO_3$ along with other cell components, it is necessary to develop methods to sinter the material below 1400°C in a relatively oxidizing atmosphere like air. Several approaches have been investigated to enhance the sinterability of $LaCrO_3$ in oxidizing atmospheres. These approaches involve use of highly reactive powders, nonstoichiometric materials, dopants, sintering aids, and processing techniques.

E. SOFC Stack and Assembly

In the SOFC, where the components are all solid, cells and stacks can be designed and fabricated into configurations that may not be feasible with cells employing liquid electrolytes. Thus, SOFC stack designs can be different from the flat-plate type commonly used in other fuel cell systems. For other fuel cells, the components are fabricated separately and then assembled to form a cell, and the process conditions for the fabrication of each component can be selected independently. This cannot be done for the SOFC, because if the components are built up one by one, the conditions for each successive layer must be tailored to avoid altering the properties of the preceding layers. If all the components are formed together in the green (unfired) state, then all the components must be sintered under the same conditions.

Four stack configurations have been proposed and fabricated [15]: the sealless tubular design, the segmented-cell-in-series design, the monolithic design, and the flat-plate design (Fig. 20).

1. *Sealless tubular design*: The cell components are configured as thin layers on a tubular support closed at one end. For cell operation, oxidant is introduced to the cell through an injector tube near the closed end, where it traverses and exits from the open annulus between the support tube and the oxidant injector tube. Fuel flows on the outside of the support tube. Tubular tubes are bundled in series and parallel electrical connection to form a stack.

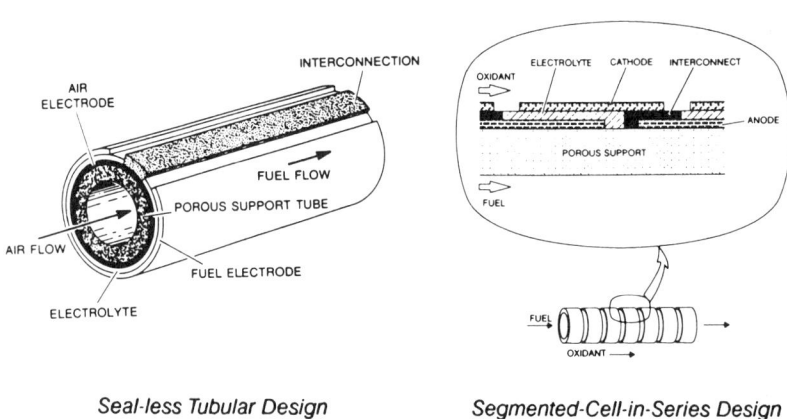

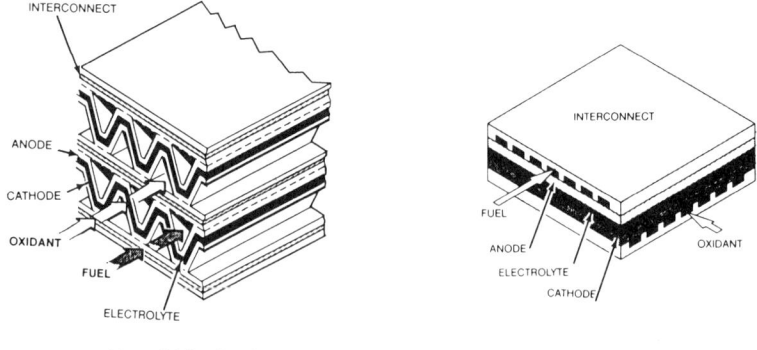

Figure 20 SOFC designs. (From Ref. 12.)

2. *Segmented-cell-in-series design*: The design consists of segmented cells connected serially in terms of electrical functions and gas flow, commonly arranged as a thin-banded structure on a porous support tube. In this design, the fuel flows from one cell to the next inside the support tube, and the oxidant flows outside. Tubes are connected to form a module for practical power generation.
3. *Monolithic design*: The monolithic configuration consists of thin cell components formed into a compact corrugated structure of gas coflow configuration. In this design, fuel and oxidant flow parallel in adjacent channels formed by corrugated anode/electrolyte/cathode and flat anode/interconnect/cathode laminates.
4. *Flat-plate design*: This design, common in other types of fuel cells, consists of a single cell and interconnect configured as thin, flat plates. The interconnect, often having ribs on both sides, forms gas channels. In the most common flow geometry, fuel flows in one face of the fuel cell stack and out the opposite face, as shown in Fig. 20. Similarly, oxidant flows in one face and out the opposite face.

The various SOFC designs differ in the extent of dissipative losses within the cells, in the manner of sealing between fuel and oxidant channels, and in the cell-to-cell electrical connections in a stack of cells. The ease of fabrication and assembly also varies among the designs.

The fabrication process selected for each SOFC design depends on the configuration of the cells within the stack. The sealless tubular design involves sequential forming of thin components on a porous support. Electrochemical vapor deposition (EVD) has been developed to produce the required dense oxide layers on the porous support tube (at elevated temperatures and reduced pressures). Cell components of the segmented-cell-in-series design are made by a process based on plasma spraying. Masking is used to make the required pattern on the support tube. The monolithic design involves forming the fuel cell in the green (unfired) state and co-firing the structure at elevated temperatures in a single step. Tape calendering (involving formation of a green ceramic tape by squeezing a softened thermoplastic ceramic powder/organic binder mix between two rolls) has been used to produce green multilayer tapes for the monolithic SOFC. The most common fabrication technique for the flat-plate design is tape casting. Fuel cell stacks are formed by stacking tape-cast single-cell and interconnect plates to the desired height.

IX. SOLID POLYMER FUEL CELL

The solid polymer fuel cell (SPFC) consists of a proton exchange membrane electrolyte such as a perfluorosulfonic acid polymer and carbon-supported platinum porous anode and cathode. Graphite plates having grooves (to distribute gas over the electrodes) are commonly used as bipolar separators in an SPFC stack. The SPFC operates at a temperature of about 80°C and a pressure of about 1–5 atm.

A. Electrolyte

The SPFC electrolyte is a solid polymer membrane, mainly made of perfluorosulfonic acid materials. Two types of membrane, often referred to as the DuPont membrane, or Nafion (registered trademark of E. I. DuPont de Nemours), and the Dow membrane, have been considered for use as SPFC electrolyte. The chemical structures of the DuPont and Dow membranes are shown in Fig. 21.

Typical Nafion membranes have an equivalent molecular weight in the range of 1100–1350. As shown in Fig. 21, the polymer consists mainly of a PTFE backbone and a perfluorinated vinyl polyether side chain. The chain carries a terminal sulfonic acid group that provides the proton exchange capacity. Nafion exhibits exceptionally high thermal and chemical stability; the material is resistant to chemical attack by strong bases, strong oxidizing and reducing acids, and other chemicals at temperatures up to 125°C. The polymer can absorb water into its molecular structure, promoting a high degree of proton dissociation and, thus, enhancing the material ionic conductivity. The conductivities of Nafion (equivalent weight = 1100) are about 0.065 Ω^{-1} cm^{-1} at 30°C (saturated water vapor) and 0.17 Ω^{-1} cm^{-1} at 80°C (immersed in water), and they depend on the water content. Consequently, operating temperatures are often limited to below 100°C to prevent dehydration of the membrane and the attending loss of conductivity.

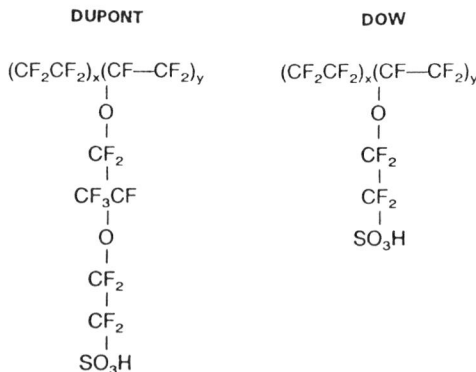

Figure 21 Structural characteristics of DuPont and Dow membranes.

The Dow membrane has a lower equivalent weight (550–950) and a shorter side chain than Nafion. As a result, this polymer has more sulfonic acid groups per CF_2 and thus better conductivity and water retention capability. The conductivities of Dow membrane (equivalent weight = 800) are 0.12 Ω^{-1} cm^{-1} at 30°C (saturated water vapor) and 0.23 Ω^{-1} cm^{-1} at 80°C (immersed in water). Cells constructed with Dow membranes have been reported to achieve higher performance than that obtained with Nafion.

During the operation of the SPFC, water is transported through the membrane with hydrogen ions because of the electroosmotic effect. About 1 to 2 water molecules are dragged with each proton, and the water drag increases at high current densities. The transport of substantial amounts of water through the membrane presents a water-management problem: there is the tendency for the anode to dry, with resulting conductivity loss, and for the cathode to flood with pore blocking. The water management in the membrane electrolyte and electrode assembly of the SPFC is complex and requires dynamic control to match the varying operating conditions of the fuel cell. The most common approach is to humidify the incoming reactant gases and to operate the cell with a different pressure, using a higher pressure on the cathode side.

B. Cathode

The SPFC cathode consists of two layers: a catalyst layer and an electrode backing (carbon cloth or paper). The catalyst layer is a gas-diffusion electrode made of Pt supported on carbon. A polymeric material such as PTFE is frequently added to serve as a binder for carbon-supported Pt particles. Since the SPFC operates at low temperatures, platinum is the most active and preferred catalyst for the oxygen reduction, the cathode typically containing about 10–20% Pt. The Pt loading used in the cathode may vary, depending on the electrode structure. For example, cathodes using PTFE binders require a Pt loading of about 4 mg/cm^2 to exhibit adequate performance. This loading can be reduced to 0.4 mg/cm^2 when the cathode is impregnated with a proton conductor such as dissolved Nafion. The improved performance at this low loading has been ascribed to increased contact area between the electrolyte and the Pt particles, resulting in an increased reactive area and thus a considerable improvement in oxygen reduction kinetics. The cathode is generally bonded to the surface of the electrolyte membrane by hot pressing at a temperature usually between the glass transition and the thermal degradation temperatures of the membrane. These pressing conditions provide an intimate contact at the cathode/electrolyte membrane interface.

The oxygen reduction at the SPFC cathode is considerably slower than the hydrogen oxidation. At 100°C, typical cathode polarization at about 200 mA/cm^2 is about 250 mV. An increase in oxygen partial pressure has been found to significantly reduce cathode polarization. The cathode, when operating on air at high current densities, shows sharp performance drop, indicating a significant effect of the gas transport limitation. Thus, pressurized operation (up to 5 atm) is often required to attain high power densities in the SPFC when air is used as the oxidant.

C. Anode

The SPFC anode, like the cathode, is composed of a carbon-supported Pt electrode and a carbon backing. At the 80°C operating temperature, the anode is insensitive to CO_2 (up to several percent), but very sensitive to CO poisoning because of the adsorption of CO on active Pt sites. The presence of CO at the 100 ppm level in the fuel causes severe loss in cell performance, and even low levels of CO (10 ppm) have a significant effect. Lowering high levels of CO in the fuel (to about 10 ppm) can be achieved in a selective oxidation process based on a Pt/Al_2O_3 catalyst. Several approaches have been considered to reduce the poisoning effect of low CO levels. One approach is to use alloy catalysts, e.g., Pt/Sn or Pt/Ru. Another approach is to bleed a small amount of oxygen or air into the anode to oxidize adsorbed CO to CO_2. Figure 22 shows that a performance equivalent to that obtained on pure hydrogen can be achieved by this approach [16]. Operating on high CO_2 (e.g., 75% H_2 + 25% CO_2), SPFC anodes of high Pt loadings suffer significant performance losses. Anodes based

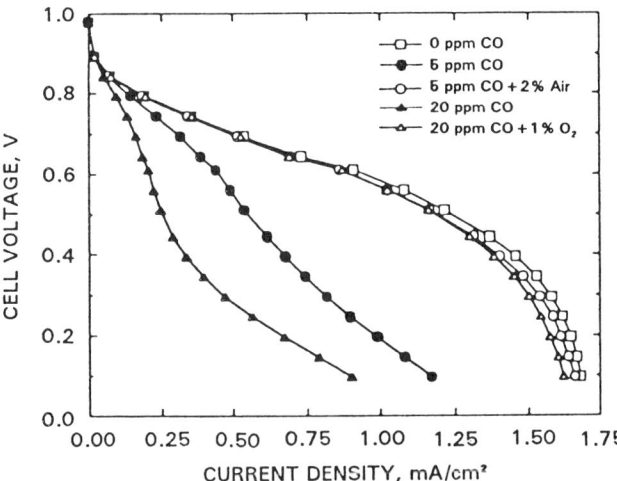

Figure 22 Polarization curves obtained for SPFC cells operating on fuel gases containing CO without and with air or oxygen bleed (0.14 mg/cm² Pt loading, 80°C). (From Ref. 16.)

on Pt/Ru alloys, on the other hand, exhibit little polarization, even when operating on hydrogen with 25% CO_2 (Fig. 23).

Methanol is a preferred fuel for the SPFC in transportation applications. However, methanol cannot be used directly in the SPFC, because under typical operating conditions methanol poisons the anode reaction and causes high polarization losses. The alcohol also causes significant crossover current due to its high solubility in the polymer membrane. Methanol is often steam re-formed and passes through a catalyst to remove CO before being fed to the fuel cell.

D. Separator

The SPFC separator is commonly made of graphite. Grooves may be machined into graphite separator plates to provide gas flow channels.

E. SPFC Stack and Assembly

The SPFC stacks consist of single cells (including electrode backings) and separator plates connected in series. The separator has grooves on both sides to serve as passages for fuel and oxidant gases. Cooling is accomplished by using a heat transfer fluid, usually water, which is pumped through integrated coolers within the stack. Figure 24 is a photograph of a 5-kW SPFC stack containing 35 cells, each 232 cm^2 in area.

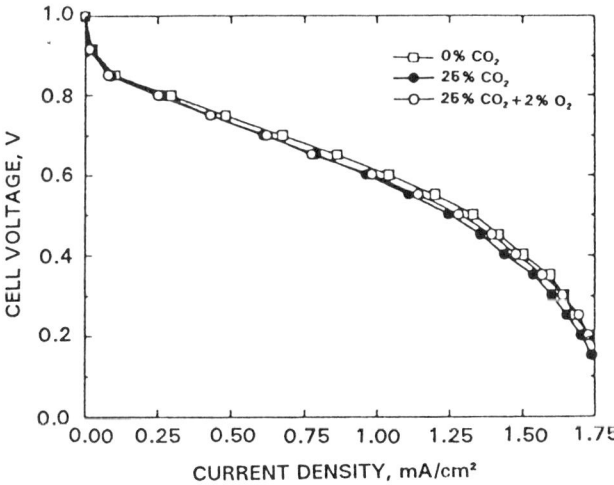

Figure 23 Polarization curves for SPFC cells (with Pt/Ru anodes) operating on fuel gases containing CO_2 (0.19 mg/cm^2 Pt/Ru loading, 80°C). (From Ref. 16.)

Figure 24 5-kW SPFC stack (35 cells). (From Ref. 3.)

A typical SPFC stack (consisting of the membrane electrolyte with electrodes attached and separator, cooling plates, and end plates) is assembled in a plate and frame or filter press arrangement. The electrodes are fabricated by painting or spraying required materials followed by sintering. To achieve intimate contact, the electrodes are generally bonded to the surface of the electrolyte membrane by hot pressing at a temperature that is usually between the glass transition and the thermal degradation temperatures of the membrane.

X. ALKALINE FUEL CELL

The alkaline fuel cell (AFC) uses an aqueous KOH solution (typically, with normalities of 6–12) as the electrolyte. Unlike the other types of fuel cell, electrode materials used in the AFC vary depending on the design and operating conditions. The common cathode catalysts are Pt, Ag, and Pt/Au, and the common anode catalysts are Pt, Pd/Rh, and Ni. The AFC operates at a temperature of 60–100°C (typically 70°C) and a pressure of 1–10 atm (typically 1 atm).

A. Electrolyte

KOH has been selected over other alkaline hydroxides because of its high conductivity and extremely low vapor pressure. However, at the fuel cell operating conditions, KOH solution is susceptible to CO_2 poisoning. Thus, an efficient AFC operation requires the complete removal of CO_2 from the fuel and oxidant gases. The presence of CO_2 in the gas streams forms a carbonate:

$$CO_2 + 2OH^- = CO_3^{2-} + H_2O \tag{29}$$

The carbonate reduces the hydroxide concentration and electrolyte conductivity and eventually causes precipitation of carbonate salts in the pores of the porous electrodes.

In the AFC, the KOH solution electrolyte is either immobile or mobile. Immobile electrolytes are usually contained in a matrix such as asbestos or potassium titanate. Mobile electrolytes are pumped through the cell. During operation, the water product has to be constantly removed from the electrolyte so that the KOH concentration remains within the desired range. In the immobile electrolyte arrangement, the cells must contain an electrolyte reservoir plate to compensate for dilution during operation. (This reservoir plate is a porous nickel or graphite sheet in contact with the anode.) In the mobile electrolyte arrangement, water is removed by using a given gas circulation rate to entrain water vapor into the gas outlet stream. The dilute electrolyte can also be pumped through an electrolyte regenerator or evaporator to restore it to its normal concentration.

B. Cathode

Oxygen reduction kinetics in the alkaline electrolyte is relatively rapid; consequently, the use of nonnoble metal catalysts is possible. In addition to noble metals, various metal oxides (e.g., Fe- and Co-based perovskites), macrocycles (porphyrins, phthalocyanines), and even carbon are electrocatalysts for oxygen reduction in KOH electrolytes. However, the most common electrocatalysts for the cathode are Pt, Ag, and Pt/Au. Dopants may be used in the catalyst as sintering inhibitors; an example is titanium-doped Ag. These catalyst particles are often deposited on carbon supports or are part of metallic electrodes (current collectors), generally based on nickel substrates.

When the catalyst is part of the metallic electrode, the catalyst particles are usually mixed with a binder and other ingredients and pressed into a supporting nickel screen. When the catalyst is supported on carbon, supported catalyst particles are often bonded with PTFE. This latter cathode may be designed to be a multilayer structure. For example, the cathode may consist of a backing layer, a gas diffusion layer, and a catalyst layer. The backing layer is a carbon cloth or paper with about 50 wt % PTFE. The high PTFE content ensures high gas permeability, good structural strength, and corrosion resistance, while maintaining a reasonable electrical conductivity. The diffusion layer is made of carbon and about 30–50 wt % PTFE to provide an adequate rate of gas diffusion. The catalyst layer contains C-supported catalysts with 10–20 wt % PTFE. The low PTFE content in the catalyst layer ensures that the catalyst particles are only partially wetted, and a stable three-phase contact area (electrode/electrolyte/gas) is formed.

The AFC cathode performs better in oxygen than in air, and as expected, the electrode performance increases with increasing temperature and pressure. The presence of CO_2 in the oxidant increases cathode polarization, leading to degradation in cell performance (Fig. 25) [17]. Despite the relatively low operating temperature of the AFC, catalyst sintering may occur over long-term operation, although it is usually minor.

C. Anode

The common anode catalysts are Pt, Pd/Rh, and Ni. Dopants can be added to the anode catalyst to inhibit sintering (e.g., titanium-doped Ni). Like the cathode, the anode structure is either a catalyst on a metallic electrode, or it is a PTFE-bonded, C-supported catalyst. In general, the AFC anode polarization is relatively small. The presence of CO_2, CO, and sulfur are again known to have a detrimental effect on the anode performance, and these impurities must be removed from the fuel (down to the ppm level) to minimize the poisoning effect.

D. Separator

Several materials have been used as separator for the AFC. Examples are nickel, a composite of 30% carbon black + 70% propylene, and conducting plastics.

E. AFC Stack and Assembly

As in other types of fuel cells, stacks in the AFC can use bipolar separators to connect the cells in electrical series. However, several designs use monopolar

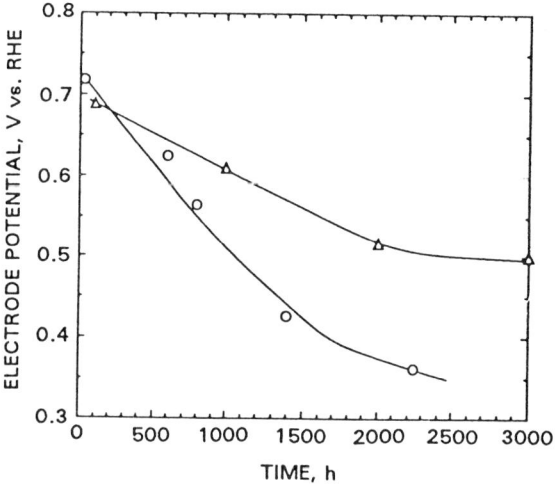

Figure 25 Effect of CO_2 on the air performance of carbon-supported platinum cathodes (0.2 mg/cm² Pt loading, 6 N KOH solution, 50°C). ○: CO_2-containing air; △: CO_2-free air. (From Ref. 17.)

arrangements with edge current collection. In the case of immobile electrolytes, cooling compartments are included in the stack design to remove by-product heat. In mobile electrolytes, the by-product heat is removed with the circulating electrolyte.

A variety of methods can be used to fabricate AFC electrodes. Electrode material powders are usually mixed and then pressed or calendered into layers. Sedimentation as well as spraying techniques followed by sintering can also be used. Assembling an AFC stack consists of two steps: framing the electrodes and other components, and then integrating the frames. The framing can be done by injection molding, and the frames can be pressed together by filter pressing or welding.

XI. SUMMARY

Fuel cells are a different way of making electrical power from a variety of fuels. Fuel cells allow the clean and efficient use of fossil fuels and provide several important operational and economical advantages. Commercial-size fuel cell systems have been operated for thousands of hours and have demonstrated desired performance characteristics with very high reliability. The interest in fuel cell technology can only widen as there is a continuing need to develop cleaner and more efficient means of converting energy sources into useful forms. As fuel cell technology becomes more mature and commercial applications increase, costs are expected to fall. Each fuel cell type, with its inherent properties and operation features, will have its own market niche. In the future, fuel cells will be an integral element in electric power generation from variety of fuels.

EXERCISES

1. A regenerative fuel cell is a fuel cell system in which the fuel cell products (e.g., water in a hydrogen/oxygen fuel cell) are reconverted into reactants (e.g., hydrogen and oxygen in a hydrogen/oxygen cell) by a thermal, chemical, photochemical, or electrical method. Consider the thermally regenerative fuel cell schematically shown in Fig. 26. In this system, A and B are the fuel cell reactants; AB is the product; T_2 is the low temperature at which the fuel cell reaction is carried out; and T_1 is the high temperature at which AB is decomposed. The operation of the system involves four stages: fuel cell reaction (stage I, temperature T_2); increase of temperature of fuel cell products (stage II, temperature $T_2 \rightarrow$ temperature T_1); regeneration of reactants (stage III, temperature T_1); and decrease of temperature of fuel cell reactants (stage IV, temperature $T_1 \rightarrow$ temperature T_2). Assuming that the reactions are carried out reversibly, isothermally, and at a constant pressure,

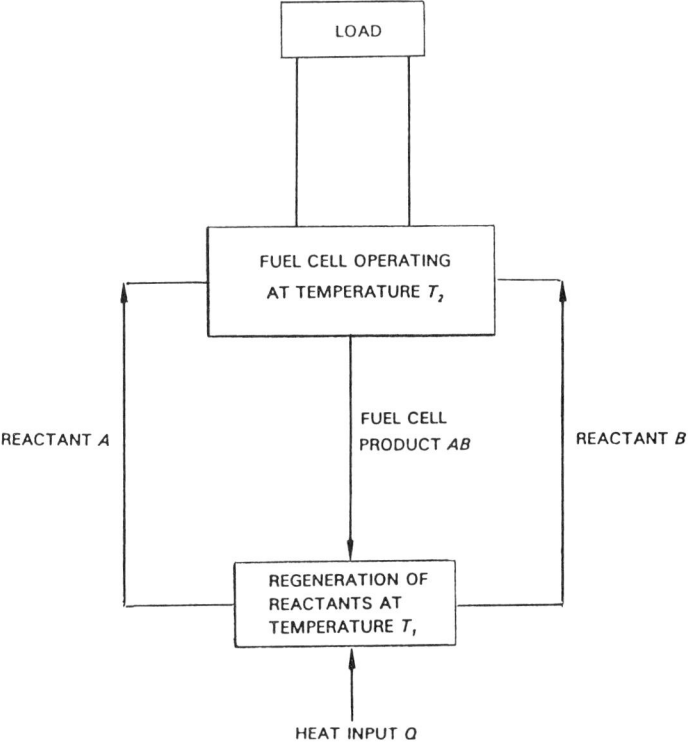

Figure 26

derive a general equation for the efficiency of the system (from useful work outputs and heat inputs) in terms of temperature and specific heat.

2. Consider the heat balance for the fuel cell system given in Fig. 27. This fuel cell is operated at voltage V with an electrochemical efficiency of ε_c (ε_c is the product of thermodynamic efficiency ε_T, the voltage efficiency ε_V and the current efficiency or fuel utilization ε_J). From the heat rejected (due to heat of reaction and heat of resistance), calculate the heat release per mole of fuel in the feed gas (df/dt is the molar flow rate of the feed gas).

3. Write down balanced electrode reactions and the overall cell reactions for phosphoric acid (PAFC), molten carbonate (MCFC), solid oxide (SOFC), solid polymer (SPFC), and alkaline (AFC) fuel cells. Derive the Nernst equation for each fuel cell. Typical operating temperatures of the PAFC, MCFC, SOFC, SPFC, and AFC are 200, 650, 1000, 80, and 100°C, respectively. Use data given in JANAF Thermochemical Tables to calculate the thermodynamic voltage of each fuel cell for the following fuel and oxidant compositions:

Fuel Cells

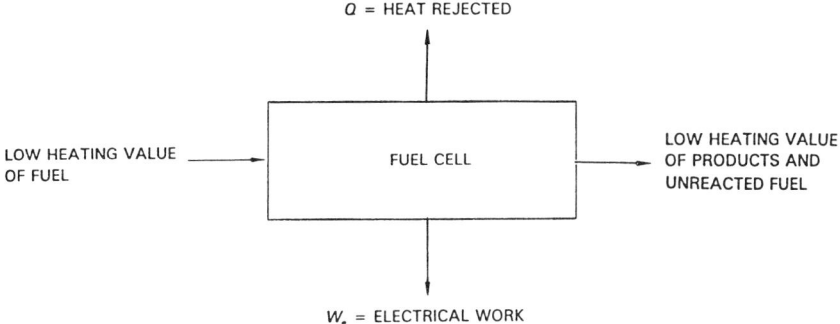

Figure 27

	Fuel	Oxidant
PAFC	64% H_2, 20% H_2O, 16% CO_2	94% air, 6% H_2O
MCFC	19% H_2, 8% H_2O, 12% CO_2, 14% CO, 47% N_2	70% air, 30% CO_2
SOFC	97% H_2, 3% H_2O	air
SPFC	100% H_2	97% air, 3% H_2O
AFC	97% H_2, 3% H_2O	97% O_2, 3% H_2O

4. The following data are available for the anode and cathode reactions in an alkaline fuel cell:

	E^0, V (vs. Ref. Elect.)	j_0, A/cm²	Tafel slope, mV⁻¹
Anode	−0.83	3×10^{-2}	1/40
Cathode	0.40	3×10^{-8}	1/60

Assuming an ohmic drop through the electrolyte and electrodes of 50 mV, estimate the cell voltage for a current density of 300 mA/cm². If the fuel used is hydrogen, calculate the energy production (kWh) per kg of fuel.

5. Consider a fuel cell operated under conditions in which the current/potential relation at each electrode is in the Tafel region (the mass transfer and ohmic polarizations are assumed to be negligible). Derive an equation for maximum power in terms of the exchange current densities and transfer coefficients. Simplify the equation assuming that activation polarization is

predominant at the cathode and discuss the influence of exchange current density on the maximum power.

REFERENCES

1. *Fuel Cells*, Report DOE/METC-85/0223, U.S. Department of Energy, Morgantown Energy Technology Center, Morgantown, West Virginia, 1984.
2. Proceedings of the Fourth Annual Fuel Cells Contractors Review Meeting, Report DOE/METC-92/6127, U.S. Department of Energy, Morgantown Energy Technology Center, Morgantown, West Virginia, 1992, p. 5.
3. D. S. Watkins, in *Fuel Cell Systems* (L. J. M. J. Blomen and M. N. Mugerwa, eds.), Plenum Press, New York, 1993, p. 493.
4. W. L. Lundberg, in Proceedings of the 25th IECEC, Vol. 3, American Institute of Chemical Engineers, New York, 1990, p. 218.
5. R. Anahara, in *Fuel Cell Systems* (L. J. M. J. Blomen and M. N. Mugerwa, eds.), Plenum Press, New York, 1993, p. 271.
6. K. Kinoshita, *Electrochemical Oxygen Technology*, Wiley-Interscience, New York, 1992.
7. W. M. Vogel and K. A. Klinedinst, *Electrochim. Acta 22*:1385 (1977).
8. *Fuel Cells*, Report DOE/METC-89/0266, U.S. Department of Energy, Morgantown Energy Technology Center, Morgantown, West Virginia, 1988.
9. J. R. Selman and L. G. Marianowski, in *Molten Salt Technology* (D. G. Lovering, ed.), Plenum Press, New York, 1982, p. 323.
10. N. Q. Minh, *J. Power Sources 24*:1 (1988).
11. General Electric, *Development of Molten Carbonate Fuel Cell Power Plant*, Report DOE/ET/17019-20 (Vol. 2), U.S. Department of Energy, Washington, D.C. 1985.
12. N. Q. Minh, in *Science and Technology of Zirconia V* (S. P. S. Badwal, M. J. Bannister, and R. H. J. Hannink, eds.), Technomic, Lancaster, Pennsylvania, 1993, p. 652.
13. J. H. Kuo, H. U. Anderson, and D. M. Sparlin, *J. Solid State Chem. 87*:55 (1990).
14. L. Groupp and H. U. Anderson, *J. Am. Ceram. Soc. 59*:449 (1976).
15. N. Q. Minh, *J. Am. Ceram. Soc. 76*:563 (1993).
16. M. S. Wilson, C. R. Derouin, J. A. Valerio, and S. Gottesfeld, in Proceedings of the 28th IECEC, Vol. 1, American Chemical Society, Washington, D.C., 1993, p. 1.1203.
17. K. Kordesch, J. Gsellmann, and B. Kraetschmer, in *Power Sources 9* (J. Thompson, ed.), Academic Press, New York, 1983, p. 379.

Index

Accelerometers, 31
Adhesion (*see also* Wafer priming), 205, 497
Adhesion failure, 77
Adhesion promoter, 212, 428
Adsorption, 99, 112, 113, 134, 478
Aluminum, 185–188
Aluminum corrosion, 77
Aluminum films
   aluminum hydride precursor, 188
   beta-hydride elimination, 186
   carbon contamination, 187
   plasma etching with CCl_4, 195
   reaction with silicon, 190
   triisobutylaluminum precursor, 186
Amorphous film growth
   boundary layer, 140
   crystallite, 160
   diffusive transport, 139
   kinetics of, 138
   Knudsen number, 139
   metastable, 131
   Peclet number, 140
   properties of, 164

[Amorphous film growth]
   sticking coefficient, 139
   tetrahedral bonding in, 134
   thermodynamics of, 131
Amorphous films
   a-C, 131, 164–166
   a-C:H, 128, 130, 131
   a-G:H, 129, 130
   a-Ge:H, 130, 131, 164–166
   a-Si:H, 14, 128, 130, 131, 164–166
   a-SiGe:H, 131
   applications, 129–131
   bandgap structure of, 165–167
   diamondlike, 131
   hydrogen binding energy of, 136
   networks, 137
   solar cells, 131, 166
   Urbach energy, 137
Amorphous thin film transistor, 15
Anisotropic
   etching, 285
   profiles, 253
Anisotropy, 446

Antireflective layers
  amorphous silicon as, 217
  applications of, 215
  bottom layers (BARL), 216
  organic BARL, 217
  silicon trinitride as, 217
  top layers (TARL), 216
Application-specific integrated circuits, 4
Aspect ratio, 180, 291
Atomic layer etching, 304–306
Auger electron spectroscopy (AES), 312
  depth profiling, 40
  display formats, 398
  electron beam density effects, 401
  fluorine desorption, 400
  lateral resolution, 394
  material considerations, 400
  titanium nitride spectra, 398–400
  titanium silicide line spectrum, 398
Autodoping, 109

Bandgap, 2
Bardeen, Cooper, Schrieffer theory, 512, 521
BARL (bottom antireflective layers)
  amorphous silicone as, 217
  organic BARL, 217
  purpose, 216
  silicon trinitride as, 217
Bathtub curve, 71, 341
Beta-spodumene, 492
Binding energy, 134, 136, 403
Bipolar transistor, 9
Bleaching, 207, 229–330
Boundary layer, 107, 108, 140–141

Ceramic chemistry
  carbon oxidation, 496
  greensheet compatibility, 476
  high temperature glass chemistry, 494
  kinetics and thermodynamics of, 493–496

[Ceramic chemistry]
  Lewis acid-base, 475
  metal paste-greensheet interaction, 480
  paste solvent/binder interactions, 482–484
  polymer binder interaction with SiO_2, 479
  polymer solubility parameters, 482–484
Ceramic materials, 461
Ceramics
  conductive paste, 472
  desirable properties, 462
  differences between glasses and, 462
  glass ceramic system, 491–493
  greensheets, 471
  interconnecting metals, 467
  liquid-phase glass system, 489
  metal-ceramic adhesion, 497
  packaging technology, 463–468
  particle surface chemistry, 474–476
  polymer binder, 479
  screenable paste, 472
Chemical analysis, 372, 393
Chemical vapor deposition, 86, 127, 538–539
  binding energy of hydrogen, 134, 136
  boundary layer, 107, 108, 140–141
  dangling bonds, 134
  definition, 127
  diffusive transport in, 139
  doping, 155–156
  effect of growth rate, 138
  free molecular growth, 139
  Gibbs free energy, 132
  hydrogen adsorption, 112, 113, 134
  hydrogen partial pressure effect, 135
  kinetics, 138
  Knudsen number, 139
  laser doping, 156–158
  mass transport limited, 138

Index

[Chemical vapor deposition]
  Peclet number, 140
  photolysis, 133, 146
  pyrolysis, 133
  rate determining, 138
  reactors, 141–147
  sticking coefficient, 139
  transport, 139
  types of films, 131
Chemical vapor deposition films
  a-C:H, 152
  a-Ge:H, 155
  a-Si:H, 153–155
  aluminum, 185–188
  amorphous hydrogenated, 128–131, 152
  applications, 129–131
  copper, 189–190
  desired characteristics, 128
  diamond, 130, 147–149
  effect of hydrogen partial pressure, 135
  energy requirements, 133
  GaAs, 89, 110, 115
  metals, 173–174
  polycrystalline silicon, 146–147, 173
  silicides, 173–174
  silicon epitaxy, 90
  Si_3N_4, 130, 151–152, 164
  SiO_2, 130, 149–151, 163–164
  titanium nitride, 192–194, 398–400
  tungsten, 181–185
Chemical vapor deposition reactors
  cold wall atmospheric, 142
  hot–filament, 144–145
  hot wall, low pressure, 141
  photochemical, 145–146
  plasma, 142–144
Chip packaging, 65–71
Chip passivation, 64
Chips, 36
Chlorosilanes, 51, 52
Class 1, 44

Cleanliness needs, 44
Compact disc (CD) technology, 25
Compound semiconductors, 13
Contamination
  bacteria growth, 47
  carbon incorporation, 114, 134
  class 10, 44
  intentional impurities, 47
  ionic, 46
  metallic, 46–47
  nonparticulate chemical, 393
  oxygen on silicon surface, 111
  particle, 43–46
  sensitivity limits, 368
  sensitivity requirements, 368
  smoker's breath, 45
  sodium, 46
  surfaces, 212
  wafer, 57–59
Cooper pairing, 512, 521, 551
Copolymers, 481
Copper, 189–190
Copper films
  hexafluoroacetylacetonate (hfa), 189
  plasma etching of, 196
  selective growth, 189
Corrosion, 324
  accelerated testing, 340
  atmospheric tests of, 337
  chemical free energy, 326
  driving force of, 326
  effect of cleaning, 348
  effect of grain boundaries, 349–350
  effects of environment, 334
  electromotive force, 328
  failure mechanism of Al-Cu, 360
  Gouy-Chapman layer, 327
  Helmholtz plane, 327
  humidity effect on, 338–339
  inhibitors for copper, 350
  life projection, 341, 357, 361
  Nernst equation, 329
  Pourbaix diagrams, 330

[Corrosion]
  protection for magnetic media, 351–359
  reactions of, 325
  resistance, 329, 348, 352
  temperature-humidity tests, 341
  testing of, 334
Corrosion mechanisms
  dendrite formation, 359
  effect of Cl-containing plasmas, 348
  effect of miniaturization, 361
  galvanic corrosion, 343–346
  "mouse" holes, 343
  of Al-Cu, 346, 360
  of conductors, 358
  of thin-film disks, 355
Corrosion reaction kinetics
  metal dissolution, 331
  rate of, 332, 335
  role of water, 338
Crystal growth, 51–52
Crystalline films
  density, 159
  properties of, 158–162

Deep UV lithography, 201
Deep UV resist
  acid-catalyzed reaction, 231
  bleachable eliphatic diazoketones, 229–230
  chemical amplification (CAM), 230
  cross-linking in, 234
  deprotection mechanism, 232
  dissolution inhibition, 236
  electrophilic aromatic substitution, 235
  negative tone CAM, 234
  performance of CAM, 238
  sensitivity to airborne contamination, 233
Defect density, 85
Defects, 43–44, 71, 85, 104–105, 111, 369

Deposition processes, 84
Desorption, 99, 134
Device capacitance, 17–18
Device crosstalk, 17
Device fabrication processes
  chemical vapor deposition, 62, 63, 83–122, 127–129, 131–155, 173–174, 181–190, 192–194
  chip passivation, 64
  doping, 59–62, 113–116, 155–159
  epitaxy, 85–122
  metal films, 171–172, 177–190
  oxidation, 62
  pattern etching, 200, 251–256
  photopatterning, 63–64
  physical vapor deposition, 62
  thin-film deposition, 62–63
  wafer cleaning, 57–58
Device failure
  analysis of, 73
  bathtub curve, 71, 341
  failure in time, 71
  mechanisms, 72–73
  reliability, 71–72
  wearout, 71
Device failure analysis
  general considerations, 409
  methodology, 407, 408
  physical unlayering, 410, 411
Device failure modes
  adhesion, 77
  aluminum corrosion, 77, 78
  electromigration, 17, 79, 177, 178
  electrostatic discharge, 75–77
  hermeticity, 77
  radiation, 79
Device miniaturization, 41
Device operation, 6–7
Device packaging
  hermetic, 68–70
  multichip, 70, 71
  plastic, 67, 68
  single chip, 65–71

Index   613

Devices
  ASIC, 4
  CD-ROM, 25–26
  definition, 6
  diode, 7
  discrete, 3
  DRAM, 5
  EPROM, 6
  integration of, 4
  lasers, 23
  logic, 4
  magnetic disk storage, 27
  microprocessor, 4
  optoelectronic, 22
  power, 31
  PROM, 6
  ROM, 5
  sensors, 29–31
  solar cells, 14
  SQUID, 19
  SRAM, 5
  superconductors, 19, 501–512
  transistor, 9
Diamond, 130, 159
Diamondlike films, 136, 160
Die, 36
Dielectric constant
  Al_2O_3, 490
  conventional polyimides, 440
  nonfluorinated polyimides, 440
  polyimides, 422
  SiO_2, 489
Dielectric films
  device quality, 162–163
  properties of, 162–164
  Si_3N_4, 151, 163, 164
  SiO_2, 149–151, 164
Diffraction
Diffusion, 59
Diffusion barriers, 190–191
  for aluminum, 190
  for copper, 191
  titanium nitride, 191–195

[Diffusion barriers]
  for tungsten, 191
Diffusive transport, 139
Diode
  LED, 22–23
  operation of, 7–8
Discrete devices, 3
Doping, 7, 47, 114, 134, 155–158, 406, 409
Doping depth profile, 157
Doping effect on Si growth rate, 156
Doping techniques
  ion implantation, 60, 155
  laser, 156–157
  liquid phase, 59
  vapor phase, 115–116, 156
Dow membrane, 598
DRAM, 5
Dry etching (see Plasma etching)
Dupont membrane, 598

E-beam lithography, 201
Electrical properties of silicon, 52–55
Electrical test, 64
Electrochemical reactions
  corrosion, 324
  electroetching, 324
  etching, 324
  plating, 324
Electrolyte
  alkaline carbonate mixture, 583–584
  KOH, 602
  phosphoric acid, 577
  polymer membrane, 598
  stabilized ZrO_2, 591
Electrolyte support
  $LiAlO_2$ matrix, 583
Electromechanical sensors, 29
Electromigration, 17, 77, 78, 177, 178
Electronic device dimensions, 37
Electronic package fabrication
  casting, 471

[Electronic package fabrication]
  co-firing, 469
  firing (*see* sintering)
  metal network, 488
  milling process, 470
  post-firing, 469
  screenable paste, 472
  sintering, 474, 484, 488, 497, 603
  stacking, 473
Electrostatic discharge, 75
Endpoint detection, 310
Energy dispersive x-ray microanalysis (EDX)
  concentration detection limits, 394
  display options, 396
  Planck's equation, 395
  x-ray emission nomenclature, 395
Energy transfer function, 259
Environmental requirements, 42
Epitaxial doping
  atomic incorporation, 98
  autodoping, 109
  carbon incorporation, 115, 116
  dopants, 59, 115–116, 156
  intentional, 116
  purpose for, 59, 113, 114
  unintentional, 115
Epitaxial growth
  adsorption, 99
  atomic incorporation, 98, 99
  boundary layer, 108
  carbon incorporation, 115, 116
  definition, 85
  desorption, 99
  doping in, 114, 115
  edge morphology, 120
  elementary steps in, 86–87
  fluid flow, 106, 108
  Frank–van der Merve, 102
  generation of misfit dislocation, 104
  hydrogen adsorption, 112, 113
  Knudsen number, 107

[Epitaxial growth]
  laminar flow, 107
  lattice-match, 103
  law of mass action, 101
  layer-by-layer, 102
  mass balance, 91
  mass transport, 106
  misfit dislocation in, 104
  oxygen surface contamination, 112
  pressure effect on boundary layer, 141
  process, 87
  pseudomorphic, 104
  rate limited, 110
  step-flow, 102
  Stranski-Krastanov, 104
  surface migration, 86
  surface nucleation, 101, 102
  transport limited, 112
  unimolecular decomposition, 96
Epitaxy
  AlAs, 103
  Al_xGa_{1-x} As, 111
  AlP, 103
  AlSb, 103
  film purity limitation, 114
  GaAs, 89, 110, 115
  GaP, 103
  GaSb, 103
  kinetics of, 92
  heteroepitaxy, 86, 101–106
  homoepitaxy, 102
  low pressure, 112
  MBE, 86
  MOCVD, 110
  properties of, 86
  selective, 117–122
  Silane, 96
  $SiCl_xH_{4-x}$, 90
  Si_xGe_{1-x}, 86, 103, 113
  silicon, 90
  thermodynamics, 89

Equilibrium coefficient, 93
Equilibrium constant, 90
Etching
    anisotropic, 253, 297–299
    anisotropy, 285
    etch rate, 255
    physical characteristics of, 257–264
    Piranha etch, 57
    plasma chemistry, 264–266
    polycrystalline silicon, 253
    radiation damage, 256
    selectivity, 256, 294
    uniformity, 256

Failure analysis, 73–74
Failure in time, 71
Field effect transistors, 9, 10–11
Focused-ion beam (FIB)
    applications, 391–392
    low temperature oxide analysis, 392
    tool functions, 391
Frank–van der Merve growth, 102
Fuel cell components
    anode, 581, 588, 593, 600, 604
    cathode, 577, 585, 592, 599, 603
    electrolyte, 577, 583, 591, 598, 602
    separator, 582, 588, 594–595, 601, 604
Fuel cell polarization
    diffusion, 571
    ohmic, 571
Fuel cells
    carbon as electrode substrate, 578–581
    components, 575–605
    CO poisoning, 582
    energy conversion efficiency of, 563, 574
    environmental compatibility of, 563
    features of, 563
    lithiation, 585
    modularity of, 564
    multifuel capability of, 564

[Fuel cells]
    operating principles, 561
    polarization in, 570–572
    power generation, 573
    Pt loading, 580
    reaction, 568
    reaction in, 568, 587–588, 602
    size of, 565
    stack designs, 582
    stacks of, 562
    thermodynamic of, 568
    types, 566–567
Fuel cell types
    alkaline, 567, 602–605
    molten carbonate, 567, 583 – 590
    phosphoric acid, 567, 576–583
    solid oxide, 567, 590–597
    solid polymer, 567, 597–602

GaAs
    carbon incorporation, 115, 116
    epitaxy, 89, 103, 109–112, 114–117
    heat of reaction, 90
    MOCVD, 110–112
    triethyl gallium precursor, 114
    trimethyl gallium precursor, 89, 110–112
Gas phase chemistry, 266–279
Gas phase chemistry analysis methods
    absorption, 311
    actinometry, 308
    coherent anti–stokes Raman spectroscopy, 148
    endpoint detection, 310
    IR tunable diode laser spectroscopy, 148
    laser induced fluorescence, 148–311
    mass spectrometry, 148, 311, 312
    optical emission spectroscopy, 307
Germanium, 2
Gibbs free energy, 89, 132, 326, 493, 574, 568
Gigascale integration, 5

Glass ceramics, 463, 466, 491–493
Greensheets
  interaction with binder, 478
  interaction with metal–paste, 480
  interaction with slurries, 476
  polymers in, 480
Growth front, 87
Growth model, 86–87, 101–105, 128–129

Hermetic seal, 68
Heteroepitaxial growth, 86, 101–106
High-temperature glass chemistry, 494
Homoepitaxial growth, 102
Hydrogen binding energy
  carbon, 134
  germanium, 136
  silicon, 136
Hydrogenated amorphous films
  carbon, 128
  germanium, 129
  silicon, 128

Impurity concentration, 406, 409
Integrated circuits
  dielectric interconnect, 418
  electrical interconnect, 177–179
  functions of metals, 171–172
  metal contacts, 175–177
  metal silicides, 172–174
  optical interconnect, 20, 419
Interconnects, 17
Ionic contamination, 46
Ion implantation, 60
Isotropic etching, 253

Josephson junctions, 525

Kinetics, 92–101, 138–139
Knudsen number, 107, 139

Lambert-Beer law, 206
Langmuir-Hinshelwood mechanism, 281
Lasers, 23
Lattice imaging, 388, 389
Lattice-matched growth, 103
Law of mass action, 101
Lewis acid-base, 475
Light emitting diodes (LED), 22
Lithography
  antireflective layers, 215
  depth of focus, 203
  Dill parameters, 207, 229
  e-beam lithography, 201
  g-line, 201
  i-line, 201
  Lambert-Beer law, 206
  latent image, 205
  photolithography, 63, 200–202
  production requirements, 203–205
  projection printing, 209
  proximity printing, 208
  Raleigh equations, 203
  resolution capability, 203, 208
  step and repeat projection printing, 209
  step and scan systems, 210
  swing curve, 204
  x-ray lithography, 201
Lithography processes
  bleaching, 207, 229–230
  film development, 214
  multifunctional silylation, 244
  multilayer systems, 239
  phase-shifting mask, 247
  post exposure bake, 211
  prebake, 211
  selective silylation, 242
  silylation, 240–241, 244
  single-layer lithography, 210
  spin coating, 213
  substrate preparation, 212
  surface priming, 212–213
Logic devices, 4

Index   617

Magnetic disk storage technology, 27
Magnetic storage devices, 33
Magneto optical storage technology, 26
Mark-Houwink equation, 431–432
Mask, 63, 251
Mass balances, 91
Massively parallel systems, 4
Material analysis
  arsenic implants, 406
  Auger transitions, 398
  backscattering electron imaging, 380
  chemical methods, 372, 393
  contamination sensitivity limits, 368
  contamination sensitivity requirements, 368
  CMOS structure, 390
  cross sections for, 387
  crystallographic defects, 369
  depth profiling, 157, 400, 406, 409
  device failure, 407, 408
  MOS transistor, 388
  optical methods, 372
  physical unlayering, 410, 411
  plan view sections, 384, 385
  secondary electron imaging, 379, 380
  "soft" x-rays, 402
  spectral chemical shift, 403, 404
  structural methods, 372, 381
  surface methods, 312–313
  TiN analysis, 398–400
  tripod polisher, 385–387
  x-ray emission nomenclature, 395
Material analysis tools
  Auger electron spectroscopy, 312, 397, 398
  electrical resistance, 513
  electron energy loss spectroscopy, 514
  electron spin resonance spectroscopy, 513
  energy-dispersive x-ray spectroscopy, 394

[Material analysis tools]
  ESCA (*see* x-ray photoelectron spectroscopy)
  focused-ion beam, 390
  magnetically modulated microwave absorption, 513
  modern analytical tools, 370
  nuclear magnetic resonance, 513
  optical microscopy, 372
  photoemission spectroscopy, 514
  scanning electron microscopy, 376
  secondary ion mass spectroscopy, 157, 404, 405
  transmission electron microscopy, 381
  x-ray crystallography, 513
  x-ray photoelectron spectroscopy, 312, 402, 514
Material electrical conductivity, 50, 51
Maxwellian distribution, 260
Meisner effect, 502
Metal films
  aluminum, 185–188
  chemical vapor deposition, 173–174, 181–190
  copper, 189–190
  plasma etching, 195–196
  selective deposition, 184
  silicides, 172–174
  sputter deposition, 173–174, 179–180
  step coverage, 180, 183, 192, 194
  stress in, 185–186
  titanium nitride, 192–194
  tungsten, 181–185
  uniform film growth, 181
  wet etching, 195
Metallic contamination, 46
Metallurgical grade silicon, 51
Metal-organo chemical vapor deposition (MOCVD)
  aluminum hydride, 188
  group III precursors, 110

[Metal-organo chemical vapor deposition (MOCVD)]
  tetrakis (dialkylamido) titanium, 192
  triisobutylaluminum, 186
  trimethyl gallium, 110
Metal properties
  aluminum, 178
  copper, 179
  silver, 177
  tungsten, 178
Metal reactions
  decomposition of tetrakis (dialkylamido) titanium, 192–194
  hydrogen catalysis, 185
  hydrogen reduction, 183
  silane with tungsten hexafluoride, 185
  silicon catalysis, 184
  silicon with aluminum, 190
  silicon with titanium, 173
  SiO_2 with titanium, 176
Metal silicides, 172–174, 179
  as electrical interconnect, 179
  cobalt disilicide, 174
  formation of, 173–174
  platinum silicide, 174
  titanium disilicide, 173
  tungsten disilicide, 173
Microelectronic
  device manufacturing discipline, 41–50
  dimensions, 37
Microelectronic industry, 1
  products, 9–31
  reliability, 71–73
Microelectronic materials
  amorphous, 14
  compound semiconductors, 13
  dimensions in, 6, 37
  integrated circuit "die," 36
  manufacturing discipline, 41–50
  metals, 17
  particle control in, 43–46

[Microelectronic materials]
  polycrystalline, 13
  semiconductor grade silicon, 51
  silicon, 10
  substrate, 50
Microelectronics, 3, 36
Microprocessor, 4
Mid UV lithography, 200–201
Mid UV resist
  dyed, 227
  formulation, 225
  Novolak, 219–221
  performance, 226
Misfit dislocations, 104
MOCVD, 186–190, 192–195
Molecular beam epitaxy (MBE), 86
Mott-Hubbard insulator, 516
MOVPE, 110–112, 115, 116
Multichip modules, 466
Multilayer ceramic, 468

Nernst equation, 329, 568

Optical
  analysis, 372
  emission spectroscopy, 307
  interconnects, 20
  lithography, 201–210
  transparency, 422, 442
Optical microscopy
  applications, 376
  basic microscope, 374
  magnification limits, 372
  phase microscopy, 375
  resolution limits, 375
Optoelectronic devices, 22–24, 416
Organic insulators, 416
Oxidation, 62

Particle control, 43
Patterning of surfaces, 63
Peierls transitions, 518
Periodic table, 36

Index

Phase-shifting mask, 247
Photoresist, 218–239
Photoresist mask, 251
Physical analysis, 372, 381
Plasma chemistry
   anisotropy mechanism, 285
   electron energy distribution function, 268–273
   electron impact reactions, 268
   effect of free fluorine on silicon, 294–297
   etch rate of silicon, 293, 294
   etch rate of silicon dioxide, 293
   gas additives, 302–303
   gas-phase, 266
   heavy particle reactions, 278–279
   ion-assisted, 282
   ion-enhanced, 283
   ion-induced, 282
   Maxwell-Boltzmann distribution (*see* Maxwellian)
   molecular process, 265
   polymer formation, 303
   product volatility, 265
   reactions of, 266, 281–282
   selectivity, 294–297
   sputtering, 280
   surface chemistry, 280
Plasma diagnostics
   absorption, 311
   ellipsometry, 313
   end point detection, 310
   gas-phase diagnostics, 307–312
   laser-induced fluorescence, 311
   mass spectrometry, 311
   optical emission spectroscopy, 307
   surface diagnostics, 312–313
Plasma etching
   aspect ratio (external variable), 291
   chemistry of, 264–267
   Debye length in, 260
   diode reactor, 257–258
   electron diffusion length, 274

[Plasma etching]
   electron heating, 259
   electron temperature, 273
   energy transfer function, 259
   frequency (external variable), 288
   gas flow rate (external variable), 287
   high-density plasmas, 303–304
   loading (external variable), 289
   Maxwellian velocity distribution, 260
   parallel plate reactor, 257
   power (external variable), 286
   pressure (external variable), 286
   scaling parameters, 275–277
Plasma etching of
   aluminum, 300
   compound semiconductors, 301
   polymers, 302
   silicon, 292–295
   silicon dioxide, 293–294
   silicon nitride, 297
   tungsten, 300–301
Plasma reactor
   barrel, 258
   heat in, 260
   parallel plate, 257
   symmetric, 264
Plastic package, 67, 68
Polycrystalline
   materials, 13
   silicon, 146–147, 173
Polyimide films
   aromatic, 422
   as dielectric interconnect, 418
   as optical interconnect, 419
   as protective overcoat, 419
   for alpha-ray shielding, 419
   for solar cell application, 419
   imidization, 425
   Kapton, 416
   PIQ, 417
   poly (amic acid) structure, 417
   processability, 421

[Polyimide films]
  process considerations, 426
  Pycalin, 416
  reaction mechanism to form, 425
  synthesis, 422–423
Polyimide properties
  chemical and physical properties, 421–422
  degradation temperature, 437
  dielectric constant, 422
  mechanical stability, 422
  optical transparency, 422, 442
  thermal coefficient of expansion, 437
  thermal stability, 422, 437
Polyimide property modifications, 447
  copolymerization, 448
  effect of fluorine content, 452, 453
  of near infrared transparency, 452–453
  of visible transparency, 451
  physical properties, 449
Polyimides
  adhesion of, 428–431
  copolymerization, 448–449
  film planarity, 433
  material properties, 436–454
  moisture sensitivity, 434
  photosensitivity, 435
  process considerations, 426
  property modification, 447–454
  refractive index, 445
  viscosity, 431
Power devices, 31
Process materials requirements
  chemicals, 48
  gases, 50
  ultrapure water, 47
Properties of
  amorphous films, 164–167
  carbon films, 159–160
  ceramics, 462
  fuel cells, 563

[Properties of]
  polyimides, 421–422
  resist films, 226–229, 236–238
  Si_3N_4, 164
  SiO_2, 163, 164
  silicon, 53–55
  superconductors, 551
Proximity printing, 208

Radiation damage, 79
Reaction mechanism
  amorphous silicon, 154
  diamond films, 148
  silicon dioxide, 149–151
  silicon nitride, 151–152
Reactive ion etching (*see* Plasma etching)
Refractive index
  effect of fluorine, 444
  Lorentz-Lorenz formula, 446
  optical anisotropy, 446
Resist films
  chemical interactions of components, 224–225
  deep UV, 229–238
  formulation, 225
  mid UV, 218–229
  photoactive compound (PAC), 206, 220–224
  positive tone resist, 218–219
  Novolak, 219–221
Rideal mechanism, 282

Scanning electron microscopy (SEM)
  backscattering electron imaging, 379, 380
  backscattering electrons, 378
  electron–material interaction, 377
  secondary electron imaging, 379
  secondary electrons, 377
  specimen magnification, 376
  with energy dispersive x-ray spectrometer, 380

Index

Secondary ion mass spectroscopy (SIMS)
  arsenic implant profile, 406
  boron implant profile, 157, 409
  detection limits, 405–406
Selective
  area doping, 59
  deposition, 184
  epitaxy, 116 121
Semiconductor grade silicon, 51
Sensors, 29
Sidewall, 180, 253, 256, 280, 299
Silicides, 173–174
Silicon
  amorphous, 13, 128
  Czochraski crystal growth, 52
  devices, 3–13
  dichlorosilane precursors, 90
  doping, 7, 54
  electrical property modification, 53–55
  float zone crystal growth, 52
  growth kinetics, 98–101
  hydrogen binding energy, 136
  intentional doping, 116
  intrinsic, 54
  low-pressure growth of, 112–113
  metallurgical grade, 51
  polycrystalline, 13, 146–147
  semiconductor grade, 51
  silane, 93–95, 100, 101
Silicon dioxide, 62, 130, 149–151
Silicon nitride, 130, 151–152
Silver, 177
Silylation, 240–241, 244
SIMS, 157
Sintering
  co-fired copper, 488
  effects of additives, 492
  inhibiter, 603
  parameters effecting, 484
  particle size effect on, 490
  profile, 486

[Sintering]
  pyrolysis, 487
  thermodynamics of, 493–496
Solar cell efficiencies, 16
Solar cells, 14, 15
Solid state electronics, 1, 7
Spectroscopy (*see* Material analysis tools)
Sputter deposition
  advantages, disadvantages, 180
  aspect ratio, 180
  metals, 173–174, 179–180
  step coverage, 180
SQUID, 19, 527, 549
SRAM, 5
Step coverage, 180, 183, 192, 194
Stranski-Krastanov growth, 104
Substrate, 50
Superconductor fabrication methods
  thin films, 548
  wires, 549–551
Superconductor preparation chemistry
  chemical modifications, 556
  chemical vapor deposition, 538–539
  evaporation methods, 537
  of charge transfer salts, 543–545
  of films, 536
  of fullerides, 545–547
  of inorganic, 528–539
  of mercury-containing cuprates, 435
  of organic, 539–545
  pulsed laser ablation methods, 538
  structural modifications, 555
  synthesis of tetrachalcagenofulvalene, 540
  synthesis of tetratellurafulvalene, 541
  synthesis of tetrathiafulvalene, 540
  synthesis routes, 429–435
Superconductors
  alkali metal fullerides, 511
  carbon allotrope buckminster fullerene, 503
  coherence lengths in, 522–523

[Superconductors]
    composition of cuprate, 506
    copper containing metal oxides, 503
    crystal structure of cuprate, 506–507
    electrical resistance, 552
    electronic structure, 514–520
    flux tubes in, 523–525
    high Tc oxide, 505
    inorganic, 505–507
    junctions in, 525
    Meisner effect, 502
    nomenclature of cuprate, 507
    organic, 509–512
    Peierls transitions, 518
    preparation chemistry, 528–547
    pressure effects on, 554
    SQUID, 9, 527, 549
    structure determination, 513–514
    structure/property relationships, 551
    superconducting state, 520
    synthetic-metal, 503, 504–505
    tunneling junctions in, 525
    Type I, Type II, 523
Surface chemistry, 280–286
Surface diagnostics
    attenuated total internal reflection, 313
    Auger electron spectroscopy, 312
    Ellipsometry, 313
    quartz crystal microbalance, 313
    x-ray photoelectron spectroscopy, 312

Tafel equation, 571
Tape casting, 589
TARL (top antireflective layers)
Thermal coefficient of expansion
    Al_2O_3, 490
    inorganic materials, 438
    metals, 438
    PMDA/ODA, 438
    polyimides, 438, 439
    SiO_2, 489

[Thermal coefficient of expansion]
    thermomechanical analysis, 438
Thermodynamics, 89
Thin-film deposition, 62
Thin-film disks, 354
Thin-film heads, 352
Titanium nitride
    as barrier films, 190–195
    Auger spectra, 400
    formation of, 192
    step coverage, 192, 194
    tetrakis (dialkylamido) titanium as precursor, 192–194
Top surface imaging
    process considerations, 245
    selective silylation, 242
Transistor
    bipolar, 9
    field effect (FET), 9
    integrated, 3
    operation of, 11
    performance of, 6
Transistor elements, 9
Transmission electron microscopy (TEM)
    cross-section analysis, 385
    MOS logic transistor, 388
    plan view section analysis, 384
    resolution, 381
    sample thickness limits, 383
    tripod polisher, 385
Transport, 101, 139
Tungsten, 181–185
Tungsten films
    blanket deposition of, 184
    chemical vapor deposition, 181–186
    effect of silicon, 184
    enhanced growth of, 185
    film stress, 185–186
    reaction with silane, 185
    selective deposition, 184
    step coverage, 183
    wet etching of, 195

[Tungsten films]
   WF_6 deposition source, 183
Tungsten hexafluoride, hydrogen reduction of, 183

Ultra large scale integration (ULSI), 19, 56
Ultrapure water, 47
Ultrapurity, 47–50

Very large scale integration (VLSI), 19, 416

Wafer cleaning, 57–59, 403
Wafer priming, 212–213
Wet etching, 253

X-ray lithography, 201
X-ray photoelectron spectroscopy (XPS)
   method, 312, 402, 514
   "soft" x-rays, 402
   spectral chemical shift, 403–404
   surface cleaning processes, 403
   surface depth range, 403